# Introduction to Engineering
# Fluid Mechanics

# Introduction to Engineering Fluid Mechanics

Marcel Escudier

OXFORD
UNIVERSITY PRESS

Great Clarendon Street, Oxford, OX2 6DP,
United Kingdom

Oxford University Press is a department of the University of Oxford.
It furthers the University's objective of excellence in research, scholarship,
and education by publishing worldwide. Oxford is a registered trade mark of
Oxford University Press in the UK and in certain other countries

Published in the United States of America by Oxford University Press
198 Madison Avenue, New York, NY 10016, United States of America

British Library Cataloguing in Publication Data

Data available

Library of Congress Control Number: 2016954938

ISBN 978-0-19-871987-8 (hbk.)
ISBN 978-0-19-871988-5 (pbk.)

Printed and bound by
CPI Group (UK) Ltd, Croydon, CR0 4YY

*To my wife Agnes, our son Stephen, and the memory of my Mother and Grandmother*

# Preface

A fluid is a material substance in the form of a liquid, a gas, or a vapour. The most common examples, to be found in both everyday life and in engineering applications, are water, air, and steam, the latter being the vapour form of water. The flow (i.e. motion) of fluids is essential to the functioning of a wide range of machinery, including the internal-combustion engine, the gas turbine (which includes the turbojet, turbofan, turboshaft, and turboprop engines), wind and hydraulic turbines, pumps, compressors, rapidly rotating discs (as in computer drives), aircraft, spacecraft, road vehicles, and marine craft. This book is concerned primarily with Newtonian fluids, such as water and air, for which the viscosity is independent of the flow. The quantitative understanding of fluid flow, termed fluid dynamics, is based upon the application of Newton's laws of motion together with the law of mass conservation. To analyse the flow of a gas or a vapour, for which the density changes in response to pressure changes (known as compressible fluids), it is also necessary to take into account the laws of thermodynamics, particularly the first law in the form of the steady-flow energy equation. The subject of fluid mechanics encompasses both fluid statics and fluid dynamics. Fluid statics concerns the variation of pressure in a fluid at rest (as will be seen in Chapter 4, this limitation needs to be stated more precisely), and is the basis for a simple model of the earth's atmosphere.

This text is aimed primarily at students studying for a degree in mechanical engineering or any other branch of engineering where fluid mechanics is a core subject. Aeronautical (or aerospace), chemical, and civil engineering are all disciplines where fluid mechanics plays an essential rôle. That is not to say that fluid flow is of no significance in other areas, such as biomedical engineering. The human body involves the flow of several different fluids, some quite ordinary such as air in the respiratory system and water-like urine in the renal system. Other fluids, like blood in the circulatory system, and synovial fluid, which lubricates the joints, have complex non-Newtonian properties, as do many synthetic liquids such as paint, slurries, and pastes. A brief introduction to the rheology and flow characteristics of non-Newtonian liquids is given in Chapters 2, 15, and 16.

As indicated in the title, this text is intended to <u>introduce</u> the student to the subject of fluid mechanics. It covers those topics normally encountered in a three-year mechanical-engineering-degree course or the first and second years of a four-year mechanical-engineering-degree course, as well as some topics covered in greater detail in the final years. The first ten chapters cover material suitable for a first-year course or module in fluid mechanics. Compressible flow, flow through axial-flow turbomachinery blading, internal viscous fluid flow, laminar boundary layers, and turbulent flow are covered in the remaining eight chapters. There are many other textbooks which cover a similar range of material as this text but often from a much more mathematical point of view. Mathematics is essential to the analysis of fluid flow but can be kept to a level within the capability of the majority of students, as is the intention here where the emphasis is on understanding the basic physics. The analysis of many flow situations rests upon a small number of basic equations which encapsulate the underlying physics. Between these fundamental equations and the final results, which can be applied directly to the solution of engineering problems, can be quite extensive mathematical manipulation and it is all too easy to lose sight of the final aim. A basic understanding of vectors is required but not of vector analysis. Tensor notation and analysis is also not required and the use of calculus is kept to a minimum.

The approach to certain topics may be unfamiliar to some lecturers. A prime example is dimensional analysis, which we suggest is approached using the mathematically simple method of sequential elimination of dimensions (Ipsen's method). The author believes that this technique has clear pedagogical advantages over the more widely used Rayleigh's exponent method, which can easily leave the student with the mistaken (and potentially dangerous) idea that any physical process can be represented by a simple power-law formula. The importance of dimensions and dimensional analysis is stressed throughout the book. The author has also found that the development of the linear momentum equation described in Chapter 9 is more straightforward to present to students than it is via Reynolds transport theorem. The approach adopted here shows very clearly the relationship with the familiar $F = ma$ form of Newton's second law of motion and avoids the need to introduce an entirely new concept which is ultimately only a stepping stone to the end result. The treatment of compressible flow is also subtly different from most texts in that, for the most part, equations are developed in integral rather than differential form. The analysis of turbomachinery is limited to flow through the blading of axial-flow machines and relies heavily on Chapters 3, 10, and 11.

'Why do we need a fluid mechanics textbook containing lots of equations and algebra, given that computer software packages, such as FLUENT and PHOENICS, are now available which can perform very accurate calculations for a wide range of flow situations?' To answer this question we need first to consider what is meant by accurate in this context. The description of any physical process or situation has to be in terms of equations. In the case of fluid mechanics, the full set of governing equations is extremely complex (non-linear, partial differential equations called the Navier-Stokes equations) and to solve practical problems we deal either with simplified, or approximate, equations. Typical assumptions are that all fluid properties remain constant, that viscosity (the essential property which identifies any material as being a fluid) plays no role, that the flow is steady (i.e. there are no changes with time at any given location within the fluid), or that fluid and flow properties vary only in the direction of flow (so-called one-dimensional flow). The derivation of the Navier-Stokes equations, and the accompanying continuity equation, is the subject of Chapter 15. Exact analytical solution of these equations is possible only for a handful of highly simplified, idealised situations, often far removed from the real world of engineering. Although these solutions are certainly mathematically accurate, due to the simplifications on which the equations are based they cannot be said to be an accurate representation of physical reality. Even numerical solutions, however numerically accurate, are often based upon simplified versions of the Navier-Stokes equations. In the case of turbulent flow, the topic of Chapter 18, calculations of practical interest are based upon approximate equations which attempt to model the correlations which arise when the Navier-Stokes equations are time averaged. It is remarkable that valuable information about practical engineering problems can be obtained from considerations of simplified equations, such as the one-dimensional equations, at minimal cost in terms of both time and money. What is essential, however, is a good physical understanding of basic fluid mechanics and a knowledge of what any computer software should be based upon. It is the aim of this text to provide just that.

Already in this brief **Preface** the names Navier, Newton, Rayleigh, Reynolds, and Stokes have appeared. In Appendix 1 we provide basic biographical information about each of the scientists and engineers whose names appear in this book and indicate their contributions to fluid mechanics.

# Acknowledgements

The author gratefully acknowledges the influence of several outstanding teachers, both as a student at Imperial College London and subsequently as a Research Associate at the Massachusetts Institute of Technology. My interest in, and enjoyment of, fluid mechanics was sparked when I was an undergraduate by the inspiring teaching of Robert Taylor. Brian Spalding, my PhD supervisor, and Brian Launder are not only internationally recognised for their research contributions but were also excellent communicators and teachers from whom I benefitted as a postgraduate student. As a research associate at MIT I attended lectures and seminars by Ascher H. Shapiro, James A. Fay, Ronald F. Probstein, and Erik Mollo-Christensen, all inspiring teachers. Finally, my friend Fernando Tavares de Pinho has given freely of his time to answer with insight many questions which have arisen in the course of writing this book.

Marcel Escudier
Cheshire, August 2016

# Contents

# Notation

Each Roman, Greek, and mathematical symbol is followed by its meaning, its SI unit, and its dimension(s).

**Lower-case Roman symbols**

| | | | |
|---|---|---|---|
| $a$ | acceleration | m/s$^2$ | L/T$^2$ |
| $c$ | blade chord length | m | L |
| $c$ | concentration | kg/m$^3$ | M/L$^3$ |
| $c$ | soundspeed | m/s | L/T |
| $c$ | wetted perimeter | m | L |
| $c_f$ | skin-friction coefficient | – | – |
| $c_0$ | speed of light in vacuum | m/s | L/T |
| $d$ | diameter | m | L |
| $e$ | energy | J | ML$^2$/T$^2$ |
| $\dot{e}_{xx}$ | extensional strain rate in $x$-direction | 1/s | 1/T |
| $f$ | non-dimensional velocity | – | – |
| $f_x$ | body force per unit mass acting in the $x$-direction | m/s$^2$ | L/T$^2$ |
| $f_D$ | Darcy friction factor | – | – |
| $f_F$ | Fanning friction factor | – | – |
| $\overline{f_F}$ | average Fanning friction factor | – | – |
| $g$ | acceleration due to gravity | m/s$^2$ | L/T$^2$ |
| $g_0$ | acceleration due to gravity at sea level ($z = z' = 0$) | m/s$^2$ | L/T$^2$ |
| $h$ | height | m | L |
| $h$ | spacing of parallel plates | m | L |
| $h$ | specific enthalpy | kJ/kg | L$^2$/T$^2$ |
| $h_0$ | specific stagnation enthalpy | kJ/kg | L$^2$/T$^2$ |
| $h_{0,REL}$ | relative stagnation enthalpy | kJ/kg | L$^2$/T$^2$ |
| $i$ | angle of incidence | ° or rad | – |
| $j$ | number of independent dimensions | – | – |
| $k$ | number of non-dimensional groups | – | – |
| $k$ | radius of gyration | m | L |
| $k$ | specific turbulent kinetic energy | m$^2$/s$^2$ | L$^2$/T$^2$ |
| $\overline{k}$ | time-averaged specific turbulent kinetic energy | m$^2$/s$^2$ | L$^2$/T$^2$ |
| $k_B$ | Boltzmann constant | J/K | ML$^2$/T$^2$K |
| $l$ | length | m | L |
| $l_K$ | Kolmogorov length scale | m | L |
| $l_M$ | mixing length | m | L |

| | | | |
|---|---|---|---|
| $m$ | mass | kg | M |
| $m$ | wedge-flow exponent | – | – |
| $m_A$ | added mass | kg | M |
| $\dot{m}$ | mass flowrate | kg/s | M/T |
| $n$ | amount of substance | kmol | M |
| $n$ | number of physical quantities | – | – |
| $n$ | power-law exponent in power-law viscosity model | – | – |
| $p$ | static pressure | Pa | $M/LT^2$ |
| $p_G$ | gauge pressure | Pa | $M/LT^2$ |
| $p_H$ | hydrostatic pressure | Pa | $M/LT^2$ |
| $p_{REF}$ | reference pressure | Pa | $M/LT^2$ |
| $p_T$ | total pressure | Pa | $M/LT^2$ |
| $p_V$ | vapour pressure | Pa | $M/LT^2$ |
| $p_0$ | stagnation pressure | Pa | $M/LT^2$ |
| $p_{0,REL}$ | relative stagnation pressure | Pa | $M/LT^2$ |
| $\bar{p}$ | average static pressure | Pa | $M/LT^2$ |
| $p'$ | fluctuating component of static pressure | Pa | $M/LT^2$ |
| $p'$ | intermediate static pressure | Pa | $M/LT^2$ |
| $p^*$ | non-dimensional static pressure | – | – |
| $\dot{q}$ | heat transfer rate | W | $ML^2/T^3$ |
| $\dot{q}'$ | heat transfer rate per unit length | W/L | $ML/T^3$ |
| $r$ | radial distance | m | L |
| $s$ | arc length | m | L |
| $s$ | cascade-blade spacing (or pitch) | m | L |
| $s$ | distance along a streamline | m | L |
| $s$ | specific entropy | $m^2/s^2 \cdot K$ | $L^2/T^2\theta$ |
| $s_0$ | specific stagnation entropy | $m^2/s^2 \cdot K$ | $L^2/T^2\theta$ |
| $t$ | elapsed time | s | T |
| $t$ | temperature | °C | $\theta$ |
| $\tilde{t}$ | non-dimensional time | – | – |
| $t^*$ | non-dimensional time | – | – |
| $u$ | specific internal energy | kJ/kg | $L^2/T^2$ |
| $u$ | velocity component in $x$-direction | m/s | L/T |
| $\bar{u}$ | time-averaged value of velocity component $u$ | m/s | L/T |
| $u'$ | fluctuating component of velocity component $u$ | m/s | L/T |
| $u^*$ | non-dimensional value of velocity component $u$ | – | – |
| $u^+$ | velocity component $u$ normalised by $u_\tau$ | – | – |
| $u_P$ | velocity of plastic plug | m/s | L/T |
| $u_0$ | centreline velocity | m/s | L/T |
| $u_\tau$ | friction velocity | m/s | L/T |
| $v$ | specific volume | $m^3/kg$ | $L^3/M$ |
| $v$ | velocity component in $y$- or $r$-direction | m/s | L/T |
| $\bar{v}$ | time-averaged value of velocity component $v$ | m/s | L/T |

| | | | |
|---|---|---|---|
| $v'$ | fluctuating component of velocity component $v$ | m/s | L/T |
| $v^+$ | velocity component $v$ normalised by $u_\tau$ | – | – |
| $v_K$ | Kolmogorov velocity scale | m/s | L/T |
| $w$ | specific weight | N/m$^3$ | M/L$^2$T$^2$ |
| $w$ | velocity component in $z$- or $\theta$-direction | m/s | L/T |
| $\overline{w}$ | time-averaged value of velocity component $w$ | m/s | L/T |
| $w'$ | fluctuating component of velocity component $w$ | m/s | L/T |
| $w^+$ | velocity component $w$ normalised by $u_\tau$ | – | – |
| $x$ | distance along or parallel to a surface/streamwise distance | m | L |
| $X$ | length | m | L |
| $y$ | distance normal to a surface | m | L |
| $y^+$ | distance $y$ normalised by $u_\tau$ and $v$ | – | – |
| $z$ | blade height (or length) | m | L |
| $z$ | depth (i.e. distance measured vertically downwards) | m | L |
| $z'$ | height (i.e. distance measured vertically upwards) | m | L |
| $z'$ | geometric altitude | m | L |
| $z'_G$ | geopotential altitude | m | L |
| $z_C$ | depth of centroid | m | L |
| $z_P$ | depth of centre of pressure | m | L |

**Upper-case Roman symbols**

| | | | |
|---|---|---|---|
| $A$ | cross-sectional area | m$^2$ | L$^2$ |
| $A$ | surface area | m$^2$ | L$^2$ |
| $A^*$ | choking (or sonic) area | m$^2$ | L$^2$ |
| $A_E$ | nozzle exit area | m$^2$ | L$^2$ |
| $A_T$ | nozzle throat area | m$^2$ | L$^2$ |
| $B$ | barometric (or atmospheric) pressure or external pressure | bar | M/LT$^2$ |
| $B$ | log-law constant | – | – |
| $Bi$ | Bingham number | – | – |
| $C_D$ | coefficient of discharge | – | – |
| $C_D$ | drag coefficient | – | – |
| $C_F$ | average friction factor | – | – |
| $C_L$ | lift coefficient | – | – |
| $C_P$ | pressure coefficient | – | – |
| $C_P$ | specific heat at constant pressure | m$^2$/s$^2$·K | L$^2$/T$^2\theta$ |
| $C_V$ | specific heat at constant volume | m$^2$/s$^2$·K | L$^2$/T$^2\theta$ |
| $D$ | diameter | m | L |
| $D$ | drag (or drag force) | N | ML/T$^2$ |
| $\overline{D}$ | mean diameter | m | L |
| $D_H$ | hydraulic diameter | m | L |
| $D_T$ | nozzle throat diameter | m | L |
| $D'$ | drag force per unit length of surface | N/m | M/T$^2$ |

| | | | |
|---|---|---|---|
| $E$ | energy released | J | $ML^2/T^2$ |
| $E$ | Young's modulus | Pa | $M/LT^2$ |
| $Eu$ | Euler number | – | – |
| $F$ | force | N | $ML/T^2$ |
| $F$ | non-dimensional stream function | – | – |
| $F_B$ | buoyancy force | N | $ML/T^2$ |
| $F_\theta$ | function in Thwaites' method | – | – |
| $Fr$ | Froude number | – | – |
| $G$ | mass velocity | $kg/m^2 \cdot s$ | $M/L^2T$ |
| $G$ | shear modulus (fluid) | Pa | $M/LT^2$ |
| $G$ | modulus of rigidity (solid) | Pa | $M/LT^2$ |
| $H$ | height or depth | m | L |
| $H$ | horizontal component of force | N | $ML/T^2$ |
| $H$ | boundary-layer shape factor | – | – |
| $He$ | Hedstrom number | – | – |
| $H_{12}$ | boundary-layer shape factor | – | – |
| $I$ | second moment of area | $m^4$ | $L^4$ |
| $I_C$ | second moment of area about an axis through the area's centroid | $m^4$ | $L^4$ |
| $I_{xy}$ | product of inertia | $m^4$ | $L^4$ |
| $K$ | bulk modulus of elasticity | Pa | $M/LT^2$ |
| $K$ | consistency index in power-law viscosity model | $Pa \cdot s^n$ | $M/LT^{2-n}$ |
| $K$ | loss coefficient | – | – |
| $K$ | turbomachine stagnation-pressure loss coefficient | – | – |
| $Kn$ | Knudsen number | – | – |
| $1/K$ | compressibility | 1/Pa | $LT^2/M$ |
| $L$ | length | m | L |
| $L$ | lift (or lift force) | N | $ML/T^2$ |
| $L^*$ | choking length | m | L |
| $M$ | Mach number | – | – |
| $M$ | molar mass | kg/kmol | – |
| $M$ | momentum | $kg \cdot m/s$ | $ML/T$ |
| $\mathcal{M}$ | molecular weight | kg/kmol | – |
| $M_{REL}$ | relative Mach number | – | – |
| $\dot{M}$ | momentum flowrate | $kg \cdot m/s^2$ | $ML/T^2$ |
| $\dot{M}'$ | momentum flowrate per unit width of duct | $kg/s^2$ | $M/T^2$ |
| MG | metacentric height | m | L |
| $N$ | molecular number density | $1/m^3$ | $1/L^3$ |
| $N$ | number of molecules | – | – |
| $N$ | rotational speed | rps | 1/T |
| $N_A$ | Avogadro number | 1/kmol | 1/M |
| $N_P$ | turbomachine power-specific speed | – | – |
| $N_S$ | turbomachine specific speed | – | – |
| $P$ | piezometric pressure | Pa | $M/LT^2$ |

| | | | |
|---|---|---|---|
| $P$ | power | W | $ML^2/T^3$ |
| $Po$ | Poiseuille number | – | – |
| $Pr$ | Prandtl number | – | – |
| $\dot{Q}$ | volumetric flowrate | $m^3/s$ | $L^3/T$ |
| $\dot{Q}'$ | volumetric flowrate per unit width | $m^2/s$ | $L^2/T$ |
| $R$ | radius | m | L |
| $R$ | reaction force | N | $ML/T^2$ |
| $R$ | resultant force | N | $ML/T^2$ |
| $R$ | specific gas constant | $m^2/s^2 \cdot K$ | $L^2/T^2\theta$ |
| $R_E$ | mean radius of the earth | m | L |
| $R_H$ | hydraulic radius | m | L |
| $R_I$ | inner radius of annulus | m | L |
| $R_O$ | outer radius of annulus | m | L |
| $\mathcal{R}$ | molar gas constant (universal gas constant) | $kJ/kmol \cdot K$ | $L^2/T^2\theta$ |
| $Re$ | Reynolds number | – | – |
| $Re_x$ | Reynolds number based upon length $x$ | – | – |
| $Re_\delta$ | Reynolds number based upon length $\delta$ | – | – |
| $Re_C$ | critical Reynolds number | – | – |
| $Re_D$ | Reynolds number based upon pipe diameter | – | – |
| $Re_H$ | Reynolds number based upon hydraulic diameter | – | – |
| $Rep$ | Reynolds number based upon plastic viscosity | | |
| $S$ | fluid-structure interaction force | N | $ML/T^2$ |
| $St$ | Strouhal number | – | – |
| $T$ | absolute temperature | K | $\theta$ |
| $T$ | skin-friction coefficient $= \theta\tau_S/\mu U_\infty$ | – | – |
| $T$ | surface-tension force | N | $ML/T^2$ |
| $T$ | thrust (or thrust force) | N | $ML/T^2$ |
| $T$ | time interval | s | T |
| $T$ | torque | $N \cdot m$ | $ML^2/T^2$ |
| $T_0$ | stagnation (or total) temperature | K | $\theta$ |
| $T_{0,REL}$ | relative stagnation temperature | K | $\theta$ |
| $Ta$ | Taylor number | – | – |
| $U$ | free-stream velocity | m/s | L/T |
| $U_0$ | scaling velocity | m/s | L/T |
| $U_\infty$ | free-stream velocity | m/s | L/T |
| $V$ | velocity | m/s | L/T |
| $V$ | vertical component of force | N | $ML/T^2$ |
| $V_B$ | buoyancy force | N | $ML/T^2$ |
| $V_D$ | vertically downwards force | N | $ML/T^2$ |
| $V_U$ | vertically upwards force | N | $ML/T^2$ |
| $V_\infty$ | terminal velocity | m/s | L/T |
| $\overline{V}$ | average (bulk-mean) velocity | m/s | L/T |
| $\tilde{V}$ | non-dimensional velocity | – | – |
| $\overline{V}^+$ | average velocity $\overline{V}$ normalised by $u_\tau$ | – | – |

| | | | |
|---|---|---|---|
| $\boldsymbol{\mathcal{v}}$ | volume | m³ | L³ |
| $\boldsymbol{\mathcal{v}}_C$ | critical volume for validity of continuum hypothesis | m³ | L³ |
| $\boldsymbol{\mathcal{v}}_D$ | displaced volume | m³ | L³ |
| $\boldsymbol{\mathcal{v}}_S$ | submerged volume | m³ | L³ |
| $V_\infty$ | $y$-direction velocity at edge of boundary layer | m/s | L/T |
| $W$ | relative velocity | m/s | L/T |
| $W$ | weight | N | ML/T² |
| $W$ | width | m | L |
| $W$ | work | J | ML²/T² |
| $\dot{W}$ | rate of work input (power input) | W | ML²/T³ |
| $We$ | Weber number | – | – |
| $X$ | length | m | L |
| $Y$ | boundary-layer thickness | m | L |
| $Y$ | surface tension | N/m | M/T² |
| $Z$ | depth of liquid | m | L |

**Lower-case Greek symbols** (English word in parentheses)

| | | | |
|---|---|---|---|
| $\alpha$ (alpha) | angle of attack | ° or rad | – |
| $\alpha$ | absolute flow angle | ° or rad | – |
| $\alpha$ | conical gap angle | ° or rad | – |
| $\alpha$ | non-dimensional constant in Blasius' equation | – | – |
| $\alpha'$ | constant in shock-structure analysis | m²/s | L²/T |
| $\beta$ (beta) | oblique shock angle | ° or rad | – |
| $\beta$ | relative-flow angle | ° or rad | – |
| $\beta$ | wedge angle | ° or rad | – |
| $\gamma$ (gamma) | ratio of specific heats | – | – |
| $\dot{\gamma}$ | shear rate | 1/s | 1/T |
| $\dot{\gamma}_{xy}$ | shear rate corresponding to $\tau_{xy}$ | 1/s | 1/T |
| $\delta$ (delta) | angle of deflection or deviation | ° or rad | – |
| $\delta$ | boundary-layer thickness | m | L |
| $\delta$ | radial gap width | m | L |
| $\delta A$ | element of area | m² | L² |
| $\delta F$ | element of force | N | ML/T² |
| $\delta h$ | infinitesimal height difference | m | L |
| $\delta H$ | element of horizontal force | N | ML/T² |
| $\delta m$ | element of mass | kg | M |
| $\delta p$ | infinitesimal change or difference in pressure | Pa | M/LT² |
| $\delta s$ | infinitesimal change of distance | m | L |
| $\delta t$ | infinitesimal change in time | s | T |
| $\delta V$ | element of vertical force | N | ML/T² |
| $\delta \boldsymbol{\mathcal{v}}$ | element of volume | m³ | L³ |

| | | | |
|---|---|---|---|
| $\delta W$ | element of weight | N | $ML/T^2$ |
| $\delta x$ | element of streamwise or $x$-direction distance | m | L |
| $\delta y$ | element of distance normal to a surface or $y$-direction distance | m | L |
| $\delta z$ | element of depth or $z$-direction distance | m | L |
| $\delta z'$ | element of height | m | L |
| $\delta^*$ | boundary-layer displacement thickness | m | L |
| $\delta_{SUB}$ | thickness of viscous sublayer | m | L |
| $\delta_1$ | boundary-layer displacement thickness | m | L |
| $\delta_2$ | boundary-layer momentum-deficit thickness | m | L |
| $\epsilon$ (epsilon) | turbulent kinetic energy dissipation rate | $m^2/s^3$ | $L^2/T^3$ |
| $\epsilon$ | upwash or downwash angle | ° or rad | – |
| $\varepsilon$ (epsilon) | eccentricity | m | L |
| $\varepsilon$ | non-dimensional annular gap with | – | – |
| $\varepsilon$ | surface-roughness height | m | L |
| $\varepsilon^+$ | surface-roughness height normalised by $u_\tau$ and $\nu$ | – | – |
| $\eta$ (eta) | dynamic viscosity | Pa · s | M/LT |
| $\eta$ | boundary-layer similarity variable | – | – |
| $\theta$ (theta) | angle | ° or rad | – |
| $\theta$ | boundary-layer momentum-deficit thickness | m | L |
| $\theta$ | contact angle | ° | – |
| $\theta$ | turning angle | ° | – |
| $\dot{\theta}$ | angular velocity | rad/s | 1/T |
| $\ddot{\theta}$ | angular acceleration | $rad/s^2$ | $1/T^2$ |
| $\kappa$ (kappa) | lapse rate | K/m | $\theta/L$ |
| $\kappa$ | von Kármán's constant | – | – |
| $\kappa$ | wavenumber | 1/m | 1/L |
| $\lambda$ (lamda) | time constant | s | T |
| $\lambda$ | pressure-gradient parameter | – | – |
| $\lambda$ | Pohlhausen's pressure-gradient parameter | – | – |
| $\lambda$ | wavelength of turbulence | m | L |
| $\lambda_P$ | Poiseuille-flow pressure-gradient parameter | – | – |
| $\lambda_\theta$ | boundary-layer pressure-gradient parameter | – | – |
| $\mu$ (mu) | dynamic viscosity | Pa · s | M/LT |
| $\mu$ | Mach angle | ° or rad | – |
| $\mu_{EFF}$ | effective viscosity | Pa · s | M/LT |
| $\mu_P$ | viscosity of plastic plug | Pa · s | M/LT |
| $\mu_T$ | eddy viscosity | Pa · s | M/LT |
| $\mu_\infty$ | infinite-shear-rate viscosity | Pa · s | M/LT |
| $\nu$ (nu) | kinematic viscosity | $m^2/s$ | $L^2/T$ |
| $\nu$ | Prandtl-Meyer function | ° | – |
| $\nu_T$ | kinematic eddy viscosity | $m^2/s$ | $L^2/T$ |

| | | | |
|---|---|---|---|
| $\xi$ (xi) | blade stagger angle | ° or rad | – |
| $\xi$ | non-dimensional distance | – | – |
| $\xi$ | turbomachine enthalpy-loss coefficient | – | – |
| $\xi_P$ | non-dimensional radius of plastic plug | – | – |
| $\rho$ (rho) | density | kg/m$^3$ | M/L$^3$ |
| $\sigma$ (sigma) | density ratio | – | – |
| $\sigma$ | relative density | – | – |
| $\sigma$ | surface tension | N/m | M/T$^2$ |
| $\sigma_{xx}$ | normal stress in $x$-direction | Pa | M/LT$^2$ |
| $\tau$ (tau) | characteristic time | s | T |
| $\tau$ | shear stress | Pa | M/LT$^2$ |
| $\tau_K$ | Kolmogorov time scale | s | T |
| $\tau_S$ | surface shear stress | Pa | M/LT$^2$ |
| $\overline{\tau_S}$ | average surface shear stress | Pa | M/LT$^2$ |
| $\tau_Y$ | yield stress | Pa | M/LT$^2$ |
| $\tau_{xy}$ | shear stress acting in $y$-direction | Pa | M/LT$^2$ |
| $\phi$ (phi) | angle | ° or rad | – |
| $\phi$ | blade camber angle | ° or rad | – |
| $\phi$ | turbomachine flow coefficient | – | – |
| $\chi$ (chi) | blade angle | ° or rad | – |
| $\chi$ | boundary-layer scale factor | – | – |
| $\psi$ (psi) | stream function | 1/s | 1/T |
| $\psi$ | hydraulic machine pressure-change coefficient | – | – |
| $\omega$ (omega) | angular velocity | rad/s | 1/T |

**Upper-case Greek symbols**

| | | | |
|---|---|---|---|
| $\Gamma$ (gamma) | circulation | m$^2$/s | L$^2$/T |
| $\Gamma$ | lapse rate | °C/km | $\theta$/L |
| $\Gamma_{AD}$ | adiabatic lapse rate | °C/km | $\theta$/L |
| $\Delta$ (delta) | finite change or difference | – | – |
| $\Delta$ | scaling length | m | L |
| $\Delta_S$ | shock thickness | m | L |
| $\Delta p$ | finite pressure difference | Pa | M/LT$^2$ |
| $\Delta p_0$ | reduction in stagnation pressure | Pa | M/LT$^2$ |
| $\Delta Z$ | finite depth difference | m | L |
| $\Delta\rho$ | density difference | kg/m$^3$ | M/L$^3$ |
| $\Theta$ (theta) | dilation | 1/s | 1/T |
| $\tilde{\Theta}$ | ratio $\theta/\delta$, where $\theta$ = boundary-layer momentum-deficit thickness | – | – |
| $\Lambda$ (lamda) | degree of reaction | – | – |
| $\Lambda$ | molecular mean free path | m | L |
| $\Pi$ (pi) | non-dimensional group | – | – |
| $\Pi$ | shock strength | – | – |

| | | | |
|---|---|---|---|
| $\Pi$ | wake parameter | – | – |
| $\Sigma$ (sigma) | summation | – | – |
| $\tilde{\Phi}$ (phi) | ratio $\delta^*/\delta$ | – | – |
| $\Omega$ (omega) | angular velocity | rad/s | 1/T |

## Mathematical symbols

| | | | |
|---|---|---|---|
| *div* | vector operator of divergence | $1/m$ | $1/L$ |
| $\nabla$ | del (or gradient) operator | $1/m$ | $1/T$ |
| $\nabla^2$ | Laplacian operator | $1/m^2$ | $1/L^2$ |

## Lower-case Roman subscripts

| | |
|---|---|
| $f$ | friction |
| $r$ | radial direction |
| $t$ | throat |
| $x$ | $x$-direction |
| $y$ | $y$-direction |
| $z$ | $z$-direction |

## Upper-case Roman subscripts

| | |
|---|---|
| $A$ | actual |
| $B$ | back (pressure) |
| $C$ | centroid or critical |
| $E$ | exhaust |
| $F$ | fluid or fuel or full scale |
| $G$ | centre of gravity or gas |
| $H$ | based on hydraulic diameter |
| $H_2O$ | water |
| $I$ | inlet or inner surface |
| $L$ | laminar or liquid or lower surface |
| $M$ | manometer |
| $M$ | model |
| $O$ | outer surface |
| $P$ | centre of pressure |
| $REF$ | reference condition |
| $S$ | isentropic or solid or submerged or surface |
| $T$ | total or turbulent |
| $TH$ | theoretical |
| $U$ | upper surface |

## Lower-case Greek subscript

| | |
|---|---|
| $\theta$ | $\theta$-direction |

## Numerical subscripts

| | |
|---|---|
| 0 | stagnation or reference conditions |
| 1 | conditions upstream of a shockwave |
| 2 | conditions downstream of a shockwave |

## Superscripts

| | |
|---|---|
| $T$ | isothermal |
| $*$ | choking (or critical or sonic) condition |

# 1 Introduction

Why do students of many branches of engineering need to study fluid mechanics? First and foremost, the answer is 'design'. It can be argued that the principal purpose of engineering is **engineering design**, and it is frequently the case that considerations of fluid flow are crucial to the engineering-design process. It would be inappropriate here to discuss in detail what is meant by engineering design. Suffice to say, design is sometimes confused with **styling**, which refers primarily to the external appearance of a device or machine, whereas engineering design is concerned with its functioning and invariably involves calculations based upon the laws of physics. In this introductory chapter we indicate the wide and diverse range of practical situations where fluid mechanics plays a central role, often together with such related subjects as heat transfer, thermodynamics, and combustion. Although the emphasis in this book is on applications of fluid mechanics in mechanical, aeronautical, and civil engineering, other examples could be taken from biomedical, building, chemical, and environmental engineering. Within this book we also mention many of the natural phenomena for which fluids, and the way they flow, play a fundamental role. Although the origins of fluid mechanics can be traced to ancient Greek (Archimedes) and Roman (Frontinus) times, and important contributions were made in the 15$^{th}$ (da Vinci), 16$^{th}$, 17$^{th}$ (Newton, Pascal), 18$^{th}$ (Bernoulli and Euler), and 20$^{th}$ centuries (Prandtl, Taylor), most of the major developments in the subject were made by engineers, mathematicians, and physicists in the 19$^{th}$ century (including Kelvin, Mach, Navier, Rankine, Rayleigh, Reynolds, and Stokes). Many effects, functions, equations, non-dimensional parameters (see Chapter 3), etc., are named after these pioneers and other major contributors to fluid mechanics: brief biographies are included in Appendix A.

A thorough understanding of the contents of this book should enable the student to

- use the results of **dimensional analysis** (Chapter 3) to scale up the results of wind-tunnel model tests[1]. A typical example is in the analysis of wind-tunnel data for the aerodynamic behaviour of a **Formula 1 racing car**, as shown in Figure 1.1 (to illustrate the point, we could just as well have chosen, e.g. a fighter aircraft or a bridge).
- specify the characteristics of a **centrifugal pump**, as illustrated in Figure 1.2, required to handle large quantities of oil, based upon small-scale tests with water, again guided by dimensional analysis
- calculate the flowspeed in a wind tunnel using a **Pitot-static tube** and a **U-tube manometer**, as shown in Figure 1.3 (the size of the manometer relative to the Pitot tube is

---

[1] Where the aerodynamic characteristics of an aircraft, a car, a locomotive, or any other vehicle are to be investigated in a wind or water tunnel, it is usual for the vehicle to be fixed in position with the fluid flowing around it. This change is known formally as a **Galilean transformation**. In a wind tunnel used to investigate vehicles in contact with a road, the surface in contact with the vehicle usually moves at the same speed (and direction) as the working fluid. Such an arrangement is referred to as a **rolling road**. Note too that the flow direction in all figures in this book is from left to right, a convention adopted in the majority of fluid mechanics textbooks.

*Introduction to Engineering Fluid Mechanics*. Marcel Escudier.
© Marcel Escudier 2017. Published 2017 by Oxford University Press.

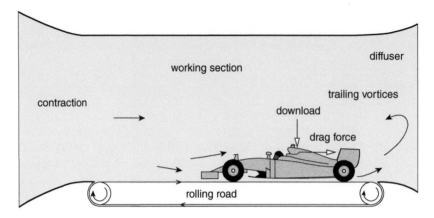

**Figure 1.1** Wind-tunnel test of a racing car

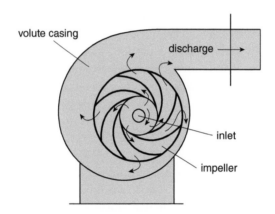

**Figure 1.2** Centrifugal-pump testing

much reduced in the diagram). This calculation involves both **hydrostatics** (Chapter 4) and **Bernoulli's equation** (Section 8.9).

- using the principles of hydrostatics (Section 8.5), calculate the resultant force exerted by the water in a reservoir on the face of a dam, as shown by $R$ in Figure 1.4[2]
- use the principles of hydrostatics to design a **floating boom** to contain an **oil slick,** as shown in Figure 1.5
- use Bernoulli's equation (see Chapter 8) to calculate the **lift force** resulting from the airflow over the surfaces of an aerofoil, as shown in Figure 1.6. A qualitative discussion of the underlying physical phenomena which explain lift is given in Section 17.7.
- use Bernoulli's equation to determine the flowrate at which **internal boiling** occurs at room temperature as a consequence of reduced pressure (so-called **cavitation,** discussed in Section 8.11) in the flow of a liquid through a constriction, such as a valve or, as illustrated in Figure 1.7, a **convergent-divergent nozzle**

---

[2] The inverted triangle is used to identify a free surface.

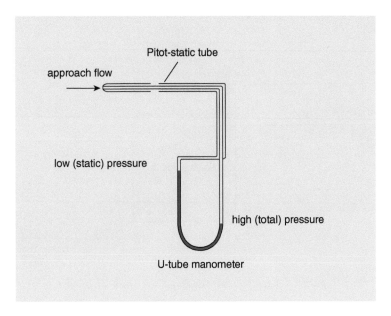

**Figure 1.3** Pitot-static tube and U-tube manometer

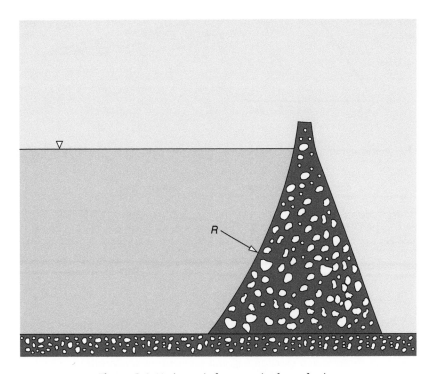

**Figure 1.4** Hydrostatic force on the face of a dam

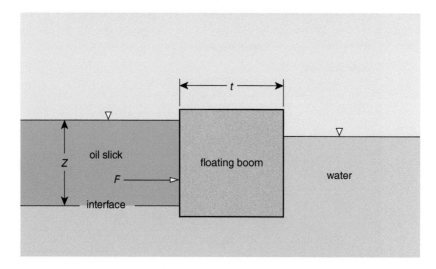

**Figure 1.5** Floating boom designed to contain an oil slick

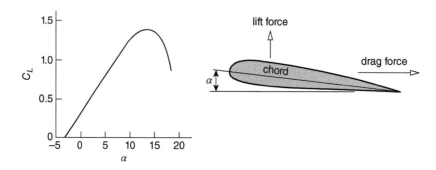

**Figure 1.6** Aerodynamic lift generated by an aerofoil: Lift coefficient $C_L$ versus angle of attack $\alpha$

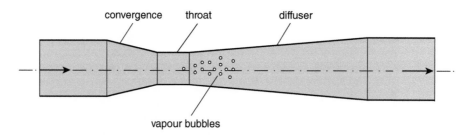

**Figure 1.7** Cavitation in water flow through a convergent-divergent nozzle

- use the **mass-conservation** (Section 6.8) and **momentum-conservation** (Chapter 9 and Section 10.4) **equations** to calculate the thrust developed by a turbofan engine, such as that shown schematically in Figure 1.8, which is a simplified version of Figure 14.1
- use the continuity and momentum equations, together with Bernoulli's equation, to calculate the power output of a **Pelton hydraulic turbine** (Section 10.11), as shown in Figure 1.9

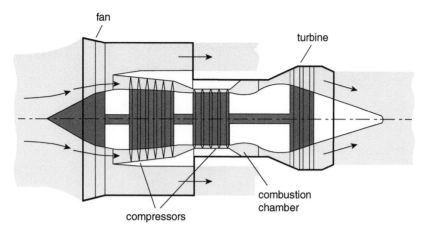

**Figure 1.8** Schematic cross section of a turbofan engine

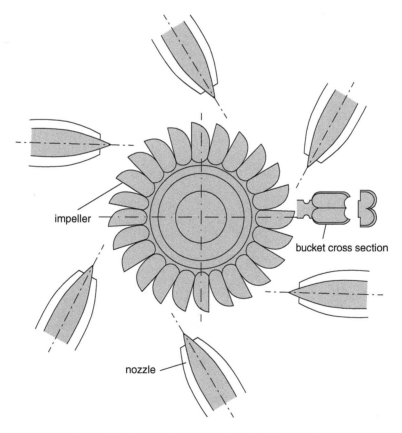

**Figure 1.9** Schematic cross section of a Pelton turbine

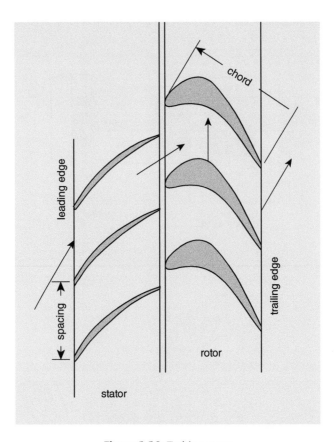

**Figure 1.10** Turbine stage

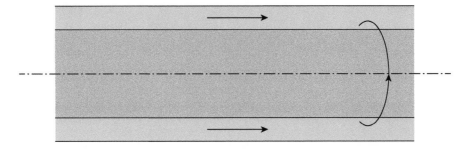

**Figure 1.11** Viscous flow through a concentric annulus with centrebody rotation

- use the mass- and momentum-conservation equations, the **steady-flow energy equation**, and the **perfect-gas law** to calculate the power output of a **turbine (or compressor) stage** (Section 14.8), as shown in Figure 1.10
- use the mass- and momentum-conservation equations, together with **Newton's law of viscosity**, to calculate the flow of a viscous fluid through a **concentric annulus** with **centrebody rotation** as shown in Figure 1.11 (Section 16.5)

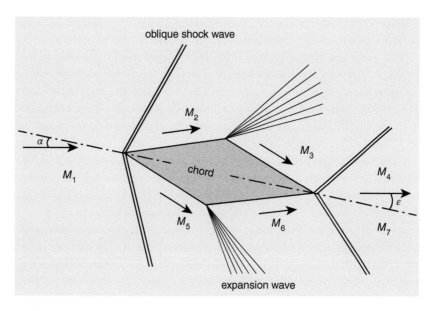

**Figure 1.12** Supersonic flow with shock and expansion waves over a diamond-shaped aerofoil

- use the tabulated solution of the **Blasius equation** for a laminar boundary layer (Section 17.3) to calculate the drag force on a thin flat plate immersed in a viscous fluid flow
- use the Virginia Tech **Compressible Aerodynamics Calculator** (see Section 11.3) to calculate the **shock** and **expansion waves**, the **Mach number** and **pressure distributions**, and lift force (Section 12.3) of supersonic perfect-gas flow over a **diamond-shaped aerofoil**

The **thickness** of the **shock waves** in Figure 1.12 is greatly exaggerated (see Section 11.8). The subscripted $M$'s indicate the Mach numbers in each region of the flow.

The foregoing is just a selection of the engineering applications of fluid mechanics considered in this textbook. As we emphasise in the remainder of this chapter, there are few areas of life, whether man-made or natural, in which fluids and fluid mechanics do not play a vital role.

## 1.1  What are fluids and what is fluid mechanics?

Without salt-free water to drink, we die within about ten days, and become brain dead within about four minutes without the oxygen which makes up about 21% by volume of the air we breathe (the rest is mainly nitrogen, 78%). Water is a **liquid**, air is a **gas**, and both are what we call **fluids**. The total mass of air in the **atmosphere** which surrounds the earth (see Section 4.13) is estimated to be about $5.3 \times 10^{18}$ kg (or 5.3 petatonnes[3]), and the total mass of

---

[3] Peta- and exa- are two of the 20 approved prefixes of The International System of Units (SI) presented in Section 3.2.

water in all the oceans, lakes, rivers, etc., the so-called **hydrosphere**, is about $1.4 \times 10^{21}$ kg (or 1.4 exatonnes). Given their abundance, and their importance to our very existence, it is hardly surprising that water and air are the two fluids encountered most commonly in fluid mechanics. There are, of course, many other familiar 'everyday' fluids: methane, ethane, hydrogen, helium, oxygen, and nitrogen are all gases which behave much like air; similarly, natural (as opposed to synthetic) fluids such as oil, petrol, mercury, honey, glycerine, and alcohol are all relatively simple liquids much like water, but with different densities, viscosities, and other properties (Chapter 2 is concerned with fluid properties and what makes fluids different from solids). We call these simple fluids, with viscosities independent of their motion (though not their temperature), **Newtonian**. Blood, synovial fluid (which lubricates our joints), custard, mayonnaise, salad cream, ketchup, hair gel, toothpaste, drilling fluid, fracking (or fracturing) fluids, freshly mixed cement slurry, and paint are all liquids but with viscous properties and flow behaviour very different from those of water. These differences arise primarily because such liquids have either a complex molecular structure or consist of a mixture of a simple liquid (such as water) and many tiny (often in the micron range) suspended particles. Because of the complexity of their viscosities, these liquids are termed **non-Newtonian**. The study of the viscous properties of non-Newtonian liquids is a subject in itself, called **rheology**. There is a brief account of non-Newtonian liquids in Section 2.10. Simple models for such liquids and their flow are discussed in Sections 15.5 and 16.6.

We know from everyday experience that liquids flow. Water flows from the mains supply when we open the tap. Water flows from the sink or bath into the drainage system. Tea flows from a teapot. Beer flows into our digestive system from a glass, bottle, or can, and then, usually after a biological/chemical transformation and temporary storage, flows out again from our urinary system. Blood flows through our arteries and veins, pumped by a natural or artificial heart. Air flows into our lungs, and carbon dioxide flows out into the atmosphere. Liquid or gaseous fuel flows into the engines of passenger vehicles, trains, aircraft, and ships, while exhaust gases flow out, again into the atmosphere. Town gas, a mixture consisting primarily of hydrogen, methane, and carbon dioxide, flows to our cookers and boilers, and products of combustion flow out. Air flows around us as we walk, run, or ride our bicycles. It flows over the bodywork of our cars, over the wings and fuselages of the aircraft in which we fly, and through the blades of wind turbines, causing them to rotate and generate electrical power. Oil, gas, brine, and drilling fluid flow from deep in the earth to the surface when we drill for oil or gas. Water flows from rivers into reservoirs, lakes, and the sea and from reservoirs through **hydraulic turbines** again to generate electrical power. It also flows into the boilers of power-generating steam turbines where it is converted into steam, a vapour. It flows around the hull of a ship or submarine. Lava, a non-Newtonian liquid, flows from an active volcano.

We should also be aware that some substances can exist in more than one state (or **phase**). Water, for example, can exist as ice (a solid), water (a liquid), or steam. The latter exhibits some of the characteristics of a gas, particularly at very high temperatures, and is termed a **vapour**. Many gases, including air, can be **liquefied** by subjecting them to very high pressure and/or low temperature.

Engineering fluid mechanics is concerned with analysing fluid flows, such as those mentioned above, in order to calculate the rates at which they flow, the changes in pressure as they flow, and the stresses and forces they exert on the machines and surfaces through and over which they flow. The **law of conservation of mass**, **Newton's laws of motion**, and the **laws of thermodynamics** (principally the **first law** in the form of the **steady-flow energy equation**), together with appropriate representation of fluid properties, form the basis of the analysis. Before we go into further detail, it is useful to expand the catalogue of situations where fluid mechanics plays an essential role.

## 1.2 Fluid mechanics in nature

The height of the **atmosphere**, that is the altitude beyond which we are in the vacuum of outer space, is usually taken to be about 80 km. For many purposes, the atmosphere can be taken as a series of stationary spherical layers of air with the temperature variation shown in Figure 1.13.

We consider this **hydrostatic** model of the atmosphere in some detail in Section 4.13. We know, of course, that the atmosphere, especially the part of it we inhabit, is very often far from static; **meteorology** is the branch of fluid mechanics devoted to the study of its motion. Anyone who has seen time-lapse images of clouds knows that, in addition to being swept along by winds, they are in constant motion due to **thermals** (finite packets of warm air moving upwards which allows gliders to rise to altitudes up to about 15 km), **evaporation**, **condensation**,

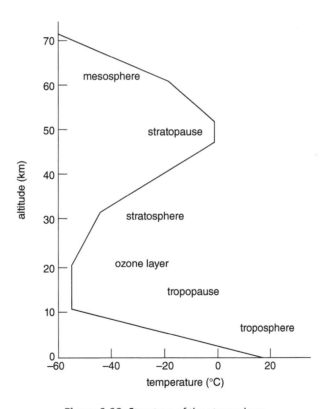

**Figure 1.13** Structure of the atmosphere

and **shearing** (which gives rise to the **clear-air turbulence** often experienced by passenger aircraft). We should also mention dust devils, tornadoes, and hurricanes which are examples of the intense, often violent, **swirling motion** which can arise in the lower atmosphere due to combined thermal and shearing effects.

While it is essential that the earth is surrounded by a layer of air, it is just as important for humans (at least in our current state of evolution) that all the water in the hydrosphere is not distributed uniformly over the planet's surface. Were that the case, the water layer would be about 2.7 km deep. Instead, this water actually covers about 71% of the earth's surface, with regions of the deepest ocean being about 10 km deep, almost equal in magnitude to the height of Mount Everest. As with the atmosphere, much can be learned about the oceans, reservoirs, lakes, etc., by considering them to be at rest. Chapters 4 and 5 are devoted to **hydrostatics**— the study of fluids at rest—with a considerable fraction concerned with the forces exerted on surfaces, such as the face of a dam, as shown in Figure 1.4. **Oceanography** is the branch of fluid mechanics which deals with tides, currents, waves, **stratification** (water-density variations due to salinity and temperature changes with depth), and other phenomena associated with water motion in the oceans. Related topics involving fluid mechanics are erosion, sedimentation, whirlpools, river flows, and also the flow in canals and sewers, although the latter are man-made rather than natural systems. In principle, we could also include here the fluid mechanics associated with the wave-like body motion which fish, eels, aquatic mammals, and sperm use to swim.

Undeniably natural are the flows of lava from an active volcano and of hot water and steam from a geyser. The flow of formation fluids (oil, methane, hydrogen sulphide, brine, etc.), as well as drilling mud, from an oil well represents a mixture of man-made and natural phenomena. There would be no flow were it not for the man-made well, but the flow of formation fluids through porous rock involves natural fluids flowing through naturally occurring channels in a natural medium. Here again, however, in **hydraulic fracturing** (commonly referred to as '**fracking**') we are dealing with a combination of man-made and natural processes.

The study of flow in the circulatory, respiratory, urinary, and other biological systems is termed **biofluid mechanics**. As with all natural systems, an additional difficulty is that the geometry of the flow channels is not well defined and often not fixed. For example, arteries and veins are flexible and so change in cross section as blood pressure increases and decreases with every beat of the heart. To further complicate matters, blood is not a homogeneous liquid but consists mainly of red corpuscles, which are thin discs about 8 $\mu$m in diameter with a thick rim, suspended in plasma. As a consequence of this composition, the effective viscosity (see Section 2.10) of blood decreases with shearing (relative tangential movement) and is slightly elastic (**viscoelastic**) in character, i.e. blood is a **non-Newtonian fluid**. At rest, blood has an effective viscosity about 100 times that of water, although this factor decreases to about five in the arteries. In any event, the saying 'blood is thicker than water' is entirely accurate. Although synovia, the fluid which lubricates our joints, is a homogeneous liquid, it is again a non-Newtonian fluid with **shear-thinning**, **viscoelastic** properties, in this instance because it has a polymer-like molecular structure.

## 1.3 External flows

As engineers, we are concerned primarily with fluids which flow either through or around man-made devices, which we term **internal** and **external flows**, respectively. In either case, **viscosity** (or to be more precise, **dynamic viscosity**) is the key fluid property which determines the details of the flow. Wherever velocity gradients occur in a flowing fluid, the fluid property viscosity leads to shear stresses and forces. A fundamental concept in fluid mechanics is the **no-slip condition** according to which, in the immediate vicinity of a solid surface, a consequence of viscosity is that the fluid is brought to rest (or, more generally, if the surface is itself moving, to the same velocity as the surface so that the relative velocity is zero). In essence, the fluid adheres to the surface. For an external flow, the major effects of viscosity are confined to a relatively thin region close to the surface called the **boundary layer**, the subject of Chapter 17.

In the case of buildings, smoke stacks (or chimneys), bridges, wind turbines, windmills, off-shore structures such as drilling platforms, etc., the external flow (there may be quite separate internal flows, such as exhaust gases) is provided by nature. The damage which sometimes occurs to these and other structures when high windspeeds arise tells us that the wind can impose massive forces on their surfaces. In certain circumstances, even at relatively low speed, a steady wind can excite vibrations (**flow-induced vibrations**) which can be of sufficient amplitude to cause structural damage. Huge plate-glass windows have been known to pop out of their frames due to wind-induced torsional oscillations of skyscrapers, as happened to the 241 m high **John Hancock Tower** opened in Boston in 1976. The best known example of wind-induced vibration was the complete destruction in 1940 of the **Tacoma Narrows Bridge** in Washington State, USA. Remarkably, in both instances, the vibration was initiated at wind-speeds no greater than about 70 kph. In order to design structures which are safe, we need to calculate both the steady and periodic forces due to the wind, either from fundamental theory or, more likely, from a combination of theory and experimental data obtained from tests carried out in a wind or water tunnel. The use of experimental data, generalised using **dimensional analysis** (Chapter 3), is termed **empiricism**. **Environmental fluid mechanics** also concerns the **dispersion of pollutants** in the atmosphere and in the sea, rivers, lakes, etc.

Some of the most advanced theoretical and experimental work in fluid mechanics has been associated with the development of aircraft, spacecraft, and missiles. There have been remarkable advances in aviation since December 1903, when Orville Wright flew a powered, heavier-than-air, machine some 260 m in 59 s. For example, we now take for granted passenger aircraft such as the turbofan-powered Airbus A380-800 with a passenger-carrying capacity up to about 850, a maximum take-off weight of 575,000 kg, a wingspan of 80 m, a cruising speed of 945 km/h (just below soundspeed), and a range of 15,700 km. Although taken out of service in 2003, just as impressive was the performance of the **turbojet**-powered (see Section 10.3) British Aerospace Corporation/Aérospatiale supersonic transport aircraft, **Concorde**, which routinely carried about 130 passengers at twice **soundspeed** (a flight speed of about 2130 km/h) in the **stratosphere** (see Section 4.13). Although, as we see from Figure 1.10, the atmospheric temperature at **cruise altitude** (about 18 km) is about –56.5 °C, the skin of Concorde reached a temperature of about 120 °C, due to **frictional heating**, causing the length of the aircraft to increase by about 0.3 m. Modern combat aircraft, such as the **Lockheed Martin F-22 Raptor**, again turbofan powered, can fly at **Mach numbers** above two (about 2500 km/h). Although manned flights into space are now regarded as almost routine, in reality each flight represents an extraordinary engineering achievement. For example, the speed required to escape the earth's gravitational pull is about 11 km/s (i.e. 40,000 km/h or a Mach number above 30) and, on re-entry into the earth's atmosphere, the air surrounding the space shuttle becomes so hot (6000 °C plus) that the craft is surrounded by a glowing plasma.

One of the ways we distinguish between different flight regimes is through the Mach number, which is the ratio of the flight speed of an aircraft to the speed of sound at the flight altitude (discussed further in Section 3.12 and Chapter 11). As the Mach number increases, the fluid mechanics becomes more complicated because an increasing number of physical phenomena have to be taken into account. If the Mach number is considerably less than unity (0.3 is the value usually taken), changes in fluid density are negligible and the flow is said to be **incompressible**. For higher Mach numbers, **compressibility** effects (i.e. density changes) become increasingly important but can be accounted for in a relatively straightforward way using the **perfect-gas law** to relate temperature, pressure, and density (see Section 2.4), together with the first law of thermodynamics (Chapter 11). Once the Mach number exceeds about five, however, very high temperatures develop near surfaces, and the air properties change due to chemical breakdown of the molecules and the subsequent reaction of free atoms. At this point, physical chemistry also comes into play, but beyond the scope of this book.

The preceding paragraphs suggest an important aspect of the subject of fluid mechanics which students often find difficult to understand: even at relatively low flowspeeds, there are few problems we can solve completely, usually because the mathematics involved becomes far too complicated, even if we understand all the physics involved and know the relevant equations. To a degree, computers can take over at some stage in the analysis of a problem to provide a numerical rather than an analytical (i.e. algebraic) solution. Unfortunately, even the largest and fastest computers available at the present and in the foreseeable future are inadequate to solve most practical problems and we have no choice but to introduce approximations, assumptions, and simplifications. In fact, this 'engineering' approach represents common sense. For example, if we are dealing with a low-speed gas flow where we know that the fluid density remains practically constant, there is no point in making our task more difficult (and more expensive) than necessary by not introducing this simplification from the outset. Of course, it is usually a matter of experience, or even hindsight, which tells us what simplifications are justified. In this textbook, we approach problems using the simplest possible physics and mathematics, with the aim of deriving approximate solutions which provide some insight into the interplay between fluid properties, flow geometry, and flowspeed. The reader needs to bear in mind that our approach often represents only a start to, rather than a complete treatment of, the solution of problems of fluid flow.

Even land vehicles have now reached speeds where air-density variations must be accounted for. The land-speed record, held by the turbofan-powered car **Thrust SSC** since 1997, is 1228 km/h, which corresponds to a Mach number of 1.018, i.e. just **supersonic** (Mach numbers in the range close to unity are termed **transonic**). A new turbofan-powered car, **Bloodhound SSC**, is being developed with a target speed of about 1700 km/h (Mach 1.9). Somewhat slower is the Japanese Tōhoku Shinkansen '**Bullet Train**' which has a top speed of about 320 km/h or 89 m/s, corresponding to a Mach number of 0.26, so that compressibility effects are largely insignificant. However, some racing cars can achieve speeds where compressibility effects cannot be neglected: the highest speed reached at the California Speedway track in Fontana, California, is about 400 km/h or 111 m/s, which corresponds to a Mach number of 0.33. Although this figure is close to the 0.3 'cutoff', it must be the case that on the bodywork of the cars there would have been regions where the airspeed was considerably higher. It has to be said that normal cars, buses, and lorries have considerably lower top speeds and the airflow around them can safely be considered to be incompressible (i.e. to have constant density).

Although the speeds of even the fastest marine vehicles are much lower than for most land vehicles, the fluid mechanics involved is complicated by **wave motion** which arises due to the tendency for gravitational pull to overcome any disturbance to a water surface. We are all familiar with the surface **gravity waves** which propagate radially outwards when we throw a stone into a pond, whereas the forward movement of a ship creates a vee-shaped pattern of surface waves. Although invisible to the eye, a submarine travelling deep below the surface also generates gravity waves as it disturbs water layers of different densities which occur due to variations with depth of salt content and temperature. The energy required to generate waves has to be provided by the propulsion system of the ship or submarine and so corresponds to an additional contribution to the drag force, so-called **wave drag**.

## 1.4 **Internal flows**

Most of the flow situations dealt with in this textbook are concerned with internal flows through **pipes, ducts, nozzles**, engines, **turbomachines**, etc. In one sense, internal flows are easier to deal with than external flows because the flow is confined within solid boundaries unlike the flow over an **aerofoil** (Figure 1.6), for example, where the region of flow is practically unlimited.

The most common man-made device through which flow occurs is a metal, plastic, or glass pipe of circular cross section. Pipes of this kind allow oil and gas to flow to the earth's surface from reservoirs which may be many kilometres below, often deep below the seabed, and then hundreds of kilometres across land, directly to refineries or to ports for transfer to ships. Oil and gas pipelines, and also the pipes which convey water into the turbines of a **hydroelectric power plant**, may be a metre or more in diameter. The enormous capital cost involved means that careful consideration has to be given to the design of such pipelines including all the associated **valves, bends, contractions, expansions, pumps**, monitoring equipment, etc. Smaller diameter pipes connect the pumps, separators, boilers, distillation columns, burners, filters, etc., of oil refineries and other chemical-processing plant. Such pipes allow gas and water to be transported to the homes where we live and to the offices and factories where we work. Fluid flow through a straight pipe is resisted by **friction** between the fluid and the internal surface of the pipe, which arises due to the viscosity of the fluid (see Section 2.8) and has to be overcome by a pressure difference created by a pump or compressor, or by gravitational effects. Friction also causes the fluid temperature to rise, the fluid density to decrease, and the average fluid velocity to increase. Much like the situation of an external flow, a boundary layer develops and grows in thickness with downstream distance so that, in an internal flow, if the flow channel is long enough, it is inevitable that eventually fluid across the entire cross section of the channel is affected by viscosity (see Chapter 16). Pipe flow of compressible fluids is considered in detail in Chapter 13. Due to surface friction or external heating, in the case of a gas, the fluid velocity in a pipe may even reach the speed of sound, causing an effect called **choking** (see Chapters 11 and 13), which limits the volume of gas which can be pumped through the pipe. Clearly, even a flow which at first sight probably appears to be the simplest we can think of turns out to be rather complicated. In fact, the situation is even more complicated than we have indicated so far because it is only for low flowrates or small-diameter pipes or highly viscous fluids (all of these influences are accounted for by a non-dimensional parameter termed the **Reynolds number**, which we discuss further in Chapters 3 and 15 to 18) that the flow remains smooth and steady (so-called **laminar flow**) and we are able to analyse it completely. The majority of flows of engineering interest exhibit a high degree of random unsteadiness which we call **turbulence** (see Chapter 18) and, even today, we are able to calculate turbulent pipe flow from first principles only through the use of supercomputers. Fortunately, the principles of dimensional analysis apply whether a flow is laminar or turbulent, and this enables us to generalise experimental data for use in engineering-design calculations.

In industrial applications, pipes rarely stay straight or keep the same diameter for long (see Section 18.11). Often more important than understanding the details of the flow within a pipe or pipe system is the ability to calculate the **hydrodynamic forces** which arise when a pipe changes direction and, perhaps, also diameter, as illustrated by the pipe bend in Figure 1.14 (see Section 10.7).

The calculation of hydrodynamic forces is one of the main topics of Chapter 10, which brings together many of the concepts and principles introduced in previous chapters, particularly those in Chapters 6, 7, and 9.

Combustion chambers, furnaces, boilers, jet pumps, control valves, guidevanes, cyclone separators, radiators, oil coolers, fuel-injection systems, carburettors, rocket engines, and the coolant channels within the core of a nuclear reactor or the block of a petrol or diesel engine

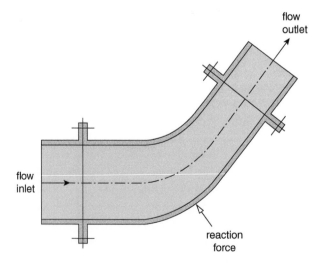

**Figure 1.14** Hydrodynamic reaction force exerted on a pipe bend

are all examples involving internal fluid flow. As we show in Chapter 10, the flow characteristics which underlie the design of many of these devices, including the **rocket engine**, **jet pump**, and **cascade of guidevanes**, shown in Figures 1.15, 1.16, and 1.17, respectively, can be determined using the principles of fluid mechanics that we cover in this textbook. The analysis of most of the other cases requires more advanced aspects of fluid mechanics and may also involve considerations of heat transfer, thermodynamics, and chemistry, all of which are beyond the scope of this book.

The turbojet and turbofan engines shown in Figures 10.3(a) and 1.8, respectively, are examples of a class of devices called **turbomachines**, derived from the Latin word *turbo*, which has the meanings 'whirlwind' and 'spinning top'. Other examples of turbomachines are pumps, fans, compressors, steam turbines, gas turbines, hydraulic turbines (see Figure 1.9), turbochargers, and superchargers. A common feature of all turbomachines is a central rotating

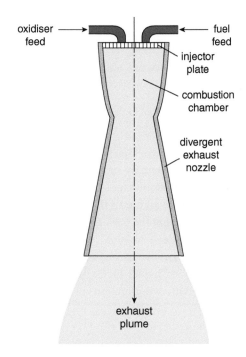

**Figure 1.15** Thrust of a liquid-propellant rocket

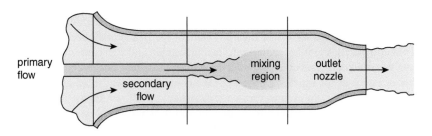

**Figure 1.16** Performance of a jet pump

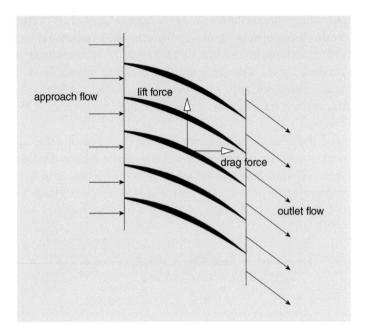

**Figure 1.17** Hydrodynamic forces on a cascade of guidevanes

shaft which carries blades (the **rotor** or **impeller**) to transfer momentum and work either to or from the fluid which flows through them by causing changes in the direction of fluid flow. Most turbomachines also incorporate stationary blades (called **stators** or **nozzle rings**) attached to the casing to guide the flow to and from the rotor stages. As we show in Chapter 10, we can learn a considerable amount about the performance of these complex machines simply by considering the state of flow at inlet and outlet. The basic flow within a **stator** or **rotor stage** can be analysed in much the same way as that through a stationary **cascade of guidevanes**; but, to take the analysis further, as we do in Chapter 14, requires that we use more advanced aspects of fluid mechanics, often together with considerations of thermodynamics.

 1.5 SUMMARY

In this chapter, we have indicated the wide array of engineering devices, from the kitchen tap (a valve) to supersonic aircraft, for which the basic design depends upon considerations of the flow of gases and liquids. Much the same is true of most natural phenomena, from our weather to ocean waves and the movement of sperm and other bodily fluids. This textbook introduces a number of the concepts, principles, and procedures which underlie the analysis of any problem involving fluid flow. In this **Introduction**, we have selected a number of examples for which, by the end of the book, the student should be in a position to make practically useful engineering-design calculations. We emphasise that simply attending lectures or reading this book is not sufficient: it is absolutely essential for the student to spend at least twice the amount of lecture time attempting to solve the self-assessment problems which follow most chapters.

# 2

# Fluids and fluid properties

Wet. Sticky. Viscous. Viscid. Gelatinous. Slippery. Greasy. Oily. Lubricious. Slimy. Oleaginous. Oozy. Soapy. Thick. Thin. Runny. Syrupy. Treacly. Tacky. Claggy. Muddy. Gummy. Gooey. Mucilaginous. Glutinous. These are among the many adjectives commonly used to describe liquids, to convey something about how liquids feel, how they flow, or how they respond to being stirred or mixed. The list of words available to describe gases is far more limited: viscous, viscid, heavy, and dense. We could also include smelly in both lists, although in the case of liquids what is sensed is the vapour form. In contrast to these adjectives, which primarily give us a qualitative tactile impression, in this chapter we introduce the properties used to quantify the physical characteristics of liquids and gases: **dynamic** and **kinematic viscosity**, **density**, **specific volume**, **relative density**, **bulk modulus of elasticity** and **compressibility**, **speed of sound** (or **soundspeed**), **vapour pressure**, and **surface tension**, together with the **perfect-gas law** and an **equation of state for liquids**. We discuss how and why fluids and solids are different both on a molecular and on a macroscopic scale. We show that central to the definition of the physical properties of fluids, and the way in which we go on to analyse fluid flow, is the **continuum hypothesis**, which allows us to define properties on a scale which is far smaller than any scale of engineering interest but still far larger than the underlying molecular scale.

## 2.1 Fluids and solids

The state of any substance can be classified as **solid** or **fluid**, with the term fluid including **liquids**, **gases**, and **vapours**. From an engineering viewpoint, the essential difference between a fluid and a solid is the way in which the substance resists **shear stress**. In the case of a solid, the shear stress is resisted by a static deformation, the magnitude of which (for a given shear stress) depends upon a material property called the **modulus of rigidity**. For a fluid, no matter how low the shear stress, the deformation increases without limit as long as the shear stress is applied. The rate of deformation of a fluid is determined by a property called the **dynamic viscosity** (or just the **viscosity**). A fluid for which the viscosity is zero is said to be **inviscid**, whereas a fluid with non-zero viscosity is said to be **viscous**[4]. A fluid with vanishingly small viscosity is also termed a **perfect fluid**, the only known example of which is liquid helium cooled to 2.17 K, at which critical temperature a fraction of the liquid becomes an inviscid **superfluid**.

We can begin to quantify the statements in the first sentences of the preceding paragraph as follows. Suppose we have a solid rectangular block subjected to a shear (i.e. tangential) force $F$,

---

[4] The term viscid is also used.

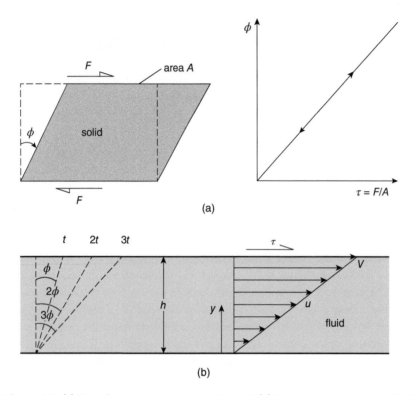

**Figure 2.1** (a) Shear force applied to an elastic solid (b) Shear stress applied to a fluid

as illustrated in Figure 2.1(a). Unless the magnitude of the force is so great that the material fractures or deforms plastically (in a sense, behaving like a liquid), the solid resists the force $F$ by a static deformation which we can measure by the angle $\phi$ (the Greek letter *phi*). In the case of an elastic solid, according to **Hooke's law**, the deformation is proportional to the applied force, so we can write

$$\frac{F}{A} = \tau = G\phi \tag{2.1}$$

where $A$ is the surface area over which $F$ is distributed, $\tau$ (the Greek letter *tau*) is the shear stress (i.e. the shear force per unit area), and the constant of proportionality $G$ is called the **modulus of rigidity** or **shear modulus**.

Consider now the situation illustrated in Figure 2.1(b), which shows a fluid between two parallel plates separated by a short distance $h$, with the lower plate stationary and the upper plate moving at velocity $V$. A fundamental concept of the flow of a viscous fluid, called the **no-slip condition** (see Sections 6.4 and 15.3), is that fluid in contact with a solid surface adheres to it and moves at the speed of the surface. Thus, the fluid in the immediate vicinity of the upper surface moves forwards at velocity $V$, the fluid in contact with the lower surface is at rest, and the fluid in-between moves as though in infinitesimally thin layers with velocity $u$, which increases progressively with distance $y$ from the lower surface, i.e.

$$u = \frac{Vy}{h}. \tag{2.2}$$

If we imagine a line normal to the plate surfaces and marking the fluid at some instant of time, at time $t$ later the line will have rotated through an angle $\phi$, as shown in Figure 2.1(b), so that

$$\tan \phi = \frac{Vt}{h}. \tag{2.3}$$

If the time $t$ is short, the angle $\phi$ will be small and negligibly different (measured in radians) from $\tan \phi$, so that

$$\phi = \frac{Vt}{h}, \tag{2.4}$$

from which we see that if $t$ doubles, $\phi$ also doubles; if $t$ triples, $\phi$ also triples; and so on. Rather than think of progressive deformation in this way, it is far more convenient to think in terms of the **rate of change of deformation**, which is given by

$$\frac{d\phi}{dt} = \frac{V}{h}. \tag{2.5}$$

From equation (2.2) and Figure 2.1(b) we can see that the quantity $V/h$ is the gradient of the velocity $u$ with respect to distance $y$, i.e.

$$\frac{du}{dy} = \frac{V}{h}. \tag{2.6}$$

Because gradients of velocity within a fluid occur due to the effects of shear stress, the **rate of deformation** $du/dy$ is referred to as the **shear rate**. For a fluid, the statement equivalent to equation (2.1) can now be written as

$$\tau = \mu \frac{d\phi}{dt} = \mu \frac{du}{dy} \tag{2.7}$$

where the symbol $\mu$ (the Greek letter *mu*) represents the fluid property known as **dynamic viscosity** (usually just referred to as the **viscosity**). In some books, the symbol $\eta$ (the Greek letter *eta*) is used rather than $\mu$. Viscosity is the principal property which distinguishes a fluid from a solid, and many of the adjectives listed at the beginning of this chapter are qualitative descriptions of the viscous nature of fluids. For many simple fluids, including air and water, $\mu$ is a thermodynamic property which depends only upon temperature and pressure but not on the shear rate. As mentioned in Chapter 1, such fluids are known as **Newtonian**. One of Newton's many contributions to scientific understanding was the recognition that the resistance to relative motion between two 'layers' of a fluid is proportional to the velocity difference between the layers, as represented by equation (2.7).

It is easy to find descriptive distinctions between the four states (solid, liquid, gas, vapour) in which matter occurs. Solids are hard and not easily deformed. A liquid has no inherent shape and is so easily deformed that under the influence of gravity it takes up the shape of any container into which it is poured without a change in volume. A gas is even easier to deform than a liquid and increases in volume without limit unless constrained by a closed container, which it then fills completely. The volume of a fixed mass of gas is decreased by any increase in pressure, whereas to decrease the volume of a liquid by a measurable amount requires very high pressures (see Section 2.6). These and other differences between the gas, liquid, and solid states can be explained on the basis of their molecular structures. Movement of the molecules of a solid is highly restricted because they are closely packed in a fixed lattice structure with large **intermolecular cohesive forces** between them. The molecules of a liquid have freedom of movement and are further apart (though the typical spacing is still only $10^{-10}$ m or 0.1 nm) so the intermolecular forces are smaller. In fact, the molecules are in a continual state of inter-action with their neighbours and never move very far. Gas molecules, on the other hand, move about randomly but in straight lines at high speed (about 1.2 to 1.5 times the speed of sound), occasionally colliding with each other or the surfaces of a confining container. For both liquids and gases the continual bombardment of any surface by molecules gives rise to a stress which is normal to the surface and which we call **pressure**.

Since many substances can exist in any one of the three basic states, the differences in mo-lecular structure are largely a matter of degree, and there is the possibility of transition between these states. For example, the volume of a fixed mass of gas is easily decreased by increasing its pressure, a process termed **compression**, while **expansion** is the opposite process. At very high levels of compression the gas molecules are forced so close together that the gas becomes indistinguishable from a liquid and is said to liquefy. **Liquefaction** can also be achieved by cooling a gas to a temperature below its **critical temperature**. The free surface of any liquid is always in contact with its gaseous state, called a **vapour**. At sufficiently high temperature many solids melt and become liquid and, with further increase in temperature, increasing amounts of vapour are produced until all the material is in the gaseous state. These different states are identified thermodynamically as **phases** which represent forms of matter which are physically and chemically stable.

## 2.2 **Fluid density** $\rho$

The **density** $\rho$ (the Greek letter *rho*) of a fluid (or a solid), sometimes referred to as its **mass density**, is the ratio of the mass $m$ of a given volume of that substance to its volume $\mathcal{V}$, i.e.

$$\rho = \frac{m}{\mathcal{V}}. \tag{2.8}$$

In the SI system of units, which we use exclusively in this textbook and present in some detail in Chapter 3, the unit of mass is the kilogram (symbol kg), that of volume is the cubic metre (m$^3$), and the unit of density is kilogram per cubic metre (kg/m$^3$). As we indicated in Section 2.1, we can decrease the volume of a fixed mass of gas by increasing its pressure. According to equation (2.8), the consequence of compression is an increase in the gas density. The pressure of the air flowing through the core of a jet engine, such as that illustrated in Figure 1.8, is increased progressively as the air passes through the compressor stages and so the air density also increases (there is an accompanying increase in temperature).

As may be evident, our definition of density in the previous paragraph is incomplete: the idea that the density of air can vary with location as it flows through a compressor implies that we regard density as having a value at a given point, as is the case for all fluid properties. A more complete definition of density requires that the volume $\mathcal{V}$, and hence the mass $m$, is so small that there is no appreciable variation of density within it. At the extreme, we could define a volume so small that at any instant of time it contained a single molecule but this

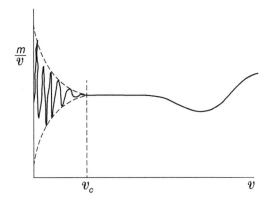

**Figure 2.2** Variation of the ratio mass: volume with volume

does not lead to a sensible definition of density, not least because the molecule would not have a fixed location. However, by progressively increasing the volume above such a low value, we eventually reach a situation where, although molecules are continuously moving into and out of the volume at its boundary, the net number of molecules within the volume at any instant is practically constant. The effect on the ratio $m:\mathcal{V}$ of progressively increasing $\mathcal{V}$ is shown schematically in Figure 2.2. The horizontal scale is compressed to the right of the vertical line, and expanded to its left. Once $\mathcal{V}$ exceeds the critical value $\mathcal{V}_C$, we can define a density as a thermodynamic property (i.e. a physical property which depends only upon temperature and pressure) which is independent of volume and which can vary smoothly and continuously throughout the entire body of fluid. We shall quantify the order of magnitude of $\mathcal{V}_C$ in Section 2.5. The densities and other properties of pure water and dry air at a pressure of 1 atm are given in Tables A.3 and A.4, respectively[5]. For other fluids of engineering interest the physical properties are given in Tables A.5 (liquids) and A.6 (gases), also at a pressure of 1 atm.

There are two principal ways in which the density of a fluid influences flow. The most important stems from **Newton's second law of motion**, which tells us that the acceleration of a given mass is proportional to the net force applied to it. We shall discuss in detail the application of Newton's second law to fluid flow in many of the chapters in this book. For the time being it is sufficient to realise that to produce a change in the velocity of a high-density fluid, such as a liquid, involves much larger forces (per unit volume) than is the case for a fluid of low density, such as a gas. For example, the power required to propel a submerged **submarine** would be about a thousand times greater than for an **airship** of the same size and speed flying through the air. The second way in which density plays a role involves gravity and the associated decrease in atmospheric pressure with altitude or increase in pressure with liquid depth. These and other **hydrostatic** effects are the subject of Chapters 4 and 5. **Compressible flow**, in which there can be very large, and even discontinuous, changes in density, is the subject of Chapters 11, 12, and 13.

---

[5] Table A.1 lists some atomic and molecular weights, and Table A.2, some universal constants. The physical properties of the 1976 Standard Atmosphere (see Section 4.13) are given in Table A.7. Tables A.1 to A.7 form Appendix 2.

## 2.3 Atoms, molecules, and moles

All matter is made up of a limited number of elementary substances, the **chemical elements** (as of November 2011, 118 had been identified: 94 naturally occurring and 24 synthetic). So far as this textbook is concerned, the basic building block for any chemical substance is the **atom**, a tiny (typically with a radius less than 1 nm) particle which cannot be split without losing the properties of the element. Each element has a **relative atomic mass (atomic weight)** based on a scale in which the mass of the carbon-12 ($^{12}C$) atom, the most abundant (almost 99%) isotope of carbon, is 12. Most substances consist of **molecules** in which atoms are bound together by interatomic forces. In a way similar to that of atomic weight, the **molecular weight** $\mathcal{M}$ (**relative molecular mass**, or **molecular mass**), with the units kg/kmol, of these compounds is defined relative to the mass of $^{12}C$. The atomic weights and molecular weights of some common substances are listed in Table A.1 in Appendix 2, together with the symbols used for atoms or the molecular formulae for molecules.

Although molecular weight is defined as a ratio, and so is a non-dimensional quantity (see Chapter 3) which has no units, it is useful to express molecular weights in terms of a unit called the **mole** (symbol mol), 1 mol being the amount of a substance in grams numerically equal to its molecular weight, or the **kilomole** (symbol kmol), which is the amount of substance in kilograms. In the case of methane, for example, $\mathcal{M} = 16.04$ kg/kmol. The number of molecules in 1 kmol of any substance is given by the **Avogadro number**, $N_A$, a fundamental physical constant the value of which is $6.022 \times 10^{26}$ molecules/kmol. If we have $N$ molecules of a substance with molecular weight $\mathcal{M}$, the amount of that substance $n = N/N_A$ kmol, and the corresponding mass is $m = n\mathcal{M} = N\mathcal{M}/N_A$ kg.

## 2.4 Perfect-gas law

At very high temperatures (above about 1000 °C) the molecular structure of a gas breaks down (a process known as **dissociation**) and at very high pressures or low temperatures, as we have already indicated in Section 2.1, gases can liquefy. Away from these extremes, most gases are in good agreement with a **thermal equation of state** known as the **perfect-gas[6] law**

$$p = \rho RT, \tag{2.9}$$

where $p$ is the gas pressure in pascal (Pa = N/m$^2$), $T$ is the absolute temperature of the gas in degrees kelvin (K = 273.15 + °C), and $R$ is a constant of the gas called the **specific gas constant** (with units m$^2$/s$^2 \cdot$ K or kJ/kg $\cdot$ K). A gas which obeys the equation of state $p = \rho RT$ is a **thermally perfect gas[7]**. The unit m$^2$/s$^2 \cdot$ K suggests a connection between $R$ and a speed which we shall show in Section 2.12 is that for the propagation of sound through the gas, i.e. the **speed of sound**. The specific gas constant is related to the **universal** (or **molar**) **gas constant** $\mathcal{R}$ as follows

---

[6] The term perfect gas should not be confused with perfect fluid, which is an idealised fluid lacking both viscosity and thermal conductivity.

[7] Such a gas is sometimes termed an **ideal gas** rather than a perfect gas.

$$\mathcal{R} = \mathcal{M}R, \tag{2.10}$$

where $\mathcal{M}$ (with the unit kilogram per kilomole) is the molecular weight of the gas. The universal gas constant is defined in terms of the **Boltzmann constant** $k_B$ and the Avogadro number $N_A$ as

$$\mathcal{R} = k_B N_A. \tag{2.11}$$

Boltzmann's constant has the value $1.3807 \times 10^{-23}$ J/K, and the universal gas constant has the value 8.31451 kJ/kmol · K (or 8314.51 J/kmol · K).

The specific gas constant $R$ is equal to the difference between the **specific heats** at constant pressure $C_P$ and constant volume $C_V$, i.e.

$$R = C_P - C_V. \tag{2.12}$$

For a range of gases, values for the molecular weight $\mathcal{M}$, the specific gas constant $R$, and the ratio of the specific heats,

$$\gamma = \frac{C_P}{C_V} \tag{2.13}$$

are tabulated in Table A.6. A perfect gas for which $C_P$ and $C_V$, and hence $\gamma$, are constant is called a **calorically perfect gas**. It is usual to refer to a calorically perfect and thermally perfect gas obeying $p = \rho RT$ simply as a **perfect gas**. The quantities in Table A.6 play an important role in compressible-flow theory (see Chapters 11, 12, and 13). Although values for the corresponding gas density $\rho$ at STP (20 °C, 1 atm) are also tabulated, this is not essential, since the density of any of the gases listed can be calculated from equation (2.9).

---

**ILLUSTRATIVE EXAMPLE 2.1**

Calculate the density of nitric oxide (NO) at 20 °C and 1 atm and also at 500 °C and 5 bar.

Solution

$\mathcal{M} = 30.01$ kg/kmol (from Table A.6); $p_1 = 1.01325 \times 10^5$ Pa; $T_1 = 293.15$ K; $p_2 = 5 \times 10^5$ Pa; $T_2 = 773.15$ K.
From equation (2.10)

$$R = \mathcal{R}/\mathcal{M} = 8314.51/30.01 = 277.1 \, \text{m}^2/\text{s}^2.\text{K}.$$

From equation (2.9)

$$\rho_1 = \frac{p_1}{RT_1} = \frac{1.01325 \times 10^5}{277.1 \times 293.15} = 1.248 \, \text{kg/m}^3$$

and

$$\rho_2 = \frac{p_2}{RT_2} = \frac{5 \times 10^5}{277.1 \times 773.15} = 2.334 \, \text{kg/m}^3.$$

**Comments:**

(a)  It is generally unnecessary to carry so many significant figures (s.f.) in an engineering calculation; 4 s.f. for $\mathcal{R}$ and 3 for other quantities are usually sufficient.

(b)  As they should be, the values calculated here for $R$ and $\rho_1$ for NO are precisely the same as those in Table A.6.

(c)  The first step in the solution was to restate the data given (in this case for temperature and pressure) in standard SI units. The student should develop the habit of converting given data to standard SI form in this way.

---

## 2.5 Continuum hypothesis and molecular mean free path

In Section 2.1 we discussed some of the qualitative differences between the molecular structures of liquids and gases. As we shall now see, these differences have a direct influence on the size of the critical volume $\mathcal{V}_C$ introduced in Section 2.4.

We consider first a gas with molecular weight $\mathcal{M}$ which obeys the perfect-gas law, to calculate the average number of molecules contained in a cube (the choice of a cube is arbitrary, and we could just as well have chosen another shape, such as a sphere) of gas of side length $L$ m. If the fluid density is $\rho$, from equation (2.8) the mass of the cube will be $\rho L^3$, since the cube volume $\mathcal{V} = L^3$. From equations (2.9) and (2.10) we have

$$\rho = p\mathcal{M}/\mathcal{R}T \tag{2.14}$$

so that the mass of our cube is given by

$$m = \rho \mathcal{V} = p\mathcal{M}\mathcal{V}/\mathcal{R}T. \tag{2.15}$$

From Section 2.3 we know that the molecular weight $\mathcal{M}$ is the mass in kg of 1 kmol of that substance, so that our cube contains $p\mathcal{V}/\mathcal{R}T$ kmol (the unit kmol is often written as kg mole). Since the number of molecules in 1 kmol of any substance is given by the **Avogadro number**, $N_A$, the value of which is $6.022 \times 10^{26}$ molecules/kmol, we see that the average number of molecules $N$ in the cube must be given by

$$N = p\mathcal{V}N_A/\mathcal{R}T. \tag{2.16}$$

Equation (2.16) can be rearranged as

$$\mathcal{V} = N\mathcal{R}T/pN_A \tag{2.17}$$

from which we can calculate the volume $\mathcal{V}$, which contains $N$ molecules of a gas at temperature $T$ (K) and pressure $p$ (Pa).

In terms of the gas density $\rho$, equation (2.17) becomes

$$\mathcal{V} = NM/\rho N_A. \tag{2.18}$$

Equation (2.16) shows that, since $\mathcal{R}$ is a universal constant, the same for all gases, the average number of molecules $N$ in a volume $\mathcal{V}$ of any gas depends only upon its pressure $p$ and

**Table 2.1** Number $N$ of gas molecules in a volume $\mathcal{V}$.

| $L$ | $\mathcal{V}$ $(m^3)$ | $N$ |
|---|---|---|
| 1 mm | $10^{-9}$ | 2.7 E+16 |
| 1 $\mu$m | $10^{-18}$ | 2.7 E+7 |
| 100 nm | $10^{-21}$ | 2.7 E+4 |
| 50 nm | $1.25 \times 10^{-22}$ | 3362 |
| 33.4 nm | $3.73 \times 10^{-23}$ | 1000 |
| 20 nm | $9 \times 10^{-24}$ | 215 |
| 10 nm | $10^{-24}$ | 27 |
| 3.34 nm | $3.73 \times 10^{-26}$ | 1 |

absolute temperature $T$. This equation can therefore be regarded as a quantitative form of Avogadro's law: equal volumes of two gases, at the same temperature and pressure, contain the same number of molecules. Table 2.1 shows the results obtained for $N$ using equation (2.16), for 0 °C and 1 atm (standard temperature and pressure, or STP).

As we shall see in Section 4.13, the air density in the atmosphere decreases with altitude. At the lower limit of the **stratosphere** (an altitude of about 20 km), according to Table A.7, the temperature is about 217 K, and the pressure is 5475 Pa (the corresponding density is 0.0880 kg/m$^3$), while at the outer limit of the mesosphere (about 80 km) the values are 196.7 K and 0.886 Pa, respectively, so that, according to equation (2.17), a cube of air containing 1000 molecules would have a side length of 81.7 nm at an altitude of 20 km, and 0.797 $\mu$m at 80 km, both of which are negligibly small compared with the dimensions of any object likely to be flying at such altitudes. Equation (2.18) also shows that the density of air would have to fall to $4.8 \times 10^{-14}$ kg/m$^3$ (which would correspond to an altitude of about 1600 km) for the cube size to reach 1 mm.

We cannot give a precise value but would probably not want the number of molecules over which to form an average to be any lower than 1000 and so conclude that for a gas at STP the concept of fluid density begins to fail if the cube size $\mathcal{V}_C$ is below about 30 nm (i.e. 0.3 $\mu$m or $3 \times 10^7$m). To put this in perspective, the diameter of a human hair is typically about 100 $\mu$m, and the wavelength of visible light is about 589 nm. There are few, if any, practical situations involving gas-flow channels with dimensions which come anywhere close to 30 nm. Even devices known as **microchannels** typically have dimensions in the range 1 to 500 $\mu$m. Gases for which the number of molecules in a 1 $\mu$m cube fall below about 1000 are said to be **rarified** and are normally encountered only in outer space.

Because the molecular structure of a liquid is generally more complex than that of a gas, the number of molecules per unit volume, $N$, which is termed the **molecular number density**, varies from liquid to liquid. For a cube of side length $L$ of liquid with density $\rho$ the mass is again given by $m = \rho L^3$. The number of kilomoles of liquid is then $\rho L^3 / \mathcal{M}$, and the number of molecules is $\rho L^3 N_A / \mathcal{M}$. Table 2.2 shows values of $N$ for several liquids, with $L = 1$ $\mu$m ($\mathcal{V} = 10^{-18}$ m$^3$).

**Table 2.2** Number $N$ of molecules in a liquid cube of side length $L = 1\ \mu$m

| Liquid | $N$ |
| --- | --- |
| Petrol, $C_8H_{18}$ | 4.4 E+9 |
| Carbon tetrachloride, $CCl_4$ | 6.3 E+9 |
| Liquid oxygen, $O_2$ | 7.7 E+9 |
| Pure glycerol, $C_3H_8O_3$ | 8.2 E+9 |
| Ethyl alcohol, $C_2H_5OH$ | 1.0 E+10 |
| Water, $H_2O$ | 3.35 E+10 |
| Mercury, Hg | 4.07 E+10 |

If we compare Table 2.2 with Table 2.1, we see that the molecular number density $N$ for liquids far exceeds that for gases. We conclude from the foregoing that, except in extreme circumstances, $\mathcal{V}_C$ will always be far smaller than any volume of engineering interest and can be regarded as defining what we mean by a **point** in a fluid. Although we have specifically discussed the property density, the same considerations apply to any physical property and enable us to define point values of these properties, which vary smoothly and continuously throughout a fluid. Although these 'large-scale' (or **macroscopic**) properties reflect the underlying molecular structure, it is generally the case that we can treat the majority of problems of fluid flow without the need to consider molecular structure directly. The idea that both fluid properties and flow properties can be treated in this way is known as the **continuum hypothesis**.

If molecules in a fluid are considered to be hard spheres of effective diameter $\sigma$ in random motion constantly colliding with each other elastically, **kinetic theory** leads an approximate expression for the average distance $\Lambda$ between successive collisions

$$\Lambda = \frac{1}{\sqrt{2}\pi N_V \sigma^2} \tag{2.19}$$

where $N_V$ is the number of molecules per unit volume or the **volume number density**. The quantity $\Lambda$ is termed the **molecular mean free path**, and equation (2.19) is usually attributed to James Clerk Maxwell. The equation for $\Lambda$ can be written in terms of other quantities as follows. The number of moles $n$ in a mass of gas $m$ of molecular mass (or molecular weight) $\mathcal{M}$ is given by

$$n = m/\mathcal{M} \tag{2.20}$$

and so the number of molecules in the mass of gas $N_M$ is

$$N_M = nN_A = mN_A/\mathcal{M} \tag{2.21}$$

where $N_A$ is the Avogadro number (see Section 2.3).
It follows that

$$N_V = N_M/\mathcal{V} = mN_A/\mathcal{M}\mathcal{V} = \rho N_A/\mathcal{M} \tag{2.22}$$

and, from equation (2.19),

$$\Lambda = \mathcal{M}/\sqrt{2}\pi\rho N_A \sigma^2. \tag{2.23}$$

Since $\mathcal{M}$ and $\sigma$ are fixed for a given gas, and $N_A$ is a universal constant, equation (2.23) leads to the conclusion that $\Lambda$ is inversely proportional to the gas density $\rho$. As we shall see in Section 4.13, the density of the air in the earth's atmosphere decreases with altitude (as can be seen from Table A.7). Although at an altitude of 71 km $\rho$ has fallen to about 0.01% of its value at sea level, the corresponding value of $\Lambda$ is still less than 1 mm.

If we introduce the perfect-gas equation (2.9), $p = \rho RT$, equation (2.23) may be written as

$$\Lambda = \mathcal{R}T/\sqrt{2}\pi p N_A \sigma^2 \tag{2.24}$$

where we have made use of equation (2.10) to introduce the universal gas constant $\mathcal{R}$. Since the ratio $\mathcal{R}/N_A$ defines the Boltzmann constant $k_B$, the equation for $\Lambda$ may also be written as

$$\Lambda = \frac{k_B T}{\sqrt{2}\pi p \sigma^2}. \tag{2.25}$$

Table 2.3 includes values of the effective molecular diameter $\sigma$ and molecular mean free path $\Lambda$ for some common gases at 0 °C and 1 atmosphere[8].

**Table 2.3** Effective molecular diameter $\sigma$ and molecular mean free path $\Lambda$ for some common gases at 0 °C and 1 atmosphere

| Gas | $\sigma$ (pm) | $\Lambda$ (nm) |
|---|---|---|
| Air | 366 | 69.1 |
| Argon | 342 | 62.6 |
| Carbon dioxide | 390 | 39.0 |
| Carbon monoxide | 371 | 58.6 |
| Chlorine | 440 | 27.4 |
| Ethylene | 423 | 34.3 |
| Helium | 258 | 173.6 |
| Hydrogen | 297 | 110.6 |
| Methane | 380 | 48.1 |
| Neon | 279 | 124.0 |
| Nitrogen | 375 | 58.8 |
| Nitrous oxide | 388 | 38.7 |
| Oxygen | 354 | 63.3 |
| Sulphur dioxide | 429 | 27.4 |

[8] With the exception of those for air, the values for $\sigma$ and $\Lambda$ have been taken from Kaye and Laby online. The values for air are from the *CRC Handbook of Chemistry and Physics*. Many of the values from these two sources differ by as much as 20%.

For the gases in the table, the arithmetic average value for $\Lambda = 66.8$ nm. A cube with this side length 66.8 nm would have a volume of about $3 \times 10^{-22}$ m$^3$ and so contain about 8000 molecules, another indication of the validity of the continuum hypothesis.

---

**ILLUSTRATIVE EXAMPLE 2.2**

Calculate the molecular mean free path for a gas with molecular weight 28.96 kg/kmol, density 1.28 kg/m$^3$, and effective molecular diameter 366 pm.

**Solution**

$\mathcal{M} = 28.96$ kg/kmol; $\rho = 1.28$ kg/m$^3$; $\sigma = 3.66 \times 10^{-10}$ m; $N_A = 6.022 \times 10^{26}$ molecules/kmol. We use equation (2.23) to find $\Lambda$

$$\Lambda = \mathcal{M}/\sqrt{2}\pi\rho N_A \sigma^2$$

$$= \frac{28.96}{\sqrt{2} \times \pi \times 1.28 \times 6.022 \times 10^{26} \times \left(3.66 \times 10^{-10}\right)^2}$$

$$= 6.33 \times 10^{-8} \text{ m or } 63.3 \text{ nm.}$$

**Comment:**

The value for $\Lambda$ calculated from equation (2.23) represents the result of kinetic theory for a gas with the properties of dry air. This value differs by about 8% from the experimentally determined value of 69.1 nm.

---

## 2.6 Equation of state for liquids

Although equation (2.12) for the ratio of specific heats $\gamma$, has no generally valid equivalent applicable to liquids, it is usually adequate to assume that $\rho$ = constant, and $C_P = C_V$ = constant, so that

$$\gamma = \frac{C_P}{C_V} = 1. \tag{2.26}$$

An approximate equation, cited by Batchelor (2000), for the influence of extreme pressure (typically in excess of 1000 bar) on the density of water is

$$\ln\left(\frac{\rho}{\rho_0}\right) = \frac{1}{n} \ln\left(\frac{p/p_0 + C}{1 + C}\right) \tag{2.27}$$

where $p$ is the static pressure measured in bar, $\rho$ is the corresponding density, $p_0 = 1$ bar (i.e. approximately equal to atmospheric pressure), $\rho_0 = 1000$ kg/m$^3$, $C = 3000$, and $n = 7$. For $p/p_0 = 1000$, approximately equal to the pressure at a water depth of 10 km, the equation gives $\rho/\rho_0 \approx 1.04$, confirming that the effect of pressure on water density can be considered practically negligible.

A more general equation for the influence of pressure on the density of a range of liquids is the modified **Tait equation**

$$1 + \frac{\rho}{\rho_0} = A \ln \left( \frac{p/p_0 + D}{1 + D} \right) \tag{2.28}$$

where $p_0$ is a low pressure (usually the barometric pressure $B$ or 1 bar), $\rho_0$ is the liquid density at pressure $p_0$, and $A$ and $D$ are constants for the given liquid.

## 2.7 Specific volume $v$, relative density $\sigma$, and specific weight $w$

In thermodynamics it is often more convenient to work in terms of **specific volume $v$** than density $\rho$. The word 'specific' here means 'per unit mass', i.e.

$$v = \frac{v}{m} \tag{2.29}$$

from which we see that the unit of $v$ is m$^3$/kg.

**Relative density** $\sigma$ (Greek letter *sigma*) is the ratio of the density of a fluid to that of a standard **reference fluid** $\rho_{REF}$, i.e.

$$\sigma = \frac{\rho}{\rho_{REF}} \tag{2.30}$$

Because it is defined as the ratio of two physical quantities with the same unit, relative density[9] has a purely numerical value without unit and is again non-dimensional (see Chapter 3).

For liquids, the reference fluid is usually taken to be pure water at 4 °C and 1 atm when it has a density $\rho_{REF} = 1000$ kg/m$^3$. Water shows anomalous behaviour in that between 0 °C and 4 °C its density increases to a maximum of 999.972 kg/m$^3$ at 4 °C. Below 0 °C water solidifies to become ice. The temperature for the reference fluid is sometimes taken as 20 °C at which the density of water is 998.20 kg/m$^3$.

For gases the reference fluid is usually pure air (although hydrogen is sometimes used), which has a density of 1.204 kg/m$^3$ at 20 °C and 1 atm. In practice, relative density is little used for gases.

**Specific weight $w$**, which should not be confused with specific gravity (i.e. relative density), is the weight per unit volume of a substance. Since density $\rho$ is mass per unit volume, it follows that

$$w = \rho g \tag{2.31}$$

where $g$ is the acceleration due to gravity and has the value 9.807 m/s$^2$, usually rounded to three significant figures as 9.81 m/s$^2$. The units of $w$ can be shown to be N/m$^3$ because, as we shall see in Chapter 3, 1 newton (symbol N) = 1 kg · m/s$^2$.

---

[9] Particularly in older texts, the term **specific gravity** is sometimes used instead of relative density.

---

**ILLUSTRATIVE EXAMPLE 2.3**

Calculate the density and specific weight for liquid oxygen, which has a relative density of 1.46 at −252.7 °C and 1 atm.

**Solution**

$\sigma = 1.46 = \rho/\rho_{REF}$; so, with $\rho_{EF} = 1000$ kg/m³, for a liquid, $\rho = \sigma\rho_{REF} = 1.46 \times 1000 = 1460$ kg/m³.

$w = \rho g = 1460 \times 9.81 = 14\,320$ N/m³.

**Comments:**

(a)  In any problem where either relative density or specific weight is specified, the first step should always be to calculate the fluid density in SI units.
(b)  It is almost always advisable to work through any problem using algebraic symbols and to substitute numerical values as late as possible.

---

## 2.8  **Dynamic viscosity (viscosity)** $\mu$

In Section 2.1, we introduced dynamic viscosity (symbol $\mu$) as the property which provides the link between the shear stress applied to a fluid and the resulting rate of deformation. For the simple case of a fluid confined between two parallel plates, one fixed, the other moving, we showed that the rate of deformation was equal to the velocity gradient within the fluid. In most flows the spatial variation of velocity is more complicated than the linear variation shown in Figure 2.1. In more general situations, such as that shown in Figure 2.3, the continuum hypothesis allows us to relate the shear stress $\tau$ at any point in a fluid to the velocity gradient (often termed the **shear rate**) $du/dy$ at that point according to

$$\tau = \mu \frac{du}{dy}. \tag{2.32}$$

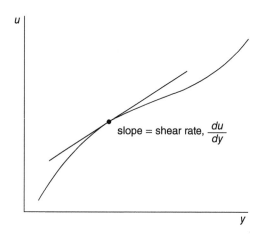

**Figure 2.3**  Velocity versus normal distance to illustrate velocity gradient

As we shall see in Chapters 15 to 18, the fluid velocity $u$ is usually a function of more than one spatial variable, and the ordinary derivative $du/dy$ must be replaced by the partial derivative $\partial u/\partial y$. In even more general situations, $u$ is only one component of the total velocity in a flow, so that $\tau$ also depends upon the derivatives of the other velocity components. The quantity $\mu$ has the units $Pa \cdot s$ (= $N \cdot s/m^2$) and is properly known as either the **absolute coefficient of viscosity** or the **dynamic viscosity** but is usually referred to simply as the viscosity. The reciprocal of $\mu$ (i.e. $1/\mu$) is called the **fluidity**.

The viscosity of a Newtonian liquid is commonly measured using an instrument such as the concentric-cylinder **viscometer**[10] illustrated schematically in Figure 2.4. The liquid is introduced into the annular gap between an inner cylinder, which rotates at angular velocity $\Omega$ (unit rad/s), and an outer stationary cylinder. According to the no-slip condition mentioned in Section 2.1 and discussed in more detail in Section 15.3, the fluid velocity at the surface of the outer cylinder is zero while that at the surface of the inner cylinder $V$ is $\Omega R$. By making the width $\delta$ of the annular gap between the two cylinders negligibly small in comparison with

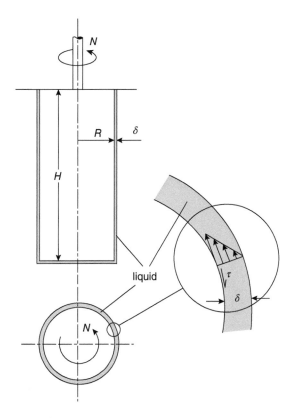

**Figure 2.4** Concentric-cylinder viscometer

---

[10] An instrument designed to measure viscosity and other mechanical properties of a viscous liquid is called a **rheometer**.

the cylinder radius $R$, the effect of curvature becomes unimportant, and the flow geometry is essentially the same as that of two parallel plates. The velocity gradient within the fluid is then

$$\frac{du}{dy} = \frac{\Omega R}{\delta}$$

and the shear stress is

$$\tau = \mu \frac{du}{dy} = \frac{\mu \Omega R}{\delta}.$$

The total torque $T$ exerted on the inner cylinder is given by $T = \tau A R$, where $A$ is the surface area of the inner cylinder. If the length of the cylinder is $H$, we have $A = 2\pi R H$ and so

$$T = \frac{2\pi R^3 H \Omega \mu}{\delta}.$$

Since $\delta$, $R$, and $H$ are known dimensions, $\mu$ can be determined by measuring the torque $T$ and the rotational speed $N$ (rps) from

$$\mu = \frac{15\delta T}{\pi^2 N R^3 H} \tag{2.33}$$

where we have used the relationship $V = \Omega R = 2\pi R^2 N$.

Further discussion of viscometers and rheometers is given in Section 16.7.

---

**ILLUSTRATIVE EXAMPLE 2.4**

A concentric-cylinder viscometer is used to measure the viscosity of an oil. The dimensions of the viscometer are $R = 30$ mm, $H = 75$ mm, and $\delta = 100$ $\mu$m. At a rotational speed of 300 rpm, the measured torque is 0.1 N $\cdot$ m. Calculate the dynamic viscosity of the oil.

Solution

$R = 3 \times 10^{-2}$ m; $H = 7.5 \times 10^{-2}$ m; $\delta = 10^{-4}$ m; $N = 5$ rps; and $T = 0.1$ Nm.
We have

$$\mu = \frac{15\delta T}{\pi^2 N R^3 H} = \frac{15 \times 10^{-4} \times 0.1}{\pi^2 \times 5 \times \left(3 \times 10^{-2}\right)^3 \times 7.5 \times 10^{-2}} = 1.50 \text{ Pa.s.}$$

---

Values for the viscosities of a wide range of Newtonian fluids are given in Tables A.5 (liquids) and A.6 (gases) at standard temperature and pressure. Values for pure water and dry air at 1 atm are tabulated in Tables A.3 and A.4, respectively. Some of the viscosities from Tables A.3 to A.6 are plotted in Figure 2.5, which shows the strong dependence on temperature, particularly for liquids (note the logarithmic ordinate). The dependence on pressure is generally negligible up to 10 bar for gases and 100 bar for liquids.

The temperature dependence for gases is well represented by **Sutherland's formula**

$$\mu = \frac{KT^{3/2}}{T + C} \tag{2.34}$$

where $T$ is the absolute temperature (in K), and $K$ and $C$ are constants characteristic of the particular gas concerned. If $\mu$ has the value $\mu_{REF}$ at a specified reference temperature $T_{REF}$, then at any other temperature $T$ we have

$$\mu = \left(\frac{T_{REF} + C}{T + C}\right)\left(\frac{T}{T_{REF}}\right)^{3/2} \mu_{REF}. \tag{2.35}$$

A useful result, based upon the **kinetic theory of gases** (see Section 2.5), is

$$\mu = \sqrt{\frac{2}{\pi \gamma}} \rho c \Lambda, \tag{2.36}$$

where $\mu$ is the dynamic viscosity of the gas, $\rho$ is its density, $c$ is the speed of sound, $\gamma$ is the ratio of specific heats, and $\Lambda$ is the molecular mean free path[11]. As we shall see in Section 11.8, equation (2.36) leads to a criterion for the validity of the continuum hypothesis in the analysis of shockwave structure.

If equation (2.36) is combined with equation (2.23) for $\Lambda$, we find

$$\mu = \frac{\mathcal{M}c}{\sqrt{\pi^3 \gamma N_A \sigma^2}} \tag{2.37}$$

where $\sigma$ is the effective molecular diameter and $N_A$ is the Avogadro number. According to equation (2.37), $\mu$ is independent of pressure, a prediction which is consistent with experimental observations up to about 10 bar. As we shall show in Section 2.12, $c \propto \sqrt{T}$ so that according to equation (2.37) $\mu \propto \sqrt{T}$, which is less satisfactory: according to Sutherland's formula $\mu \propto T^{3/2}/(T + C)$. A more accurate prediction requires a more sophisticated analysis which is beyond the scope of this book.

---

**ILLUSTRATIVE EXAMPLE 2.5**

For air, the constant $C$ in Sutherland's formula has the value 110.4 K, and a viscosity at 20 °C of $1.8 \times 10^{-5}$ Pa · s. Calculate the viscosity of air at 400 °C.

**Solution**

$T_{REF} = 20 + 273 = 293$ K; $\mu_{REF} = 1.8 \times 10^{-5}$ Pa · s; $C = 110.4$ K; $T = 400 + 273 = 673$ K. From Sutherland's formula

$$\mu = \left(\frac{293 + 110.4}{673 + 110.4}\right)\left(\frac{673}{293}\right)^{1.5} = 3.23 \times 10^{-5} \text{ Pa.s.}$$

**Comment:**

The value calculated for $\mu$ at 400 °C in this example is within 1% of the value given in Table A.4 ($3.32 \times 10^{-5}$ Pa · s).

---

[11] The factor 2/3 sometimes appears in equation (2.36), depending upon the assumptions made in its derivation.

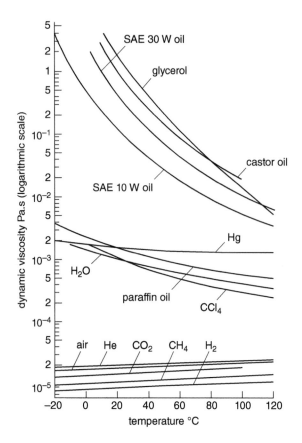

**Figure 2.5** Dynamic viscosity of common fluids as a function of temperature

For liquids the temperature dependence of viscosity can be approximated by the formula

$$\mu = \exp\left[C\left(\frac{T_{REF}}{T} - 1\right)\right]\mu_{REF} \tag{2.38}$$

where $\mu_{REF}$ is the viscosity at a reference absolute temperature $T_{REF}$, typically 293 K, and $C$ is a numerical constant for the particular liquid. The presence of the exponential function (i.e. exp) reflects the strong temperature dependence for liquids evident in Figure 2.5. For example, for water the viscosity just above the freezing point at a pressure of 1 bar, 0 °C, is $1.787 \times 10^{-3}$ Pa · s, compared with $2.818 \times 10^{-4}$ Pa · s just below the boiling point 100 °C, corresponding to a decrease of 84% or a ratio of 6.3:1. The value of $C$ in this case is 6.9.

---

**ILLUSTRATIVE EXAMPLE 2.6**

An engine oil has a viscosity of 1.0 Pa · s at 20 °C. If the constant $C$ in the formula for the viscosity of a liquid is 17 in this case, calculate the viscosity of the oil at 150 °C.

**Solution**

$C = 17$; $T_{REF} = 20 + 273 = 293$ K; $\mu_{REF} = 1.0$ Pa $\cdot$ s; $T = 150 + 273 = 423$ K.
From equation (2.38),

$$\mu = \exp\left[17 \times \left(\frac{293}{423} - 1\right)\right] \times 1.0 = 5.38 \times 10^{-3} \text{ Pa.s.}$$

---

## 2.9 **Kinematic viscosity $\nu$**

The kinematic viscosity of a fluid $\nu$ (Greek letter *nu*), defined by

$$\nu = \frac{\mu}{\rho} \qquad (2.39)$$

is frequently used instead of the dynamic viscosity $\mu$ because, in many problems, $\mu$ and the density $\rho$ occur only in the combination $\mu/\rho$. An interesting consequence of combining $\mu$ and $\rho$ in this way is that the kinematic viscosity of many gases is higher than that of many liquids, a trend which becomes more pronounced as the temperature increases, because the dynamic viscosities of liquids decrease whereas those for gases increase. The term kinematic is associated with the units of $\nu$, which are m$^2$/s, and so involves only metres and seconds, like the terms displacement (m), velocity (m/s), and acceleration (m$^2$/s), which are descriptive terms not directly involving the dynamics (i.e. the mass, stresses, and forces) of a problem. If equation (2.32) is rewritten as

$$\tau = \nu \frac{d(\rho u)}{dy} \qquad (2.40)$$

we see that the shear stress $\tau$ is proportional to the **gradient of fluid momentum** $\rho u$, the constant of proportionality being $\nu$. Equation (2.39) has the form of a diffusion equation, and $\nu$ is sometimes referred to as the **viscous diffusivity**. The diffusive nature of viscosity is also apparent from the units of $\nu$ which are also those of thermal diffusivity, the property which determines the rate at which thermal energy is transported at a molecular level (i.e. diffused) because of a gradient in energy concentration.

## 2.10 **Non-Newtonian liquids**

Whether a fluid is termed Newtonian or **non-Newtonian**, the relationship (by definition) between the applied shear stress $\tau$ and the resulting shear rate $\dot{\gamma}$ ($\gamma$ is the Greek letter *gamma*; the dot above $\gamma$ indicates rate), is the same, i.e.

$$\tau = \mu\dot{\gamma} \text{ (spoken as } tau \text{ equals } mu \text{ } gamma \text{ dot).} \qquad (2.41)$$

In Section 2.1 we identified Newtonian fluids as those for which the viscosity $\mu$ can be regarded as a thermodynamic property which may depend upon temperature and pressure but is independent of any deformation of the fluid as it flows, i.e. $\mu$ independent of the shear rate. To be more precise, a Newtonian fluid must not only have a viscosity independent of any fluid deformation but also not exhibit such properties as elasticity. Gases have a simpler molecular structure than liquids and always exhibit Newtonian characteristics. However, as we indicated in Chapter 1, the molecular structures of most synthetic liquids, as well as such naturally occurring liquids as blood and synovial fluid, are complex and in consequence the viscosities (sometimes the term **apparent viscosities** is used) of these liquids change not only with temperature and pressure but also with the shear rate itself. As we shall illustrate, $\mu$ can both increase and decrease with increases in the shear rate. Such liquids are correctly termed non-Newtonian.

Blood, cement slurry, yoghurt, toothpaste, mud (either natural mud or synthetic drilling fluid), salad cream, and many other non-Newtonian liquids of practical importance are **shear thinning**, which means that their viscosities decrease with increase in the shear rate. The term **pseudoplastic** is also used for shear-thinning liquids. Salad cream and tomato ketchup are often 'reluctant' to leave the bottle until it is shaken vigorously after which, because the viscosity has decreased, the liquid flows easily, sometimes too easily with unfortunate consequences. The shear-thinning character of these condiments is due to small amounts (typically less than 1%) of an additive, often xanthan gum. The shear-thinning effect is used in a more subtle way in ballpoint pens. The viscosity of the ink is decreased locally when the ball rotates, thereby allowing the ink to flow onto the paper. High viscosity is recovered and the ink flow stops (or should) when the ball is again stationary so that a ballpoint pen doesn't leak in the pocket.

In the case of salad cream and ketchup, another effect called **thixotropy** is also present, this being the term used to describe liquids which take time to adjust to the state of shear. If it were not for this, once the shaking ceased, the liquid would instantaneously return to its high-viscosity state and no longer flow. In addition to being shear thinning, many water-soluble polymers have the effect of reducing the resistance to turbulent fluid motion even in concentrations so low (parts per million) that there is no measurable change in the viscosity of the base solvent. This effect, called **drag reduction**, is still not fully understood but is associated with **viscoelasticity**, another property of some non-Newtonian liquids.

Some non-Newtonian liquids are **shear thickening** (or **dilatant**), which is to say that the viscosity increases with shear rate. Starch- or cornflower-based liquids, such as egg-free custard made from powder (e.g. Bird's Custard Powder), are shear thickening. The surface of a thick paste of custard powder and milk appears to be almost solid if 'stabbed' with the point of a spoon, though the spoon will sink gradually into the paste under its own weight. If the spoon is moved rapidly through the paste, the liquid surface appears to fracture then flow gradually back together.

Other materials, such as certain gels, lubricating greases, ice cream, and margarine, appear to be solid when subjected to low shear stress. Even unconfined, they maintain their shape without deformation due to gravity but become fluid once the shear stress exceeds a certain threshold level, called the **yield stress** $\tau_Y$.

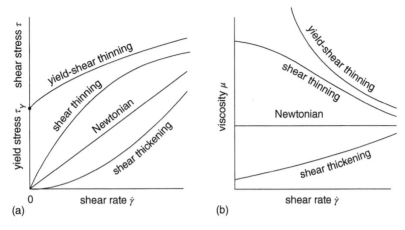

**Figure 2.6** Variation of (a) shear stress and (b) dynamic viscosity with shear rate for Newtonian and non-Newtonian liquids

Typical variations in shear stress and viscosity with shear rate are shown in Figure 2.6 for four basic liquid types

- Newtonian:        $\mu$ constant independent of $\dot{\gamma}$
- shear thinning:   $\mu$ decreases with increasing $\dot{\gamma}$
- shear thickening: $\mu$ increases with increasing $\dot{\gamma}$
- yield stress:     $\tau < \tau_Y, \dot{\gamma} = 0$ (i.e. there is no flow and from a practical point of view the material is solid)

    $\tau > \tau_Y, \dot{\gamma} > 0$ (i.e. flow occurs, the material is liquid and initially shear thinning)

An important subset of non-Newtonian liquids for which the viscosity is dependent upon the shear rate but the fluid is inelastic and not time dependent is the so-called **generalised Newtonian fluid**. Models for a number of generalised Newtonian fluids are presented in Section 15.5, and their flow between infinite parallel plates is analysed in Section 16.6.

Since the flow behaviour of non-Newtonian fluids is complicated and still not completely understood, particularly for those which are elastic or thixotropic, it is fortunate for us that all gases and many liquids of engineering interest are Newtonian in character at most temperatures and pressures. We should not forget, however, that numerically there are more non-Newtonian liquids (predominantly synthetic) than Newtonian. In terms of total volume, however, the reverse is probably true, although even gases cease to be Newtonian at very high temperatures or low pressures.

## 2.11  **Bulk modulus of elasticity *K* and compressibility**

We have seen already that viscosity is the property which relates the rate of change of shape of a fluid to applied shear stress. The property which relates the change in volume to a change in pressure is the **bulk modulus of elasticity** $K$, i.e. $K$ is the property which characterises the **compressibility** of a fluid. Although all fluids are compressible to some extent, gases are far more so than liquids and we sometimes distinguish between them primarily on this basis. However, as we discuss briefly in Section 7.5, in many practical gas-flow problems the variation in density is sufficiently small that we can treat the gas as incompressible. In Chapters 11, 12, and 13 we deal with gas flows where compressibility is the dominant influence.

The bulk modulus of elasticity is defined by the equation

$$K = -\frac{\delta p}{\delta \mathcal{V}/\mathcal{V}} = -\mathcal{V}\frac{\delta p}{\delta \mathcal{V}} \tag{2.42}$$

where $\delta\mathcal{V}$ is the change in the fluid volume $\mathcal{V}$ due to a pressure change $\delta p$. From the definition, it can be seen that the units of $K$ are the same as those of pressure, i.e. Pa. Since an increase in pressure (i.e. $\delta p > 0$) causes a decrease in volume (i.e. $\delta\mathcal{V} < 0$), a minus sign is introduced into the defining equation to ensure that $K$ is a positive quantity. In the limit of an infinitesimal change in pressure, with a resulting infinitesimal change in volume, we can write

$$K = -\mathcal{V}\frac{dp}{d\mathcal{V}} \tag{2.43}$$

where $dp/d\mathcal{V}$ is the derivative (or gradient) of the $p - \mathcal{V}$ curve during a compression (or expansion) process. If we consider a fixed mass of fluid $m$, then $m = \rho\mathcal{V}$ so that, since there is no change in mass during compression or expansion,

$$\frac{dm}{dp} = \mathcal{V}\frac{d\rho}{dp} + \rho\frac{d\mathcal{V}}{dp} = 0 \tag{2.44}$$

from which we have

$$\frac{dp}{d\mathcal{V}} = -\frac{\rho}{\mathcal{V}}\frac{dp}{d\rho} \tag{2.45}$$

and, finally,

$$K = \rho\frac{dp}{d\rho}. \tag{2.46}$$

This is a far more satisfactory definition of $K$ than that involving the arbitrary volume $\mathcal{V}$ because, just like density, the bulk modulus of elasticity is a thermodynamic property defined at a point. The reciprocal of $K$ (i.e. $1/K$) is called the **compressibility**, with units $Pa^{-1}$.

In general, compression and expansion are thermodynamic processes involving an increase or decrease in pressure accompanied by a corresponding change in the temperature of the fluid and other properties, such as specific entropy (see Section 11.2). To define completely a property which quantifies the compressibility of a fluid, especially gases, we need to specify either the temperature or the thermodynamic process itself. A process in which the temperature remains constant is called **isothermal**, and one in which there is no heat transfer is called **adiabatic**. If there is neither heat transfer nor friction, the thermodynamic property known as entropy remains constant, and the process is said to be **isentropic**.

Values of $K$ for a range of commonly encountered liquids are listed in Table A.5. For a perfect gas, the pressure–density relationship for an isentropic process can be shown to be (see Section 11.2)

$$\frac{p}{\rho^{\gamma}} = \text{constant} \qquad (2.47)$$

where $\gamma$ is the ratio of the specific heats at constant pressure ($C_P$) and constant volume ($C_V$), i.e. $\gamma = C_P/C_V$ (see Section 2.4). We then have

$$\frac{dp}{d\rho} = \frac{\gamma p}{\rho}$$

and the **isentropic modulus of elasticity** for a perfect gas is

$$K_S = \gamma p, \qquad (2.48)$$

where the subscript $S$ denotes that the expansion or compression process is isentropic. The **isentropic compressibility** is $1/\gamma p$.

## 2.12 **Speed of sound $c$**

Sound travels through a fluid in the form of small-amplitude pressure fluctuations or waves, a process which can be regarded as isentropic. The speed at which such waves propagate through a fluid is called the **speed of sound** (or **soundspeed**) $c$ and can be shown to be given by

$$c^2 = \frac{dp}{d\rho}\bigg|_S = \frac{K_S}{\rho}. \qquad (2.49)$$

In Section 2.5 the equation of state for a perfect gas was given as $p = \rho RT$ and, in Section 2.11, we showed that, for an isentropic process, $K_S = \gamma p$. If we combine the three equations, we have

$$c = \sqrt{\gamma RT}, \qquad (2.50)$$

i.e. since the ratio of specific heats $\gamma$ and the specific gas constant $R$ are constant for a given gas, the speed of sound $c$ is proportional to the square root of its absolute temperature $T$. Thus, the speed of sound is lower on a cold day than on a hot one: between early morning and early afternoon the air temperature in the Black Rock Desert in Nevada might increase from 0 °C (273 K) to 40 °C (313 K). The corresponding increase in the speed of sound is from 331 m/s to 355 m/s, which partly explains why it was advantageous for the successful attempt of the **Project Thrust supersonic car** to break through the sound barrier (i.e. to exceed the speed of sound) to take place early in the day (the actual air temperature was reported to be between 5 and 8 °C).

Since $R = \mathcal{R}/\mathcal{M}$, where $\mathcal{R}$ is the universal gas constant, we have

$$c = \sqrt{\gamma RT} = \sqrt{\frac{\gamma \mathcal{R} T}{\mathcal{M}}} \qquad (2.51)$$

and we see that the speed of sound will be much higher for gases with low molecular weight $\mathcal{M}$, such as hydrogen ($\mathcal{M} = 2.02$ kg/kmol, $c = 1332$ m/s at 20 °C) and helium (4.00 kg/kmol, 1007 m/s), than for heavier gases, such as air (28.965 kg/kmol, 343 m/s), carbon dioxide (44.01 kg/kmol, 268 m/s), and the electrically insulating gas sulphur hexafluoride (146.06 kg/kmol, 133 m/s).

**ILLUSTRATIVE EXAMPLE 2.7**

Calculate the speed of sound for air ($\gamma = 1.4$, $\mathcal{M}_{AIR} = 29$ kg/kmol), helium ($\gamma = 1.63$, $\mathcal{M} = 4$ kg/kmol), and water ($\rho = 998$ kg/m$^3$, $K_S = 2.19 \times 10^9$ Pa) at 20 °C. The universal gas constant $\mathcal{R}$ has the value 8314.5 J/kmol·K.

**Solution**

$T = 20 + 273 = 293$ K.

For air: $\gamma_{AIR} = 1.4$, $\mathcal{M}_{AIR} = 29$ kg/kmol, so $R_{AIR} = \mathcal{R}/\mathcal{M}_{AIR} = 8314.5/29 = 286.7$ J/kg·K, and

$$c = \sqrt{1.4 \times 286.7 \times 293} = 342.9 \text{ m/s}.$$

For helium: $\gamma_{He} = 1.63$, $\mathcal{M}_{He} = 4$ kg/kmol, so $R_{He} = \mathcal{R}/\mathcal{M}_{He} = 8314.5/4 = 2079$ J/kg·K, and

$$c = \sqrt{1.63 \times 2,079 \times 293} = 996.4 \text{ m/s}.$$

For water: $\rho = 998$ kg/m$^3$, $K_S = 2.19 \times 10^9$ Pa, so that, from equation (2.49),

$$c = \sqrt{\frac{2.19 \times 10^9}{998}} = 1481 \text{ m/s}.$$

**Comments:**

(a) In the two calculations for gases we made use of the relation 1 J = 1 N·m, which is a consequence of the mechanical equivalent of heat, and also the definition of the newton, N = kg·m/s$^2$ (see Chapter 3).
(b) In the calculation for water we made use of the definition of the pascal, Pa = 1 N/m$^2$.
(c) Only in the case of helium is the calculated value for $c$ significantly different from the values in Table A.6 and, even for helium, the difference is only 1.1%.

## 2.13 Vapour pressure $p_V$, boiling, and cavitation

It is a common observation that water, in a container open to the atmosphere, evaporates. What we mean by this is that the liquid molecules just below the liquid surface have sufficient momentum to overcome **intermolecular cohesive forces** (see Section 2.1) and escape in vapour form into the atmosphere. If the same liquid is placed in a closed container, and the space above the liquid surface evacuated (i.e. any air is pumped out and the pressure reduced), the rate of evaporation of the liquid rises until an equilibrium is reached when as many molecules leave the surface to create vapour (in the case of water, the vapour is called **steam**) as return to the liquid. Under these equilibrium conditions, the vapour is said to be saturated, and the pressure is the property termed the **saturated vapour pressure** $p_V$, usually referred to simply as the **vapour pressure**. The corresponding temperature is called the **saturation**

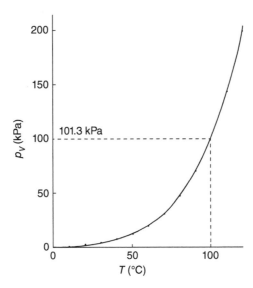

**Figure 2.7** Variation of saturated vapour pressure with temperature for water

**temperature.** The tendency of a liquid to **evaporate** (or **vaporise**) is termed **volatility**. Since molecular activity increases with temperature, the vapour pressure also increases with temperature. The variation of vapour pressure with temperature for water is given in Table A.3 and shown in graphical form in Figure 2.7.

If the pressure within a body of liquid equals the vapour pressure corresponding to the liquid temperature, vapour bubbles are produced within the liquid until all the liquid has become vapour. This is the process we call **boiling.** In a closed container, the production of vapour at a given temperature increases the pressure until equilibrium conditions are reached, corresponding to a point on the $p_V(T)$ curve (Figure 2.7). As we shall see in Chapters 7 and 8, for a subsonic flow, the pressure in a fluid stream decreases if the fluid velocity increases, for example, in flowing through a valve or nozzle. If the pressure within a liquid stream falls below the vapour pressure corresponding to the liquid temperature, internal boiling will initiate an undesirable phenomenon called **cavitation** (see Section 8.11). Since the vapour pressure increases with temperature, the danger of cavitation also increases in, for example, poorly designed domestic-heating systems.

**ILLUSTRATIVE EXAMPLE 2.8**

The atmospheric pressure at the summit of Mount Everest (height 8848 m) is 31 kPa. At what temperature does water boil at this altitude? What pressure would be required for the boiling point to be raised to 100 °C?

**Solution**

$p_1 = 31\,\text{kPa} = 3.1 \times 10^4$ Pa; so, from Table A.3, $T_1 = 70\,°\text{C}$.
$T_2 = 100\,°\text{C}$; so, from Table A.3, $p_2 = 1.013 \times 10^5$ Pa = 1 atm = 1.013 bar.

**Comment:**

The temperature 100 °C corresponds to what we normally think of as the boiling point of water at normal ambient conditions. Due to the low pressure at high altitude, the boiling point is reduced by 30 °C. The pressure has to be increased to 1 atm (slightly higher than 1 bar) for the boiling point to return to 100 °C. A pressure cooker is used to ensure food normally boiled at 100 °C is properly cooked at high altitude.

---

## 2.14 Surface tension $\sigma$ and contact angle $\theta$

The discussion in Section 2.3 was limited to situations in which molecules of fluid in the interior of a fluid interacted with molecules of the same fluid in the same thermodynamic state. As we have just seen, at the surface of a liquid in contact with either a gas or its own vapour, we need to take into account the fact that molecules constantly cross the surface. **Interface** is a more general term for the surface which separates two fluids, such as a liquid and a gas or two immiscible liquids, such as water and mercury or oil and water. For liquid in a tube, the curved interface is also called a **meniscus**. Although the chemistry and physics are complex, for most practical purposes such an interface can be treated as a skin or membrane in tension and this leads to the identification of the fluid property called **surface tension, $\sigma$**, defined by[12]

$$\sigma = \text{interfacial (tensile) force per unit length.} \tag{2.52}$$

From the definition it can be seen that $\sigma$ must have the units N/m.

Figure 2.8 shows a curved line drawn in the free surface of a liquid. The arrows which are everywhere tangential to the surface and normal to the curve, but pointing away from it, represent the force due to surface tension. For an infinitesimal element of surface of length $\delta s$, the force is $\sigma \delta s$. Surface tension for a pure liquid decreases almost linearly with increasing temperature and also depends upon whether the liquid is in contact with its own vapour or with air (or some other gas). An important, but difficult to quantify, influence on surface temperature is contamination of the liquid, due either to unwanted impurities or detergents which markedly decrease $\sigma$. The values of surface tension listed in Tables A.3 (water) and A.5 are for pure liquids.

The shape of small liquid drops and of small soap bubbles can be explained using the concept of surface tension. Figure 2.9(a) shows a section through a segment of a spherical drop of liquid of radius $R$. Consider the circular surface segment of radius $r$ which subtends the angle $\phi$ at the centre of the drop. If the external pressure is barometric, $B$, and the internal pressure is $p_I$, such that the pressure difference between the inside of the surface and the outside is $\Delta p = p_I - B$, then for the segment to be in static equilibrium requires

$$2\pi r\sigma \sin\phi - \pi r^2 \Delta p = 0.$$

---

[12] The symbol $Y$ is sometimes used.

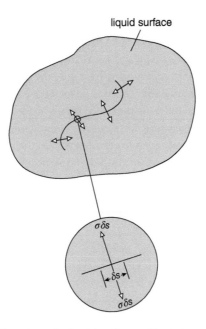

**Figure 2.8** Plan view of a liquid surface to illustrate surface tension

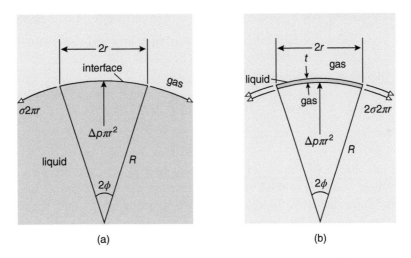

**Figure 2.9** Forces acting on (a) a liquid droplet and (b) a soap bubble

From the geometry of the situation, $\sin \phi = r/R$ so that the static-equilibrium equation reduces to

$$\Delta p = \frac{2\sigma}{R} \tag{2.53}$$

from which we see that the internal pressure increases as the drop size decreases. It is because the internal pressure becomes so large for very small drops (called droplets), that their shape is negligibly affected by gravity and they become spherical in shape, as we assumed at the outset. As $R$ increases, gravity causes a drop to distort and become increasingly non-spherical. A suspended drop develops a tear shape and eventually breaks free.

In the case of a liquid bubble surrounded by a gas, the equivalent to the equation for a droplet is $\Delta p = 4\sigma/R$. The increase in $\Delta p$ by a factor of 2 is a consequence of the fact that the surface of a bubble has a finite thickness (typically a few hundred nm) and therefore an inner and an outer surface, each of which exerts a surface-tension force. Because the gas which fills a bubble generally has a much lower density than the liquid which forms its surface, bubbles can reach much larger sizes than drops before gravitational effects have a significant influence.

---

**ILLUSTRATIVE EXAMPLE 2.9**

The surface tension for petrol is $2.16 \times 10^{-2}$ N/m. Calculate the pressure in the interior of a petrol droplet, 2 $\mu$m in diameter, created by a fuel-injection nozzle if the external pressure is 2.5 bar.

**Solution**

The surface tension $\sigma = 2.16 \times 10^{-2}$ N/m; $R = 10^{-6}$ m; $p_E = 2.5 \times 10^5$ Pa.
From equation (2.46)

$$\Delta p = p_I - p_E = \frac{2\sigma}{R} = \frac{2 \times 2.16 \times 10^{-2}}{10^{-6}} = 4.32 \times 10^4 \text{ Pa or 0.432 bar}$$

so that the internal pressure $p_I$ is given by

$$p_I = p_E + \Delta p = 2.5 \times 10^5 + 4.32 \times 10^4 = 2.93 \times 10^5 \text{ Pa or 2.93 bar.}$$

---

We started this chapter with the word 'wet'. Liquids are said to be wet because surface tension causes them to adhere to solid surfaces. Whether or not a solid surface is wetted by a liquid depends upon the extent to which there is an attraction between the liquid molecules and the surface molecules. The degree of attraction is measured by the angle, called the **contact angle $\theta$**, at which the liquid meets the surface, a quantity which depends on the same factors as surface tension and, in addition, upon the nature of the surface and the surrounding fluid (normally a gas). For $\theta < 90°$ the liquid is said to be **wetting** and, for $\theta > 90°$, **non-wetting**.

Water on a clean, grease-free glass surface has a contact angle practically equal to zero while for mercury the value is about 130°. The combined effects of surface tension and contact angle thus determine the shape of a liquid drop on a horizontal surface, as shown in Figure 2.10. As we shall show in Section 4.8, a wetting liquid is drawn upwards into a vertical small-diameter tube due to surface tension, whereas the surface of a non-wetting liquid is depressed. This effect

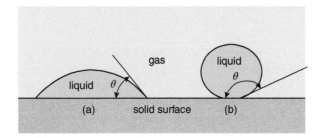

**Figure 2.10** Drops of (a) a wetting and (b) a non-wetting liquid on a horizontal surface

is known as **capillarity**. For a liquid in a container or large-diameter tube, surface tension is also responsible for the curvature of the meniscus where the liquid contacts the solid surface.

 2.15 SUMMARY

In this chapter we have shown that the differences between solids, liquids, and gases have to be explained at the level of the molecular structure. The continuum hypothesis allows us to characterise any fluid and ultimately analyse its response to pressure difference $\Delta p$ and shear stress $\tau$ through macroscopic physical properties, dependent only upon absolute temperature $T$ and pressure $p$, which can be defined at any point in a fluid. The most important of these physical properties are density $\rho$ and viscosity $\mu$, while some problems are also influenced by compressibility, vapour pressure $p_V$, and surface tension $\sigma$. We showed that the bulk modulus of elasticity $K_s$ is a measure of fluid compressibility and determines the speed at which sound propagates through a fluid. We also introduced the perfect-gas law and derived an equation for the soundspeed $c$.

The student should be able to

- state what is meant by the continuum hypothesis
- calculate the number of molecules in a given volume of any fluid
- calculate the molecular mean free path for any gas
- define density as

$$\rho \equiv m/\mathcal{V}$$

- define dynamic viscosity as

$$\mu \equiv \tau/\dot{\gamma}$$

  where $\dot{\gamma}$ is the shear rate
- calculate the soundspeed from $c^2 = K_S/\rho$ and for a perfect gas $c = \sqrt{\gamma \mathcal{R} T/\mathcal{M}}$
- state what is meant by saturated vapour pressure $p_V$
- define surface tension as

$$\sigma = \text{interfacial (tensile) force per unit length}$$

- use the tables in Appendix 2 to look up values for fluid properties and use them in calculations

**2.16  SELF-ASSESSMENT PROBLEMS**

**2.1**  Calculate the sizes of cubes which contain one billion (i.e. $10^9$) molecules of the following substances at STP: a perfect gas, water, Freon 12, and mercury. The Avogadro number $N_A$ has the value $6.02214 \times 10^{26}$ molecules/kmol, and the universal gas constant $\mathcal{R}$ has the value 8314.5 J/kmol $\cdot$ K.
(Answers: 3.42 $\mu$m; 0.31 $\mu$m; 3.42 $\mu$m; 0.53 $\mu$m; 3.42 $\mu$m)

**2.2**  Calculate the molecular mean free path for oxygen at 0 °C and 1 atm. The Boltzmann constant $k_B$ has the value $1.380658 \times 10^{-23}$ J/K.
(Answer: 66.8 nm)

**2.3**  Calculate the relative density and specific weight of air at 500 °C, 1 bar, and of methanol at 20 °C, 1 bar.
(Answers: 0.379; 4.48 N/m³; 0.792; 7760 N/m³)

**2.4**  Calculate the specific gas constant, the density, and the speed of sound for sulphur hexafluoride (SF$_6$) at 100 °C and 3 bar. Take the molecular weight of SF$_6$ as 146 kg/kmol and the ratio of its specific heats as 1.085.
(Answers: 56.9 m²/s² $\cdot$ K; 14.1 kg/m³; 152 m/s)

**2.5**  Calculate the density of water at a depth of 5000 m, where the pressure is 5000 bar, given that the relationship between density $\rho$ and pressure $p$ (in bar) for water is

$$ \ln\left(\frac{\rho}{\rho_0}\right) = \frac{1}{n} \ln\left(\frac{p/p_0 + C}{1 + C}\right) $$

where $p_0 = 1$ bar, $\rho_0 = 1000$ kg/m³, $C = 3000$, and $n = 7$.
(Answer: 1150 kg/m³)

**2.6**  A concentric-cylinder viscometer has the dimensions $R = 25$ mm, $H = 80$ mm, and $\delta = 150\,\mu$m. If the fluid in the annular gap is ethylene glycol at STP, calculate the shear stress and the torque exerted on the inner cylinder if the rotation speed of the inner cylinder is 12 rpm.
(Answers: 41.7 Pa; 0.013 N $\cdot$ m)

**2.7**  The viscosity of a non-Newtonian liquid is 0.5 Pa $\cdot$ s for a shear rate $\dot{\gamma}$ of 10 s$^{-1}$. At very high shear rates the viscosity falls to a constant value of 0.2 Pa $\cdot$ s. Calculate the yield stress $\tau_Y$, assuming that the shear stress $\tau$ obeys the equation $\tau = \tau_Y + C\dot{\gamma}$, where $C$ is a constant.
(Answer: 3 Pa)

**2.8**  Calculate the soundspeed for petrol at STP.
(Answer: 1187 m/s)

**2.9**  An experiment is being designed in which water has to boil at the normal body temperature of 37 °C. What pressure is required?
(Answer: 6.44 kPa)

**2.10**  The mass of liquid used to create a soap bubble 100 mm in diameter is 100 $\mu$g. The surface tension of the liquid is 0.03 N/m, its density is 1000 kg/m³, and its molecular weight is 18 kg/kmol. Calculate the pressure difference between the inside and the outside of the bubble and the thickness of the soap film. Is the continuum hypothesis satisfied?
(Answers: 2.4 Pa; 3.2 nm; just)

# 3

# Units of measurement, dimensions, and dimensional analysis

This chapter is about the **dimensions** and **units** of physical quantities and how they can give us insight into physical problems, simplify the representation of the solutions to problems, and provide a partial check on the correctness of any resulting formulae. Although the illustrative examples are limited primarily to fluid flow, it is important to realise that the principles introduced here apply to many branches of physics and engineering. At first sight **dimensional analysis** may appear abstract and mystifying, perhaps because it involves no difficult mathematics and quite elementary physical ideas, yet leads to far-reaching consequences. It involves little more than simple algebra and the basic principle that each term in any equation involving a combination of physical quantities must have the same overall dimensions.

We start by discussing the units of measurement and the dimensions which are essential to the specification of any physical quantity. We then introduce those units, according to **The International System of Units** (abbreviated as **SI**), and the corresponding dimensions for physical quantities that are involved in the description of fluid flow. We show how units can be multiplied and divided (but not added or subtracted), the same applying to dimensions, and introduce the underlying principle of **dimensional homogeneity**: the overall dimensions of each term in any formula or relationship involving physical quantities have to be the same. We then demonstrate how this principle leads to a systematic procedure which allows the quantities which describe any physical problem to be combined so producing a smaller number of terms which we call **non-dimensional groups**. Some of these groups are particularly significant, appear repeatedly, and are given names, such as the Reynolds and Mach numbers, after the scientists who made important contributions to understanding the flows where these groups arise. The chapter concludes with the topics of **dynamic similarity** and **scaling** which allow us, for example, to predict the aerodynamic performance of a full-scale racing car from a reduced-scale wind-tunnel test of a geometrically similar model.

## 3.1 Units of measurement

The measure or value of any physical quantity, such as acceleration, force, pressure, density, or viscosity, is practically meaningless unless its units are also stated. With few exceptions, all physical quantities have units and dimensions—the two always go together. One of the exceptions to this general rule is the plane angle, which can be thought of as the ratio of two lengths. The quantity $\pi$ (= 3.141592654 . . . . . . .), for example, which plays an important role in plane geometry and, in turn, many branches of engineering and physics, is defined as the ratio of the circumference of a circle to its diameter. More generally, the angle $\theta$ (in radians) subtended by the arc of any circle is defined as the arc length $s$ divided by the circle radius $R$, i.e. $\theta = s/R$.

*Introduction to Engineering Fluid Mechanics*. Marcel Escudier.
© Marcel Escudier 2017. Published 2017 by Oxford University Press.

The ratio of any two quantities with the same units (or dimensions) has neither units nor dimensions (in dimensional analysis, the dimension of any non-dimensional quantity is 1).

In ancient times, basic units of length were often based upon the size of parts of the human anatomy, such as the hand, forearm (cubit), and foot, and extended by multiplying factors to suit particular applications, e.g. yard (3 feet), chain (22 yards), furlong (10 chains), and mile (8 furlongs). Over the centuries a wide array of units evolved, particularly for length, mass, and weight, and for closely associated quantities such as area and volume: inch, metre, pound, hundredweight, ton, gram, kilogram, poundal, slug, hectare, pint, gallon, peck, bushel, etc. With the addition of second, minute, hour, and other units of time, together with units of temperature (degree Celsius or centigrade, Fahrenheit, and kelvin), these units are sufficient to express the magnitude of all the physical quantities we shall consider in this book, and indeed all we are likely to encounter in much of engineering, with the exception of electrical quantities. As is the case with monetary units, especially pre-Euro, the preference for one system of units over another is largely a matter of history and familiarity. The inch, foot, yard, pound, hundredweight, ton, etc., are part of the **Imperial System of Units**, superseded in the UK by The International System of Units, which is now used, especially in engineering, throughout most of the world, with the exception of the United States of America.

The units of most of the quantities we encounter in engineering and science cannot be expressed in terms of length, or mass, or time, alone. Instead, they have to be expressed as combinations of these **basic units** or in terms of new units (**derived units**) defined in terms of the basic units. In the absence of a simple and well-defined system of units, the conversion between units can become complicated and prone to error, sometimes with catastrophic results. The most spectacular (and expensive) example of the latter occurred in 1999, when controllers at the Jet Propulsion Laboratory in California fired the thrusters of the **Mars Climate Orbiter** to adjust its orbit. Unfortunately, the onboard software of the thrusters specified the thrust of the rockets in pounds force ($lb_f$), whereas the control software on the ground assumed it was in newtons (N). Since 1 $lb_f \cong 4.45$ N, the controllers applied an excessively large thrust causing the Orbiter (fortunately unmanned) to disintegrate in the atmosphere of planet Mars.

---

**ILLUSTRATIVE EXAMPLE 3.1**

What tractive force $F$ $lb_f$ is required to accelerate a car of mass $m$ tons from rest to a speed $V$ miles per hour in a time of $t$ minutes?

### Solution

We recognise that this problem requires the use of **Newton's second law of motion** in the familiar form $F = ma$, where the acceleration $a = V/t$ (assuming constant acceleration) so that $F = mV/t$. With $m$ in tons, $V$ in mph, and $t$ in minutes, the units of $F$ will be ton · mile/h · min. It is quite clearly not very useful to have a force expressed in such peculiar units so we need to introduce appropriate conversion factors in the hope of producing a more familiar and practical unit of force. We have 1 (long) ton = 2240 pounds mass ($lb_m$), 1 mile = 5280 ft, 1 h = 3600 s, and 1 min = 60 s. The result is then

$$F = 2240 \, m \times \frac{5280 \, V}{3600} \times \frac{1}{60 \, t} = \frac{54.76 \, mV}{t} \frac{\text{lb}_\text{m} \cdot \text{ft}}{\text{s}^2}$$

which is not much of an improvement.

Since the weight in $\text{lb}_\text{f}$ of a mass $m$ is given by $m \cdot g$, where $g$ is the gravitational acceleration (approximately 32.2 ft/s$^2$), it becomes apparent that $1 \, \text{lb}_\text{m} \times 32.2 \, \text{ft/s}^2$ must be equivalent to 1 $\text{lb}_\text{f}$, so that

$$F = \frac{54.76 \, mV}{32.2 \, t} = \frac{1.702 \, mV}{t} \, lb_f.$$

If this example doesn't convince the reader that a more coherent system of units would be preferable, it's unlikely that anything will.

## 3.2  The International System of Units (SI)

The units of measurement now preferred for practically all engineering and science applications are those of the International System of Units (**Le Système International d'Unités** or, simply, SI). There are seven **base units**: the **kilogram** (symbol kg) for mass, **metre** (m) for length, **second** (s) for time, **kelvin** (K) for absolute thermodynamic temperature, **mole** (mol) for amount of substance, **ampere** (A) for electric current, and **candela** (cd) for luminous intensity. There are numerous **derived units** which are products of powers of base units. Examples are area (m$^2$), volume (m$^3$), velocity (m/s), and density (kg/m$^3$). A small proportion of these derived units have special names and symbols. So far as fluid mechanics and thermodynamics[13] are concerned, these include **newton** ($\text{N} \equiv \text{m} \cdot \text{kg} \cdot \text{s}^{-2}$) for force, **pascal** ($\text{Pa} \equiv \text{N/m}^2$) for pressure or stress, **joule** ($\text{J} \equiv \text{N} \cdot \text{m}$) for energy or work, **watt** ($\text{W} \equiv \text{J/s}$) for power, and **hertz** for frequency ($\text{Hz} \equiv 1/\text{s}$). Other derived units, such as pascal second ($\text{Pa} \cdot \text{s}$) for dynamic viscosity and joule per kilogram kelvin ($\text{J}/(\text{kg} \cdot \text{K})$) for specific heat capacity, are combinations of the derived units. Although obviously named after great scientists and engineers, the names of these derived units are never capitalised when written out in full, whereas the symbols always are.

To avoid very large or very small numbers with many zeros, the SI system also specifies twenty prefixes for the decimal multiples and submultiples of SI units in the range $10^{-24}$ (**yocto**, symbol y) to $10^{24}$ (**yotta**, Y). For the most part, successive prefixes differ by the factor $10^{\pm3}$, those normally encountered in engineering applications being **pico** (p, $10^{-12}$), **nano** (n, $10^{-9}$), **micro** ($\mu$, $10^{-6}$), **milli** (m, $10^{-3}$), **kilo** (k, $10^3$), **mega** (M, $10^6$), **giga** (G, $10^9$), and **tera** (T, $10^{12}$). Examples of the use of these prefixes are pm (picometre), GW (gigawatt), and THz (terahertz). Prefixes which may also be encountered include **centi** (c, $10^{-2}$), **deci** (d, $10^{-1}$), **deca** (da, $10^1$), and **hecto** (h, $10^2$) but these are not common or recommended in engineering practice.

Full details of the SI system of units are to be found at http://physics.nist.gov/Pubs/SP330/sp330.pdf (2008).

---

[13] Some aspects of fluid mechanics, such as compressible fluid flow (Chapters 11–13), also require consideration of **thermodynamics**, and the two subjects, together with **heat transfer**, are often treated as a single subject called **thermofluids**.

## 3.3 Dimensions

While there is an almost unlimited choice of units of measurement, the same is not true of dimensions which are far more fundamental in character. For present purposes, the dimensions of any physical quantity can be expressed in terms of the dimensions **mass** (symbol M), **length** (L), **time** (T), and **temperature** ($\theta$). Other choices are possible, such as incorporating a dimension for force instead of mass, but M, L, T, and $\theta$ corresponding with four of the basic SI units is a logical choice. Since it may be difficult to remember the dimensions of such quantities as dynamic viscosity $\mu$ (M/LT) and power $P$ (ML$^2$/T$^3$), it is worth remembering that there is a one-to-one correspondence between the basic SI units (i.e. kg, m, s, and K) and the dimensions M, L, T, and $\theta$, so the units of any quantity expressed in terms of basic rather than derived units can always be used to work out its dimensions, as we demonstrate in Illustrative Example 3.2.

It is conventional to use **square brackets** around any quantity to denote that only its dimensions are involved. Thus $[P] = ML^2/T^3$ is an equation indicating that the dimensions of power $P$ are $ML^2/T^3$.

---

### ILLUSTRATIVE EXAMPLE 3.2

Convert the derived units for pressure $(p)$, dynamic viscosity $(\mu)$, and power $(P)$ to basic SI units and hence find the dimensions of these three quantities.

### Solution

The derived SI unit for pressure $p$ is the pascal

$$Pa = \frac{N}{m^2} = \frac{kg.m}{s^2}.\frac{1}{m^2} = \frac{kg}{m.s^2}$$

and the dimensions of pressure must be

$$[p] = \frac{M}{LT^2}.$$

Similarly, for dynamic viscosity $\mu$, for which the derived SI unit is Pa · s,

$$Pa.s = \frac{kg}{m.s^2}.s = \frac{kg}{m.s}$$

and so

$$[\mu] = \frac{M}{LT}.$$

Finally, for power $P$, the derived SI unit is the watt (W)

$$W = \frac{J}{s} = \frac{N.m}{s} = \frac{kg.m}{s^2}.\frac{m}{s} = \frac{kg.m^2}{s^3}$$

and

$$[P] = \frac{ML^2}{T^3},$$

as we stated above.

___

One quantity which often causes difficulties for students working out dimensional problems is **rotational speed** $N$, for which the non-SI unit rpm (revolutions per minute) is commonly used. Since one complete revolution corresponds to $2\pi$ rad and there are 60 s in a minute, it can be seen that the corresponding **angular velocity** $\omega$ in rad/s is given by $\omega = 2\pi N/60$. Since $2, \pi$, and 60 are pure numbers, and so non-dimensional, the dimension of $N$ must be the same as that of $\omega$, i.e. $[\omega] = [N] = 1/T$. It is a common mistake to assume that angular velocity is no different dimensionally from linear velocity. To emphasise the point, recall that the linear velocity $V$ of a point on the circumference of a wheel of radius $R$ rotating at angular velocity $\omega$ is given by $V = \omega R$, from which we have

$$[\omega] = \left[\frac{V}{R}\right] = \frac{L}{T} \cdot \frac{1}{L} = \frac{1}{T}.$$

The symbols, their meaning, units, and dimensions of all the physical quantities we shall encounter in this book, and, to a large extent, in engineering fluid mechanics generally, are tabulated in the **Notation** section at the beginning of the book. To assist the reader, the English word form of each of the Greek symbols has also been included in the table.

## 3.4 **Combining dimensions and combining units**

If two or more physical quantities are combined, either by multiplication or division, then the dimensions of the resulting quantity are obtained from the dimensions of the original quantity by the same arithmetic process. For example, if $m$ is mass and $a$ is acceleration, then $[ma] = [m][a] = M \times L/T^2 = ML/T^2$. If we refer to the **Notation**, we see that $ML/T^2$ is in fact the dimension of force $F$. Since $F = ma$ is a common form of Newton's second law of motion, this outcome is just what we should have expected. As with normal arithmetic, we can cancel dimensions (as we did in Section 3.3 to obtain the dimensions of $\omega$), multiply powers of them together by adding indices, and divide by subtracting indices. It should be self-evident that dimensions, just like units, can be neither added nor subtracted: it makes no sense, for example, to add L and T or m and s. We note that it is generally less confusing and so 'safer' to combine groups of units or dimensions by multiplication rather than division. For example, if $a$ is acceleration and $V$ is velocity, we find the dimensions of $a/V$ as follows

$$\left[\frac{a}{V}\right] = \frac{[a]}{[V]} = \frac{L}{T^2} \times \frac{T}{L} = \frac{1}{T}$$

where instead of dividing $L/T^2$ by $L/T$ we have multiplied $L/T^2$ by $T/L$.

Since there is a one-to-one correspondence between units and dimensions, it follows that all of the principles outlined in the previous paragraph apply equally to the manipulation of units. An important point with regard to combinations of SI units is that it is essential to separate individual units when combined by multiplication by a dot, as we do here, or by a space. In this way we avoid confusion between ms meaning millisecond and m·s (or m s) for metre second. For units in the denominator we can use either a solidus (i.e. a slash /) or negative indices; for example, m/s$^2$ can also be written as m · s$^{-2}$.

If the combination of dimensions or units for a physical quantity is unity, that quantity is said to be **non-dimensional**[14].

---

**ILLUSTRATIVE EXAMPLE 3.3**

Show that if $p$ is pressure, $\rho$ is density, and $V$ is velocity, the quantity $p/\rho V^2$ is non-dimensional.

**Solution**

The first step is to write down the dimensions of $p$, $\rho$ and $V$ as follows

$$[p] = \frac{M}{LT^2} \qquad [\rho] = \frac{M}{L^3} \qquad \text{and} \qquad [V] = \frac{L}{T},$$

from which we have

$$\left[\frac{p}{\rho V^2}\right] = \frac{M}{LT^2} \times \frac{L^3}{M} \times \frac{T^2}{L} = 1$$

so demonstrating that the quantity $p/\rho V^2$ is non-dimensional.

If instead of working in terms of dimensions we use units, we arrive at the same conclusion: the unit of pressure is the pascal, or in base units kg/m · s$^2$, those of density are kg/m$^3$, and those of velocity m/s, so that the units of $p/\rho V^2$ are (kg/mv · s$^2$) · (m$^3$/kg) · (s$^2$/m$^2$) = 1, which again demonstrates that $p/\rho V^2$ must be non-dimensional.

---

---

**ILLUSTRATIVE EXAMPLE 3.4**

Calculate the value of $p/\rho V^2$ if $p$ = 7 bar, $\rho$ = 2 kg/m$^3$, and $V$ = 10 m/s. Carry out the calculation using first SI units then Imperial units.

**Solution**

In **consistent** SI units, we have $p = 7 \times 10^5$ Pa, $\rho = 2$ kg/m$^3$, and $V = 10$ m/s. Thus,

$$\frac{p}{\rho V^2} = \frac{7 \times 10^5}{2 \times 10^2} = 3.5 \times 10^3.$$

---

[14] As an alternative to 'non-dimensional' the term **'dimensionless'** is in common use.

The only conversion needed was for the units of pressure since bar (= $10^5$ Pa), though in common use, is not an SI unit but is accepted for use with SI.
We now repeat the exercise for $p$ in psi (i.e. $lb_f/in^2$), $\rho$ in $lb_m/ft^3$, and $V$ in ft/s.

$$7 \text{ bar} = 7 \times 101.2 \text{ psi} \quad 2 \text{ kg/m}^3 = 0.125 \text{ lb}_m/\text{ft}^3 \quad \text{and} \quad 10 \text{ m/s} = 32.81 \text{ ft/s}$$

so that

$$\frac{p}{\rho V^2} = \frac{101.2 \times 144 \times 32.2}{0.125 \times 32.81^2} = 3.49 \times 10^3.$$

Comment:

(1)  The factor 144 had to be introduced to convert psi to $lb_f/ft^2$.
(2)  The factor 32.2 $lb_m \cdot ft/lb_f.s^2$ was needed to convert $lb_f$ to $lb_m \cdot ft/s^2$.
(3)  The final result using Imperial units is not precisely $3.5 \times 10^3$, as it should be, but $3.49 \times 10^3$ as a consequence of the accumulation of small errors in each of the conversions from SI to Imperial units.

## 3.5  The principle of dimensional consistency (or homogeneity)

Each additive term in any physical equation must have the same overall dimensions. This statement of the **principle of dimensional consistency** is the basis of dimensional analysis. It follows that each additive term in any physical equation must have the same overall units. It is, of course, essential that consistent, ideally base SI, units are used for each term in such an equation. We state again, that although this book is concerned with fluids and fluid flow, it should be apparent that any statement about dimensions or units applies in any branch of engineering or physics.

**ILLUSTRATIVE EXAMPLE 3.5**

Show that each of the following equations is dimensionally consistent

(a) $e = mc_0^2$; (b) $v = u + at$; (c) $s = ut + \frac{1}{2}at^2$; (d) $p = B + \rho gz$; (e) $T = 2\pi \sqrt{(l/g)}$;

(f) $D = 6\pi \mu VR + \frac{9}{4}\pi \rho V^2 R^2$; (g) $\dot{Q} = \frac{\pi R^4 \Delta p}{8\mu L}$.

Solution

(a)  We recognise $e = mc_0^2$ as a result of **Einstein's theory of relativity**. In this equation, known as **Einstein's mass-energy relation**, $e$ is the energy released by matter if its **rest mass** reduces by an amount $m$, and $c_0$ is the **speed of light** in a vacuum[15]

$$[e] = \frac{ML^2}{T^2} \qquad \text{and} \qquad [mc_0^2] = M \times \left(\frac{L}{T}\right)^2,$$

[15] Note that in this instance $m$ represents mass reduction, not mass.

and we are relieved to find that Einstein's relation is dimensionally consistent. Note that the units and dimensions of the change or difference in any quantity are the same as those of the quantity itself.

(b) $v = u + at$ is a simple kinematic equation relating the velocity $(v)$ of an object accelerating at constant acceleration $(a)$ for a time $(t)$ from an initial velocity $(u)$. The dimensions of the three terms in this equation are

$$[v] = \frac{L}{T} \qquad [u] = \frac{L}{T} \qquad \text{and} \qquad [at] = \frac{L}{T^2} \times T = \frac{L}{T}$$

which demonstrates that the equation is dimensionally consistent because each additive term has the overall dimensions of velocity, L/T.

(c) $s = ut + \frac{1}{2}at^2$ is a kinematic equation corresponding to $v = u + at$ for the distance $s$ travelled by the object during the time $t$. In this case,

$$[s] = L \qquad [ut] = \frac{L}{T} \times T = L \qquad \text{and} \qquad \left[\frac{1}{2}at^2\right] = 1 \times \frac{L}{T^2} \times T^2 = L$$

again demonstrating that the equation is dimensionally consistent. Note that the numerical factor 1/2 is non-dimensional and highlights an important general principle: no consideration of dimensions can tell us anything about the correctness (or absence) of a purely numerical factor in an equation since it will always have the dimension unity.

(d) For a **simple pendulum** of length $l$, the period of small-amplitude swing $T$ is given by $T = 2\pi\sqrt{(l/g)}$, $g$ again being the acceleration due to gravity

$$2\pi\sqrt{\frac{l}{g}} = 1 \times 1 \times \sqrt{L \times \frac{T^2}{L}} = T$$

which corresponds with the dimension of the period $T$ (i.e. a time).

We note that the first four examples have nothing to do with fluid mechanics, thereby illustrating the observation made at the start of this section that dimensional considerations apply to any branch of engineering or physics.

(e) As we shall see in Section 4.3, the equation $p = B + \rho gz$ gives the pressure $p$ at depth $z$ below the surface of a liquid of constant and uniform density $\rho$ with barometric pressure $B$ at the surface (i.e. at $z = 0$), $g$ being the acceleration due to gravity, also assumed to be constant. We have

$$[\rho gz] = \frac{M}{L^3} \times \frac{L}{T^2} \times L = \frac{M}{LT^2}$$

which corresponds with the dimensions of $p$ and $B$, both of which are pressures, and again we find that the equation is dimensionally consistent.

(f) The **drag force** (or just **drag**) $D$ exerted on a sphere of radius $R$ moving at a low constant velocity $V$ through a fluid of uniform and constant density $\rho$ and dynamic viscosity $\mu$ is given by **Oseen's formula**

$$D = 6\pi\mu VR + \frac{9}{4}\pi\rho V^2 R^2.$$

The formula is approximate and its validity depends upon the value of a non-dimensional parameter termed the **Reynolds number** *(Re)* being small (typically $\ll 1$). The Reynolds number, which is commonly regarded as the most important non-dimensional parameter in fluid mechanics, will be discussed later in this chapter (Section 3.12). The dimensions of each term in the equation are as follows

$$[D] = \frac{ML}{T^2} \qquad [6\pi\mu VR] = 1 \times 1 \times \frac{M}{LT} \times \frac{L}{T} \times L = \frac{ML}{T^2}$$

$$\text{and} \qquad \left[\frac{9}{4}\pi\rho V^2 R^2\right] = 1 \times 1 \times \frac{M}{L^3} \times \frac{L^2}{T^2} \times L^2 = \frac{ML}{T^2}.$$

Note that the first term on the right-hand side of Oseen's formula is linear in $V$ and $R$ while the second is quadratic. It is also the case that the first term includes $\mu$ but not $\rho$ while the second includes $\rho$ but not $\mu$, but the equation is still dimensionally consistent as all three terms have the overall dimensions $ML/T^2$.

(g) The final example involves the **Hagen-Poiseuille formula** for the pressure drop $\Delta p$ along a circular tube of radius $R$ and length $L$ for the flow of a fluid of density $\rho$ and dynamic viscosity $\mu$ with volumetric flowrate $\dot{Q}$

$$\dot{Q} = \frac{\pi R^4 \Delta p}{8\mu L}.$$

While there are restrictions on the applicability of this formula, these are not relevant to our consideration of dimensions. Since the formula concerns the volumetric flow rate $\dot{Q}$, which has the units $m^3/s$, rather than mass flow rate with units kg/s, we have

$$[\dot{Q}] = \frac{L^3}{T} \quad \text{and} \quad \left[\frac{\pi R^4 \Delta p}{8\mu L}\right] = 1 \times L^4 \times \frac{M}{LT^2} \times 1 \times \frac{LT}{M} \times \frac{1}{L} = \frac{L^3}{T}.$$

Note that the dimensions of **pressure** <u>**difference**</u> $\Delta p$ are the same as those of pressure.

---

## 3.6 Dimensional versus non-dimensional representation

We have seen already that certain combinations of quantities have no overall dimensions, i.e. they are non-dimensional. In fact, if any two quantities have the same overall dimensions, their quotient (i.e. ratio) is non-dimensional. As an example, if we divide through Oseen's formula in Illustrative Example 3.5 (f) by $6\pi\mu VR$, we have

$$\frac{D}{6\pi\mu VR} = 1 + \frac{3}{8}\frac{\rho VR}{\mu}. \tag{3.1}$$

Since the first term on the right-hand side is now unity, it is immediately obvious that equation (3.1) is non-dimensional. We have already mentioned the Reynolds number and the combination $\rho VR/\mu$ is precisely that. From equation (3.1) we see that as the Reynolds number approaches zero, Oseen's formula reduces to

$$\frac{D}{6\pi\mu VR} = 1 \tag{3.2}$$

which is known as **Stokes' formula**. The advantage of the non-dimensional version of Oseen's formula is that its five dimensional 'constituents' can be combined into two non-dimensional groups so that the formula can be represented graphically by a single curve of $D/6\pi\mu VR$ versus $\rho VR/\mu$.

We cannot emphasise too strongly that it is always advantageous to convert a dimensional equation or formula to non-dimensional form. As a second illustration, we use the equation of Illustrative example 3.5(c)

$$s = ut + \frac{1}{2}at^2. \tag{3.3}$$

We can easily calculate the distance $s$ for any values of the initial velocity $u$, the acceleration $a$, and the time $t$. Should we wish to, we could plot $s$ versus $t$ for different combinations of $u$ and $a$. Even for such a simple formula, this is a tedious exercise because for every value of $u$ there is an infinite choice of values for $a$ and to cover even a limited range for $u$ and $a$ we would need to plot a large number of curves. Figure 3.1 shows $s$ plotted versus $t$ for just five values of $a$ with $u = 1$ m/s.

Suppose we now divide each of the terms in equation (3.3) by $u^2$ and multiply each by $a$. Then we have

$$\frac{sa}{u^2} = \frac{at}{u} + \frac{1}{2}\left(\frac{at}{u}\right)^2 \tag{3.4}$$

and we observe that instead of three entirely different terms, our equation now has only two, $sa/u^2$ and $at/u$, both of which are non-dimensional

$$\left[\frac{sa}{u^2}\right] = \mathrm{L} \times \frac{\mathrm{L}}{\mathrm{T}^2} \times \frac{\mathrm{T}^2}{\mathrm{L}^2} = 1 \quad \text{and} \quad \left[\frac{at}{u}\right] = \frac{\mathrm{L}}{\mathrm{T}^2} \times \mathrm{T} \times \frac{\mathrm{T}}{\mathrm{L}} = 1.$$

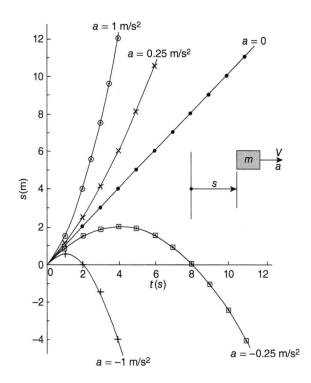

**Figure 3.1** Curves of $s$ versus $t$ for $s = ut + at^2/2$ with $u = 1$ m/s and $a = 0, \pm 0.25$ m/s$^2$, and $\pm 1.0$ m/s$^2$

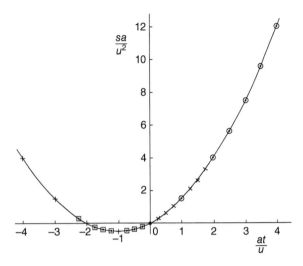

**Figure 3.2** Curve of $sa/u^2$ versus $at/u$ corresponding to $s = ut + at^2/2$

Just as in the example of Oseen's formula, the advantage of our non-dimensional equation over the dimensional version is that all combinations of the physical variables (in this case $s, u, a,$ and $t$) can be represented by a single curve of $sa/u^2$ plotted versus $at/u$ (Figure 3.2). It should be obvious that it takes far less effort to generate one curve rather than five (or more). The benefits of non-dimensionalising a more complicated equation are even greater, as are the advantages of plotting experimental data in non-dimensional form. A major benefit of non-dimensional representation is that information is presented in a very compact form. In the case of experimental data, far fewer experiments need be performed. Arguably even more important, in situations where we have experimental results but limited or no theoretical guidance, a non-dimensional plot is more likely to reveal any underlying relationship between the variables than a dimensional representation.

## 3.7 Buckingham's Π (pi) theorem

In the second example of the previous section, we reduced a problem involving four individual dimensional quantities ($s, u, a,$ and $t$) to one involving just two non-dimensional groups of quantities ($sa/u^2$ and $at/u$). In any problem of **dimensional analysis**, as this mathematical process is called, in the absence of an equation or formula, it is convenient to know in advance how many non-dimensional groups will result from the set of physical quantities thought to describe a physical process. This information is provided by **Buckingham's Π (pi) theorem**

If a physical process involves $n$ dimensional quantities (or variables) which can be described in terms of $j$ independent dimensions, then this process can be represented by $k$ non-dimensional combinations of the dimensional quantities, where

$$k = n - j. \tag{3.5}$$

Although throughout this book we place little emphasis on memorising formulae, equation (3.5) is one of the few that the student should commit to memory.

Since we are limiting ourselves to problems involving physical quantities with dimensions M, L, T, and $\theta$, $j$ can only take on the values 1, 2, 3, or 4. It is immediately obvious that the number of **non-dimensional quantities** $k$ (often called **non-dimensional groups** or **non-dimensional numbers**) is always less than the number of dimensional variables $n$. For example, in the case of the formula $s = ut + at^2/2$, we see that there are four variables ($s, u, a$, and $t$) and two independent dimensions (L and T) so $n = 4$, $j = 2$, and $k = n - j = 2$, which confirms what we found previously in Section 3.6.

Although called Buckingham's $\Pi$ theorem, the symbol $\Pi$, which is the capital version of the Greek letter pi, has nothing to do with the familiar numerical constant $\pi = 3.1415927\ldots$ but is simply the symbol chosen by Buckingham to represent a non-dimensional combination of dimensional quantities. In our example, we can write

$$\Pi_1 = \frac{sa}{u^2} \qquad \text{and} \qquad \Pi_2 = \frac{at}{u}.$$

In a more general case, the $n$-dimensional variables would reduce to $k = n - j$ combinations of those variables, $\Pi_1, \Pi_2, \Pi_3, \ldots \ldots \ldots \Pi_k$ or $\Pi_1 = f(\Pi_2, \Pi_3, \ldots \ldots \ldots \Pi_k)$, which means that the non-dimensional group $\Pi_1$ is a function of (i.e. depends upon) $\Pi_2, \Pi_3, \ldots \ldots \ldots \Pi_k$.

## 3.8 Sequential elimination of dimensions (Ipsen's method)

Although with experience it is often possible to identify the non-dimensional groupings (i.e. the $\Pi$'s) in any problem by **inspection**, it is usually preferable to use a systematic approach (see Section 3.10). Although not the most common procedure in use (which is the exponent method, presented in Section 3.9), the method of **sequential elimination of dimensions** (also known as **Ipsen's step-by-step method**) presented here is an essentially foolproof 'recipe' which requires only elementary mathematics.

We illustrate the method by reference once again to the sphere drag part of Illustrative Example 3.5(f) but pretend now that we know (or postulate) only that $D$ depends upon $V, R, \rho$, and $\mu$ but we do not know the formula $D = 6\pi \mu VR + 9\pi \rho V^2 R^2/4$. We start by writing

$$D = f(V, R, \rho, \mu), \tag{3.6}$$

which simply means that $D$ is a function of (i.e. depends upon or is determined by) $V, R, \rho$, and $\mu$. The quantity $D$ is called the **dependent variable**, while $V, R, \rho$, and $\mu$ are the **independent variables** (i.e. the variables under our control).

In this case, then, we have five physical variables, so $n = 5$. It is vital in any problem of dimensional analysis not to forget the dependent variable, in this case $D$, when counting the number of physical variables.

The dimensions of the physical variables are

$$[D] = \frac{ML}{T^2} \qquad [V] = \frac{L}{T} \qquad [R] = L \qquad [\rho] = \frac{M}{L^3} \qquad \text{and} \quad [\mu] = \frac{M}{LT}$$

so we have just three dimensions (i.e. M, L, and T), and $j = 3$. From Buckingham's theorem, $k = n - j = 2$, so we expect to find two non-dimensional groups (i.e. two $\Pi$'s).

Our aim is to eliminate the three dimensions M, L, and T systematically by multiplying or dividing each of the variables by any one of the others (or a power of any one of them). It is important to realise that, although we can start the elimination process with any variable, the end result will always be correct, although not the same. Suppose we choose to eliminate M first using the variable $\rho$ (we could just as well have chosen $\mu$ to eliminate M). Then we have

$$\left[\frac{D}{\rho}\right] = \frac{ML}{T^2} \times \frac{L^3}{M} = \frac{L^4}{T^2} \qquad \text{and} \qquad \left[\frac{\mu}{\rho}\right] = \frac{M}{LT} \times \frac{L^3}{M} = \frac{L^2}{T}$$

and we can rewrite our original equation (3.6) as

$$\frac{D}{\rho} = f_1\left(V, R, \frac{\mu}{\rho}\right) \tag{3.7}$$

in which we have written $f_1(\ldots)$ to indicate that the dependence of $D/\rho$ on $V$, $R$, and $\mu/\rho$ is not the same as the dependence of $D$ on $V$, $R$, $\rho$, and $\mu$. At this stage we have already reduced the number of variables from four to three (i.e. $D/\rho$, $V$, $R$, and $\mu/\rho$), and the number of dimensions from three to two (i.e. L and T), so that $k$ is still equal to 2, as it should be.

We now choose $R$ to eliminate the dimension L from $D/\rho$, $V$, and $\mu/\rho$, as follows

$$\left[\frac{D}{\rho R^4}\right] = \frac{L^4}{T^2} \times \frac{1}{L^4} = \frac{1}{T^2} \qquad \left[\frac{V}{R}\right] = \frac{L}{T} \times \frac{1}{L} = \frac{1}{T} \quad \text{and} \quad \left[\frac{\mu}{\rho R^2}\right] = \frac{L^2}{T} \times \frac{1}{L^2} = \frac{1}{T}$$

and we can write our equation as

$$\frac{D}{\rho R^4} = f_2\left(\frac{V}{R}, \frac{\mu}{\rho R^2}\right) \tag{3.8}$$

i.e. we now have just three variables ($D/\rho R^4$, $V/R$, and $\mu/\rho R^2$) and one remaining dimension (T).

Finally, we choose $V/R$ to eliminate the dimension T, as follows

$$\left[\frac{D}{\rho R^4}\left(\frac{R}{V}\right)^2\right] = \left[\frac{D}{\rho V^2 R^2}\right] = \frac{1}{T^2} \times T^2 = 1 \quad \text{and} \quad \left[\frac{\mu}{\rho R^2} \cdot \frac{R}{V}\right] = \frac{\mu}{\rho VR} = \frac{1}{T} \times T = 1$$

so that our two non-dimensional groups are $D/\rho V^2 R^2$ and $\mu/\rho VR$, i.e. we can write

$$\Pi_1 = \frac{D}{\rho V^2 R^2} \qquad \text{and} \qquad \Pi_2 = \frac{\mu}{\rho VR}$$

and the end result is

$$\Pi_1 = F(\Pi_2) \qquad \text{or} \qquad \frac{D}{\rho V^2 R^2} = F\left(\frac{\mu}{\rho VR}\right). \tag{3.9}$$

Note that dimensional analysis tells us only that $D/\rho V^2 R^2$ depends upon $\mu/\rho VR$, assuming our original assumption that $D = f(V, R, \rho, \mu)$ was itself correct, but can give us no further information as to the form of the dependence (or, in the case of a single non-dimensional group, its value; this is the situation in Illustrative Example 3.6). The final result here is not at first sight consistent with the non-dimensional form of Oseen's equation

$$\frac{D}{6\pi \mu VR} = 1 + \frac{3}{8}\frac{\rho VR}{\mu}. \tag{3.1}$$

However, if we divide through this equation by $\rho VR/\mu$ we find

$$\frac{D}{6\pi \rho V^2 R^2} = \frac{\mu}{\rho VR} + \frac{3}{8} \tag{3.10}$$

which is entirely consistent with the result of dimensional analysis. The two constants $6\pi$ and 3/8 arise from a solution of the governing **Navier-Stokes equations** (see Chapter 15) and cannot be determined from dimensional analysis. We should also recognise that, if instead of using $\rho$ to eliminate M we had chosen $\mu$, the final result of dimensional analysis would have been

$$\frac{D}{\rho \mu^2} = F_1\left(\frac{\rho VR}{\mu}\right) \tag{3.11}$$

which is easily shown to be consistent with

$$\frac{D}{\rho V^2 R^2} = F\left(\frac{\mu}{\rho VR}\right) \text{ or } \frac{D}{\rho V^2 R^2} = F_2\left(\frac{\rho VR}{\mu}\right). \tag{3.12}$$

We should also consider the possibility that other physical variables might influence the drag on the sphere. For example, if the fluid is a gas, we should include a variable that accounts for compressibility. As we saw in Chapter 2, such a variable is the speed of sound $c$ (an alternative would be the **isentropic bulk modulus**, $K_S$, or its inverse, the **isentropic compressibility**). Our starting point for dimensional analysis would then be

$$D = f(V, R, \rho, \mu, c) \tag{3.13}$$

and the end result

$$\frac{D}{\rho V^2 R^2} = F_3\left(\frac{\rho VR}{\mu}, \frac{V}{c}\right). \tag{3.14}$$

Adding one more physical variable has led to an additional non-dimensional group because we changed $n$ from 5 to 6, and $k$ increased from 2 to 3 because $j$ remained equal to 3.

We should be aware that this example applies generally to any problem concerning the fluid-dynamic drag on a body of any shape immersed in a fluid flow, provided we replace the sphere radius $R$ by a characteristic dimension $l$ of the body.

Later in this chapter we shall see that $D/\rho V^2 R^2$, $\rho VR/\mu$, and $V/c$ have special places in fluid mechanics: the first is a **drag coefficient**, the second is called the **Reynolds number**, and the third is the **Mach number**. Other physical quantities which play a role in flow problems include the acceleration due to gravity ($g$) and surface tension ($\sigma$), and these also lead to 'named' non-dimensional groups: the **Froude number** $V/\sqrt{gl}$ and the **Weber number**, $\rho V^2 l/\sigma$ (see Section 3.12).

If, as is frequently the case, we are unable to fully analyse a problem from basic principles, then a non-dimensional representation of experimental data is of great value in guiding us how best to establish a correlation, not least because it always reduces, sometimes significantly, the number of variables we need to deal with independently.

---

**ILLUSTRATIVE EXAMPLE 3.6**

In 1945 the first test of an **atomic bomb** took place in New Mexico, the so-called **Trinity Test**. Photographic images were released in 1947 showing the expansion of the fireball with time $t$ (0.1 to 62 ms) after the instant of initiation. If it is assumed that the fireball radius $R$ depends upon $t$, the atmospheric density $\rho$ (taken as constant), and the energy released[16] in the explosion $E$, show that

$$\frac{Et^2}{\rho R^5} = \text{constant.}$$

Solution:

Step 1: The functional dependence in terms of the dimensional physical variables may be written as

$$R = f(t, \rho, E)$$

so that the number of independent physical variables $n = 4$.

Step 2: The dimensions of each of the physical variables are

$$[R] = L, \quad [t] = T, \quad [\rho] = \frac{M}{L^3}, \quad \text{and} \quad [E] = \frac{ML^2}{T^2}$$

and the number of dimensions $j = 3$.

Step 3: According to Buckingham's $\Pi$ theorem,

$$k = n - j = 1$$

and we expect the four dimensional quantities will combine together to produce a single non-dimensional group, $\Pi_1$, such that $\Pi_1 = \text{constant}$.

Step 4: We select $\rho$ to eliminate the dimension M from $E$

$$\left[\frac{E}{\rho}\right] = \frac{ML^2}{T^2} \times \frac{L^3}{M} = \frac{L^5}{T^2}$$

[16] The energy released in an explosion is known as the **yield**.

so that

$$R = f_1\left(\frac{E}{\rho}, t\right).$$

**Step 5:** Use $t$ to eliminate T from $E/\rho$

$$\left[\frac{Et^2}{\rho}\right] = \frac{L^5}{T^2} \times T^2 = L^5$$

so that

$$R = f_2\left(\frac{Et^2}{\rho}\right).$$

**Step 6:** Use $R$ to eliminate the remaining dimension L from $Et^2/\rho$

$$\left[\frac{Et^2}{\rho R^5}\right] = L^5 \times \frac{1}{L^5} = 1$$

so that finally

$$\frac{Et^2}{\rho R^5} = \text{constant}$$

or

$$R = C\left(\frac{E}{\rho}\right)^{1/5} t^{2/5}$$

where $C$ is a non-dimensional constant. A graph of the fireball radius $R$ versus the time from initiation of the explosion $t$, in logarithmic coordinates, thus has a slope of 2/5 and an intercept on the $R$–axis of $\ln C + \ln(E/\rho)/5$.

From a more detailed analysis, the British scientist Sir Geoffrey (G. I.) Taylor was able to determine the value of $C$ and so calculate the energy released in the Trinity explosion. He also showed that the values of $R$ and $t$ determined from the 1947 pictures followed closely the trend predicted by dimensional analysis. Taylor's estimate for $E$ was 71.5 TJ or 16.8 kilotonnes of TNT, while the figure stated in an official US Army report published shortly after the test was 15–20 kt of TNT.

---

**ILLUSTRATIVE EXAMPLE 3.7**

A common method for mixing large batches of liquid-food products, plastics, cement, and other viscous liquids is with a rotating-paddle mixer. The power $P$ required to rotate the paddle depends upon its rotational speed $\omega$, its radius $R$, the density of the liquid $\rho$, and its dynamic viscosity[17] $\mu$. Derive a non-dimensional form to represent this dependence.

---

[17] In practice most synthetic liquids are **non-Newtonian** in character so that $\mu$ varies with $\omega$ and other flow variables. Some consideration of non-Newtonian liquids is given in Sections 2.10 and 15.5.

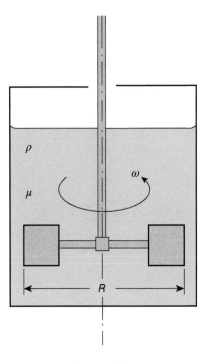

**Figure E3.7**

Solution:

Step 1: The functional dependence in terms of the dimensional physical variables may be written as

$$P = f(\omega, R, \rho, \mu)$$

so that the number of physical variables $n = 5$.

Step 2: The dimensions of each of these physical variables are

$$[P] = \frac{ML^2}{T^3} \qquad [\omega] = \frac{1}{T} \qquad [R] = L \quad [\rho] = \frac{M}{L^3} \quad \text{and} \quad [\mu] = \frac{M}{LT}$$

and the number of dimensions $j = 3$.

Step 3: According to Buckingham's Π theorem,

$$k = n - j = 2$$

and we expect the five dimensional quantities will combine together to produce two non-dimensional groups, $\Pi_1$ and $\Pi_2$, such that $\Pi_1 = F(\Pi_2)$.

Step 4: To find Π and $\Pi_2$, we select $\rho$ to eliminate the dimension M from $P$ and $\mu$

$$\left[\frac{P}{\rho}\right] = \frac{ML^2}{T^3} \times \frac{L^3}{M} = \frac{L^5}{T^3} \qquad \text{and} \qquad \left[\frac{\mu}{\rho}\right] = \frac{M}{LT} \times \frac{L^3}{M} = \frac{L^2}{T}$$

so that

$$\frac{P}{\rho} = f_1\left(\omega, R, \frac{\mu}{\rho}\right).$$

**Step 5:** Use $R$ to eliminate L from $P/\rho$ and $\mu/\rho$

$$\left[\frac{P}{\rho R^5}\right] = \frac{L^5}{T^3} \times \frac{1}{L^5} = \frac{1}{T^3} \text{ and } \left[\frac{\mu}{\rho R^2}\right] = \frac{L^2}{T} \times \frac{1}{L^2} = \frac{1}{L}$$

so that

$$\frac{P}{\rho R^5} = f_2\left(\omega, \frac{\mu}{\rho R^2}\right).$$

**Step 6:** Use $\omega$ to eliminate the remaining dimension T from $P/\rho R^5$ and $\mu/\rho R^2$

$$\left[\frac{P}{\rho R^5 \omega^3}\right] = \frac{1}{T^3} \times T^3 = 1 \quad \text{and} \quad \left[\frac{\mu}{\rho R^2 \omega}\right] = \frac{1}{T} \times T = 1$$

so that finally we have two non-dimensional groups

$$\Pi_1 = \frac{P}{\rho R^5 \omega^3} \quad \text{and} \quad \Pi_2 = \frac{\mu}{\rho R^2 \omega}$$

and

$$\Pi_1 = F(\Pi_2) \quad \text{or} \quad \frac{P}{\rho R^5 \omega^3} = F\left(\frac{\mu}{\rho R^2 \omega}\right).$$

Once again this is a perfectly valid result but only one of several possibilities determined by the sequence in which each of the dimensions (here, M, L, and T) was eliminated and which physical quantities were chosen to carry out the elimination procedure. Alternative non-dimensional groups can be formed by multiplying or dividing the groups, or powers or roots of the groups, which are the 'natural outcome' of dimensional analysis. Some of the other non-dimensional groups we might have obtained in the present example are $\rho R^2 \omega/\mu$, $P\rho^2 R/\mu^3$, $P/\mu\omega^2 R^3$, and $P\sqrt{(\rho^3/\omega\mu^5)}$.

---

## 3.9 Rayleigh's exponent method

The **exponent** (or **indicial**) **method** of dimensional analysis is attributed to the English physicist Lord Rayleigh and also associated with the American physicist Percy Williams Bridgman. This method is based on the rather sophisticated idea that any mathematical function can be expressed as an infinite power series, each term of which, according to the principle of dimensional consistency (see Section 3.5), must have the same overall dimensions.

We can illustrate the exponent method using the aerodynamic-drag example. Since we have

$$D = f(V, l, \rho, \mu) \tag{3.6}$$

it must be that

$$D = kV^a l^b \rho^c \mu^d + k'V^{a'} l^{b'} \rho^{c'} \mu^{d'} + \ldots\ldots\ldots \tag{3.15}$$

where $k, k', \ldots\ldots$ are numerical constants and $a, b, c, d, a', b', c', d', \ldots..$ are the exponents (i.e. constants or powers). Dimensional analysis can be used to determine the values of the exponents but not the numerical constants. According to the principle of dimensional consistency, all terms in the series must have the same dimensions as the dependent variable $D$, i.e.

$$[D] = \left[ kV^a l^b \rho^c \mu^d \right] \tag{3.16}$$

and

$$\frac{ML}{T^2} = 1 \times \left( \frac{L}{T} \right)^a \times L^b \times \left( \frac{M}{L^3} \right)^c \times \left( \frac{M}{LT} \right)^d. \tag{3.17}$$

The key point is to recognise that, since this is a dimensional equation, we require that each dimension balances separately. For example, in the case of L,

$$L^1 = L^a \times L^b \times L^{-3c} \times L^{-d} = L^{a+b-3c-d} \tag{3.18}$$

which in turn means that the exponents must balance, i.e.

$$1 = a + b - 3c - d. \tag{3.19}$$

Similarly, from considerations of M,

$$1 = c + d \tag{3.20}$$

and, for T,

$$-2 = -a - d. \tag{3.21}$$

As an observation, we note that the number of unknown exponents (i.e. $a, b, c,$ and $d$) is the same (i.e. four) as the number of **independent** physical quantities (i.e. $V, l, \rho,$ and $\mu$) while the number of equations equals the number of dimensions (i.e. M, L, and T). With only three equations we cannot determine all four unknowns; the best we can do is to write three of the unknown exponents in terms of the fourth. For the latter, we can choose any one of the four exponents. If we choose $d$, from equation (3.20) we have

$$c = 1 - d, \tag{3.22}$$

from equation (3.21),

$$a = 2 - d, \tag{3.23}$$

from equation (3.19),

$$b = 1 - a + 3c + d = 2 - d \tag{3.24}$$

and we have now found $a, b,$ and $c$ in terms of $d$.

We return to the infinite series, equation (3.15), which can now be written as

$$D = kV^{2-d} l^{2-d} \rho^{1-d} \mu^d + \ldots.. \tag{3.25}$$

$$= k\rho V^2 l^2 \left( \frac{\mu}{\rho V l} \right)^d + \ldots\ldots\ldots\ldots\ldots \tag{3.26}$$

In the final version of our infinite series, we have separated the independent variables into those having (known) pure-number exponents and those involving the unknown exponent $d$. If we now divide through by $\rho V^2 l^2$, we have

$$\frac{D}{\rho V^2 l^2} = k \left( \frac{\mu}{\rho V l} \right)^d + \dots \dots \dots \dots \dots \tag{3.27}$$

or

$$\frac{D}{\rho V^2 l^2} = F \left( \frac{\mu}{\rho V l} \right) \tag{3.28}$$

which means that the non-dimensional group $D/\left(\rho V^2 l^2\right)$ is a function of $\mu/\left(\rho V l\right)$ which, as it must be, is also non-dimensional. In fact, the inverse of $\mu/\left(\rho V l\right)$, i.e. $\rho V l/\mu$, is the special non-dimensional group, the **Reynolds number**, which arises in the majority of viscous fluid-flow problems (see Section 3.12).

It is crucially important for the reader to realise that the final result in the form of equation (3.27) is not simply

$$\frac{D}{V^2 \rho l^2} = k \left( \frac{\mu}{\rho V l} \right)^d \tag{3.29}$$

even though, if it were, this would be a very simple and convenient formula to use once $k$ and $d$ were known, e.g. from experiment or a complete analysis of the flow problem. Unfortunately, the exponent method is sometimes presented with the vitally important '+ $\dots\dots\dots\dots$' omitted from equation (3.27), and the unwary reader forgets, or is never told, that he or she is dealing with only one term of an infinite series.

Of course, if carried out correctly, both the exponent method and the sequential-elimination process produce the same result. The exponent method is probably the most commonly used but, in the author's opinion, **Ipsen's sequential-elimination process** (Section 3.8) is more straightforward and less likely to lead the inexperienced user into difficulty or misunderstanding.

## 3.10 Inspection method

In many instances, with experience, it becomes quite straightforward to write down the appropriate non-dimensional groups for any given problem, essentially from memory or inspection. For example, if the flow velocity $V$ and fluid viscosity $\mu$ are involved in a problem, there is a very good chance that the non-dimensional group $\rho V l/\mu$ will be one of the non-dimensional groups (obviously, the fluid density $\rho$ and a length $l$ are also required). Since an approach of this kind is more ad hoc than systematic, it is not to be recommended for the inexperienced.

## 3.11 Role of units in dimensional analysis

Since there is a one-to-one correspondence between the basic SI units, m, kg, s, and K here, and the primary dimensions M, L, T, and $\theta$, it should be clear that, in principle, any problem in dimensional analysis can be worked through using basic units rather than dimensions, although this is not recommended. However, it is important to realise that the units of any quantity can be used to determine its dimensions. This is particularly useful in cases for quantities where the dimensions may be difficult to remember, such as viscosity and power. The units of practically all physical properties of engineering significance can be found in thermodynamic tables, such as those by Rogers and Mayhew (1994), or any of the mechanical-engineering handbooks which many university departments make available to students. As mentioned in Section 3.3, the **Notation** section at the beginning of this book lists all the physical properties and other quantities which appear in this book, together with their units and dimensions.

---

**ILLUSTRATIVE EXAMPLE 3.8**

The power $P$ required to drive a centrifugal pump depends upon the volumetric flowrate $\dot{Q}$ it delivers, the pressure difference $\Delta p$ imposed between the outlet and inlet of the pump, the density of the liquid $\rho$ and its viscosity $\mu$, the rotational speed of the pump $\omega$, and the impeller radius $R$. Put the preceding sentence into the form of a non-dimensional equation.

**Solution**

As always, the first step is to write down the functional dependence

$$P = f(\dot{Q}, \Delta p, \omega, R, \rho, \mu).$$

To illustrate the point being made in this section, we now write down the units of each of the eight quantities involved

$$P\,(\text{W}); \quad \dot{Q}\,(\text{m}^3/\text{s}); \quad \Delta p\,(\text{Pa}); \quad \omega\,(1/\text{s}); \quad R\,(\text{m}); \quad \rho\,(\text{m}^3/\text{s}); \text{ and } \mu\,(\text{Pa.s});$$

from which we can state or derive the dimensions of each quantity, where necessary first converting the derived units (W, Pa, and N) into basic units

$$W = \frac{J}{s} = \frac{N.m}{s} = \frac{kg.m}{s^2}.\frac{m}{s} = \frac{kg.m^2}{s^3}; \quad Pa = \frac{N}{m^2} = \frac{kg.m}{s^2}.\frac{1}{m^2} = \frac{kg}{m.s^2}$$

and 
$$Pa.s = \frac{kg}{m.s^2}.s = \frac{kg}{m.s};$$

so that

$$[P] = \frac{ML^2}{T^3}; \quad [\dot{Q}] = \frac{L^3}{T}; \quad [\Delta p] = \frac{M}{LT^2}; \quad [\omega] = \frac{1}{T}; \quad [R] = L; \quad [\rho] = \frac{M}{L^3}; \text{ and } [\mu] = \frac{M}{LT}.$$

We now use the sequential elimination method to perform the dimensional analysis, starting with the density $\rho$ to eliminate the dimension M

$$\left[\frac{P}{\rho}\right] = \frac{ML^2}{T^3} \times \frac{L^3}{M} = \frac{L^5}{T^3}; \quad \left[\frac{\Delta p}{\rho}\right] = \frac{M}{LT^2} \times \frac{L^3}{M} = \frac{L^2}{T^2}; \quad \text{and} \quad \left[\frac{\mu}{\rho}\right] = \frac{M}{LT} \times \frac{L^3}{M} = \frac{L^2}{T}$$

and at this stage we can write

$$\frac{P}{\rho} = f_1\left(\dot{Q}, \frac{\Delta p}{\rho}, \omega, R, \frac{\mu}{\rho}\right).$$

We now use $R$ to eliminate the dimension L

$$\left[\frac{P}{\rho R^5}\right] = \frac{L^5}{T^3} \times \frac{1}{L^5} = \frac{1}{T^3}; \quad \left[\frac{\Delta p}{\rho R^2}\right] = \frac{L^2}{T^2} \times \frac{1}{L^2} = \frac{1}{T^2}; \quad \left[\frac{\mu}{\rho R^2}\right] = \frac{L^2}{T} \times \frac{1}{L^2} = \frac{1}{T}$$

and $\quad \left[\dfrac{\dot{Q}}{R^3}\right] = \dfrac{L^3}{T} \times \dfrac{1}{L^3} = \dfrac{1}{T}$

so that

$$\frac{P}{\rho R^5} = f_2\left(\frac{\dot{Q}}{R^3}, \frac{\Delta p}{\rho R^2}, \omega, \frac{\mu}{\rho R^2}\right).$$

Finally, we use $\omega$ to eliminate the dimension T (noting that $[\omega] = 1/T$)

$$\frac{P}{\rho R^5 \omega^3} = F\left(\frac{\dot{Q}}{\omega R^3}, \frac{\Delta p}{\rho R^2 \omega^2}, \frac{\mu}{\rho \omega R^2}\right).$$

In this case the seven physical quantities have produced four non-dimensional groups, as we would expect from Buckingham's Π theorem since with $n = 7$ and $j = 3$, we have $k = 4$.

For a fluid with relatively low viscosity, such as water, viscosity plays only a minor role in determining pump performance except at very low speeds. In practice this means that viscosity is of little importance if $\rho R^2 \omega / \mu \gg 1$. $\rho R^2 \omega / \mu$, of course, is the inverse of our fourth non-dimensional group and is a rotational form of the Reynolds number we have mentioned previously.

## 3.12 **Special non-dimensional groups**

In engineering fluid mechanics we have to take account of the influence on flow of those fluid properties relevant to the specific problem under consideration. These properties may include the fluid density $\rho$, either the fluid dynamic viscosity $\mu$ or the fluid kinematic viscosity $v$ (defined by $v = \mu / \rho$), surface tension $\sigma$, and soundspeed $c$. The latter quantity is included to take account of the compressibility of the fluid (i.e. the increase or decrease in density produced by an increase or decrease in pressure). We could also have included here the acceleration due to gravity $g$, which in most instances plays no role, although there are obvious exceptions, for example, in determining the vertical variation of pressure in a body of fluid at rest (see Section 4.2) or in problems where there is a free surface or an interface between two immiscible liquids. If we are concerned with flow through a machine or object (**internal flow**, see Chapter 16) or around a body surrounded with fluid (**external flow**), it is necessary to select a velocity $V$ and a length $l$ that characterises the object's scale (or size). In the case of a pipe or **duct**[18] through which there is flow, this length is likely to be its diameter or radius (or more generally its **hydraulic diameter or radius**[19]) although the pipe length $L$ may also be of importance as may be the average height of any **surface roughness** $\varepsilon$ (see Section 18.9). For an

---

[18] The term duct is used to mean any pipe, tube, channel, nozzle, etc., through which there is fluid flow.

[19] The concept of hydraulic diameter and radius is explained in Section 16.2.

external flow, the characteristic length may be the diameter or radius of an object, for example in the case of a smoke stack, or the wing span in the case of an aircraft. In a given problem, it is most unlikely that all of the fluid properties, lengths, etc., play a significant role and it is usual to include in any analysis only those likely to do so. For example, liquids are practically incompressible so the soundspeed would only be included if the fluid concerned was a gas in which the occurrence of major pressure variations was likely. Collectively, the quantities which influence a flow are termed the **independent variables**.

The **dependent variables** (i.e. the quantities which are determined by the independent variables) might be $D$, the drag force for external flow around an object, $f$ the frequency of the periodic disturbances which sometimes occur in the wake of the object, and $\Delta p$ a pressure difference, as discussed further below. This being the case, we may write

$$D \left(\text{or} f \text{ or } \Delta p\right) = F\left(V \left\{\text{or } \omega\right\}, \rho, \mu, \sigma, c, g, l\right) \tag{3.30}$$

where $V$ is a characteristic velocity of the flow. In the case of an external flow, $V$ is usually the velocity of the approach flow relative to the object. In the case of a boundary layer, the **free-stream velocity** $U_\infty$ is used. For an internal flow, an average velocity $\bar{V}$ across the flow cross section is the usual choice. So far as turbomachinery, such as a pump, compressor, or turbine, is concerned, $\omega$ is the rotational speed (in rad/s).

If we apply the method of sequential elimination of dimensions to equation (3.30), we arrive at

$$\frac{D}{\rho V^2 l^2} \left(\text{or} \frac{fl}{V} \text{ or } \frac{\Delta p}{\rho V^2}\right) = F_1 \left(\frac{\mu}{\rho V l}, \frac{\sigma}{\rho V^2 l}, \frac{gl}{V^2}, \frac{c}{V}\right). \tag{3.31}$$

Additional quantities, such as the lengths $L$ and $\varepsilon$, could easily be added on the right-hand side of equation (3.30). So far as dimensional analysis is concerned, this would merely add $L/l$ and $\varepsilon/l$ to the list of non-dimensional groups in equation (3.31).

Although perfectly valid as they are, each of the non-dimensional groups in equation (3.31) (with the exception of $fl/V$) is usually modified as follows

$$\frac{D}{\frac{1}{2}\rho V^2 A}, \frac{\Delta p}{\frac{1}{2}\rho V^2}, \frac{\rho V l}{\mu}, \frac{\rho V^2 l}{\sigma}, \frac{V}{\sqrt{gl}}, \text{ and } \frac{V}{c}. \tag{3.32}$$

We now consider each of these special non-dimensional groups.

### 3.12.1 Drag coefficient $C_D$

The first group in the list above defines the **drag coefficient**

$$C_D = \frac{D}{\frac{1}{2}\rho V^2 A}. \tag{3.33}$$

As explained in Section 7.5, the quantity $\rho V^2/2$ is termed the **dynamic pressure** and arises from considerations of energy conservation along a streamline: it represents the kinetic energy per unit volume of fluid. Since the factor 1/2 is non-dimensional, its inclusion in the definition of $C_D$ does not affect its 'dimensionality'. The quantity $A$, which has replaced $l^2$, is a characteristic (or representative) area for the problem considered. In many instances, this is taken as the frontal area or silhouette area of the object on which the drag force is being exerted or the **planform area** in the case of a wing. The drag coefficient can be regarded as the ratio of the drag force $D$ to the force the dynamic pressure $\rho V^2/2$ would exert on an area $A$.

Further consideration of the drag coefficient for various objects is discussed in Section 18.15.

### 3.12.2  Lift coefficient $C_L$

If $L$ is the lift force exerted by fluid flow on an object, such as an aerofoil, then, in a similar way to a drag coefficient, we can define a **lift coefficient**

$$C_L = \frac{L}{\frac{1}{2}\rho V^2 A}. \tag{3.34}$$

Lift coefficients for many standard aerofoil sections have been compiled by Abbot and von Doenhoff (1959).

### 3.12.3  Euler number *Eu*, cavitation number *Ca*, pressure coefficient $C_P$

The **Euler number** is defined as

$$Eu = \frac{\Delta p}{\rho V^2}. \tag{3.35}$$

There are several variants on the non-dimensional combination of the quantities $\Delta p$ and $\rho V^2$, largely dependent upon how the pressure difference $\Delta p$ arises. For example, if $\Delta p = p_{REF} - p_V$, where $p_{REF}$ is a reference pressure, such as the prevailing barometric (or atmospheric) pressure, and $p_V$ is the vapour pressure of a liquid, $\Delta p/\rho V^2$ may be used as a measure of the propensity for **cavitation** to occur (see Section 8.11 for a discussion of cavitation) and is then also referred to as a **cavitation number, *Ca*.**

For flow around an object, $\Delta p$ is usually taken as the difference between the local **static pressure** $p$, e.g. at a point on the surface of an aerofoil, and a reference pressure, such as the undisturbed static pressure upstream of the object, $p_\infty$. Here again, it is also conventional to introduce the factor 1/2 so that

$$C_P = \frac{\Delta p}{\frac{1}{2}\rho V^2} \tag{3.36}$$

which is known as the **pressure coefficient**.

For internal flows, the **dynamic pressure** $\rho \bar{V}^2/2$ is used to 'normalise' (i.e. make non-dimensional) the pressure drop in a duct $\Delta p$, and we have the **friction factor**

$$c_f = \frac{\Delta p}{\frac{1}{2}\rho \bar{V}^2} \tag{3.37}$$

where $\bar{V}$ is the flow velocity averaged over the duct cross section. The factor 1/2 is often moved to the left-hand side so that the friction factor is defined as $c_f/2 = \Delta p/\rho V^2$.

Pressure reduction in a simple, straight duct is a consequence of wall shear stress $\tau_S$ (the subscript $S$, for surface, is often replaced by $W$, for wall). More generally, in a pipe system, pressure losses also occur for a number of other reasons, such as increases, particularly sudden increases, or decreases, in cross-sectional area (see Sections 10.5 and 18.11).

### 3.12.4 Fanning friction factor $f_F$, Darcy friction factor $f_D$, and skin-friction coefficient $c_f$

Friction factors for duct flow based directly upon the surface shear stress $\tau_S$ are also commonly encountered, for example the **Fanning friction factor**, $f_F \equiv 2\tau_S/\rho\bar{V}^2$, and the **Darcy friction factor**, $f_D = 8\tau_S/\rho\bar{V}^2$ (see Chapters 16 and 18).

It is easily shown that $f_F$ and $c_f$ for fully developed duct flow through a cylindrical duct of length $L$ and hydraulic diameter $D_H$ (see Section 16.2) are related by

$$c_f = \frac{4L}{D_H}f_F. \tag{3.38}$$

For external flows, a **boundary-layer friction factor** (or **skin-friction coefficient**) is defined by

$$\frac{c_f}{2} = \frac{\tau_S}{\rho U_\infty^2} \tag{3.39}$$

which is similar to the definition of the Darcy friction factor for duct flow except for the 1/2-factor being on the left-hand side (see Chapters 17 and 18).

At first sight, all these related but slightly different definitions may seem confusing. They have been introduced independently by different people over many decades and retained in their areas of applicability. It is obviously vital for the user of any formula involving one of these non-dimensional groups to be aware of which definition is relevant.

### 3.12.5 Reynolds number *Re* and Poiseuille number *Po*

In Chapter 2, we identified viscosity $\mu$ as the material property of a fluid which distinguishes it from a solid. It is hardly surprising, therefore, that the non-dimensional group which incorporates viscosity, the **Reynolds**[20] **number**, plays a role in the majority of flow problems. The Reynolds number is defined by

$$Re = \frac{\rho Vl}{\mu} \tag{3.40}$$

where $l$ is a **characteristic (or representative) length** and $V$ is a **characteristic velocity**. Depending upon the problem, $l$ may be a diameter, a radius, a wing span, a height, etc. While the choice is relatively unimportant, if it is stated, for example, that 'the critical Reynolds number is 2100' it is crucial that it is known what length and velocity have been used to define the Reynolds number.

If the Reynolds number for an external flow is very small compared with unity, fluid inertia is of minor importance and there is a balance between viscous and pressure forces. Such flows are referred to as **creeping flows**, an example of which is **Stokes' flow** (see Section 3.6). If the Reynolds number is very large, much of the flow represents a balance between inertia and

[20] Reynolds number is sometimes written as Reynold's number, even though this is completely incorrect since the surname of the man it is named after is Reynolds not Reynold. In fact, if it were named after someone called Reynold, it would still be incorrect to write Reynold's as it is conventional to adopt forms such as Mach number (not Mach's number), etc.

pressure forces, with viscous effects being negligible. In the latter case, viscosity still dominates flow in the region immediately adjacent to any solid surface, known as the **boundary layer** (see Chapters 17 and 18), where the fluid velocity progressively approaches that of the surface (i.e. zero if the surface is at rest). The Reynolds number plays an important role in characterising a viscous flow as either laminar (see Chapters 16 and 17) or turbulent (Chapter 18). For cylindrical pipe flows, the Reynolds number is based upon either the internal diameter or radius while, for duct flows in general, the **hydraulic diameter** $D_H$ is often chosen as the length scale.

Much of Chapter 16 is concerned with **fully-developed** laminar **flows** through **cylindrical ducts** where, as for Stokes' flow, fluid inertia plays no role. The important flow parameter is the **Poiseuille**[21] **number** defined as

$$Po = \frac{2\tau_S D_H}{\mu \bar{V}} \tag{3.41}$$

from which it is easily shown that $Po = f_F\,Re_H$ although, as pointed out in Section 16.2, the Reynolds number (here based upon $D_H$) has been introduced artificially.

### 3.12.6   Mach number *M*, Cauchy number *Ca*, and Knudsen number *Kn*

The **Mach number**[22] is a non-dimensional number named in honour of Ernst Mach, at the suggestion of the Swiss scientist Jakob Ackeret, and defined as the ratio of a flowspeed (or the speed of an object moving through a stationary fluid), $V$, to the soundspeed of the fluid $c$

$$M = \frac{V}{c}. \tag{3.42}$$

It should be noted that $c$ is not a constant for any given fluid but depends primarily upon its temperature (in the case of a gas, $c \sim \sqrt{T}$, where $T$ is the absolute temperature). Provided the Mach number is less than about 0.3, a flow can be considered as practically incompressible, which is why many gas flows can be treated (with the appropriate values for $\rho$ and $\mu$) in exactly the same way as a liquid flow. For higher values of $M$, compressibility becomes increasingly important. Up to about $M = 0.75$, the effects of density changes can be accounted for by applying a compressibility correction to the results of incompressible theory. For values of $M$ close to unity, the flow is termed **transonic**, typically with some regions remaining **subsonic** ($M < 1$) while others are **supersonic** ($M > 1$) and shockwaves begin to appear. In supersonic flow, abrupt decreases in velocity, stagnation pressure, and Mach number occur across shockwaves, with corresponding increases in temperature, pressure, and density. These changes become increasingly strong as the upstream Mach number becomes increasingly greater than unity. Flows for which $M > 3$ are termed **hypersonic**.

A more general non-dimensional number which can be used to characterise compressible flow is the **Cauchy number** defined as

$$Ca = \frac{\rho V^2}{K} \tag{3.43}$$

---

[21] The pronunciation of Poiseuille is 'pwazoy'.
[22] The symbol *Ma* is frequently used for the Mach number.

where $K$ is the isentropic bulk modulus of the fluid. It can be shown that, for an isentropic flow process, $Ca = M^2$. Compressible fluid flow is the subject of Chapters 11, 12, and 13.

In Section 2.5 we showed how the validity of the continuum hypothesis is related to the molecular mean free path $\Lambda$. The continuum hypothesis breaks down for flows where the characteristic length scale $L$ has a magnitude approaching $\Lambda$, such that $\Lambda/L = O(1)$ or greater. This **Knudsen regime** includes flows involving a particle moving through the lower atmosphere or a satellite in the exosphere, and flow through the channels of microfluidic devices. For a gas with molecular weight $\mathcal{M}$ and effective molecular diameter $\sigma$, according to equation (2.23), we have

$$\Lambda = \mathcal{M}/\sqrt{2}\pi\rho N_A\sigma^2, \tag{2.23}$$

where $N_A$ is the Avogadro number.

The Knudsen number $Kn$ is defined as

$$Kn = \frac{\Lambda}{L}. \tag{3.44}$$

From equation (2.23), gas flows with $Kn > 1$ will thus arise where the gas density $\rho$ is very low, as in the outer regions of the atmosphere (see Section 4.13), and are referred to as **rarefied flows**.

The Knudsen number $Kn$ is related to the Mach $M$ and Reynolds numbers $Re$ as follows

$$Kn = \frac{\Lambda}{L} = \frac{3}{2}\sqrt{\frac{\pi\gamma}{2}}\frac{\mu}{\rho cL} = \frac{3}{2}\sqrt{\frac{\pi\gamma}{2}}\frac{M}{Re} \tag{3.45}$$

where $M = V/c$, and $Re = \rho VL/\mu$, $V$ being a typical velocity for the flow. Equation (3.45) is based upon equation (2.36) for $\mu$

$$\mu = \frac{2}{3}\sqrt{\frac{2}{\pi\gamma}}\rho c\Lambda. \tag{2.36}$$

### 3.12.7 **Weber number *We***

The Weber number, defined by

$$We = \frac{\rho V^2 l}{\sigma}, \tag{3.46}$$

represents the ratio of inertia forces to surface-tension forces. Surface-tension effects in fluid flow are only important if the Weber number is of order unity or smaller. This can be the case for small droplets or bubbles, capillary flows, and flows of very shallow water. For $We \gg 1$, or if there is no free surface, surface-tension effects are negligible or non-existent.

### 3.12.8 **Froude number**

It can be shown that the speed of propagation of small-amplitude waves on the surface of a liquid layer of depth $h$ is $\sqrt{gh}$. For a free-surface flow with flow velocity $V$, the **Froude number** is defined as the ratio of $V$ to $\sqrt{gh}$

$$Fr = \frac{V}{\sqrt{gh}}. \tag{3.47}$$

Such a flow with $Fr < 1$ is said to be **subcritical** (i.e. the flowspeed is below the **wavespeed** so that small disturbances move faster than the flow) while a flow with $Fr > 1$ is termed **supercritical**. From this it becomes apparent that the Froude number is rather like the Mach number, which distinguishes between subsonic ($M < 1$) and supersonic ($M > 1$) flow.

Where there are two fluids of different densities, such as at the interface between two immiscible liquids, a so-called **densimetric Froude number** can be defined as

$$Fr = \frac{V}{\sqrt{gh|\Delta\rho|/\bar{\rho}}} \tag{3.48}$$

where $|\Delta\rho|$ is the magnitude of the density difference and $\bar{\rho}$ is the average density. The densimetric Froude number also arises in the analysis of **buoyant jets**, also known as **plumes**, when the appropriate length scale is the initial diameter.

### 3.12.9 Strouhal number *St*

The Strouhal number, defined as

$$St = \frac{fl}{V} \tag{3.49}$$

is used to characterise the periodic (i.e. fixed-frequency) disturbances (so-called **æolian tones**) which arise in the wake of an object such as a circular cylinder immersed in a steady flow. If the cylinder diameter is $d$, and the crossflow velocity is $V$, the frequency $f$ is given by $St = fd/V \approx 0.2$ in the Reynolds-number range, $400 < \rho Vd/\mu < 3 \times 10^5$. **Self-excited flow oscillations** of this type can feed energy into the structure, leading, in turn, to **(flow-induced) structural vibrations** which can reach dangerously high levels if the frequency is close to a natural frequency of the structure. The collapse of the **Takoma Narrows suspension bridge** in Washington State, USA, in 1940 was a consequence of this effect, as was the vibration of the 241 m-high **John Hancock Tower** opened in Boston in 1976, which led to large plate-glass windows falling from their frames. Remarkably, in both instances the vibration was initiated at windspeeds no greater than about 70 kph. The helical strakes which are wound around tall chimneys are designed to suppress such periodic flow behaviour.

## 3.13 Non-dimensional groups as force ratios

It is useful to find a physical interpretation of non-dimensional groups, as ratios of quantities with the same overall dimensions. As we have just seen, the quantity $\rho V^2/2$ (or just $\rho V^2$), which represents the kinetic energy per unit volume of fluid and also its dynamic pressure, occurs in the numerators of several of the **dependent** non-dimensional groups. The quantity $\rho V^2 l^2$, where $l$ is a characteristic length of the flow problem under consideration ($l^2$ could just as well be replaced by $A$, a characteristic area), is often referred to as an **inertia force** (it is easily shown that it has the dimensions $ML/T^2$, the same as those of force). The drag coefficient $C_D$ can then be regarded as the ratio of the drag force to the fluid inertia force, while the Euler number $Eu$, the pressure coefficient $C_P$, and the friction factor $c_f/2$ are the ratio of a pressure force ($\Delta pA$) to the inertia force. In a similar way, the Fanning and Darcy friction factors represent the ratio of a friction force to the inertia force.

We turn now to the **independent** non-dimensional groups, starting with the Reynolds number which can be written as

$$Re = \frac{\rho Vl}{\mu} = \frac{\rho V^2 l^2}{\mu Vl}. \tag{3.50}$$

The shear stress at a point in a fluid is proportional to the velocity gradient at that point, according to $\tau = \mu \, du/dy$ so that $\tau \sim \mu V/l$ (where $u$ is the local velocity, and $y$ is the normal distance from a boundary) and the denominator of $Re$, $\mu Vl \sim \tau l^2$, which is the product of a shear stress and an area ($l^2$) and again has the dimensions of force. The Reynolds number can therefore be viewed as the ratio of inertia force to shear force.

We have already seen that the Mach number is the square root of the Cauchy number, the latter defined as $Ca = \rho V^2/K$, $K$ being the isentropic bulk modulus of the fluid. The pressure difference $\Delta p$ required to change the volume $\mathcal{V}$ of a fluid by an amount $\Delta \mathcal{V}$ is given by $\Delta p = -K \, \Delta \mathcal{V}/\mathcal{V}$ (the negative sign is introduced because a decrease in volume requires a positive pressure difference) so that $Ca \sim \rho V^2 l^2/\Delta p l^2$, i.e. the ratio of inertia force to compressive (or pressure) force.

In the case of the Weber number, $\rho V^2 l$ appears in the numerator and it is immediately apparent that this number represents the ratio of inertia force to surface-tension force, the latter being proportional to $\sigma l$.

The Froude number can be written as $\sqrt{\rho} Vl/\sqrt{\rho g l^3}$, where the numerator can be seen to be the square root of the inertia force, and the denominator is the square root of the gravity force.

Not all non-dimensional groups can be interpreted as the ratio of a force to inertia force, for example the Strouhal number. Other non-dimensional groups, which arise in areas such as heat transfer, can be identified as ratios of such physical phenomena as viscous and thermal diffusion (e.g. the **Prandtl number**).

## 3.14 **Similarity and scaling**

There would be little point in carrying out experimental studies on scaled-down (or even scaled-up) models if we did not known how the results could be translated (i.e. **scaled**) to full size. Fortunately, this is just the information provided by dimensional analysis. We require two things

- **geometric similarity**, which means that the model and full size (or prototype) differ only in size (or scale) but not in shape. The ratio of full scale to model scale is termed the **scale factor**.
- **dynamic similarity**, which requires that each independent non-dimensional group has the same value for the model (M) and full scale (F), i.e. $\Pi_{2M} = \Pi_{2F}$, $\Pi_{3M} = \Pi_{3F}$, etc.

We have seen already that a major simplification resulting from dimensional analysis is that the number of separate variables we need to deal with is always reduced by the number of independent dimensions involved (usually three).

Although it was not stated at the time, an implicit assumption in dimensional analysis is that we are considering geometrically similar situations. For example, if we write

$$D = f(V, l, \rho, \mu) \tag{3.51}$$

for the drag force $D$ exerted on a car of length $l$, we intuitively realise that, if we carry out experiments on a model car, it should be a scaled-down replica of the full-size version in all relevant respects. If the model is $1/5^{\text{th}}$ the length of the full size, then the wheels of the model should be $1/5^{\text{th}}$ the diameter of those of the full size, the width $1/5^{\text{th}}$, etc. Certain aspects of the design of a car, such as the car's interior, play no role in determining its drag and so are not relevant in dimensional analysis. Other features, such as gaps between the doors and body panels, or the trim, or the windscreen wipers, may have a minor influence on drag but would normally be too difficult or expensive to reproduce accurately on a model.

The requirement of dynamic similarity in scaling model tests is the same as what is required to duplicate the results of any experiment, i.e. to ensure that each of the independent non-dimensional groups has the same value for the model and the full scale and, in consequence, so do the dependent non-dimensional groups. As we have already seen, the power of dimensional analysis is that the number of independent non-dimensional quantities is always less, often significantly so, than the number of dimensional quantities.

In our aerodynamic-drag example, we have

$$D = f(V, l, \rho, \mu) \tag{3.52}$$

which tells us that if we carry out an experiment on a model (or full-size) car to measure the drag $D$ for given values of $V, \rho, \mu$, and $l$ and repeat the experiment, for exactly the same values of the dimensional independent variables $V, \rho, \mu$, and $l$, we shall obtain, within experimental uncertainty, the same value for the dependent variable $D$.

In the non-dimensional representation, the equation for $D$ transforms to

$$\frac{D}{\rho V^2 l^2} = F\left(\frac{\rho V l}{\mu}\right) \tag{3.53}$$

which tells us that, for every value of the non-dimensional group of independent variables $\rho V l/\mu$ (i.e. the Reynolds number), there will be a corresponding value of the dependent non-dimensional group $D/\rho V^2 l^2$ (i.e. the drag coefficient, although, as mentioned above, the denominator is conventionally replaced by $\rho V^2 A/2$, where $A$ is the car's silhouette area). The beauty of this result is that it is only the values of the two non-dimensional groups $\rho V l/\mu$ and $D/\rho V^2 l^2$ which matter, not the values of their 'constituents'. In other words, we can quite freely change the values of $V, \rho, \mu$, and $l$, for example by changing the fluid (e.g. using water instead of air), but, if $\rho V l/\mu$ stays the same, then so will $D/\rho V^2 l^2$.

As a final point, here, we note that if we have **geometric** and **dynamic similarity**, then we shall also have **kinematic similarity**. What this means is that within the flowfields for the model and full scale, the ratio of velocities at corresponding points will be the same as the ratio of the reference velocities $V_M$ and $V_F$, where the subscripts $M$ and $F$ indicate model and full scale, respectively. The velocities will also have the same vector directions at corresponding points in the two cases.

**ILLUSTRATIVE EXAMPLE 3.9**

A sports car designed for a top speed of 356 kph is being developed for the **24 Heures du Mans** endurance race. The prevailing atmospheric conditions are assumed to correspond to an air density of 1.2 kg/m$^3$ and a dynamic viscosity of $1.8 \times 10^{-5}$ Pa · s (i.e. the values for pure air at normal temperature and pressure, 20 °C and 1 atm). Calculate the wind-tunnel speed for tests to be carried out on a quarter-scale model car in a pressurised and cooled wind tunnel in which the air density is 4.7 kg/m$^3$ and the dynamic viscosity $1.7 \times 10^{-5}$ Pa · s (i.e. the property values at 0 °C and 3.7 bar). If the model test gives a drag force of 1334 N, what is the corresponding drag for the full-size car and the tractive power required, assuming dynamic similarity between the wind-tunnel and full-scale situations?

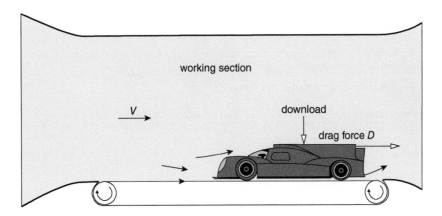

**Figure E3.9**

Solution

We would normally start by carrying out the basic dimensional analysis for this problem. However, in this case, we know the result already

$$\frac{D}{\rho V^2 l^2} = F\left(\frac{\rho V l}{\mu}\right).$$

Dynamic similarity requires that each of the two non-dimensional groups has the same value for the model and for the full-scale car, i.e.

drag coefficient $C_D = \dfrac{D_M}{\rho_M V_M^2 l_M^2} = \dfrac{D_F}{\rho_F V_F^2 l_F^2}$

and

Reynolds number $Re = \dfrac{\rho_M V_M l_M}{\mu_M} = \dfrac{\rho_F V_F l_F}{\mu_F},$

where, as earlier, the subscripts $M$ and $F$ refer to the model and full scale, respectively.

It is always advisable in problems of this kind to tabulate the known and unknown quantities using consistent SI units. For the top speed we have $V_F = 356$ kph which converts to $V_F = 98.9$ m/s; then

|  | Model | Full scale |
| --- | --- | --- |
| Speed $V$ (m/s) | ? | 98.9 |
| Length $l$ (m) | $l_F/4$ | $l_F$ |
| Air density $\rho$ (kg/m$^3$) | 4.7 | 1.2 |
| Air viscosity $\mu$ (Pa·s) | $1.7 \times 10^{-5}$ | $1.8 \times 10^{-5}$ |
| Drag force $D$ (N) | 1334 | ? |
| Power $P$ (W) | ? | ? |

Note that we have no absolute information about the size of either the model or the full-scale car, only the ratio between them which, in this instance, is sufficient.

From the Reynolds-number equality, we have

$$V_M = \frac{\rho_F}{\rho_M} \times \frac{\mu_M}{\mu_F} \times \frac{l_F}{l_M} \times V_F$$

$$= \frac{1.2}{1.7} \times \frac{1.7 \times 10^{-5}}{1.8 \times 10^{-5}} \times 4 \times 98.9 = 95.4 \frac{m}{s}.$$

From the drag-coefficient equality, we have

$$D_F = \frac{\rho_F}{\rho_M} \times \left(\frac{V_F}{V_M}\right)^2 \times \left(\frac{l_F}{l_M}\right)^2 \times D_M$$

$$= \frac{1.2}{4.7} \times \left(\frac{89.9}{95.4}\right)^2 \times 4^2 \times 1334 = 5859 \text{ N or } 5.86 \text{ Kn}$$

and the corresponding tractive power required is

$$P_F = D_F V_F = 5859 \times 98.9 = 5.79 \times 10^5 \text{ W or } 579 \text{ kW (or 777 hp)}.$$

The power required to overcome the drag of the model car $P_M$ can also be calculated as

$$P_M = D_M V_M = 1334 \times 95.4 = 1.27 \times 10^5 \text{ W}.$$

The question we should ask now is 'does power also obey the principle of dynamic similarity?' This is easily answered as follows. Using $\rho$, $V$, and $l$, we find that a non-dimensional group for the power is $P/\rho V^3 l^2$ and we should expect this group to have the same value for model and full scale.

For the model we have

$$\frac{P_M}{\rho_M V_M^3 l_M^2} = \frac{1.27 \times 10^5}{4.7 \times 95.4^3 \times (l_F/4)^2} = \frac{0.50}{l_F^2}$$

where we have again used the ratio 1:4 for the length scales, and, for the full scale,

$$\frac{P_F}{\rho_F V_F^3 l_F^2} = \frac{5.79 \times 10^5}{1.2 \times 98.9^3 \times l_F^2} = \frac{0.50}{l_F^2}$$

so the two non-dimensional groups have the same value, as they should.

---

## 3.15 Scaling complications

The scaling situation we are usually confronted with is that of a model test on a scale much smaller than full size, and this can easily lead to conflicting or impossible requirements for the model tests. We illustrate the difficulty which can arise, and suggest ways in which the corresponding conflict can be resolved by reconsidering the problem of aerodynamic drag. The observant reader will have noticed that in the sports-car example (Illustrative Example 3.9) for the wind tunnel the air density was given as 4.7 kg/m$^3$ and the dynamic viscosity as $1.7 \times 10^{-5}$ Pa $\cdot$ s. These property values, which correspond to a pressure of about 3.7 bar and a temperature of 0 °C, would be attainable only in a specially designed and expensive pressurised, cryogenic wind tunnel. In some circumstances, strict adherence to the requirements of dynamic similarity is possible only through such extreme measures and it may be necessary to accept a compromise solution.

We should, of course, ask the question 'What would be the consequences of performing the model test in a wind tunnel operating at the same temperature and pressure as would be the case for the full-scale car, so that $\rho_M$ and $\mu_M$ would have the same values as their full-scale counterparts, i.e. 1.2 kg/m$^3$ and $1.8 \times 10^{-5}$ Pa $\cdot$ s, respectively?' Since we took the speed of the full-scale car as 356 kph or 98.9 m/s, Reynolds-number equality led to

$$V_M = \frac{\rho_F}{\rho_M} \times \frac{\mu_M}{\mu_F} \times \frac{l_F}{l_M} \times V_F$$
$$= 1 \times 1 \times 4 \times 98.9 = 395.6 \, \text{m/s} = 1424 \, \text{kph}.$$

It should be apparent immediately that for a model test such an airspeed is unrealistically high and would again require a rather special (and again expensive) wind tunnel. There is, however, a more fundamental problem: since the speed of sound at 20 °C is 342 m/s, an airspeed of 395.6 m/s corresponds to a Mach number $M = 1.15$. At this Mach number, the flow approaching the model is just supersonic (i.e. $M > 1$), and changes in the airflow as it passed over the model would produce corresponding changes in pressure, density, and temperature and introduce compressibility effects, such as shockwaves, which would drastically affect aerodynamic behaviour, including drag. In fact, there would be regions in the flow where $M$ is much greater than 1.15, as well as others where it would be much lower. For the full-scale car, for which $M = 0.29$, compressibility effects would be (just about) negligible.

Clearly, something has gone wrong, and the foregoing is a reminder that when we carry out dimensional analysis (or any other theoretical analysis) it is assumed that the physical quantities we have included account for all the physical effects of importance to the problem under consideration. Compressibility would be expected to influence the aerodynamic behaviour of rockets, missiles, most jet aircraft, and even cars designed to challenge the world land-speed record (currently 1228 kph but the **Bloodhound SSC project** is aiming for 1600 kph), but not a car with a top speed of 'only' 356 kph.

In Chapter 2 we showed the isentropic bulk modulus $K$ is the appropriate property to characterise the compressibility of a fluid. However, it is more usual to use the soundspeed $c$ to characterise compressible fluid flow. If we include $c$ in the list of physical quantities that determine aerodynamic drag, we can write

$$D = f(V, \rho, \mu, l, c) \tag{3.54}$$

and the end (i.e. non-dimensional) result is

$$\frac{D}{\rho V^2 l^2} = F_1 \left( \frac{\rho V l}{\mu}, \frac{V}{c} \right) \qquad \text{or} \qquad C_D = F_1(Re, M). \tag{3.55}$$

Note that we did not need to go through the entire dimensional analysis again: since we added one more variable, $c$, and no new dimensions, Buckingham's $\Pi$ theorem tells us to look for one additional non-dimensional group to involve $c$. The conventional choice is the Mach number, $M \equiv V/c$, though in principle a second Reynolds number, based upon $c$ rather than $V$ (i.e. $\rho c l/\mu$), would be just as good.

Dynamic similarity now requires not only the Reynolds number $Re$ to have the same value for the model and the full-scale car, but also the Mach number $M$ if the drag coefficient $C_D$ is to be the same. As we saw in our example at normal temperature and pressure (i.e. 20 °C and 1 atm), Reynolds-number equality was not consistent with Mach-number equality, so that dynamic similarity was not achievable. We can see that this will always be a problem if Reynolds-number equality is enforced with the same fluid properties for both model and full scale, because

$$\frac{\rho_M V_M l_M}{\mu_M} = \frac{\rho_F F l_F}{\mu_F} \tag{3.56}$$

leads to

$$V_M l_M = V_F l_F \quad \text{or} \quad V_M = \frac{l_F}{l_M}.V_F. \tag{3.57}$$

Apart from any other considerations, this result reveals as completely erroneous the common layman's assumption that, to replicate the aerodynamic behaviour of a car or aircraft, its speed should be reduced in proportion to its size. In fact, as we see, exactly the opposite is true!

The situation is much more satisfactory for the example with high pressure (3.7 bar) and reduced temperature (0 °C) for the model. The dynamic viscosity of air is weakly dependent upon its absolute temperature ($\mu \sim \sqrt{T}$) and practically independent of pressure. Essentially, the scale difference has been compensated for by increasing the density, whereas the sound-speed, also proportional to $\sqrt{T}$ for a perfect gas, is only slightly reduced (from 343 m/s to 331 m/s) and we find

$$M_M = 95.4/331 = 0.29 \text{ and } M_F = 98.9/343 = 0.29.$$

Thus, the conditions for dynamic similarity are now completely satisfied, although the Mach number appears to be low enough for compressibility effects to be regarded as negligible (normally, an airflow can be considered incompressible if $M < 0.3$). In fact, the situation is not quite so straightforward because there will almost certainly be zones on the car's surface where the Mach number could reach significantly higher values, and compressibility effects would no longer be completely negligible (e.g. for $M = 0.5$ the density is reduced by about 11%). The more the full-scale Mach number exceeds 0.3 and approaches unity, the more important is equality of the model and full-scale Mach numbers to account for compressibility effects.

In discussing the wind-tunnel evaluation of the aerodynamic behaviour of a sports car, we have made no mention of two vital aspects (both of which are outside the scope of dimensional analysis). The first is **blockage**, associated with the fact that the flow around an object in a wind tunnel is affected by the proximity of the tunnel's walls. To some degree, this effect can be accounted for and is minimised if the tunnel cross section is far greater than that of the model. The second aspect is **ground effect**, associated with the fact that the airflow around a car moving over a stationary road is significantly different from the situation where both car and road are stationary. Modern wind tunnels used to evaluate the aerodynamics of cars use a 'rolling road' to account for ground effect.

## 3.16  **Other Reynolds-number considerations**

In general, the Reynolds number is influenced by changes in both density and dynamic viscosity. It is often more convenient, therefore, to define the Reynolds number in terms of the kinematic viscosity $\nu \equiv \mu/\rho$, i.e.

$$Re = \frac{\rho Vl}{\mu} = \frac{Vl}{\nu}. \tag{3.58}$$

We can now regard the influence of increased pressure (see Section 3.15) as a reduction in the kinematic viscosity. A reduction in $\nu$ can also be achieved by changing the model fluid from air to a liquid such as water. At first sight it may seem surprising that the effective viscosity of water is less than that of air. However, the density of water may be taken as 1000 kg/m$^3$ and its dynamic viscosity at 20 °C as $10^{-3}$ Pa · s, so that its kinematic viscosity is $10^{-6}$ m$^2$/s, which is a factor of 15 lower than the value for air, i.e. $1.5 \times 10^{-5}$ m$^2$/s. For Reynolds-number equality we now have

$$\frac{V_M l_M}{10^{-6}} = \frac{V_F l_F}{1.5 \times 10^{-5}} \qquad \text{or} \qquad V_M = \frac{1}{15} \frac{l_F}{l_M}.V_F$$

that is to say, if, as in the sports-car example, $l_F/l_M = 4$, we require $V_M = 0.27\,V_F$, or $V_M = 26.7$ m/s. Although a speed of 26.7 m/s is much too high for most water channels (10 m/s would already be considered a very high speed for a water channel) and would almost certainly introduce the new problem of **cavitation** (see Section 8.12), it is clear that the low kinematic viscosity of water will allow relatively high Reynolds numbers to be achieved at modest speeds.

We have already indicated that, if the Mach number is less than about 0.3, fluid[23] compressibility is of negligible significance and Mach-number equality is not vital. It is natural to ask if we can make a similar statement for the Reynolds number. If we calculate the Reynolds number for our sports car, assuming a length $l_F = 5$ m, we find $Re = 3.3 \times 10^7$, obviously a very large number. In fact, for most situations of practical engineering significance (we are excluding here the emerging fields of **microtechnology** and **nanotechnology**, where length scales are typically $10^{-6}$ and $10^{-9}$ m, respectively), the Reynolds numbers turn out to be quite large. It is also the case that above a critical Reynolds number, which is different for every body shape (for external flow) or channel cross section (for internal flow) but typically of the order of $10^3$, the flow becomes unsteady and increasingly random (the term chaotic is also used). Such a state of quasi-random flow is said to be **turbulent**, in contrast to that at much lower Reynolds numbers, where the flow is smooth-flowing and said to be **laminar** (the intermediate state is termed **transitional**) (see Chapters 16 and 18). In many instances, the drag coefficient in turbulent flow becomes almost constant (i.e. independent of Reynolds number, see Section 18.15) and so can be determined from model tests run at Reynolds numbers lower than full scale but still sufficiently high for the flow to be fully turbulent.

 3.17  SUMMARY

In this chapter we have explained the crucial role of units and dimensions in the analysis of any problem involving physical quantities. The underlying principle of dimensional homogeneity has been introduced, i.e. the individual terms in any equation or function which connects physical quantities must have the same overall dimensions (and units). The major advantage of collecting the physical quantities, which are included in either a theoretical analysis or an experiment, into non-dimensional groups has been shown to be a reduction in the number of quantities which need to be considered separately. Buckingham's Π theorem was introduced as a method for determining the number of non-dimensional groups (the Π's) corresponding with a set of dimensional quantities and their dimensions. The sequential elimination of dimensions was shown to be a systematic and simple procedure for identifying these groups.

The scale up from a model to a geometrically similar full-size version requires dynamic similarity, which means that each of the non-dimensional groups describing the model-scale conditions is equal to that for its full-scale counterpart. The definitions and names of the non-dimensional groups most frequently encountered in fluid mechanics have been introduced and their physical significance explained. The chapter concluded by pointing out that the requirements for dynamic similarity may be too costly, technically difficult, or physically impossible to achieve in practice, and a compromise solution has to be accepted.

The student should be able to

- write down the units and dimensions of any of the physical quantities listed in Table A.6
- convert any physical equation into non-dimensional form
- apply Buckingham's Π theorem to determine the number of non-dimensional groups corresponding with a set of dimensional quantities and their dimensions

---

[23] Practically speaking, liquids can normally be regarded as incompressible, so that the term fluid here really means a gas or vapour.

- use a systematic procedure, such as the sequential elimination of dimensions, to convert a functional dependence into an equivalent non-dimensional form
- recognise the more common non-dimensional groups which arise in fluid mechanics, such as Reynolds number, Mach number, and drag coefficient
- determine full-size physical quantities from the results of a dynamically similar model test
- recognise and resolve scaling contradictions

## 3.18  SELF-ASSESSMENT PROBLEMS

**3.1**  Determine the dimensions of the following combinations of physical quantities, where $p$ represents pressure, $\rho$ is density, $V$ is velocity, $g$ is acceleration due to gravity, $t$ is time, $h$ and $l$ represent lengths, and $\nu$ is kinematic viscosity

$$p/\rho V, \quad gt/V, \quad \sqrt{g/l}, \quad Vl/\nu, \quad \rho V^2, \quad p/gh, \quad \text{and} \quad \rho Vl.$$

**3.2**  Find the values of the exponents $a, b,$ and $c$, which make each of the following combinations of physical quantities non-dimensional

$$\frac{p}{\rho V^a}, \quad \frac{\rho VD}{\mu^b}, \quad \text{and} \quad \frac{D\sigma^c}{\rho \nu^2}.$$

In addition to the symbols in problem 3.1, $D$ represents diameter, $\mu$ is dynamic viscosity, and $\sigma$ is surface tension.
(Answers: 2; 1; 1)

**3.3**  A disc of weight $W$ and radius $R$ slides with velocity $V$ over a smooth, flat, horizontal surface. Lubricant with dynamic viscosity $\mu$ is pumped into the gap between the disc and the surface at a volumetric flowrate $\dot{Q}$. It can be shown that the drag force $D$ acting on the disc is given by

$$D = \pi V \left( \frac{\mu^2 R^4 W}{3\dot{Q}} \right)^{1/3}.$$

Show that this equation is dimensionally correct.

**3.4**  A spherical drop of liquid of diameter $D$ and density $\rho$ oscillates under the influence of its surface tension $\sigma$. Show that the frequency of oscillation is given by

$$f = k\sqrt{\frac{\sigma}{\rho D^3}},$$

where $k$ is a numerical (i.e. non-dimensional) constant.

**3.5**  (a) The wave resistance $R$ of a ship depends upon its length $L$, the water depth $d$, the water density $\rho$, the acceleration due to gravity $g$, and the ship speed $V$. Derive a non-dimensional form for the preceding sentence.

(b) A test carried out in a towing tank on a one-twentieth-scale model ship of length 1.5 m gave a value for the wave resistance of 160 N at a model speed of 4 m/s. The water density in the towing tank was 1000 kg/m$^3$ while that for the full-size ship would

be 1050 kg/m$^3$. Assuming dynamic similarity between the model and full scale, calculate the depth of water in the towing tank to correspond to a depth of 20 m for the full-scale ship and the corresponding wave resistance.
(Answers: 1 m; 1.34 MN)

**3.6** The power $P$ developed at the shaft of a wind turbine is a function of the windspeed $V$, the air density $\rho$, the rotation speed $N$, and the turbine blade rotor diameter $D$. Show that

$$\frac{P}{\rho N^3 D^5} = F\left(\frac{V}{ND}\right). \tag{3.59}$$

Tests were performed on a model turbine giving values of $P/\rho N^3 D^5$ and the corresponding values of $V/ND$. It is required to design a wind turbine to generate a specified power output for a given air density, windspeed, and rotation speed. How would you plot and use the experimental results for this purpose?

**3.7** (a) The shaft power $P$ developed by a hydraulic turbine depends upon the volumetric flowrate of water through the turbine $\dot{Q}$, the water density $\rho$, the pressure difference across (i.e. between inlet and outlet) the turbine $\Delta p$, the impeller rotation speed $N$, and the impeller diameter $D$. Derive a non-dimensional form of this statement, ensuring that $P$, $\dot{Q}$, and $\Delta p$ appear in independent non-dimensional groups.

(b) The volumetric flowrate of water through a Kaplan hydraulic turbine is 5 m$^3$/s, and the head difference 40 m (head difference $h$ is a measure of pressure difference such that $\Delta p = \rho g h$, where $g$ is the acceleration due to gravity). If the turbine rotation speed is 250 rpm, assuming conditions of dynamic similarity, calculate the flowrate and head difference for a geometrically similar, quarter-scale model turbine running at 400 rpm with a fluid of relative density 0.8. Also calculate the ratio of the power outputs for the two machines.
(Answers: 0.125 m$^3$/s; 5.12 m; 313)

**3.8** A rotating-paddle mixer with a paddle diameter of 1 m is to be designed for use with a liquid of relative density 2 and kinematic viscosity 1 m$^2$/s. The power required by a model mixer with a paddle diameter of 100 mm operating at a rotational speed of 50 rpm in a fluid of density 1200 kg/m$^3$ and dynamic viscosity 100 Pa·s is 500 W. Assuming geometric and dynamic similarity between the model and full-scale mixers, calculate the power and speed for the full-scale mixer.
(Answers: 144 kW; 6 rpm)

**3.9** (a) The velocity $V$ with which a shell can be fired from a gun barrel depends upon the shell diameter $D$, the shell mass $m$, the air density $\rho$, the soundspeed $c$, and the explosive energy of the shell $E$. Derive a non-dimensional version of this statement.

(b) In a laboratory test, a shell-shaped projectile of diameter 10 mm and mass 0.05 kg is fired at a speed of 500 m/s through a heavy gas with a density of 2 kg/m$^3$ and a soundspeed of 100 m/s. The explosive energy required is 10 kJ. Calculate the shell speed, shell mass, and explosive energy for a geometrically similar shell of diameter 100 mm fired into air with a density of 1.2 kg/m$^3$ and a soundspeed 340 m/s, assuming conditions of dynamic similarity.
(Answers: 1700 m/s; 30 kg; 69.4 MJ)

**3.10** (a) The drag force $D$ on a supersonic aircraft may be assumed to depend on its wingspan $S$, its speed $V$, the air density $\rho$, and the compressibility of the air $K$. Derive a non-dimensional form of this statement. Note that compressibility has the same units as the inverse of pressure.

(b) A fighter aircraft is designed to fly at a speed of 2500 km/h at an altitude where the air density is 0.287 kg/m$^3$ and the compressibility is $4 \times 10^{-5}$ Pa$^{-1}$. A 1/20$^{th}$- scale model of the aircraft is to be tested in a pressurised and cooled wind tunnel. If the air in the tunnel has a density of 4 kg/m$^3$ and compressibility $2.5 \times 10^{-6}$ Pa$^{-1}$, calculate the airspeed for dynamic similarity. If the model has a wingspan of 0.5 m and the drag force on the model is 1000 N, calculate the drag force on the full-size aircraft at design conditions and the required propulsive power.
(Answers: 744 m/s; $2.5 \times 10^4$ N; 17.4 MW)

**3.11** (a) The drag force $D$ exerted on a sphere moving through a fluid depends upon the sphere's radius $R$, the speed $V$ of the sphere relative to the fluid, the fluid density $\rho$, and the dynamic viscosity of the fluid $\mu$. Derive a non-dimensional form of this sentence.

(b) Measurements of the drag force on a series of spheres, each moving through a different fluid, yield the following results

| $R$ (mm) | $V$ (m/s) | $\rho$ (kg/m$^3$) | $\mu$ (Pa·s) | $D$ (N) |
|---|---|---|---|---|
| 5 | 0.1 | 1260 | 1.3 | 0.123 |
| 30 | 0.01 | 920 | 0.059 | $3.34 \times 10^{-4}$ |
| 2.5 | 0.5 | 935 | 0.12 | $2.83 \times 10^{-3}$ |
| 1 | 1 | 1000 | $810^{-3}$ | $6.3 \times 10^{-4}$ |
| 7.5 | 10 | 1.2 | $1.8 \times 10^{-5}$ | $4.25 \times 10^{-3}$ |
| 50 | 120 | 0.07 | $8 \times 10^{-6}$ | 1.588 |

Use the results of part (a) to convert these results to non-dimensional form. Plot the logarithm of one of the non-dimensional groups against the logarithm of the other and comment on the results.

**3.12** At very low Reynolds numbers the drag force $D$ exerted on a sphere of radius $R$ moving through a fluid of dynamic viscosity $\mu$ at constant velocity $V$ is given by Stokes' law

$D = 6\pi \mu R V.$

Show that a small sphere of density $\rho_S$ dropping under the influence of gravity through a fluid of density $\sigma \rho_S$, where $\sigma$ is a non-dimensional constant less than unity, reaches a terminal velocity $V_\infty$ given by

$$V_\infty = \frac{2}{9}(1-\sigma)\frac{\rho_S R^2 g}{\mu}$$

where $g$ is the acceleration due to gravity.

If the sphere has zero initial velocity, the equation governing the initial phase of motion, during which the sphere is accelerating, may be approximated as

$$mg - D - V_B = (m + m_A)\frac{dV}{dt}$$

where $t$ is the elapsed time, $m$ is the mass of the sphere, $V_B = \rho_F \mathcal{V}g$ is the buoyancy force acting on the sphere, and $m_A$ is the added mass given by $m_A = \sigma \rho_S \mathcal{V}/2$. Added mass accounts for the fact that an object accelerating through a fluid also accelerates some of the surrounding fluid. In the case of a sphere, the added mass is equal to half the mass of the displaced fluid. Show that a non-dimensional form of the equation for $V$ is

$$\frac{d\tilde{V}}{d\tilde{t}} = 1 - \tilde{V}$$

where the non-dimensional velocity $\tilde{V} = V/V_\infty$, the non-dimensional time $\tilde{t} = t/\tau$, and the characteristic time $\tau = (2 + \sigma)\, V_\infty / \left[(1 - \sigma)g\right]$.

Integrate the equation for $\tilde{V}$ and show that

$$V = V_\infty \left\{1 - e^{-(2+\sigma)V_\infty/\left[(1-\sigma)g\right]}\right\}.$$

# 4 Pressure variation in a fluid at rest (hydrostatics)

This chapter and the next are concerned with **hydrostatics**[24]: the study of fluids at rest. The word hydrostatics is derived from the Greek word *hudor* meaning water but the term applies to all fluids, gases and liquids alike. Shear stresses cannot arise in a body of fluid at rest because there is no relative motion (except at the molecular level) between fluid particles, and the only internal forces that can arise are due to changes in pressure with vertical location which result from gravitational pull. In this chapter we derive the mathematical statement of this principle, which is called the **hydrostatic equation**. The solution of many practical engineering problems involves the application of the hydrostatic equation. In the second part of this chapter we shall apply it first to the measurement of pressure using liquid-filled tubes (**manometry**) and then to analyses of pressure variations in the earth's atmosphere and in very deep water, in both instances accounting for compressibility. We conclude the chapter by extending consideration to a body of fluid which, in spite of the chapter heading, is not in fact at rest but is in steady motion, or even being accelerated, but where there is no relative motion between fluid particles.

## 4.1 Pressure at a point: Pascal's law

Any fluid in contact with a surface exerts a normal stress on that surface, i.e. a force per unit area normal (i.e. perpendicular) to the surface. This stress is what we call **pressure**, symbol $p$, unit Pa (i.e. N/m$^2$). For a gas, the lowest possible absolute pressure is zero, which defines what we mean by a **vacuum**. As we pointed out in Section 2.13, the situation for a liquid is more complex. If its absolute pressure is reduced to the value of the **vapour pressure**, vapour is produced and the liquid boils. Under no circumstance can absolute pressure become negative, i.e. tensile. Some pressure gauges measure pressure relative to atmospheric pressure $B$, i.e. $p-B$, the so-called **gauge pressure**. Confusion may arise if the absolute pressure is sub-atmospheric (i.e. $p < B$) as the gauge pressure is then negative. In older books, particularly those where Imperial units were used, gauge pressure is indicated by adding the letter '$g$' to its units, as in psig indicating pounds per square inch gauge.

An important underlying principle of hydrostatics, called **Pascal's law**, is that the pressure at any point in a body of fluid at rest is the same in all directions. To prove this law we consider a triangular wedge of fluid of infinitesimal[25] cross section, as shown in Figure 4.1.

---

[24] Instead of hydrostatics, the termed **aerostatics** is sometimes used when the fluid is a gas, especially air.

[25] The term 'infinitesimal' here implies that the dimensions of the wedge are so small that the pressure acting on any face of the wedge can be assumed to be uniform across that particular face, but may well vary from *(cont'd)*

*Introduction to Engineering Fluid Mechanics*. Marcel Escudier.
© Marcel Escudier 2017. Published 2017 by Oxford University Press.

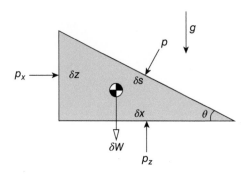

**Figure 4.1** Equilibrium of an infinitesimal wedge of fluid

We assume there is no variation of pressure normal to the triangular face shown. The wedge has thickness $l$ (into the page), a vertical face of height $\delta z$, a horizontal face of length $\delta x$, and a sloping face of length $\delta s$. The fluid density within the wedge is $\rho$ and the corresponding weight of the wedge is $\delta W$. At the outset we assume that the pressures acting normal to the three faces of the wedge, $p_x, p_z$, and $p$, as shown in the figure, are all different. Since the fluid is at rest, the net force acting on the wedge in any direction must be zero.

For the horizontal direction we have

$$p_x \delta z\, l - p\, \delta s\, l \sin\theta = 0$$

and, from the geometry of the wedge, $\delta z = \delta s\, \sin\theta$, so that

$$p_x \delta z\, l - p\, \delta z\, l = 0$$

or, after cancelling out $l\, \delta z$,

$$p_x = p. \tag{4.1}$$

For the vertical direction we have

$$p_z \delta x\, l - p\, \delta s\, l \cos\theta - \delta W = 0$$

and, again from the geometry of the wedge, $\delta x = \delta s\, \cos\theta$. Since the volume of the wedge $\delta \mathcal{V}$ is given by $\delta z\, \delta x\, l/2$, its weight is given by $\delta W = \rho\, \delta \mathcal{V} g = \rho\, \delta z\, \delta x\, l g/2$ so that

$$p_z\, \delta x\, l - p\, \delta x\, l - \frac{1}{2}\rho\, \delta z\, \delta x\, l g = 0$$

or

$$p_z = p + \frac{1}{2}\delta z\, g. \tag{4.2}$$

Equation (4.2) must hold no matter how small the wedge so that, as $\delta z$ is reduced to zero, the term including $\delta z$ must also reduce to zero, and we find

*(footnote 25 cont'd)* face to face. The concept of an infinitesimal element is commonly employed in all branches of applied mechanics (fluid mechanics, heat transfer, solid mechanics, etc.). Although the **continuum hypothesis** imposes a lower limit on the size of such an element, as we saw in Section 2.5, in practice we rarely come close to it.

$$p_z = p. \tag{4.3}$$

Taken together, equations (4.1) and (4.3) allow us to conclude that

$$p_x = p_z = p, \tag{4.4}$$

i.e. the pressure at a point in a fluid at rest is the same in all directions: we have proved **Pascal's law**.

## 4.2 **Pressure variation in a fluid at rest; the hydrostatic equation**

For a fluid at rest, the pressure is constant over any horizontal surface within the fluid but increases with depth. These statements of everyday experience are easily proved, as we do in this section, but have far-reaching implications.

Consider first a horizontal cylinder of arbitrary length $l$ and infinitesimal cross-sectional area $\delta A$, as shown in Figure 4.2. If the pressure is assumed to change from the value $p$ at one end of the cylinder to $p + \Delta p$ at the other, then the net horizontal force acting on the fluid cylinder is

$$p\,\delta A - \left(p + \Delta p\right)\delta A = -\Delta p\,\delta A.$$

This net force must equal zero unless the fluid cylinder is being accelerated to the right, which would require $\Delta p < 0$, or to the left ($\Delta p > 0$). Thus, <u>for a fluid at rest</u>, $\Delta p$ must be zero, and we conclude that <u>pressure is constant along any horizontal line within the fluid and, in consequence, over any horizontal surface.</u> What this means in practice is that, if we can connect any two points in a single body of fluid at rest without 'leaving' the fluid, the pressure at the two points will be the same. This conclusion has important consequences for pressure measurement, as we shall see in Sections 4.7 and 4.9. It also explains why the **free surface** of a liquid at rest must be horizontal.

We consider now the forces acting on a vertical cylinder of fluid of infinitesimal length $\delta z$ and infinitesimal cross-sectional area $\delta A$, as shown in Figure 4.3. For convenience the distance $z$ is measured vertically downwards so that, if measured relative to a surface at $z = 0$, it represents **depth**. The pressure changes from $p$ at the top of the cylinder to $p + \delta p$ at its base. The essential difference compared with the horizontal cylinder considered above is that the force

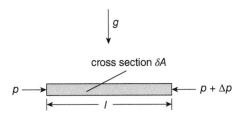

**Figure 4.2** Horizontal fluid cylinder of infinitesimal cross section

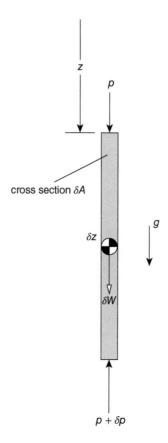

**Figure 4.3** Infinitesimal vertical fluid cylinder

balance must now include the weight of the fluid cylinder $\delta W$ acting vertically downwards (the circular symbol in Figure 4.3, with alternating black and white quadrants, represents the centre of gravity of the cylinder through which $\delta W$ acts).

The net downward vertical force acting on the fluid cylinder is

$$p\,\delta A + \delta W - (p + \delta p)\,\delta A = 0$$

so that, after cancellation of the term $p\,\delta A$ and rearrangement, we have

$$\delta p = \frac{\delta W}{\delta A}$$

i.e. the pressure increases by the amount $\delta p$ due to the weight of fluid per unit area in a layer of depth $\delta z$. The cylinder weight is given by $\delta W = \rho\,\delta \mathcal{V} g$, where $\delta \mathcal{V}$ is the cylinder volume. Since $\delta \mathcal{V} = \delta z\,\delta A$, we have $\delta p = \rho\,\delta z g$, or $\delta p/\delta z = \rho g$. If we now reduce $\delta z$ to zero, the finite-difference ratio $\delta p/\delta z$ must approach $dp/dz$, which we term the **pressure gradient**, and we have the first-order, ordinary differential equation

$$\frac{dp}{dz} = \rho g \tag{4.5}$$

which is known as the **hydrostatic equation**[26]. Since both the fluid density $\rho$ and the acceleration due to gravity $g$ are always positive, we conclude that $dp/dz$ is always positive so that the pressure in a body of fluid at rest can only increase with vertical depth $z$. We could just as well say that pressure decreases with **altitude** (or **elevation**) $-z$, altitude being the distance measured vertically upwards, usually from mean sea level. For convenience we use the symbol $z'$, with $z' = 0$ at sea level, rather than regard altitude as a negative depth, as we have done hitherto[27]. The hydrostatic equation can now be written as

$$\frac{dp}{dz'} = -\rho g \tag{4.6}$$

the minus sign appearing because $dz = -dz'$.


## 4.3  Pressure variation in a constant-density fluid at rest

In most applications it is sufficient to regard the gravitational acceleration $g$ as a constant, invariant with altitude or depth, the value $9.81$ m/s$^2$ normally being adopted for medium latitudes[28]. In many engineering calculations, particularly where the fluid is a liquid, sufficient accuracy is achieved if the fluid density $\rho$ is also assumed to be a constant (values are listed in Table A.5 for some common liquids and in Table A.6 for several gases). In these circumstances, equation (4.5) can be integrated to give

$$p = \rho g z + C \tag{4.7}$$

where $C$ is a constant of integration to be determined from a **boundary**, or **reference**, condition which provides a value for $p$ at a known depth or altitude. It is frequently convenient to take the pressure at the origin for $z$ (i.e. $z = 0$) as the reference, which is very often the **barometric** (or **atmospheric**) **pressure** $B$ (standard value $1.01325$ bar or $1$ atm), so that $C = B$ and

$$p = B + \rho g z. \tag{4.8}$$

The combination $\rho g z$ is referred to as the **hydrostatic pressure**, $p_H$. More generally, equation (4.8) suggests that any pressure difference $\Delta p$ can be expressed in terms of the vertical height $h$ of a column of liquid of density $\rho_L$ such that $\Delta p = \rho_L g h$. The height $h$ is then referred to as the **head**. Although in principle any liquid can be chosen as the reference liquid (for the density $\rho_L$), the usual choices are water, mercury, or an oil.

Equation (4.7) can also be written in terms of elevation $z'$ as

$$p + \rho g z' = C. \tag{4.9}$$

---

[26]  The term **buoyancy equation** is occasionally used instead of hydrostatic equation.

[27]  The reader should be aware that the use of the symbol $z$ to represent vertical depth is inconsistent with the Cartesian coordinate system adopted in later chapters in which $x, y$, and $z$ are orthogonal coordinates, with $y$ usually representing upward vertical distance. The use of $z$ for depth in this chapter follows common practice.

[28]  As discussed in Subsection 4.13.1, the reduction in $g$ with altitude in the earth's atmosphere becomes significant and has to be accounted for.

The combination of terms $p + \rho g z'$ is known as the **piezometric pressure**, and $p/\rho g + z'$ as the **piezometric head**. The prefix piezo stems from the Greek word for press.

As we discussed in Section 2.11, gases are far more compressible than liquids. Nevertheless, the constant-density assumption is often acceptable, for example, if pressure changes are due to relatively small altitude (ca 100 m) or velocity changes (see Chapters 7 and 8). For liquids, the equations derived above are perfectly adequate even for depths well in excess of 1000 m (see Section 4.12).

---

**ILLUSTRATIVE EXAMPLE 4.1**

As shown schematically in Figure E4.1, a boiler is supplied with water from an open tank located 8 m above it (the arrangement is similar to the central-heating system still found in older houses where the water tank, called a cistern, is installed in the loft). Calculate the pressure of the water in the boiler and also the reading of the pressure gauge (i.e. the gauge pressure) attached to the boiler. Take the atmospheric pressure to be 1.02 bar.

**Solution**

$H = 8$ m; $B = 1.02 \times 10^5$ Pa; $g = 9.81$ m/s$^2$; $\rho = 10^3$ kg/m$^3$.

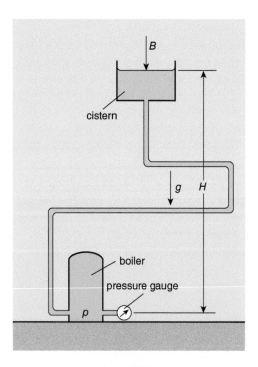

**Figure E4.1**

The boiler pressure $p$ is given by

$p = B + \rho g H$

$\quad = 1.02 \times 10^5 + 10^3 \times 9.81 \times 8$

$\quad = 1.80 \times 10^5$ Pa or 1.80 bar.

The gauge pressure $p_G$ is

$p_G = p - B = \rho g H = 1.80 - 1.02$

$\qquad\qquad\qquad = 0.78$ bar.

Comment:

Due to bends in the pipework connecting the water tank to the boiler, the length of piping may be considerably greater than the vertical height $H$. As we saw in Section 4.2, in a fluid at rest, the pressure is the same at any given horizontal location. In consequence only the vertical height difference is significant in determining the pressure difference $p - B$: the two locations at which $p$ and $B$ are determined do not need to be vertically in line.

## 4.4 Basic pressure measurement

In Section 4.4 we found that, for a static body of fluid of constant and uniform density $\rho$, the pressure increases linearly with depth $z$ according to $p = \rho g z + C$, where $C$ is a constant (equation (4.7)). If we apply this equation to a vertical tube containing a column of liquid of density $\rho_M$ and height $h$, as shown in Figure 4.4, with the origin for $z$ taken as the upper meniscus, we have

$C = p_1 \quad$ and $\quad p_2 = \rho_M g h + C$

or

$$p_2 - p_1 = \rho_M g h. \qquad\qquad (4.10)$$

This important equation provides the basis for **manometry**, which is the measurement of pressure or pressure difference using liquid-filled tubes called **manometers**. The choice of liquid is influenced by the magnitude of the pressure difference to be measured (e.g. a high-density liquid such as mercury is used if the difference is large), immiscibility with other fluids, and low volatility. Common choices are water, a light oil, and mercury.

## 4.5 Mercury barometer

If the upper end of the tube shown in Figure 4.4 is sealed and the space above the liquid evacuated, such that the pressure $p_1$ is as close to zero as practically possible, we have a device called a **barometer**, which can be used to measure the absolute pressure $p_2$. Unfortunately, this principle has a major flaw: as we discussed in Section 2.13, at room temperature many common

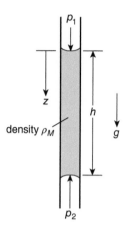

**Figure 4.4** A vertical tube containing a column of liquid

liquids tend to boil at pressures which are very low but still well above absolute zero, making them unsuitable for use as a barometer liquid. In the case of water, for example, the pressure at which boiling begins, called the **saturated vapour pressure** $p_V$, is about 1.013 bar at 100 °C and still as high as 23 mbar at 20 °C. The universal use of mercury as the barometer liquid is attributed to the Italian scientist Evangelista Torricelli, who must have realised that its vapour pressure is negligibly small for all practical purposes. As we now know, the vapour pressure for mercury is $1.1 \times 10^{-3}$ Pa or about $10^{-8}$ bar.

The basic arrangement of the classical mercury barometer is shown in Figure 4.5. The pressure above the mercury column in the vertical tube has been reduced to the saturated vapour

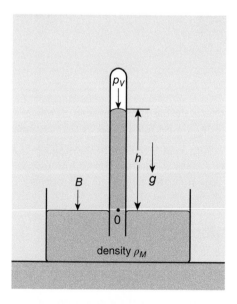

**Figure 4.5** Basic mercury barometer

pressure of mercury while the free surface of the mercury pool is exposed to the prevailing atmospheric pressure $B$ so that, from equation (4.10),

$$B = p_V + \rho_M gh. \qquad (4.11)$$

Although we have just pointed out that for mercury the value of the vapour pressure $p_V$ is negligible compared with $B$, for completeness we have included it in equation (4.11) and Figure 4.5. Note too that we have made use of the fact that the pressure at the point O (denoting the origin from which $h$ is measured) within the barometer tube, on the same horizontal level as the free surface of the mercury pool, must be precisely the same as the atmospheric pressure. The density of mercury at 20 °C is 13,546 kg/m³ so that for the **Standard Atmosphere**, for which $B$ = 1.01325 bar, with $g$ = 9.80665 m/s², equation (4.11) gives $h$ = 0.76275 m, or 30.0 in of mercury. As we pointed out in Section 4.3, any pressure difference can be converted into an equivalent height of a liquid (in this case, since one pressure is effectively zero, the pressure itself can be expressed as the height of a mercury column). The principal application of a barometer is to measure the small deviations from 1.01325 bar which are an important guide to forthcoming changes in the weather.

## 4.6 **Piezometer tube**

The **piezometer tube**, shown in Figure 4.6, is the second direct application of a column of liquid in a vertical tube being used for the measurement of pressure. In this case, the upper end of the tube is open to the atmosphere at absolute pressure $B$ (or any reference pressure)

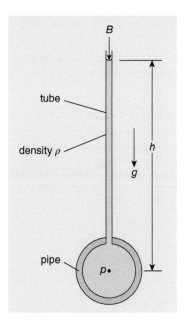

**Figure 4.6** Piezometer tube

while the lower end is attached to a vessel or pipe containing a pressurised (i.e. $p > B$) liquid which rises up the tube to a height $h$. Once again, equation (4.8) is applicable and leads to

$$p = B + \rho g h, \tag{4.12}$$

$\rho$ being the density of the liquid in the pipe and the tube. From the measured height $h$ we have $\rho g h = p - B = p_G$, the gauge pressure. Note that the liquid depth within the vessel or pipe may represent a significant contribution to the overall height $h$ and hence the gauge pressure $p_G$.

## 4.7 **U-tube manometer**

As is probably apparent to the reader, the piezometer tube is not a practical device for pressure measurement: there is the ever-present danger of liquid being blown into the environment if the tube is too short, but to cope with high pressures the vertical tube becomes excessively long—for water an absolute pressure of 2 bar (i.e. approximately 1 bar greater than atmospheric pressure) would require $h = 10.2$ m. One solution to this problem might appear to be to place a high-density liquid, such as mercury, in the vertical tube. In the case of water at 2 bar, for example, the corresponding height of a mercury column is only 0.75 m. Unfortunately, this idea is also impractical for a more fundamental reason: the interface between a heavy fluid on top of a light fluid is unstable. The mercury would simply run down into the main pipe or pressure vessel and create a major clean-up problem. Mercury is a poisonous substance and its use is best avoided wherever possible.

The **U-tube** arrangement shown in Figure 4.7 uses the heavy-liquid (density $\rho_M$) idea but avoids the stability problem since the lighter liquid (density $\rho_F$) is above the heavier one. The analysis of this and other manometer problems depends upon the two fundamental results we obtained in Sections 4.2 and 4.3

(a) in a fluid at rest, the pressure is the same for all points on the same horizontal level; and
(b) for a fluid of constant density $\rho_M$, the pressure increase due to a vertical height difference $h$ is $\rho_M g h$.

The first of these two statements tells us that, in the fluid of density $\rho_F$, the pressure at points ① and ② must be the same and, in the manometer liquid of density $\rho_M$, the pressure at points ③ and ④ must also be the same. Although we have no real interest in the intermediate pressure at the interface between the two liquids (point ③), it is convenient to give it a symbol, such as $p'$, to use in our analysis.

For the right-hand side of the manometer, we have

$$p' = B + \rho_M g H$$

and, for the left-hand side,

$$p' = p + \rho_F g h.$$

Note that in both cases we have evaluated the intermediate pressure $p'$ by working our way down the manometer, in the direction of increasing pressure, adding together the appropriate pressure and pressure difference for each fluid. This is a convenient and foolproof

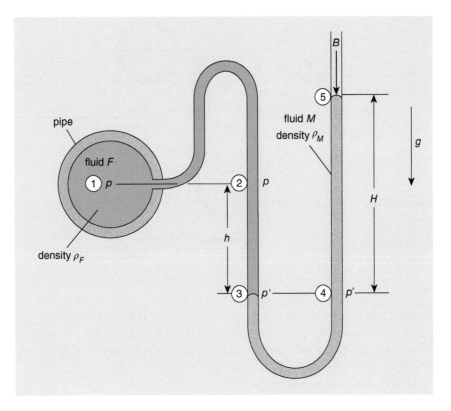

**Figure 4.7** U-tube manometer

'bookkeeping' approach to solving manometer problems which is easily extendable to any number of fluid layers. Since the right-hand side of each of the above expressions is equal to $p'$, we can equate the two and write

$$B + \rho_M g H = p + \rho_F g h \quad \text{or} \quad p - B = (\rho_M H - \rho_F h)g \tag{4.13}$$

which gives us the unknown pressure $p$, once again in the form of a gauge pressure $p - B$. In fact the pressure imposed on the free surface of the manometer liquid (point ⑤) need not be the barometric pressure $B$ but could be any known reference pressure.

---

**ILLUSTRATIVE EXAMPLE 4.2**

The U-tube manometer shown in Figure E4.2 is used to measure the pressure difference $p_1 - p_2 = \Delta p$ between two pipes on the same horizontal level. Show that

$$\Delta p = (\rho_M - \rho_F)gH$$

where $\rho_M$ is the density of the manometer liquid, $\rho_F$ is the density of the fluid in the pipes, $g$ is the acceleration due to gravity, and $H$ is the height difference between the levels of the manometer liquid in the two arms of the U-tube.

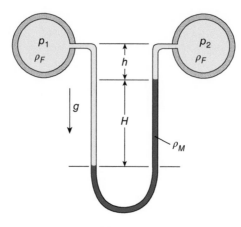

**Figure E4.2**

## Solution

In the figure the fluid pressures in the two pipes have been indicated by $p_1$ and $p_2$, where $p_1 = p_2 + \Delta p$, the height of the pipes above the manometer liquid interface on the right-hand side as $h$, and the pressure in the manometer liquid on the horizontal level of the left-hand interface as $p'$.

For the left-hand side of the manometer we have

$$p' = p_1 + \rho_F g \, (h + H)$$

and, for the right-hand side,

$$p' = p_2 + \rho_F g h + \rho_M g H.$$

If we equate these two expressions for $p'$, we have

$$p_1 + \rho_F g \, (h + H) = p_2 + \rho_F g h + \rho_M g H$$

which we can rearrange, after substituting $p_2 + \Delta p$ for $p_1$, as

$$p_2 + \Delta p + \rho_F g h + \rho_F g H = p_2 + \rho_F g h + \rho_M g H.$$

Each of the terms $p_2$ and $\rho_F g h$ appears on both sides of the equation, so they cancel out, and we have, finally,

$$\Delta p = (\rho_M - \rho_F) \, g H.$$

## Comment:

Neither the actual fluid pressures nor the height of the pipes appears in the final result, which can be derived directly from the final equation of Section 4.7 by setting $p - B = \Delta p$, and $h = H$.

The use of a U-tube manometer or a **differential pressure transducer**[29] to measure the pressure difference between two points at different heights, as shown in Figure 4.8, again relies upon the hydrostatic equation. Considering first the U-tube, we have

$$p' = p_1 + \rho_F g \left( z_1 + \frac{1}{2}h \right) = p_2 + \rho_F g \left( z_2 - \frac{1}{2}h \right) + \rho_M g h \tag{4.14}$$

from which

$$p_1 - p_2 = (\rho_M - \rho_F)\, gh + \rho_F g z_{12} \tag{4.15}$$

where $z_{12} = z_2 - z_1$.

For the pressure transducer we have

$$\Delta p = p_1 + \rho_F g z_3 - \left( p_2 + \rho_F g z_4 \right) = p_1 - p_2 - \rho_F g z_{12} \tag{4.16}$$

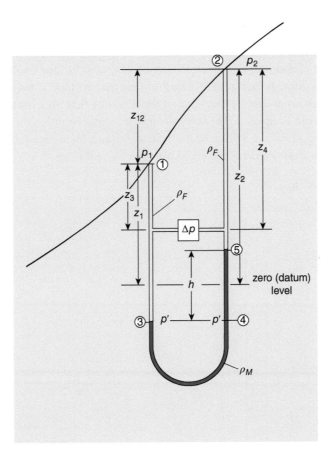

**Figure 4.8** U-tube manometer or pressure transducer used to measure pressure difference between locations at different heights

---

[29] A differential pressure transducer is an electronic device used to measure pressure difference rather than absolute pressure. Absolute pressure can be measured if one side of a differential pressure transducer is subjected to a known, reference pressure.

or

$$p_1 - p_2 = \Delta p + \rho_F g z_{12}. \tag{4.17}$$

For the U-tube manometer, equation (4.14) is based upon the recognition that the pressure $p'$ at location ③ has to be the same as that at location ④. Equation (4.16) is arrived at using the hydrostatic equation to determine the pressure on either side of the pressure transducer, which is a vertical distance $z_3$ below location ①, and $z_4$ below location ②. The distance $z_{12}$ is the vertical distance between locations ① and ②. The result of Illustrative Example 4.2 is recovered if $z_{12} = 0$.

## 4.8 Effect of surface tension

Although until now we have neglected the effect of surface tension (see Section 2.14), for liquids in small-diameter tubes this property leads to an additional pressure difference which needs to be accounted for. We consider the situation shown in Figure 4.9. The lower end of an open vertical tube of diameter $D$ is immersed in a liquid with surface tension $\sigma$ and contact angle $\theta$, measured through the liquid. We shall assume that the fluid above the liquid surface (a gas) has a much lower density than that of the liquid so that pressure differences in the surrounding fluid can be ignored. The vertically upward surface-tension force $F$ acting on the periphery of the liquid surface is given by $F = \pi D \sigma \cos \theta$. and, for static equilibrium, this force must exactly balance the weight of the liquid column of height $h$, i.e.

$$\pi D \sigma \cos \theta = \frac{\pi D^2}{4} h \rho g$$

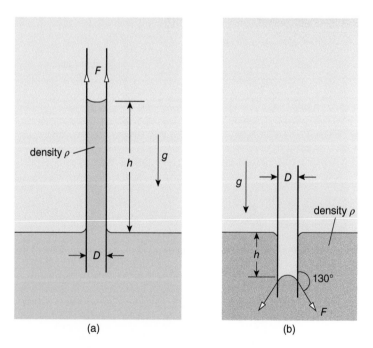

(a)             (b)

**Figure 4.9** Effect of surface tension for liquid in a vertical tube with (a) $\theta < 90°$ and (b) $\theta > 90°$

so that

$$h = \frac{4\sigma \cos\theta}{\rho g D}. \tag{4.18}$$

For pure water in contact with air the surface tension at 20 °C is $7.28 \times 10^{-2}$ N/m with a contact angle which is practically zero, so that for a tube 1 mm in diameter we find $h = 29.7$ mm, which is clearly far from negligible. Even for a 10 mm tube we have $h = 3$ mm. For mercury the surface tension is 0.472 N/m, the contact angle 130°, and the corresponding $h$ values are –9.1 mm and 0.91 mm, respectively. Note that what these contact angles tell us is that, for water in a vertical tube, surface tension produces a force pulling vertically upwards whereas for mercury the resultant force is vertically downward (due to axisymmetry there can be no radial component of force).

Although the value of $\sigma$ for mercury is about seven times that for water, its effect on the liquid level is much smaller, first because the density of mercury is about 13.6 times that of water, and second because the vertical component of the surface-tension force is reduced by the contact angle ($\cos 130° = -0.643$ compared with $\cos 0° = 1$).

Based upon the foregoing, it is clearly straightforward to account for the effect of surface tension on the liquid level in a manometer. However, this is usually unnecessary because we almost always measure changes in the liquid levels produced by changes in the applied pressure difference so that the surface-tension effect cancels out.

## 4.9 Inclined-tube manometer

A common form of manometer for teaching-laboratory use is shown in Figure 4.9. In this case, instead of a vertical U, one side of the manometer tube (cross-sectional area $a$) is inclined at an angle $\phi$ to the horizontal. This inclined tube is attached to the pipe on the right-hand side, in which there is fluid of density $\rho_F$ at a pressure $p$, to be determined. The left-hand side of the manometer tube is vertical and attached at its top to a reservoir of cross-sectional area $A$ ($\gg a$) in which the surface of the manometer fluid, density $\rho_M$, is subjected to a reference pressure $p_{REF}$. The analysis of the inclined-tube manometer is very similar to that of the U-tube manometer (Section 4.7). We measure all liquid levels from a horizontal reference level, as shown in the figure, defined such that the surface of the liquid in the reservoir is at the same level ① as the interface ④ in the inclined tube when the pressure in the pipe is $p_0$, so that

$$p_{REF} = p_0 + \rho_F g H_0, \tag{4.19}$$

$H_0$ being the vertical height of the pipe centre ⑤ above the reference level. It is evident from equation (4.19) that, unless $\rho_F g H_0$ is negligible compared with $p_{REF}$, $p_0$ may be considerably different from the reference pressure $p_{REF}$. An increase in the pipe pressure from $p_0$ to $p$ causes the interface to move down the inclined tube by an amount $L$ (i.e. from ④ to ③), and the level in the reservoir to rise above the reference level by an amount $\delta h$. Since the volume of liquid which enters the reservoir must be the same as that which moves along the inclined tube, we have $\delta h\, A = La$, so that $\delta h = La/A$, which shows that $\delta h \ll L$ since $a \ll A$.

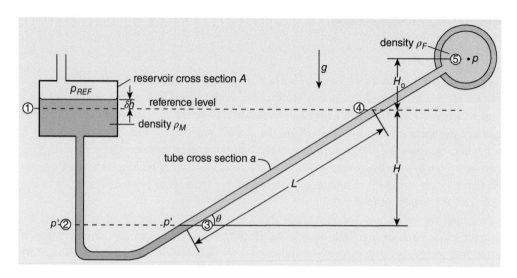

**Figure 4.10** Inclined-tube manometer

As we have stressed throughout this chapter, in manometry it is the vertical height which is significant. From Figure 4.10 we see that the interface has moved a vertical distance $H = L \sin \phi$ so that in measuring $L$ rather than $H$ we have effectively applied an amplification factor of $1/\sin \phi$. A typical value for the inclination angle $\phi$ is $15°$, which corresponds to an amplification factor of 3.86.

For the situation shown in the figure, we again make use of the fact that in a single fluid at rest the pressure must be the same at all positions on the same horizontal level, in this case, the intermediate pressure $p'$ at locations ② and ③. On the left-hand side, again working vertically downwards from the liquid surface in the reservoir to location ②, we have

$$p' = p_{REF} + \rho_M g (\delta h + H)$$

and on the right-hand side the corresponding result is

$$p' = p + \rho_F g (H_0 + H).$$

If we equate the two results, after some rearrangement, we have

$$p - p_{REF} = \rho_M g (\delta h + H) - \rho_F g (H_0 + H). \tag{4.20}$$

This equation reveals that, for a given pressure difference $p - p_{REF}$, the smaller we make $\delta h$, i.e. the smaller we make $a/A$, the larger will be $H$ and so $L$. This is an important point because it is generally easier to measure a long length than it is a short length. If we substitute $\delta h = La/A$, and $H = L \sin \phi$, in the equation for $p - p_{REF}$, we find

$$p - p_{REF} = \rho_M g \left( L \sin \phi + \frac{La}{A} \right) - \rho_F g (L \sin \phi + H_0)$$

or, after some rearrangement,

$$p - p_{REF} = (\rho_M - \rho_F) g L \sin \phi + \frac{\rho_M g L a}{A} - \rho_F g H_0. \tag{4.21}$$

A number of simplifications of the final equation for $p - p_{REF}$ are possible, depending upon the values of the density of the manometer liquid $\rho_M$, the density of the fluid in the pipe $\rho_F$, the area ratio $a/A$, and the inclination angle $\phi$. The ratio of the second term to the first in equation (4.21) can be regarded as a constant for the manometer equal to $\rho_M a/[(\rho_M - \rho_F) A \sin \phi]$. If, as is often the case, $\rho_M$ far exceeds $\rho_F$, for example mercury and a gas, or mercury and water, and the ratio $a/A$ is also small compared with unity, the second term then becomes negligible, and the manometer equation reduces to

$$p - p_{REF} \approx \rho_M g L \sin \phi - \rho_F g H_0. \tag{4.22}$$

The final term may also be negligible, or it may be that we are interested only in changes in the pressure $p$ compared with $p_0$, in which case we can substitute for $p_{REF}$ from equation (4.19) to obtain

$$p - p_0 \approx \rho_M g L \sin \phi.$$

---

**ILLUSTRATIVE EXAMPLE 4.3**

(a) The inclined manometer shown in Figure E4.3 is used to measure the difference between the pressure $p$ in a horizontal pipe and the constant reference pressure $p_R$ in the manometer reservoir. The reference level is defined by the free surface of the manometer liquid in the reservoir being at the same height as the interface between the pipe and manometer fluids in the inclined tube. The pipe axis is a vertical distance $H$ above the reference level. Derive a relationship between $p, p_R, H, g, L, \phi, \rho_M, \rho, a$, and $A$, where $g$ is the acceleration due to gravity, $L$ is the manometer reading (the distance moved from the reference level by the interface in the inclined tube), $\rho_M$ is the density of the manometer liquid, $\rho$ is the density of the fluid in the pipe, $a$ is the cross-sectional area of the inclined tube, $\phi$ is the angle of inclination of the tube measured from the horizontal, and $A$ is the cross-sectional area of the reservoir. Surface-tension effects can be neglected.

(b) For a particular manometer, $p_R = 0.5$ bar, $H = 0.5$ m, $\phi = 20°$, $\rho_M = 13,600$ kg/m³, $\rho = 800$ kg/m³, the internal diameter of the inclined tube is 5 mm, and that of the reservoir 200 mm. Calculate the manometer reading $L$ if the reservoir and pipe pressures are equal. Calculate the pressure difference $p - p_R$ if $L = 0$. Calculate the saturation vapour pressure of the liquid in the pipe if cavitation occurs within the pipe for $L = -0.9$ m.

**Solution**

(a) In the figure, the pressure at the level of the interface $p'$ is the same in both arms of the manometer.

On the left-hand side,

$$p' = p_R + \rho_M g (\delta + L \sin \phi)$$

where $\delta$ is the vertical height of the surface of the manometer liquid above the reference level. Also, $L \sin \phi$ is the vertical change in height due to movement of the manometer liquid in the sloping tube.

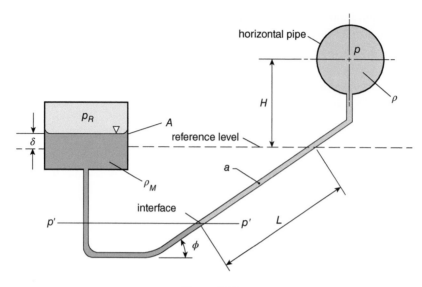

**Figure E4.3**

On the right-hand side,

$$p' = p + \rho g (H + L \sin \phi).$$

We can eliminate $p'$ to give

$$p - p_R = \rho_M g (\delta + L \sin \phi) - \rho g (H = L \sin \phi)$$
$$= (\rho_M - \rho) g L \sin \phi + \rho_M g \delta - \rho g H.$$

Since the volume of manometer liquid which moves into the reservoir must equal the volume of liquid which moves down the inclined tube, we have

$$\delta A = aL$$

so that

$$p - p_R = (\rho_M - \rho) g L \sin \phi + \left( \frac{\rho_M a L}{A} - \rho H \right) g. \qquad \text{(E4.3)}$$

(b) We have $p_R = 5 \times 10^4$ Pa; $H = 0.5$ m; $\phi = 20°$; $\rho = 800$ kg/m³; $\rho_M = 1.36 \times 10^4$ kg/m³; $d = 5 \times 10^{-3}$ m; $D = 0.2$ m; $g = 9.81$ m/s²
If the reservoir and pipe pressures are equal, $p = p_R$, and equation (E4.3) reduces to

$$(\rho_M - \rho) L \sin \phi + \frac{\rho_M a L}{A} - \rho H = 0$$

from which we find

$$L = \frac{\rho H}{(\rho_M - \rho) \sin \phi + \rho_M a/A}$$

$$= \frac{800 \times 0.5}{\left( 1.36 \times 10^4 - 800 \right) \sin 20° + 1.36 \times 10^4 \times (5/200)^2}$$

$$= 0.091 \text{ m or } 91 \text{ mm}.$$

For $L = 0$, from equation (E4.3) the pressure difference is given by

$$p - p_R = -\rho H g$$
$$= -800 \times 0.5 \times 9.81$$
$$= -3924\,\text{Pa}.$$

For $L = -0.9$ m, from equation (E4.3) the pressure difference is given by

$$p - p_R = -\left(1.36 \times 10^4 - 800\right) \times 9.81 \times 0.9 \times \sin 20^\circ$$
$$- \left[1.36 \times 10^4 \times 0.9 \times \left(\frac{5}{200}\right)^2 + 800 \times 0.5\right] \times 9.81$$
$$= -4.265 \times 10^4\,\text{Pa}$$

thus

$$p = 5 \times 10^4 - 4.265 \times 10^4\,\text{Pa} = 7.35 \times 10^3\,\text{Pa}.$$

Since this is the pressure at which we are told cavitation occurs, we conclude that the vapour pressure of the liquid in the pipe is 7350 Pa.

## 4.10 Multiple fluid layers

If there are several layers of immiscible fluid, of thickness $Z_1$ and density $\rho_1$, $Z_2$, $\rho_2$ ($> \rho_1$), $Z_3$, $\rho_3$ ($> \rho_2$), etc., the pressure increases across each layer simply add together to give the total increase in pressure with depth. Using the notation of Figure 4.11, we have

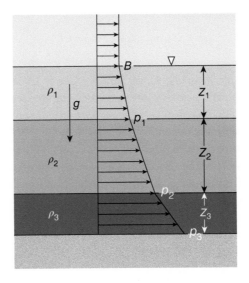

**Figure 4.11** Pressure increase through a series of fluid layers

$$p_1 = B + \rho_1 g Z_1$$
$$p_2 = B + \rho_1 g Z_1 + \rho_2 g Z_2$$
$$p_3 = B + \rho_1 g Z_1 + \rho_2 g Z_2 + \rho_3 g Z_3$$

$B$ being the pressure acting on the surface of the top layer. The linear increase in pressure $p$ with depth $z$ corresponding to $p = \rho g z + C$, $C$ being a constant, is shown schematically in each layer.

---

**ILLUSTRATIVE EXAMPLE 4.4**

As shown in Figure E4.4, a vertical cylinder of inside diameter 50 mm is sealed at the bottom and filled to a depth of 500 mm with mercury. If the barometric pressure $B$ is 1.1 bar and the mercury supports a close-fitting frictionless piston of mass 5 kg, calculate the absolute and gauge pressures at the bottom of the cylinder.

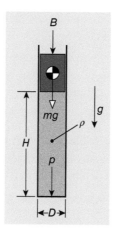

**Figure E4.4**

Solution

$D = 0.05$ m; $B = 1.1 \times 10^5$ Pa; $m = 5$ kg; $H = 0.5$ m; $\rho = 13.6 \times 10^3$ kg/m$^3$.
The statement in the problem that the piston is close-fitting implies that there is no flow of fluid (leakage) past the piston. We denote the cross-sectional area of the piston and cylinder by $A \left(= \pi D^2/4\right)$, and the absolute pressure at the bottom of the cylinder by $p$. The solution to this problem requires that we recognise that the force acting on the bottom of the cylinder $pA$ is made up of three components: the vertically downward pressure force on the top face of the piston $BA$, the weight of the piston $mg$, and the force due to the pressure difference from top to bottom of the mercury $\rho g H$. Thus, we have

$$pA = BA + mg + \rho g H A$$

so that

$$p = B + \frac{mg}{A} + \rho g H.$$

The **effective pressure difference** $\Delta p_P$ due to the weight of the piston is $mg/A$. If the density of the piston material is $\rho_P$, and the piston height is $Z_P$, then $m = \rho_P A Z_P$, so that $\Delta p_P = \rho_P g Z_P$, i.e. a hydrostatic or '$\rho g z$' term, just as for a fluid.

If we now substitute the numerical values into the equation for $p$, we have

$$p = 1.1 \times 10^5 + \frac{5 \times 9.81}{\pi \times 0.05^2/4} + 13.6 \times 10^3 \times 9.81 \times 0.5$$

$$= 2.02 \times 10^5 \text{ Pa} = 2.02 \text{ bar}$$

and the gauge pressure $p_G = p - B = 0.92$ bar.

---

## 4.11 **Variable-density fluid; stability**

Since we put no restriction on the density $\rho$ in the derivation of the hydrostatic equation (4.5) in Section 4.2, this equation must apply whether $\rho$ is constant or varies in some known way with depth $z$, i.e. $\rho = \rho(z)$. It may be that the density variation with $z$ is known, in which case the hydrostatic equation can be integrated immediately. The more usual, and more complicated, situation is that the density is related to the fluid pressure (and possibly also the fluid temperature), and finding the density variation with depth is itself part of the solution of any problem.

We consider first the situation where the density is a specified function of depth. For example, we might assume that close to the bed of a reservoir increasing amounts of silt cause the effective fluid density $\rho$ to increase linearly with depth, i.e.

$$\rho = \rho_0 + \alpha(z - z_0) = \rho_0 + \alpha z^* \tag{4.23}$$

where $\rho_0$ is the density of silt-free water, $\alpha$ is a constant which depends upon the silt concentration, and $z_0$ is the depth below which $\rho$ increases. For convenience we introduce $z^* = z - z_0$. If equation (4.23) is substituted in the hydrostatic equation, we have

$$\frac{dp}{dz^*} = \rho g = (\rho_0 + \alpha z^*) g \tag{4.24}$$

which can be written as

$$dp = \rho_0 g dz^* + \alpha g z^* \, dz^*.$$

If we integrate between the levels $z_1^*$, and $z_2^*$, we have, finally,

$$p_2 - p_1 = \rho_0 g \left(z_2^* - z_1^*\right) + \frac{1}{2}\alpha g \left(z_2^{*2} - z_1^{*2}\right) \tag{4.25}$$

i.e. the pressure increase is made up of two parts, the first due to the pure-water density $\rho_0$, and the second a quadratic (i.e. $z^{*2}$) term due to the silt.

Fresh water has a lower density (normally taken as 1000 kg/m$^3$) than salt water (density approximately $1000 + 7c$ kg/m$^3$, where $c$ is the percentage salt concentration by weight), so if one meets the other, as in an estuary, the lighter fresh water tends to form a layer above the heavier salt water. This phenomenon of layering according to progressively increasing density with depth, which can occur in both liquids and gases, is called **stratification** and can be of major practical significance. For example, under certain atmospheric and topographical conditions, as may occur in a natural basin or valley, a 'lid' of low-density warm air can settle above heavy cooler air and lead to the build-up of high levels of pollutant at ground level. In some instances the two fluids may be miscible, as is the case for gases or salt and fresh water. In others, such as oil and water, there is no mixing, and a well-defined **interface** forms between the two. The interface is **stable** if the density of the fluid above it is less than that of the fluid below it, and **unstable** if the density is higher above the interface than below. Oil spilled onto water spreads out under the influence of gravity to form a thin stable layer called a **slick**. The stability of a stratified body of liquid where the density variation is continuous can be determined as follows. If there is a density decrease with upward vertical distance $z'$, i.e. $d\rho/dz' < 0$, a fluid particle of density $\rho$ moved vertically upwards by an infinitesimal amount $\delta z'$ would find itself surrounded by fluid of lower density

$$\rho + \delta z' \frac{d\rho}{dz'}$$

and so gravity would cause the (higher-density) particle to fall back to its original position. This situation is dynamically stable, whereas if $d\rho/dz' > 0$ it is unstable. The stability of a large body of a gas, in particular the earth's atmosphere, is considered in Section 4.13.

## 4.12  Deep oceans

The **Mariana Trench** in the western Pacific Ocean is the deepest part of the world's oceans, with a depth of about 10.9 km. The corresponding pressure at that depth, calculated from the constant-density equation $B + \rho g z$ for a liquid of density 1000 kg/m$^3$, would be 1070 bar. An ideal gas subjected to such a large pressure (at constant temperature) would decrease to $1/1070^{\text{th}}$ of its volume at 1 bar, with a corresponding increase in density. How realistic, then, is our assumption of constant density for water and other liquids? As stated in Section 2.5, a good approximation for the observed behaviour of liquids is the modified **Tait equation**

$$\frac{\rho_0}{\rho} = 1 - A \ln\left(\frac{p + C}{B + C}\right) \tag{2.21}$$

where, for water at 25 °C, $A = 0.137$, $B = 1$ bar, and $C = 2996$ bar.

For a pressure of 1070 bar equation (2.25) gives a density 4.4% higher than that for 1 bar. We conclude that for water, at least, the increase in density with pressure is likely to be negligible in most practical circumstances. If a more accurate $p(z)$ relation is required than that which results from the assumption of constant density, we could integrate the hydrostatic equation (4.5) numerically with the density evaluated using equation (2.21).

## 4.13  Earth's atmosphere

The earth's atmosphere is a relatively thin region of gas held to the earth's surface by gravitational attraction. Although we rarely think of air as being heavy, it's a remarkable fact that the total mass of the atmosphere is estimated to be about $5.3 \times 10^{18}$ kg, about 80% of it contained within the **lower atmosphere**, an 11 km thick layer called the **troposphere**. The troposphere is the first of seven layers which make up the **International Standard Atmosphere 1976 (ISA)**[30], which extends from sea level to 86.0 km. Within each layer the temperature is assumed to be constant or to vary linearly with **altitude** $z'$. The ISA attempts to represent average atmospheric conditions in temperate latitudes. At sea level, the pressure (the **standard atmospheric pressure**) is taken as 101,325 Pa (1 atm), and the temperature as $(T_0)$ 15.15 °C (or 288.15 K). It should be understood that the ISA is a simplified model of the actual atmosphere, which is dynamic rather than static in character and also moist with water vapour rather than dry.

### 4.13.1 Geopotential altitude

For most practical purposes the acceleration due to gravity $g$ can be regarded as a constant but over the altitude range of the atmosphere (0 to 86 km in the ISA and up to 10,000 km if the **exosphere** is to be included) this approximation is increasingly unacceptable. According to **Newton's gravitational law**, the $g\left(z'\right)$ dependence follows an inverse square equation

$$g = g_0 \left( \frac{R_E}{R_E + z'} \right)^2 , \tag{4.26}$$

where $R_E$ is the mean radius of the earth and $g_0$ is the acceleration due to gravity at sea level $(z' = 0)$. The usual values assumed are $R_E = 6371$ km, and $g_0 = 9.80665$ m/s$^2$ (usually rounded to 9.81 m/s$^2$)[31]. Substitution for $g$ in the hydrostatic equation (4.6) then leads to

$$\frac{dp}{dz'} = -\rho g = -\rho g_0 \left( \frac{R_E}{R_E + z'} \right)^2 . \tag{4.27}$$

Since equation (4.27) is awkward to integrate analytically unless $\rho$ is constant (an unrealistic approximation within the atmosphere), it is usual in meteorology to present the properties of the atmosphere in terms of the **geopotential altitude** $z'_G$ such that the hydrostatic equation becomes

$$\frac{dp}{dz'_G} = -\rho g_0 \tag{4.28}$$

which avoids the difficulty. Division of equation (4.27) by equation (4.28) leads to

$$\frac{dz'_G}{dz'} = \frac{g}{g_0} = \left( \frac{R_E}{R_E + z'} \right)^2$$

which can be integrated to give the relationship between the **geometric altitude** $z'$ and the geopotential altitude $z'_G$

$$z'_G = \frac{R_E z'}{R_E + z'}. \tag{4.29}$$

---

[30] The ISA is one of a number of models for the earth's atmosphere. Others include the International Civil Aviation Organization (ICAO) Standard Atmosphere and the U.S. Standard Atmosphere.

[31] The standard acceleration due to gravity is specified as 9.80665 m/s$^2$ in ISO 80000-3:2006 Quantities and units— Part 3: Space and time.

From equation (4.29) with $z' = 86$ km, we find $z'_G = 84.852$ km ($\approx 0.0133\,R_E$) which corresponds with the outer surface of the final layer of the Standard Atmosphere, known as the **mesosphere** (further details are given below). Since $z'_G/R_E \approx 0.0133$ the Standard Atmosphere can be considered as a thin layer. Equation (4.29) accounts for the reduction in $g$ with increasing altitude[32].

### 4.13.2 Structure of the earth's atmosphere

According to the ISA, in the **troposphere** the temperature $T$ (in °C) decreases with $z'_G$ according to the relation

$$T = T_0 - \Gamma z'_G \tag{4.30}$$

where the **lapse rate** $\Gamma$, defined as the negative of the temperature gradient[33] $-dT/dz'_G$, is 6.5 °C/km. As we shall see in the Subsection 4.13.3, this lapse rate suggests that the troposphere is in stable equilibrium. In fact, due to natural convection caused by surface heating, the troposphere is highly unstable resulting in a great deal of mixing which manifests itself in the weather we experience, from a gentle breeze to tropical storms and hurricanes. The second layer, the **tropopause** ($z'_G = 11.0$ to $20.0$ km), is isothermal at a temperature of $-56.5$ °C. The pressure at an altitude of about 19 km, known as **Armstrong's limit**, is so low (about 0.064 kPa) that water boils at the normal temperature of the human body (37 °C).

In contrast to the troposphere, the **stratosphere** (20.0 to 47.0 km) is very stable with a progressively increasing temperature. In the **lower stratosphere** (20.0 to 32.0 km) the lapse rate is $-1$ °C/km leading to a temperature increase from $-56.5$ to $-44.5$ °C. For subsonic civil aircraft the typical cruising altitude is about 10 km, while for supersonic and combat aircraft it is more like 20 to 30 km. The low temperature at these altitudes is responsible for condensing the water vapour in an aircraft-engine exhaust to produce the white vapour trails, called **contrails** (a contraction of **condensation trails**), often visible in a clear blue sky. Most **atmospheric ozone** resides in the lower stratosphere in the so-called **ozone layer**. The temperature continues to increase in the **upper stratosphere** (32.0 to 47.0 km) with a lapse rate of $-2.8$ °C to a temperature of $-2.5$ °C. The **stratopause**, which is between 47.0 and 51.0 km, is again isothermal at a temperature of $-2.5$ °C.

The temperature decreases within the **mesosphere** (51.0 to 84.852 km), initially at a rate of 2.8 °C/km to $-58.5$ °C at an altitude of 71.8 km and then at 2.0 °C/km to $-86.3$ °C at 86.0 km, the altitude which can be taken as the upper limit of the ISA. Four further layers can be defined: the **mesopause**, which is again isothermal and extends from the top of the mesosphere to the **thermosphere** (approximately 85 to 600 km), followed by the **exosphere** or **outer thermosphere** (approximately 600 to 10,000 km), while the **ionosphere** (approximately 60 to 300

---

[32] Unless otherwise stated, the term altitude refers to the geopotential altitude.

[33] Since the lapse rate is defined as $\Gamma = -dT/dz'_G$, a positive lapse rate, as in the troposphere, corresponds to decreasing temperature and vice versa when the lapse rate is negative. We note too that the symbol $\gamma$ is frequently used to represent the lapse rate but, to avoid confusion with the ratio of specific heats for which $\gamma$ is the usual symbol, we have chosen to use $\Gamma$.

km) starts in the mesosphere and ends in the thermosphere. Solar radiation leads to temperatures in excess of 1000 °C in the outer thermosphere. The **Kármán line** at an altitude of 100 km defines the lower boundary of **outer space**.

As we have just outlined, the basic specification for the earth's atmosphere is in terms of static temperature, $T\left(z'_G\right)$. To find the variation of static pressure $p$ and density $\rho$ with altitude $z'_G$ we need to integrate the hydrostatic equation (4.28) for each layer and, in addition to $T\left(z'_G\right)$, this requires the introduction of an **equation of state**, which connects $\rho, p$, and $T$. From Section 2.4 we have the **ideal-gas law**

$$p = \rho RT \tag{2.9}$$

where $R$ is the specific gas constant, with the value for air taken as 287.1 m$^2$/s$^2$K.

We illustrate the general problem of determining $p\left(z'_G\right)$ and $\rho\left(z'_G\right)$ given $T\left(z'_G\right)$ by considering the **tropopause**, in which the temperature remains constant with altitude (i.e. the tropopause is isothermal). Since the temperature $T$ here is constant, it is convenient to use equation (2.9) to eliminate the variable density from the hydrostatic equation to give

$$\frac{dp}{dz'_G} = -\rho g_0 = -\frac{pg_0}{RT} \tag{4.31}$$

which can be rearranged as

$$\frac{1}{p}dp = -\frac{g_0}{RT}dz'_G.$$

After integration we have

$$\ln p = -\frac{g_0 z'_G}{RT} + C$$

where $C$ is a constant of integration which we can determine from the condition at the upper limit of the troposphere ($z'_T = 11$ km)[34], where the pressure is $p_T$ (0.226 bar), so that

$$C = \ln p_T + \frac{g_0 z'_T}{RT}$$

hence,

$$\ln p = -\frac{g_0 z'_G}{RT} + \ln p_T + \frac{g_0 z'_T}{RT}$$

or

$$\frac{p}{p_T} = exp\left\{-\frac{g_0\left(z'_G - z'_T\right)}{RT}\right\}. \tag{4.32}$$

Equation (4.32) shows that the air pressure in the tropopause decreases exponentially with altitude difference.

A similar analysis for the **troposphere**, in which the temperature decreases linearly with altitude, i.e. $dT/dz'_G = -\Gamma$, leads to the relation

[34] To avoid double subscripts, the geopotential altitude of the outer limit of the stratosphere is represented by $z'_T$ and that of the outer limit of the stratopause by $z'_p$.

$$p = B\left(1 - \frac{\Gamma z'_G}{T_0}\right)^{g_0/\Gamma R} \tag{4.33}$$

where $T_0$ is the temperature at sea level ($z'_G = 0$), taken as 288.15 K. Since $\Gamma = 6.5 \times 10^{-3}\,°C/m$ for the troposphere, the exponent $g_0/\gamma R = 5.26$.

---

**ILLUSTRATIVE EXAMPLE 4.5**

The temperature in the lower mesosphere ($51 < z'_G < 71$ km) decreases linearly with altitude $z'$ with lapse rate $\Gamma$. If the gas which makes up the mesosphere can be treated as a perfect gas with gas constant $R$, show that the pressure varies according to

$$\frac{p}{p_P} = \left(\frac{T}{T_P}\right)^{g_0/\Gamma R}$$

where the subscript $P$ denotes the 'top' of the stratopause (i.e. $z'_p = 51$ km).

If the values of $p_P$ and $T_P$ are 66.9 Pa and 270.7 K, respectively, and the lapse rate $\Gamma$ for the lower mesosphere is 2.8 °C/km, calculate the temperature, pressure, and density at an altitude of 60 km. The gas constant $R$ can be taken as 287 m²/s² K.

Solution

The specified temperature variation with $z'_G$ is

$$\frac{dT}{dz'_G} = -\Gamma$$

and the pressure follows the hydrostatic equation

$$\frac{dp}{dz'_G} = -\rho g_0.$$

These two equations can be combined to eliminate $z'_G$ to give

$$\frac{dp}{dT} = \frac{\rho g_0}{\Gamma}.$$

The density $\rho$ can be eliminated using the perfect-gas law $p = \rho RT$ to give

$$\frac{dp}{dT} = \frac{p g_0}{\Gamma RT}$$

which can be rearranged as

$$\frac{dp}{p} = \frac{g_0}{\Gamma R}\frac{dT}{T}.$$

This equation can be integrated to give

$$\ln p = \frac{g_0}{\Gamma R}\ln T + C.$$

The constant of integration $C$ can be determined from $p_P$ and $T_P$ using

$$\ln p_P = \frac{g_0}{\Gamma R}\ln T_P + C$$

so that, after substituting for $C$ in the equation for $p$, we have

$$\ln p = \frac{g_0}{\Gamma R}\ln T + \ln p_P - \frac{g_0}{\Gamma R}\ln T_P$$

which can be rearranged as

$$\ln p - \ln p_P = \frac{g_0}{\Gamma R}(\ln T - \ln T_P)$$

so, finally,

$$\frac{p}{p_P} = \left(\frac{T}{T_P}\right)^{g_0/\Gamma R}.$$

We note that $g_0/\Gamma R$ is a non-dimensional constant.

For the numerical part of the problem we have $p_P = 66.9$ Pa, $T_P = 270.7$ K, $R = 287$ m$^2$/s$^2\cdot$K, $g_0 = 9.81$ m/s$^2$, $z_P' = 5.1\times10^4$ m, $\Gamma = 2.8\times10^{-3}$ °C/m, and $z_G' = 6\times10^4$ m. We shall need the value of $g_0/\Gamma R$ so we calculate this first: $9.81/(2.8\times10^{-3}\times287) = 12.2$.

From $dT/dz_G' = -\Gamma$ we have $T = -\Gamma z_G' + C$ so that $T_P = -\Gamma z_P' + C$ and we have $T - T_P = -\Gamma(z_G' - z_P')$. From the last equation we find that the temperature at $z' = 60$ km is

$$T = 270.7 - 2.8\times10^{-3}\left(6\times10^4 - 5.1\times10^4\right)$$
$$= 245.5\text{ K or } -27.5\,°\text{C}.$$

The corresponding pressure is given by

$$\frac{p}{p_P} = \left(\frac{T}{T_P}\right)^{g_0/\Gamma R}$$
$$= \left(\frac{245.5}{270.7}\right)^{12.2}$$
$$= 0.3036$$

so $p = 20.3$ Pa, and the density is found from $\rho = p/RT = 20.3/(287\times245.5) = 2.88\times10^{-4}$ kg/m$^3$.

---

The properties (temperature, pressure, and density) of the International Standard Atmosphere as functions of geopotential altitude $z_G'$ are listed in Table A.7 and illustrated in Figure 4.12.

As can be seen, both the pressure and the density decrease almost exponentially with altitude, each falling to near zero at about $z_G' = 40$ km. If this value is taken as the **height of the atmosphere**, an average density $\bar\rho = B/g_0 z_G'$ can be calculated as 0.26 kg/m$^3$, i.e. only about 20% of the sea-level value at 15 °C and 1 bar.

### 4.13.3 Adiabatic lapse rate and atmospheric stability

In Section 4.11 we showed how the stability of a body of liquid depends upon the density gradient $d\rho/dz'$. To determine the stability of a region of the atmosphere where the lapse rate

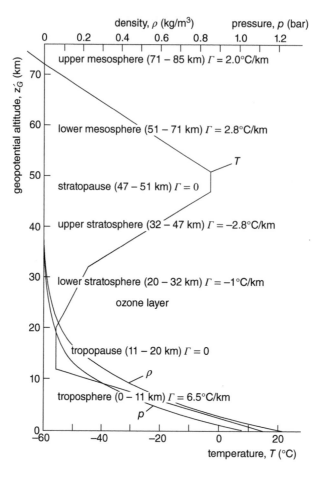

**Figure 4.12** Temperature, density, and pressure variation for the International Standard Atmosphere

is $\Gamma$, assumed constant, we first need to find the corresponding density variation with altitude. We start with the pressure variation, given by the hydrostatic equation

$$\frac{dp}{dz'_G} = -\rho g_0. \tag{4.28}$$

Assuming the air in the atmosphere is dry and follows the perfect-gas law $p = \rho RT$, then

$$\frac{dp}{dz'_G} = RT\frac{d\rho}{dz'_G} + \rho R\frac{dT}{dz'_G} = RT\frac{d\rho}{dz'_G} - \rho R\Gamma = -\rho g_0$$

from which

$$T\frac{d\rho}{dz'_G} = \rho\left(\Gamma - \frac{g_0}{R}\right). \tag{4.34}$$

We cannot simply conclude that if $d\rho/dz'_G < 0$ the atmospheric layer is stable, because now when a fluid particle is moved vertically its density changes to match the pressure at its new location. It is reasonable to assume that there is no heat transfer (adiabatic) to the particle as it moves. If the process is also reversible, then it is **isentropic** so that, from Section 2.11, the pressure and density are related as follows

$$\frac{p}{\rho^\gamma} = \text{constant}. \tag{2.38}$$

The hydrostatic equation (4.28) combined with equation (2.38) and the perfect-gas law, leads to

$$\frac{d\rho}{dz'_G} = -\frac{\rho^2}{\gamma p} = -\frac{\rho g_0}{\gamma RT}$$

or

$$T\frac{d\rho}{dz'_G} = -\frac{\rho g_0}{\gamma R}. \tag{4.35}$$

If the vertical distance moved by the particle is $+\delta z'_G$, then from equation (4.34) its density changes from $\rho$ to

$$\rho - \frac{\rho g_0}{\gamma RT}\delta z'_G. \tag{4.36}$$

From equation (4.33) the density of the surrounding air at altitude $z'_G + \delta z'_G$ is

$$\rho + \frac{\rho}{T}\left(\Gamma - \frac{g_0}{R}\right)\delta z'_G. \tag{4.37}$$

For stability, gravity must cause the particle to return to its original position, which equations (4.36) and (4.37) show will be the case if

$$\rho - \frac{\rho g_0}{\gamma RT}\delta z'_G > \rho + \frac{\rho}{T}\left(\Gamma - \frac{g_0}{R}\right)\delta z'_G.$$

After simplification, this inequality leads to

$$\frac{g_0}{R}\left(\frac{\gamma - 1}{\gamma}\right) > \Gamma \tag{4.38}$$

as the condition for stability within the atmosphere. From the perfect-gas equation it can be shown that the lapse rate for an isentropic atmosphere, where $p/\rho^\gamma$ = constant, is

$$\Gamma_{AD} = \frac{g_0}{R}\left(\frac{\gamma - 1}{\gamma}\right). \tag{4.39}$$

The quantity $\Gamma_{AD}$ is termed the **adiabatic lapse rate**, and the criterion for stability is thus $\Gamma_{AD} > \Gamma$.

With $g_0$ = 9.81 m/s$^2$, $\gamma$ = 1.402, and $R$ = 287 m$^2$/s·K, we have $\Gamma_{AD}$ = 9.800 °C/km. The lapse rates for the Standard Atmosphere are 6.5 °C/km (troposphere), 0 °C/km (tropopause), −1.0 °C/km (lower stratosphere), −2.8 °C/km (upper stratosphere), 0 °C/km (stratopause), +2.8 °C/km (lower mesosphere), +2.0 °C/km (upper mesosphere), and 0 °C/km (mesopause). We thus conclude that all segments of the Standard Atmosphere are stable.

## 4.14 Pressure variation in an accelerating fluid

Until now we have restricted consideration to the vertical variation of pressure in fluids at rest. To be more precise, the crucial restriction is that there is no relative tangential movement between fluid particles and hence the shear stress is identically zero throughout the body of fluid. It should be self-evident that there is no relative moment between fluid particles within a fluid in a container moving at constant velocity, provided sufficient time has elapsed for any motion created at the start of the process to have died out. In such circumstances everything we have said in this chapter so far still holds. In a fluid subjected to constant acceleration there may still be no relative movement between fluid particles but the situation is more complex.

We consider a horizontal cylinder of fluid of infinitesimal length $\delta x$ and infinitesimal cross section $\delta A$, as shown in Figure 4.13, with a constant component of acceleration $a_x$ in the horizontal direction. If we apply Newton's second law of motion (i.e. net force = mass × acceleration) to the fluid cylinder, we have

$$p\delta A - \left(p + \delta p\right)\delta A = \rho\delta A\,\delta x\,a_x$$

wherein we have substituted $\delta A\,\delta x$ for the volume $\delta V$ of the elemental cylinder, and $\rho\delta V = \rho\delta A\,\delta x$ for its mass. The equation simplifies to $\delta p = -\rho\delta x\,a_x$ or, in the limit as $\delta x$ approaches zero,

$$\frac{\partial p}{\partial x} = -\rho a_x. \tag{4.40}$$

The negative sign in equation (4.40) should come as no surprise: if the pressure gradient $\partial p/\partial x$ is positive (i.e. the pressure increases from left to right), the cylinder will accelerate to the left and not to the right as the acceleration vector in Figure 4.13 would suggest. For the first time in this book we have used the symbol $\partial/\partial x$ to denote a partial derivative rather than a total derivative $d/dx$. The partial derivative is appropriate here because the hydrostatic pressure $p$ can now vary both in the horizontal (i.e. $x$-) direction as well as the vertical (i.e. $z$-) direction. If we also subject the fluid to a vertically downward component of acceleration $a_z$, the hydrostatic equation (4.5) must be replaced by

$$\frac{\partial p}{\partial z} = \rho\left(-a_z + g\right). \tag{4.41}$$

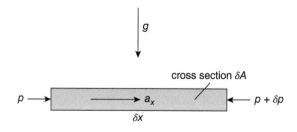

**Figure 4.13** Accelerating horizontal cylinder of fluid

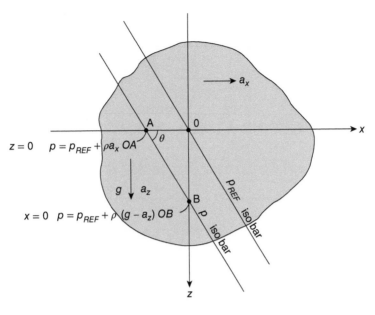

**Figure 4.14** Isobars in an accelerating body of fluid

The spatial variation (i.e. with $x$ and $z$) of the hydrostatic pressure requires the solution of both equations (4.40) and (4.41). Provided the acceleration components $a_x$ and $a_z$ are constant, both equations are easily integrated and we find

$$p = -\rho a_x x + C_1(z) \quad \text{and} \quad p = -\rho\left(a_z - g\right)z + C_2(x).$$

$C_1(z)$ and $C_1(z)$ are now functions of integration rather than the constants of integration we have for ordinary differential equations. Since the two equations for $p$ must be simultaneously valid, the final result is

$$p - p_{REF} = \rho\left(gz - a_z z - a_x x\right) \tag{4.42}$$

where $p_{REF}$ is a reference pressure (i.e. $p = p_{REF}$ at $x = 0$, and $z = 0$).

Since the terms within the brackets in equation (4.42) are constant, we see that lines for which $\left(g - a_z\right)z - a_x x$ = constant represent lines of constant pressure (called **isobars**). The slope of the isobars, as shown in Figure 4.14, is given by

$$\tan\theta = \frac{dz}{dx} = \frac{OB}{OA} = \frac{a_x}{g - a_Z}. \tag{4.43}$$

---

**ILLUSTRATIVE EXAMPLE 4.6**

A high-performance sports car is driven around a corner of radius 64 m at a constant speed of 180 kph. What angle does the free surface of the petrol in a fuel tank make with the horizontal? Calculate the pressure difference between the free surface and a point a perpendicular distance 200 mm from the free surface. Take the density of petrol as 800 kg/m$^3$, and the acceleration due to gravity as 9.81 m/s$^2$.

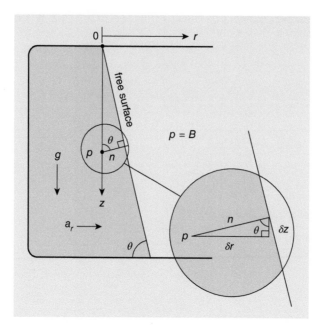

**Figure E4.6**

## Solution

Since the car speed $V$ = 180 kph = 50 m/s, the centripetal (i.e. radially inward) acceleration $a_r$ for a radius $R$ = 64 m is $a_r$ = $V^2/R$ = $50^2/64$ = 39.1 m/s$^2$ (i.e. a lateral acceleration of almost 4$g$).

In this example, the vertical acceleration is zero, so the slope of the free surface is given by equation (4.16) as $\tan \theta$ = $a_r/g$ = 3.98 from which $\theta$ = 75.9°. From Figure E4.6, we see that, for point P, 200 mm (= $n$) from the free surface, $x$ = $-n \sin \theta$ and $z$ = $n \cos \theta$ so that the pressure at P is given by

$$p - B = \rho g n \left( \frac{a_x}{g} \sin \theta + \cos \theta \right)$$

$$= 800 \times 9.81 \times 0.2(3.98 \times \sin 75.9° + \cos 75.9°)$$

$$= 6441 \text{ Pa or } 0.064 \text{ bar.}$$

 ## 4.15  SUMMARY

We started this chapter by establishing the three fundamental principles for the variation of pressure $p$ throughout a body of fluid at rest: (a) the pressure at a point is the same in all directions (Pascal's law), (b) the pressure is the same at all points on the same horizontal level, and (c) the pressure increases with depth $z$ according to the hydrostatic equation $dp/dz$ = $\rho g$. If the fluid density $\rho$ is constant, the increase in pressure over a depth increase

$h$ is $\rho gh$, a result which can be used to analyse the response of simple barometers and manometers to applied pressure changes and differences. In situations where very large changes in pressure occur, such as throughout the earth's atmosphere and in very deep water, the assumption of constant density may no longer be adequate, and an equation of state is required to relate pressure and density, together with an assumption about the fluid temperature. The hydrostatic equation is still valid but more difficult to integrate, as illustrated by consideration of the earth's atmosphere. The vertical density gradient in a body of fluid determines whether it is stable or unstable.

The student should

- be able to calculate the pressure variation with vertical depth for a fluid of constant density, including the situation of a series of fluid layers
- be able to analyse the response of a simple barometer to changes in the external (barometric) pressure
- be able to analyse the response of a U-tube or inclined-tube manometer to changes in the applied pressure difference
- be able to calculate the pressure variation with vertical depth or height for a variable-density fluid where there is a simple relationship between pressure and fluid density
- be able to determine the stability of a body of fluid given the vertical density distribution
- understand the concept of geopotential altitude
- be familiar with the series of layers which make up a model of the earth's atmosphere and be able to calculate the density and pressure distribution for a specified lapse rate

## 4.16 SELF-ASSESSMENT PROBLEMS

**4.1** If the atmosphere is assumed to have a height of 100 km and a constant density, calculate the density if the pressure at ground level is 1 bar.
(Answer: 0.102 kg/m$^3$)

**4.2** According to kinetic theory, the molecular mean free path $\Lambda$ for a gas of molecular weight $\mathcal{M}$, effective molecular diameter $\sigma$, and density $\rho$ is given by

$$\Lambda = \mathcal{M}/\sqrt{2}\pi\rho N_A\sigma^2$$

where $N_A = 6.022 \times 10^{26}$ molecules/kmol is the Avogadro number. The Knudsen number $Kn = \Lambda/L$, where $L$ is a characteristic length of an object moving through the gas. Calculate the density of air if $Kn = 1$ for values of $L$ of 1 $\mu$m, 1 mm, and 1 m and, with reference to Table A.7, identify the corresponding regions of the earth's atmosphere. The value of $\sigma$ for air is 366 pm.
(Answers: 0.808 kg/m$^3$ (lower troposphere), $8.08 \times 10^{-4}$ kg/m$^3$ (stratopause), $8.08 \times 10^{-7}$ kg/m$^3$ (lower thermosphere)).

**4.3** A tube, closed at the bottom, open at the top, of length 20 m, and inclined at an angle of 20° to the horizontal, is half full of water and half full of oil of relative density 0.8. Calculate the hydrostatic pressure at the bottom of the tube.
(Answer: 0.604 bar)

**4.4**   A vertical U-tube manometer contains two liquids, one of density $\rho_1$ and the other of lower density $\rho_0$. Show that the difference in height between the free surfaces of the two liquids when no pressure difference is applied is $h_0 (1 - \rho_0/\rho_1)$, where $h_0$ is the height of the lighter liquid above its interface with the heavier liquid.

**4.5**   Figure P4.5 shows a U-tube of cross-sectional area $A$, which is sealed on the left-hand side, open on the right-hand side, and contains a liquid of density $\rho$. The density of the gas above the liquid on the left-hand side is negligible. The solid cylinder of mass $m$ on the right-hand side is completely supported by the liquid (i.e. the cylinder is a perfect fit in the tube with no leakage or friction). Derive an expression for the absolute pressure of the gas if the external pressure acting on the cylinder is $B$.

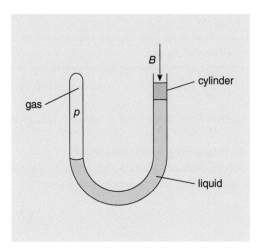

**Figure P4.5**

**4.6**   Figure P4.6 shows an inverted U-tube manometer used to measure the pressure $p$ of a gas in a pipe. The U-tube contains two liquids, of densities $\rho_1$ and $\rho_2$, as shown. Show that

$$p - B = (\rho_1 H - \rho_2 L) g$$

where $B$ is the external pressure and $g$ is the acceleration due to gravity. The gas density may be assumed to be negligible.

**4.7**   The pressure measured at the summit of a mountain is 0.31 bar. Calculate the height of the mountain, assuming the temperature decreases linearly with altitude at a rate of 6.5 °C/km. Take the pressure at sea level as 1.015 bar and the temperature as 15 °C. What would be the error in calculating the height, assuming the air density remained constant at its sea-level value?
(Answers: 8947 m, -34.6%)

**4.8**   (a) Figure P4.8 shows a manometer consisting of two vertical arms of cross-sectional area $a$ connected to form a U-tube. The open reservoirs at the top of each arm are identical and of cross-sectional area $A \gg a$. The liquid in the right arm is water with density $\rho_W$, and that in the left arm is an oil with density $\rho_O$ which is less than $\rho_W$. The manometer is used to measure the difference between the pressure $p_1$, which

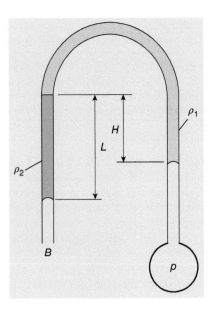

**Figure P4.6**

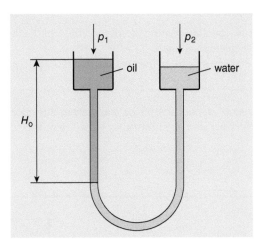

**Figure P4.8**

acts on the oil surface, and $p_2$, which acts on the water surface. Show first that when $p_1$ and $p_2$ are equal, the difference in height between the oil and water surfaces is

$$H_O \left( 1 - \frac{\rho_O}{\rho_W} \right)$$

where $H_O$ is the height of the oil column measured above the oil/water interface. Show further that a pressure difference $p_2 - p_1$ moves the interface an amount $h$ given by

$$h = \frac{p_2 - p_1}{\rho_w g \,(1 + a/A) - \rho_o g \,(1 - a/A)}.$$

(b) If the oil has a density of 800 kg/m³ and the U-tube has an internal diameter of 5 mm, calculate the reservoir diameter required if the manometer reading for a pressure difference of 200 Pa is to be 100 mm. What would the manometer read for a pressure difference of 200 Pa if the reservoirs had the same internal diameter as the manometer tube?

(Answers: 107.8 mm, 10 mm)

**4.9**  (a) An inclined manometer, as shown in Figure P4.9, is used to measure the pressure $p_F$ in a pipe containing fluid of density $\rho_F$. The measuring leg of the manometer has an internal cross-sectional area $a$ and is inclined at angle $\phi$ to the horizontal. The other side of the manometer is connected to a pipe containing a fluid of density $\rho_R$ at constant reference pressure $p_R$. The interface between the reference fluid and the manometer liquid, which has a density $\rho_M$, is maintained within a reservoir of cross-sectional area $A$. When the pressures $p_F$ and $p_R$ are equal, the manometer reading $L$ is zero, and the liquid level in the reservoir $\delta$ is also zero.

(i) If $H_F$ is the height of the pipe on the right-hand side above the zero line (i.e. $L = 0$) and $H_R$ is the height of the pipe on the left-hand side above the zero line, show that

$$\frac{H_R}{H_F} = \frac{\rho_F}{\rho_R}.$$

(ii) Derive an equation relating the rise in level $\delta$ of the fluid in the reservoir with the manometer reading $L$.

(iii) Show that when $p_F > p_R$ the pressure difference $p_F - p_R$ is given by

$$p_F - p_R = (\rho_M - \rho_F)\, gL \sin \phi + (\rho_M - \rho_R)\, gL \, a/A.$$

(b) In a particular application of the manometer, the angle of inclination $\phi$ is 20°, the inside diameter $d$ of the measuring leg is 5 mm and that of the reservoir $D$ is

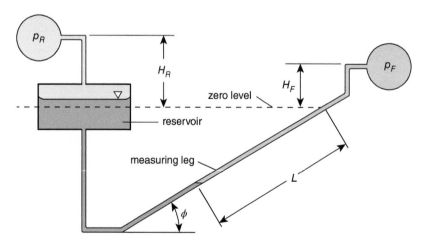

**Figure P4.9**

100 mm, and the manometer fluid is mercury with a relative density of 13.6. The reference fluid is a silicon oil which has a relative density of 0.8 and is maintained at an absolute pressure $p_R$ of 5 bar.

(i) Calculate the pressure $p_F$ of a gas with density $\rho_F$ of 10 kg/m$^3$ if the manometer reading $L$ is 350 mm.

(ii) Calculate the value of $p_F - p_R$ from the simplified equation

$$p_F - p_R = \rho_M g L \sin \phi.$$

(iii) What is the percentage error in the result for $p_F - p_R$ in (ii) compared with the full equation derived in part (a)(iii)?

(Answers: 5.16 bar, 0.62% too low)

**4.10** (a) A combat aircraft is flying in a nose-down attitude as shown in Figure P4.10. If the aircraft is accelerating at a rate $a$, show that the isobars in the fuel tanks are inclined to the horizontal at an angle $\theta$ given by

$$\tan \theta = \frac{a \cos \alpha}{g - a \sin \alpha}$$

where $\alpha$ is the pitch angle of the aircraft.

(b) If the aircraft in part (a) accelerates at $3g$, $g$ being the acceleration due to gravity, and the pitch angle is 30°, calculate the inclination of the isobars.

(c) Calculate the inclination of the isobars if the aircraft climbs with a nose-up attitude of 30° and acceleration $3g$.

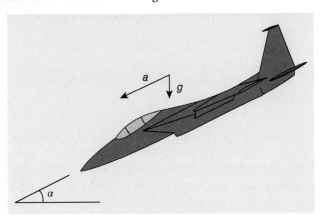

**Figure P4.10**

(Answers: −79.1°, 46.1°)

# 5

# Hydrostatic force exerted on a submerged surface

This chapter is concerned with **hydrostatic force**, which is the force exerted on a submerged body due to the hydrostatic pressure distributed over its surface (or surfaces). We start by showing that uniform pressure acting on the entire surface of a solid object results in a zero net force. The remainder of the chapter is concerned with the force exerted on a submerged surface due to the linear increase in pressure with vertical depth in a stationary body of fluid of constant density. We show that the vertical component of the net hydrostatic force on a submerged surface is equal to the weight of the fluid which occupies (or could occupy) the volume directly above the surface. It is also shown that the difference between the vertical components of hydrostatic force acting on the lower and the upper surfaces of a submerged body is the **buoyancy force of Archimedes' principle**. We then analyse the horizontal component of the hydrostatic force acting on a submerged surface. This component is less straightforward to calculate than the vertical component but is shown to equal the hydrostatic force acting on an equivalent flat vertical surface. The chapter concludes by considering the **stability** of a body either fully submerged or floating in a fluid.

## 5.1 Resultant force on a body due to uniform surface pressure

Figure 5.1(a) shows a body of arbitrary shape which is subjected to a uniform external pressure $B$ acting on its surface. An imaginary cylinder of infinitesimal cross section $\delta A$ is shown passing through the body and intersecting its surface at points $X$ and $Y$ where the surface areas are $\delta A_1$ and $\delta A_2$, respectively. Since the body is of arbitrary shape, these surface elements will be at arbitrary orientations to the line $XY$. Although the argument here applies to any three-dimensional shape, it may be easier for the reader to imagine the body has a two-dimensional shape, i.e. we are looking at a cross section which would be the same for any plane parallel to the page.

If, as shown in Figure 5.1(b), the angles between the line $XY$ and the surface normals at $X$ and $Y$ are $\theta_1$ and $\theta_2$, respectively, then

$$\delta A = \delta A_1 \cos \theta_1 = \delta A_2 \cos \theta_2, \tag{5.1}$$

i.e. the area $\delta A$ corresponds to the projection of both $\delta A_1$ and $\delta A_2$ onto a plane perpendicular to the axis of the elemental cylinder. The external pressure $B$ results in forces $B\delta A_1$ and $B\delta A_2$

*Introduction to Engineering Fluid Mechanics.* Marcel Escudier.
© Marcel Escudier 2017. Published 2017 by Oxford University Press.

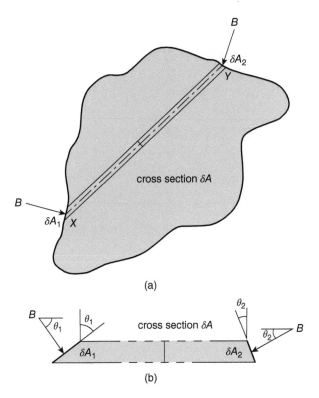

**Figure 5.1** (a) Body of arbitrary shape surrounded by uniform pressure $B$ (b) Infinitesimal cylinder within body

acting normal to $\delta A_1$ and $\delta A_2$, respectively. The net force due to $B$ acting along the line $XY$ is thus

$$(B\delta A_1)\cos\theta_1 - (B\delta A_2)\cos\theta_2 = B\,(\delta A_1\cos\theta_1 - \delta A_2\cos\theta_2)$$

which must be zero because of the area relationship, equation (5.1).

Since we can use the above argument for every part of the body surface, we conclude that  a body of arbitrary shape subjected to a uniform external pressure experiences zero net[35] force.

This conclusion has an important consequence for the calculation of not only hydrostatic forces but also hydrodynamic forces (see Chapters 9 and 10): in situations where the pressure acting on a surface varies from point to point, we can add or subtract a uniform pressure everywhere without affecting the net hydrostatic or hydrodynamic force balance. In particular, in problems for a liquid of constant density $\rho_F$, where the variation of the pressure $p$ with depth $z$ below the liquid surface is given by (see Section 4.3)

$$p = B + \rho_F gz, \tag{5.2}$$

we can subtract the uniform barometric pressure $B$ from $p$ and calculate hydrostatic forces using the **gauge pressure** $p_G = \rho_F gz$.

[35] The word net here means overall.

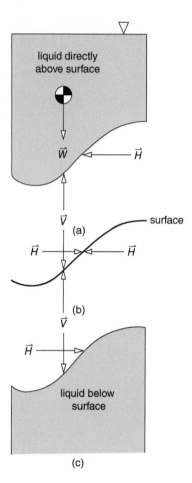

**Figure 5.2** Horizontal ($H$) and vertical ($V$) components of the hydrostatic force acting on a thin surface submerged in a liquid: (a) liquid above surface, (b) surface, (c) liquid below surface

## 5.2 Vertical component of the hydrostatic force acting on a submerged surface

Just as do solids, fluids obey Newton's laws of motion. From **Newton's third law** of action and reaction, we can state, for a body of liquid at rest, the net force exerted by the liquid on any submerged solid surface must be equal in magnitude and opposite in direction to the force exerted by the surface on the liquid. Since we need to consider the hydrostatic forces on objects or structures which may have liquid above, below, or both, the submerged surface in Figure 5.2 is shown as an infinitesimally thin, weightless sheet with an upper and a lower surface.

The first condition which must be satisfied for the liquid directly above the sheet to be in static equilibrium is that there must be a vertically upward force $V$ exerted on it by the upper surface of the sheet equal in magnitude to the weight $W$ of the liquid above it, i.e.

$$W = V = \rho_F \mathcal{V} g \tag{5.3}$$

where $\rho_F$ is the liquid density, $g$ is the acceleration due to gravity, and $\mathcal{V}$ is the volume of liquid directly above the sheet. The vertical component of the hydrostatic force exerted by the liquid on the upper surface of the sheet must be of equal magnitude to the force exerted on the liquid (i.e. $V = \rho_F \mathcal{V}g$) but act vertically downwards, as indicated in Figure 5.2(b). Finally, for the lower surface of the sheet, the hydrostatic force must again be of magnitude $|V|$ but act vertically upwards.

Each of the foregoing results could have been obtained by calculating the force due to the gauge pressure $p_G$ distributed over the upper and lower surfaces of the sheet. In the case of the upper surface, the vertical component of the force is obtained by integrating the vertical component of the force on every element of the surface $\delta A$, i.e.

$$\delta V = p_G \delta A \cos\theta = \rho_F g z \delta A \cos\theta$$

where $z$ is the depth below the liquid surface of the surface element $\delta A$, and $\theta$ is the angle between the vertical and the normal to the surface element. Since $\delta A \cos\theta$ is the area of an element of the sheet projected onto a horizontal plane, $z \delta A \cos\theta$ represents the volume $\delta \mathcal{V}$ of the vertical cylinder of liquid directly above the surface element $\delta A$. The force $V$ is thus

$$V = \int_{\mathcal{V}} \rho_F g \, d\mathcal{V} = \rho_F g \mathcal{V} \tag{5.4}$$

as before. If the surface has a simple shape (e.g. flat or cylindrical), there is a good chance the volume $\mathcal{V}$ can be calculated from well-known formulae for the volumes of rectangles, cylinders, etc. However, as we shall show in Section 5.5, if the shape is more complex, we have to evaluate the integral using the mathematical description of the shape.

So far we have considered only the magnitude of $V$ but not its line of action. The location of the latter is important because, for a system of forces acting on a body to be in static equilibrium (i.e. for a stationary body to remain at rest), we require that the forces exert no net moment on the body. As a consequence of this condition, for the liquid directly above the sheet the line of action of $V$ must pass vertically through the centroid of the liquid volume.

We note that to calculate both the magnitude of $V$ and the location of its line of action, we are concerned primarily with the geometry (i.e. the shape and size) of the volume $\mathcal{V}$.

---

**ILLUSTRATIVE EXAMPLE 5.1**

Calculate the magnitude of the hydrostatic force exerted on the upper surface of the kite-shaped plate shown in Figure E5.1, and the location of its line of action if the plate is submerged horizontally at a depth $Z$ below the surface of a liquid of density $\rho_F$.

Solution

Since the plate is horizontal, the entire hydrostatic force acting on its surface must be vertical and act downwards through the centroid of the plate surface. The force $V$ is equal to the weight of liquid directly above the plate, i.e.

$$V = \rho_F g A Z.$$

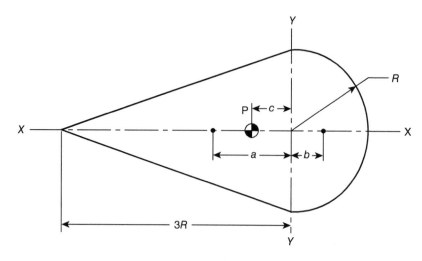

**Figure E5.1**

The surface area of the plate $A$ is given by the sum of the areas of the triangle to the left of YY and that of the semicircle to its right

$$A = 3R^2 + \frac{1}{2}\pi R^2$$

so that $V$ is given by

$$V = \left(3 + \frac{1}{2}\pi\right)\rho_F g Z R^2.$$

We can show that this result is consistent with what we obtain by considering the gauge pressure $p_G$ acting on the plate. Since the plate is horizontal, $p_G$ is constant

$$p_G = \rho_F g Z$$

and $V$ is given by

$$V = p_G A = \rho_F g Z A,$$

exactly as before.

The symmetry of the plate about XX tells us that the line of action of $V$ must pass through a point $P$ somewhere along the line XX. The location of $P$ is given by its distance $c$ from the line YY, which can be calculated by equating the moment of $V$ about YY (or any line parallel to YY) to the combined moments of the hydrostatic forces acting on the triangular section of the kite to the left of YY and on the semicircular section to the right of YY

$$\rho_F g Z\, 3R^2 a - \rho_F g Z\, \frac{1}{2}\pi R^2 b = \rho_F g Z\left(3 + \frac{1}{2}\pi\right)R^2 c.$$

The locations of the lines of action of the hydrostatic forces acting on the two sections correspond with their centroids, which we can find from Appendix 3, which includes the areas, centroid locations, and other information for a number of basic shapes as

$$a = R \quad \text{and} \quad b = \frac{4R}{3\pi}$$

so that, after dividing through by the common factor $\rho g Z R^2$, the moment equation becomes

$$3R - \frac{1}{2}\pi \frac{4R}{3\pi} = \left(3 + \frac{1}{2}\pi\right)c$$

from which

$$c = \frac{7R}{3\left(3 + \pi/2\right)}.$$

Since the plate in this example is horizontal, the only difficulty in the problem stemmed from the shape of the plate. In the following example, the surface shape is made up of two rectangles, and the difficulty arises because they are sloping rather than horizontal. As we have already remarked, later in this chapter we shall deal with a situation where the surface shape is sufficiently complex that the volume of liquid directly above it must be obtained by integration.

---

**ILLUSTRATIVE EXAMPLE 5.2**

A dam has the cross section shown in Figure E5.2, with $\tan\theta = 4$. The water depth is $H$, and the length of the dam along its top is $L$. Calculate the vertical component of the hydrostatic force acting on the face of the dam, and the horizontal distance of its line of action from the point O.

Solution

The vertical component of the force acting on the face of the dam $V$ is equal to the weight of water $W$ vertically above the face. This weight is given by

$$W = \rho_F \mathcal{V} g$$

where $\mathcal{V}$ is the volume of water directly above the two sloping surfaces, $\rho_F$ is the water density, and $g$ is the acceleration due to gravity. As shown in part (b) of the diagram, it is convenient to split $\mathcal{V}$ into three smaller volumes, for each of which the volume and centroid location can be found from Appendix 3, as follows

Volume 1: Cross section is a right-angle triangle with

$$\text{height} = \frac{2}{3}H, \quad \text{base length} = \frac{1}{6}H \text{ (since } \tan\theta = 4),$$

$$\text{centroid } \frac{1}{18}H \text{ from vertical face of triangle,}$$

$$\text{volume } \mathcal{V}_1 = \frac{1}{2}\frac{1}{6}H\frac{2}{3}HL = \frac{1}{18}H^2 L.$$

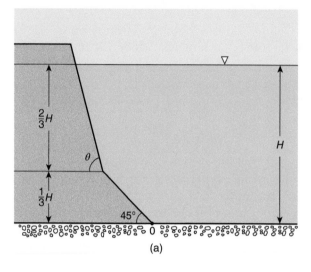

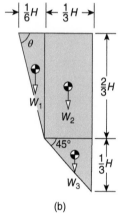

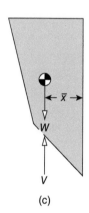

(b)                                    (c)

**Figure E5.2**

Volume 2: Cross section is a rectangle with

$$\text{height} = \tfrac{2}{3}\,H, \quad \text{width} = \tfrac{1}{3}\,H, \quad \text{centroid } \tfrac{1}{6}\,H \text{ from either vertical face,}$$

$$\text{volume } \mathcal{V}_2 = \tfrac{1}{3}H\,\tfrac{2}{3}\,H\,L = \tfrac{2}{9}\,H^2\,L.$$

Volume 3: Cross section is a right-angle triangle with

$$\text{height} = \tfrac{1}{3}\,H, \quad \text{base length} = \tfrac{1}{3}\,H, \quad \text{centroid } \tfrac{1}{9}\,H \text{ from vertical face of triangle,}$$

$$\text{volume } \mathcal{V}_3 = \tfrac{1}{2}\tfrac{1}{3}H\,\tfrac{1}{3}\,H\,L = \tfrac{1}{18}\,H^2\,L.$$

From the above,

$$\mathcal{V} = \mathcal{V}_1 + \mathcal{V}_2 + \mathcal{V}_3 = \left(\frac{1}{18} + \frac{2}{9} + \frac{1}{18}\right) H^2 L = \tfrac{1}{3}H^2\,L$$

hence,

$$V = W = \rho_F \mathcal{V} g = \tfrac{1}{3}\rho_F H^2 Lg.$$

To find the horizontal distance $\bar{x}$ of the line of action of $V$ from O, we equate the moment of $V$ about O to the combined moments of the vertical forces $V_1$, $V_2$, and $V_3$ due to the weight of the three liquid volumes $\mathcal{V}_1$, $\mathcal{V}_2$, and $\mathcal{V}_3$, i.e. $V_1 = W_1$, $V_2 = W_2$, and $V_3 = W_3$, as follows

$$V\bar{x} = \left(\frac{1}{3} + \frac{1}{18}\right)HV_1 + \frac{1}{6}HV_2 + \frac{1}{9}HV_3$$

$$= \left(\frac{1}{18}\frac{7}{18} + \frac{2}{9}\frac{1}{6} + \frac{1}{18}\frac{1}{9}\right)\rho_F H^3 Lg$$

$$= \frac{7}{108}\rho_F H^3 Lg$$

from which, if we substitute $\rho_F H^2 Lg/3$, for $V$, we have

$$\bar{x} = \frac{\frac{7}{108}\rho_F H^3 Lg}{\frac{1}{3}\rho_F H^2 Lg} = \frac{7}{36}H.$$

As we pointed out earlier in this section, the thin sheet depicted in Figure 5.2 has both an upper and a lower surface. For this weightless sheet to be in equilibrium, the net force in any direction must be zero. The hydrostatic force on the downwards-facing surface must therefore be equal in magnitude and opposite in direction to the hydrostatic force on the upwards-facing surface, and this must also apply to the vertical and horizontal components of the hydrostatic

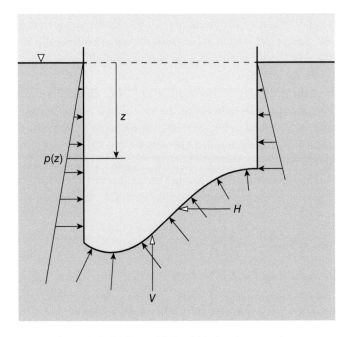

**Figure 5.3** Surface with liquid below but not above

force. While this conclusion may seem to be obvious for a submerged sheet, it may be less apparent if there is liquid beneath the sheet but not above.

The shape of the surface shown in Figure 5.3 is the same as that in Figure 5.2 but the vertical sides now prevent contact between the liquid and the upper surface of the sheet. Since the hydrostatic pressure is constant along any horizontal line within a liquid at rest, the pressure distribution over the downwards-facing surface is completely unaffected by the fact that there is no liquid above the sheet. The hydrostatic force must also be unaffected since it represents the integrated effect of the pressure distributed over a surface. We conclude that the magnitude of the vertical component of the hydrostatic force exerted on a surface submerged in a liquid is equal to the weight of the liquid which occupies the volume directly above the surface or the weight of the liquid which could occupy this volume.

---

**ILLUSTRATIVE EXAMPLE 5.3**

Figure E5.3 shows the cross section of an axisymmetric container. The upper and lower sections are both cylindrical with radii $r$ and $R$, respectively, and separated by a conical section which slants at an angle $\theta$ to the horizontal. If the container is filled with a liquid of density $\rho_F$ to a height $h$ above the top of the conical section, calculate the magnitude of the hydrostatic force exerted by the liquid on the conical section. State the direction of this force and the location of its line of action.

Solution

Since the container is symmetric about its vertical axis, the radial (i.e. horizontal) component of force arising from the pressure acting on any element of the interior surface is counteracted by a force of equal magnitude but opposite in direction arising on an identical element on the diametrically opposite side of the container. Due to the axisymmetry, therefore, the net hydrostatic force on the conical section of the container must be vertical, and its line of action coincident with the axis of symmetry. From the geometry of the conical section, it must also be that the hydrostatic force is directed vertically upwards.

The magnitude of the hydrostatic force is given by $\rho_F \mathcal{V} g$, where $\mathcal{V}$ is the volume of liquid which could occupy the space directly above the conical surface up to the level of the free surface of the liquid, as indicated in Figure E5.3 by the broken lines. From the figure, we see that we can calculate $\mathcal{V}$ by calculating first the volume of a cylinder of radius $R$ and height $h + (R - r)\tan\theta$ and then subtracting the volume of the small cylindrical section at the top, of height $h$ and radius $r$, together with the volume of the interior of the conical section which is a frustum of a cone with upper radius $r$, lower radius $R$, and height $(R - r)\tan\theta$. Thus, we have

$$\mathcal{V} = \pi R^2 \left[ h + (R - r)\tan\theta \right] - \pi r^2 h - \tfrac{1}{3}\pi \left( R^3 - r^3 \right)\tan\theta$$

and, finally, for the hydrostatic force $V$, we have

$$V = \rho_F g \pi (R - r) \left[ (R + r)h + \tfrac{1}{3}\tan\theta \left\{ 2R^2 - r(R + r) \right\} \right].$$

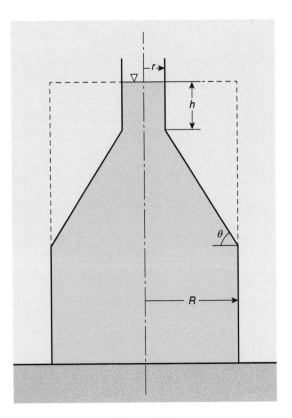

**Figure E5.3**

## 5.3 Archimedes' principle and buoyancy force on a submerged body

Although Archimedes' principle is often <u>stated</u> simply as 'the magnitude of the buoyancy force exerted on a submerged body is equal to the weight of fluid displaced by the body', it is easily <u>proved</u> using the concepts already presented in this chapter. A body of arbitrary shape submerged in a liquid, as shown in Figure 5.4, can be thought of as having an upper surface and a lower surface. From Section 5.2, the vertical component of the hydrostatic force exerted on the upper surface, $V_U$, will be a downward force equal in magnitude to the weight, $W_U$ of the liquid in the volume $\mathcal{V}_U$ directly above the upper surface, i.e.

$$V_U = W_U = \rho_F \mathcal{V}_U g. \tag{5.5}$$

Similarly, the vertical component of the hydrostatic force exerted on the lower surface, $V_L$, will be an upward force equal in magnitude to the weight, $W_L$, of the liquid which could occupy the entire volume (i.e. including the body itself), $\mathcal{V}_L$, directly above the lower surface, i.e.

$$V_L = W_L = \rho_F \mathcal{V}_L g. \tag{5.6}$$

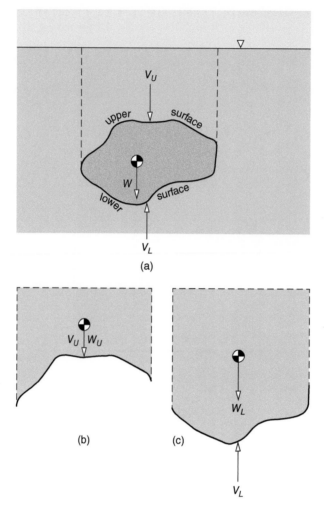

**Figure 5.4** Hydrostatic forces on a completely submerged body: (a) submerged body, (b) liquid above body, (c) liquid below body

The magnitude of the net vertical force acting on the body, $V_B$, must equal the difference between $V_L$ and $V_U$, i.e.

$$V_B = V_L - V_U = \rho_F(\mathcal{V}_L - \mathcal{V}_U)g \tag{5.7}$$

As is evident from the figure, the difference $(\mathcal{V}_L - \mathcal{V}_U) = \mathcal{V}_S$, the volume of the submerged object, so that finally

$$V_B = \rho_F \mathcal{V}_S g. \tag{5.8}$$

The subscript $B$ has been introduced as a reminder that $V_B$ is usually called the **buoyancy force**. Equation (5.8) is the mathematical representation of Archimedes' principle, which was stated in words at the start of this section.

Throughout much of this chapter we have used the word 'liquid' rather than 'fluid' because hydrostatic problems usually concern liquids rather than gases. In practice, any result in this chapter which does not involve the depth below a free surface applies to a surface or body immersed in any fluid of uniform density. Equation (5.8), for example, can be used to calculate the buoyancy force exerted by the surrounding atmosphere on a lighter-than-air balloon or airship.

Since $\mathcal{V}_L > \mathcal{V}_U$, it should be clear that the buoyancy force $V_B$ always acts vertically upwards. This conclusion can also be seen to result from the increase in hydrostatic pressure with depth, equation (5.2) (see Section 4.2), since the forces acting on the surface of a body submerged in a fluid at rest are a consequence of the pressure distributed over it. It should also be apparent that the line of action of $V_B$ must pass through the centroid of the displaced volume $\mathcal{V}_S$. If the submerged body is of uniform density $\rho_S$, the location of its centre of gravity corresponds with its centroid, but the two will not in general correspond if there is a density variation within the interior of the submerged body, for example, as would be the case for a submarine. If a body is submerged only partially rather than completely, Archimedes' principle is still valid but the volume $\mathcal{V}_S$ must be replaced by that part of the volume of the body which is below the surface of the liquid (i.e. the volume of liquid displaced, $\mathcal{V}_D = m/\rho_F$, $m$ being the mass of the body). The centroid of the submerged volume is termed the **centre of buoyancy**. Whether or not a body floats or sinks in a fluid is determined by the average density of the body, $\rho_S = m/\mathcal{V}_S$. If $\rho_S > \rho_F$, the body will become fully submerged and sink unless constrained. If $\rho_S < \rho_F$, the body will float with part of its volume, $\mathcal{V}_D$, below the surface and the rest above.

---

**ILLUSTRATIVE EXAMPLE 5.4**

A balloon in the form of a thin rigid sphere of diameter $D = 1$ m is filled with helium of density $\rho_{He} = 0.17$ kg/m$^3$. Calculate the force required to prevent the balloon from rising if the surrounding air density at ground level $\rho_{AIR} = 1.2$ kg/m$^3$, and the balloon material has negligible mass. What payload could the balloon lift to an altitude of 8500 km where the air density is 0.5 kg/m$^3$?

**Solution**

The volume of the balloon $\mathcal{V}_S$ is given by $\pi D^3/6 = 0.52$ m$^3$. The corresponding buoyancy force at ground level $V_B$ is thus

$$V_B = \rho_{AIR}\mathcal{V}_S g = 1.2 \times 0.52 \times 9.81 = 6.12\,\text{N}.$$

The weight of the filled balloon $W$ is

$$W = \rho_{He}\mathcal{V}_S g = 0.17 \times 0.52 \times 9.81 = 0.87\,\text{N}.$$

From Figure E5.4 we can see that static equilibrium at ground level requires

$$V_B - W - F = 0$$

so that the force $F$ required to prevent the balloon from rising is

$$F = V_B - W = 5.25\,\text{N}.$$

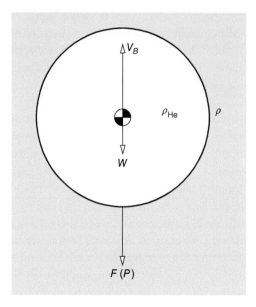

**Figure E5.4**

If the balloon is released with a payload $P < F$, it will rise to an altitude where static equilibrium requires

$$V_B - W - P = 0.$$

Since the balloon is rigid, both its volume $\mathcal{V}_S$ and its weight $W$ remain the same as at ground level. However, the buoyancy force $V_B$ decreases with altitude in direct proportion to the ambient density. For $\rho_{AIR} = 0.5\,\text{kg/m}^3$, we have

$$V_B = \rho_{AIR}\mathcal{V}_S g = 0.5 \times 0.52 \times 9.81 = 2.55\,\text{N}$$

and the payload for static equilibrium is given by

$$P = V_B - W = 2.55 - 0.87 = 1.68\,\text{N}.$$

---

    Although the pressure does not appear in equation (5.8), it is worth reminding ourselves that the buoyancy force is a direct consequence of the decrease in pressure with altitude (or increase with depth). What is quite remarkable is that, although in the atmosphere the pressure differences which arise from the vertical pressure gradient are usually very small, the resulting force can be quite substantial. For example, for a balloon of diameter 1 m the pressure difference $\Delta p$ from top to bottom is only 0.012% of 1 bar, i.e.

$$\Delta p = \rho_{AIR}Dg = 1.2 \times 1 \times 9.81 = 11.8\,\text{Pa}$$

but, as we found, the lift force at ground level is 6.12 N, which is roughly equal to the weight of a pint (just over half a litre) of beer. The term **lift force** was used quite deliberately: just as for a balloon, the lift force on an aerofoil arises because the average pressure acting on its lower surface is higher than that on its upper surface. As we shall show in Section 8.7 and discuss in more detail in Section 17.7, in the case of an aerofoil, the pressure difference is a consequence of its forward motion through the air, a much more complicated situation than the hydrostatic pressure difference due to gravity.

Before we leave the topic of balloons, it is interesting to calculate the force required to submerge the helium-filled balloon of Illustrative Example E5.4 in water. Because the density of water $\rho_{H_2O}$ is so much greater than that of air, the buoyancy force $V_B$ is also much greater, as we can see that

$$V_B = \rho_{H_2O} \mathcal{V}_S g = 10^3 \times 0.52 \times 9.81 = 5.1 \times 10^3 \text{ N}$$

and this is effectively the force which must be overcome in order to submerge the balloon since the weight (unchanged at 0.87 N) is obviously negligible compared with $V_B$. To put this result in perspective, the reader might like to compare this value with the weight of a small car, such as a BMW 320i, which is about $1.4 \times 10^4$ N.

## 5.4 Hydrostatic force acting on a submerged vertical flat plate

Figure 5.5 shows the front and side views of a flat plate of area $A$ submerged vertically in a liquid of density $\rho_F$. On the right-hand side of the figure is a graph showing the proportional increase in hydrostatic pressure $p_H$ with depth $z$, i.e. $p_H = \rho_F g z$. Since this pressure acts normal (i.e. perpendicular) to the plate at every point, the direction of the net hydrostatic force $H$ exerted by the liquid on the plate must also be horizontal.

To calculate the magnitude of $H$ we split the area $A$ into a series of infinitesimal horizontal strips, such as the strip of area $\delta A$ at depth $z$ shown in Figure 5.5. Since $p_H$ is constant along a horizontal line (see Section 4.2), the infinitesimal hydrostatic force $\delta H$ exerted on the strip is given by

$$\delta H = p_H \delta A = \rho_F g z \delta A$$

and the net force $H$ is obtained by summing all such elemental forces over the area $A$, i.e.

$$H = \sum_A \delta H = \sum_A \rho_F g z \delta A.$$

In the limit as $\delta A$ approaches zero, the summation is replaced by an integration so that, assuming $\rho_F$ and $g$ are constant with respect to the depth $z$, we have

$$H = \rho_F g \int_A z dA. \tag{5.9}$$

The integral $\int_A z dA$, called the **first moment of area**, turns out to be rather special because it is directly related to the location $z_C$ of the **centroid** of the area $A$ as follows

$$\int_A z dA = z_C A. \tag{5.10}$$

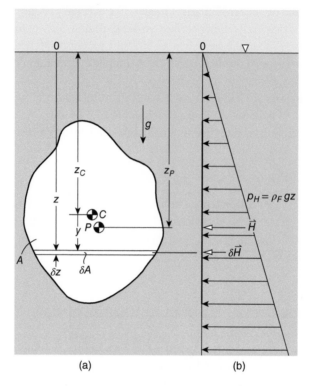

**Figure 5.5** Flat plate, submerged vertically

If the submerged plate has uniform thickness and density, it is easily shown that $z_C$ corresponds to the location of its **centre of gravity**. We can make use of the general equation for $z_C$ to write

$$H = \rho_F g z_C A = p_C A \tag{5.11}$$

where $p_C$ is the hydrostatic pressure at the centroid of $A$. For many shapes, the area $A$ and the location of the centroid $z_C$ are either well known (e.g. for a rectangle or a circle) or tabulated as in Appendix 3. More complicated shapes can often be treated as combinations of simpler ones, much like the kite shape of Illustrative Example 5.1. For shapes where this approach is not possible, the area $A$ and the integral $\int_A z\,dA$ have to be evaluated from first principles, as we show in Illustrative Example 5.5.

Although equation (5.11) shows that the hydrostatic force $H$ acting on the area $A$ is given by the product of the pressure $p_C$ at the centroid of $A$ and the area itself, as we shall now show, the line of action of $H$ is always below the centroid. The line of action of $H$ would be at a depth $z_C$ if the hydrostatic pressure $p$ acting on the plate was uniform but, as we know from the hydrostatic equation (5.2), $p$ increases linearly with depth so that the average pressure above $C$ is less than that below $C$.

To find the line of action of $H$ means finding the depth $z_P$ below the surface at which $H$ acts on the plate, and the horizontal distance $x_P$ from a vertical reference line in the plane of the plate. We start with $z_P$ by taking moments about a line in the liquid surface parallel to the plate.

For the elemental force $\delta H$ the moment is

$$\delta H\, z = p_H \delta A\, z = \rho_F g z^2 \delta A$$

and the net moment of all such elements is given by

$$\int_A \rho_F g z^2\, dA = H z_P = \rho_F g z_C A z_P \tag{5.12}$$

wherein we have equated the net moment due to all the elemental forces $\delta H$ to the moment of the net force $H$ acting at the depth $z_P$. If we cancel out the common factor $\rho_F g$, we find

$$z_P = \frac{\int_A z^2\, dA}{z_C A}. \tag{5.13}$$

Since the centroid is at depth $z_C$, we can write

$$z = z_C + y$$

where, as shown in Figure 5.5, $y$ is the vertical depth of the elemental strip below the centroid. The integral can now be written as follows

$$\int_A z^2\, dA = \int_A (z_C^2 + 2 z_C y + y^2)\, dA$$

$$= z_C^2 A + 2 z_C \int_A y\, dA + \int_A y^2\, dA.$$

The integral $\int_A y\, dA$ can be shown to be identically zero as follows

$$\int_A z\, dA = \int_A (z_C + y)\, dA = z_C A + \int_A y\, dA$$

but, as we saw earlier, $z_C$ is defined by the equation $\int_A z\, dA = z_C A$, so that $\int_A y\, dA = 0$, and we are left with

$$\int_A z^2\, dA = z_C^2 A + \int_A y^2\, dA. \tag{5.14}$$

As was the case for $\int_A y\, dA$, the integral $\int_A y^2\, dA$ is again special and is encountered in many problems of mechanical, aeronautical, structural, etc., engineering, for example in the stress analysis of beams, and the dynamics of rotating objects. It is given the symbol $I_C$ and called the second[36] moment of the area $A$ and is closely related to the moment of inertia about an axis in the plane of the area and passing through its centroid

$$I_C = \int_A y^2\, dA. \tag{5.15}$$

Appendix 3 includes the areas, centroid locations, and second moments of area for a number of basic shapes.

---

[36] It is called the second moment because the integral involves $z^2$.

As a final step, after substituting from equation (5.14), we can write equation (5.13) as

$$z_P = \frac{\int_A z^2 \, dA}{z_C A} = z_C + \frac{\int_A y^2 \, dA}{z_C A} = z_C + \frac{I_C}{z_C A}. \tag{5.16}$$

Equation (5.16) confirms what we argued earlier: P is always below C because $y^2$, and so $I_C$, is always positive.

We now turn our attention to calculating $x_P$, which, as we shall see, is a little more difficult than finding $z_P$. It might seem that we could proceed in a similar way as for $z_P$ by starting with a vertical strip of infinitesimal width $\delta x$. The difficulty that arises immediately is that the hydrostatic pressure $p_H$ is not constant along the strip, as it was for the horizontal strip, but varies with depth z. Instead of an elemental strip, therefore, we consider an elemental rectangle of width $\delta x$ and depth $\delta z$ at a location $(x, z)$ on the plate surface. The hydrostatic force on this element is $p_H \delta x \, \delta z$, and the moment about a vertical line in the plane of the plate is $p_H x \delta x \, \delta z$. The net moment is then

$$\iint_A p_H x \, dx \, dz = H x_P$$

where the double integral sign indicates that we must integrate over the surface area $A$ in both the x- and z-directions. We now substitute $\rho_F g z$ for $p_H$ so that

$$H x_P = \rho_F g \iint_A xz \, dx \, dz.$$

From equation (5.11) we have $H = \rho_F g z_C A$ so that

$$x_P = \frac{\iint_A xz \, dx \, dz}{z_C A}.$$

Once again we substitute $z = z_C + y$ so that

$$x_P = \frac{\iint_A x \, dx \, dy}{A} + \frac{\iint_A xy \, dx \, dy}{z_C A}. \tag{5.17}$$

If we specify that the vertical line, from which x is measured, passes through the centroid of the plate, then $\iint_A xy \, dx \, dy$ can be identified as the **product of inertia** of the plate, another tabulated quantity for a range of 'standard' shapes. The symbol $I_{xy}$ is often used for this quantity, i.e.

$$I_{xy} = \iint_A xy \, dx \, dy. \tag{5.18}$$

We also recognise that the first term on the right-hand side of equation (5.17) is identically zero because x is measured from the centroid, i.e.

$$\iint_A x \, dx \, dy = 0 \tag{5.19}$$

so that

$$x_P = \frac{I_{xy}}{z_C A}. \tag{5.20}$$

For convenience, we bring together the key results of this subsection

$$H = \rho_F g z_C A = p_C A, \tag{5.11}$$

$$z_P = z_C + \frac{I_C}{z_C A}, \tag{5.16}$$

and

$$x_P = \frac{I_{xy}}{z_C A}. \tag{5.20}$$

As a final point here, we note that if the plate is symmetrical about a vertical line through the centroid, then $I_{xy} \equiv 0$ and so $x_P = 0$, i.e. the line of action of $H$ is directly below the centroid.

---

### ILLUSTRATIVE EXAMPLE 5.5

A circular disc of radius $R$ is immersed vertically in a liquid of density $\rho_F$, with its centre a depth $Z$ below the surface. Calculate the hydrostatic force $H$ which the liquid exerts on one face of the disc, and the depth $z_P$ at which it acts.

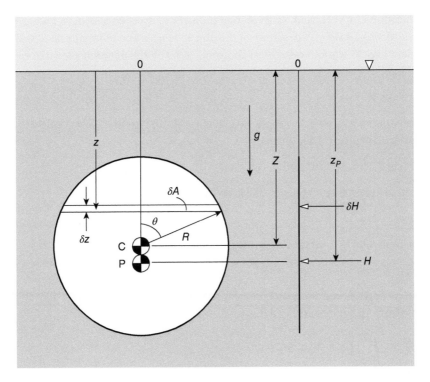

**Figure E5.5**

### Solution

We shall solve the problem in two ways. The first illustrates how we can use the information in Appendix 3.

The surface area of the disc is given by $A = \pi R^2$, and the centroid is coincident with the centre so that the depth of the centroid $z_C = Z$. The hydrostatic pressure at the centroid $p_C$ is thus given by $p_C = \rho_F g Z$, and the hydrostatic force $H$ exerted by the liquid on the disc is given by

$$H = p_C A = \rho_F g Z \, \pi R^2.$$

From equation (5.16), the depth $z_P$ at which $H$ acts is given by

$$z_P = Z + \frac{I_C}{\pi R^2 Z}.$$

From Appendix 3 the second moment of area about a horizontal axis in the plane of the disc and passing through the centroid is

$$I_C = \frac{\pi R^4}{4}$$

so that

$$z_P = Z + \frac{R^2}{4Z}.$$

For our second approach to this problem, we suppose that we do not know the area of the disc, the location of its centroid, or its second moment of area. As shown in Figure E5.5, we identify a horizontal strip on the surface of the disc at depth $z$ and of infinitesimal width $\delta z$. It is convenient here to work in cylindrical coordinates, using the notation shown in the figure, so that we have

$$z = Z - R \cos \theta.$$

If we differentiate with respect to $\theta$, we have $dz/d\theta = R \sin \theta$ so that $\delta z = R \sin \theta \, \delta \theta$, and the area of the elemental strip $\delta A$ is given by

$$\delta A = 2R \sin \theta \, \delta z = 2R^2 \sin^2 \theta \, \delta \theta.$$

The hydrostatic pressure $p_H$ at depth $z$ is given by

$$p_H = \rho_F g z = \rho_F g (Z - R \cos \theta)$$

so that the elemental hydrostatic force acting on the strip $\delta H$ is

$$\delta H = \rho_F g (Z - R \cos \theta) 2R^2 \sin^2 \theta \, \delta \theta.$$

We note that $\rho_F, g, Z$, and $R$ are all constant, and the only variable quantity is $\theta$ so that the hydrostatic force $H$ can be calculated from

$$H = 2\rho_F g R^2 \int_0^\pi (Z - R \cos \theta) \sin^2 \theta \, d\theta$$

$$= 2\rho_F g R^2 \left( Z \int_0^\pi \sin^2 \theta \, d\theta - R \int_0^\pi \cos \theta \sin^2 \theta \, d\theta \right).$$

The second integral is identically zero while the first has the value $\pi/2$, both results being obtainable from tables of standard integrals, so that

$$H = \rho_F g Z \pi R^2,$$

which is the same result as before but required a lot more effort to obtain.

To find the depth $z_P$ at which $H$ acts from first principles, we take moments about a line in the surface through O, which is parallel to the plane of the disc and vertically above C

$$H z_P = \int_{z=Z-R}^{z=Z+R} z \, dH$$
$$= 2\rho_F g R^2 \int_0^\pi (Z - R \cos\theta)^2 \sin^2\theta \, d\theta.$$

Since $H = \rho_F g Z \pi R^2$, we can rewrite the last equation as follows

$$\frac{\pi z_P}{2Z} = \int_0^\pi \left(1 - \frac{R \cos\theta}{Z}\right)^2 \sin^2\theta \, d\theta$$
$$= \frac{1}{2}\pi + \frac{1}{8}\pi \left(\frac{R}{Z}\right)^2$$

or

$$z_P = Z + \frac{R^2}{4Z},$$

once again the same result as before.

---

## 5.5 Hydrostatic force acting on a submerged curved surface

In this section we show how the results obtained so far can be used to calculate the magnitude, direction, and line of action of the horizontal and vertical components, $H$ and $V$, of the resultant hydrostatic force $R$ acting on a submerged surface of any specified shape. To illustrate the general approach, we use the example of a dam with the cross section[37] shown in Figure 5.6. The cross section is taken to be symmetrical and two dimensional, i.e. there is no curvature in any horizontal plane, so that the lines of action of $H$ and $V$ must both lie in the vertical plane of symmetry. The span of the dam is $S$, and the particular shape shown (i.e. the curve representing the surface in contact with water) is given by

$$y = \frac{Cx^2}{D} \tag{5.21}$$

where $C$ is a constant, $y$ is the upward vertical distance from the foot of the dam O, $x$ is the corresponding horizontal distance, and $D$ represents the vertical height of the dam. In the analysis below it is convenient to introduce $z$, the depth below the water surface, i.e. $z = Z - y$, where $Z$ is the total water depth. Also shown in Figure 5.6 is the line of action of $H$ at depth $z_P$, and the line of action of $V$ a horizontal distance $x_C$ from O.

---

[37] The cross section shown is similar to that of the **Hoover dam** on the border of the states of Nevada and Arizona, USA. Water from the Colorado River flows into Lake Mead on the upstream (Arizona) side of the dam. The water depth is about 180 m, and the installed Francis hydraulic turbines generate up to about 2000 MW of electrical power.

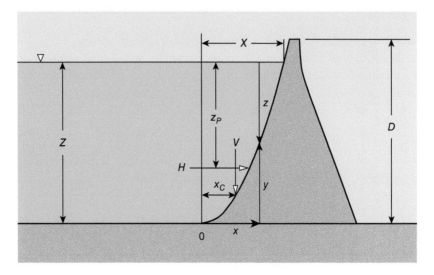

**Figure 5.6** Components of hydrostatic force acting on the face of a dam

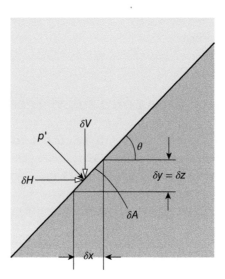

**Figure 5.7** Components of hydrostatic force on an elemental strip of the dam surface

### 5.5.1 Horizontal component of R

We consider an elemental horizontal strip of the surface of the dam a depth $z$ below the water surface, as shown in Figure 5.7. The hydrostatic pressure $p_H$ acting on the strip is given by $p_H = \rho_F g z$.

The strip is assumed to be so narrow that it may be considered to be flat. The elemental horizontal force $\delta H$ acting on the strip is given by

$$\delta H = p_H \delta A \, \sin \theta$$

where $\delta A \sin\theta$ is the area of the strip projected onto a vertical plane parallel to any horizontal line in the face of the dam, and $\theta$ is the angle between a tangent to the strip and the horizontal. We can see from Figure 5.7 that $\delta A \sin\theta = \delta y\, S$ so that

$$\delta H = \rho_F g z S \delta y$$

and the horizontal component of $R$ is thus given by

$$H = \rho_F g S \int_0^Z z\, dy \qquad (5.22)$$

$$= \rho_F g S \int_0^Z (Z - y)\, dy,$$

and, with $y(x)$ given by equation (5.21), we have

$$H = \tfrac{1}{2}\rho_F g S Z^2.$$

The final result reveals that, so far as the horizontal component of $R$ is concerned, the cross-sectional shape of the dam is irrelevant, and all that matters is the width $S$ and water depth $Z$: the magnitude of $H$ is identical to the hydrostatic force exerted on a submerged vertical rectangle of area $SZ$. This area represents the area of the shape obtained by projecting the face of the dam onto a vertical plane. What we have is a general result: the magnitude of the horizontal component of the hydrostatic force acting on a curved surface submerged in a fluid of uniform density is equal to the hydrostatic force exerted on the projection of that surface onto a vertical plane.

If the shape of the dam had not been symmetrical, it would have been necessary to resolve $H$ into two orthogonal components in a horizontal plane, $F_1$ and $F_2$, say, and then combine them using

$$H = \sqrt{F_1^2 + F_2^2}. \qquad (5.23)$$

### 5.5.2 Line of action of H

We refer again to the example of the dam in Figure 5.6. As always, we calculate the depth of the line of action of $H$ by taking moments about any convenient line. In this case we select a horizontal line through the point O at the foot of the dam and parallel to the face of the dam. The moment of the elemental force $\delta H$ about O is given by $\delta H\, y$, with $\delta H = \rho_F g z S \delta y$. The net moment of all such elemental forces is then $\int_{y=0}^{y=Z} y\, dH$ which must equal the moment of $H$ itself. If $z_P$ is the depth of the line of action of $H$, we have

$$H(Z - z_P) = \int_{y=0}^{y=Z} y\, dH \qquad (5.24)$$

$$= \rho_F g S \int_0^Z z y\, dy$$

$$\rho_F g S \int_0^Z y(Z - y)\, dy = \tfrac{1}{6}\rho_F g S Z^3.$$

From Subsection 5.5.1, we have $H = \rho_F g S Z^2/2$ so that $z_P = 2Z/3$.

We see that only the water depth $Z$ is important and the cross-sectional shape of the dam is of no consequence.

### 5.5.3  Vertical component of $R$

In Section 5.2 we showed that the magnitude of the vertical component $V$ of the hydrostatic force exerted on a submerged surface is equal to the weight of the fluid which occupies the volume $\mathcal{V}$ directly above the surface, i.e.

$$V = \rho_F g \mathcal{V}. \tag{5.4}$$

To determine $V$, therefore, we need to know $\mathcal{V}$, either from tables or by calculation from first principles. In this instance we adopt the latter approach. We consider a vertical slice of the fluid directly above the elemental strip of the curved surface of the dam, with thickness $\delta x$ (see Figure 5.7), depth $z$, and length (span) $S$. The volume of the elemental slice is given by

$$\delta \mathcal{V} = zS\delta x$$

so that the entire volume $\mathcal{V}$ is

$$\mathcal{V} = S \int_0^X z dx$$

where the symbol $X$ denotes the horizontal distance between the point O and the point in the cross section where the water surface meets the curved face of the dam (see Figure 5.6). We now have

$$V = \rho_F g S \int_0^X z dx \tag{5.25}$$

$$= \rho_F g S \int_0^X (Z - y) dx$$

and at this stage we need to connect $y$ and $x$ by introducing the shape of the curved surface (equation (5.21)), i.e.

$$y = \frac{Cx^2}{D}.$$

The final result is found to be

$$V = \frac{2}{3} \rho_F g S Z X$$

wherein we have also made use of the relationship from equation (5.21), $Z = CX^2/D$.

### 5.5.4  Line of action of $V$

As we did for $H$, we need to find the location of the line of action of $V$, i.e. the horizontal distance $x_C$ from O. Once again we take moments about a line through O as follows:

$$V x_C = \int_{x=0}^{x=X} x dV = \rho_F g S \int_0^X x (Z - y) dx \tag{5.26}$$

wherein we have substituted $(Z - y)S dx$ for $d\mathcal{V}$. The final result, after substituting $y = Cx^2/D$, $Z = CX^2/D$, and $V = 2\rho_F gsZX/3$, is

$$x_C = \frac{3}{8}x_C.$$

As we pointed out in Section 5.2, $x_C$ corresponds with the $x$-location (with respect to O) of the centroid of the volume of liquid $\mathcal{V}$ directly above the curved surface.

### 5.5.5 Resultant hydrostatic force $R$

The resultant hydrostatic force $R$ is calculated as the vector sum of $H$ and $V$, i.e.

$$R = \sqrt{(H^2 + V^2)} \tag{5.27}$$

and the angle between $R$ and the horizontal is given by $\theta = \tan^{-1}(V/H)$.

## 5.6 Stability of a fully-submerged body

To introduce the concept of stability, we consider first the behaviour of a simple pendulum which consists of a bob of weight $W$ at the end of a weightless rod, of length $l$, supported by a pivot at O, and free to swing in a vertical plane. The situation is shown in Figure 5.8. In position (a), where the centre of gravity G of the bob is vertically below O, the force $W$ is balanced by the reaction at O, the moment of $W$ about O is zero, and the pendulum is at rest in a state of static equilibrium. If given an angular displacement $\theta$ to position (b) and then released, the bob will move towards and oscillate about position (a). In practice, the oscillation is damped by friction at O and the resistance to motion due to the fluid surrounding the pendulum so that the pendulum eventually comes to rest in position (a). In the absence of damping, for small displacements $\theta$ a simple pendulum oscillates in a simple-harmonic motion with period $T = 2\pi\sqrt{l/g}$, where $g$ is the acceleration due to gravity. It can be seen from Figure 5.8(b) that the motion of the pendulum is driven by the moment $Wl\sin\theta$, which always acts to decrease $\theta$. An object which returns to a position of static equilibrium when displaced from that position is said to be in a state of stable equilibrium.

The position of the pendulum depicted in Figure 5.8(c) is also one of static equilibrium. However, this position is unstable because the moment $Wl\sin\theta$ is now such that the response of the pendulum to the slightest displacement, as shown in Figure 5.8(d), is for the displacement $\theta$ to increase. The pendulum eventually moves to a position of static and stable equilibrium, in this case again position (a).

The close analogy between the stability of a body freely floating completely submerged in a fluid is illustrated by Figure 5.9, in which G represents the centre of gravity of the body, B is the centre of buoyancy (see Section 5.3), $l$ is the distance between G and B, and $V_B$ is the buoyancy force. The condition for static equilibrium now is that G is directly below or directly above B. An angular displacement $\theta$ gives rise to a moment $V_B l\sin\theta$ and it is evident that position (b) is stable while position (d) is unstable. The analogy with the simple pendulum is not perfect because, when displaced, the body tends to roll about a horizontal axis close to G rather than a fixed pivot.

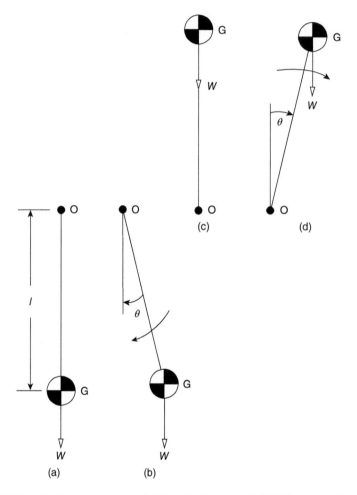

**Figure 5.8** Stability of a simple pendulum: (a) G vertically below O, (b) OG displaced by angle $\theta$, (c) G vertically above O, (d) OG displaced by angle $\theta$

## 5.7 **Stability of a freely floating body and metacentric height**

For a completely submerged body, the centre of buoyancy coincides with the centroid of the body (which is also its centre of gravity if the density of the body is uniform throughout) and this is always in the same place relative to the body. For a freely floating, partially submerged body, the buoyancy force $V_B$ must still equal the body's weight $W$ so the volume of displaced liquid remains constant (equal to $W/\rho_{Fg}$). However, as illustrated in Figure 5.10, as the position of the body changes, for example due to a rolling motion, so does the shape of the submerged volume. In consequence, the centroid of the submerged volume, which defines the centre of buoyancy, is not fixed but dependent upon the body position and this, in turn, has a critical influence on the stability of the floating body. For this reason, analysing the stability of a body which is floating partially submerged is more complicated than for one which is completely submerged.

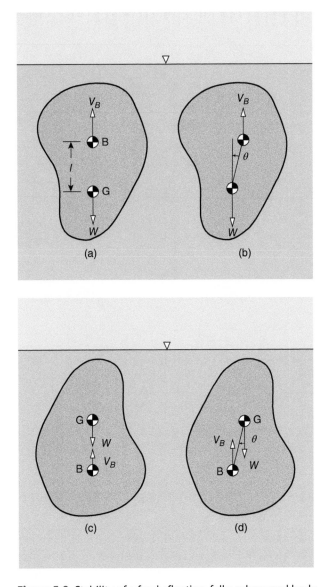

**Figure 5.9** Stability of a freely floating, fully-submerged body

The upper half of Figure 5.10 shows a body of uniform density $\rho_S$ and weight $W$ with its centre of gravity G below the centre of buoyancy B of the submerged volume. When displaced through an angle $\theta$ from the vertical, the magnitude of the submerged volume $\mathcal{V}$ must remain the same ($\mathcal{V} = W/\rho_F g$) but its centroid will move a horizontal distance $\bar{x}$ relative to G to a new location B'. In this case the couple $W\bar{x}$ exerted by $W$ and the buoyancy force $V_B$ (= $W$) acts to restore the body to its original position: situation (a) is therefore stable. From the diagram, we can see that $\bar{x} = \text{MG} \sin\theta$, MG being the distance between G and the point M where the line of action of $V_B$ intersects the extended line through BG. The point M is called the **metacentre**,

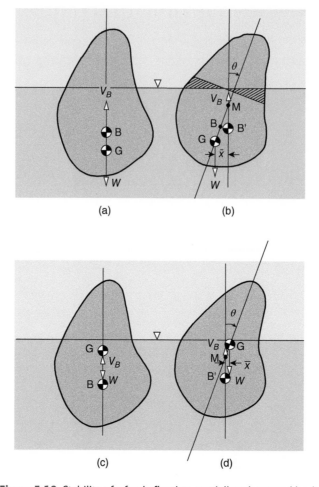

**Figure 5.10** Stability of a freely floating, partially submerged body

and the length MG the **metacentric height**. If M is above G, MG is taken as positive. For small angular displacements, MG is independent of $\theta$, i.e. the location of M is fixed.

The lower half of Figure 5.10 shows a body of the same shape, weight, and orientation as that in Figures 5(a) and 5(b). The centre of buoyancy of the submerged volume B must be in the same location as before but we now consider the situation where, due to a different internal distribution of mass within the body $\rho_S$ is no longer uniform, and G is now above B. In the case of a ship, for example, the location of the centre of gravity depends upon the way in which the cargo is distributed. As can be seen in Figure 5.10(d), the couple $W\bar{x}$ now acts to increase the angular displacement $\theta$, and situation (c) is unstable. Because the metacentre M is now below G, it is considered negative: the magnitude and sign of MG play a critical role in determining the stability of a floating body. If MG > 0, the body floats in stable equilibrium, and the larger MG the more stable the situation; the opposite applies if MG < 0.

The foregoing explains the general principles regarding the stability of a floating object. We now consider the particular case of a ship with the symmetrical cross section shown in Figure 5.11(a) (it is assumed that the sides of the ship are parallel). The length of the hull

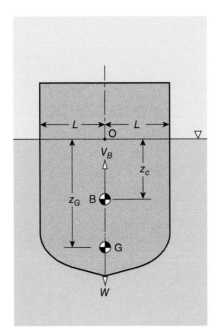
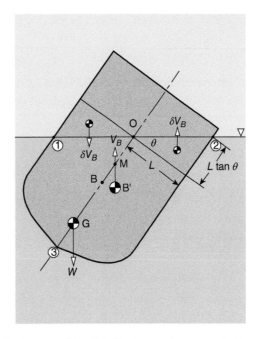

**Figure 5.11** Stability of a ship: (a) G vertically below B, (b) hull inclined at angle $\theta$ to vertical

is $S$, and the width at the water line is $2L$, while the depths below the surface of the centres of buoyancy B and gravity G are $z_C$ and $z_G$, respectively. The point O lies on the line of symmetry at the level of the water line. Figure 11(b) shows the situation if the ship is given an angular displacement $\theta$[38]. The centre of buoyancy B′ for the ship in the displaced position is obtained by calculating the centroid of the submerged volume defined by the shape ①②③①. The metacentre M is defined by the intersection of the vertical through B′ and the line of symmetry OG. The centroid of the shaded triangular volume to the right of O is a distance $2L/3$ from O. The buoyancy force $\delta V_B$ corresponding to this volume can be taken as $\delta V_B = SL^2 \tan\theta \rho_F g/2 \approx \rho_F g SL^2\theta/2$, where we have used the approximation $\sin\theta = \theta$ since $\theta$ is a small angle (note that $\theta$ must be measured in radians for this approximation to be valid). Due to the symmetry about OG, the reduction in the buoyancy force on the left-hand side is also $\delta V_B = \rho_F g SL^2\theta/2$.

To determine the location of B′ (the centroid of the volume below the water line) we take moments about O, again approximating $\sin\theta$ by $\theta$:

$$V_B OM\theta = V_B OB\theta - 2\delta V_B \frac{2}{3}L.$$

We can now substitute $V_B = W$ and, from Figure 5.11, OB = $z_C$, and OM = $z_G$ − MG, so that

$$W(z_G - \text{MG})\theta = Wz_C\theta - \frac{2}{3}\rho_F g SL^3\theta$$

---

[38] For a stability analysis we need consider only a small displacement. The angle $\theta$ shown in Figure 5.11(b) is greatly exaggerated.

which, after cancellation of $\theta$, can be rearranged to give

$$MG = \frac{2\rho_F L^3 Sg}{3W} - (z_C - z_G). \tag{5.28}$$

We have established already that a body is in stable equilibrium if $MG > 0$, which will always be the case if G is below B (i.e. $z_G > z_C$), but even if G is above B (i.e. $z_G < z_C$) stability is still seen to be possible if

$$z_C - z_G < \frac{2\rho_F L^3 Sg}{3W}. \tag{5.29}$$

---

### ILLUSTRATIVE EXAMPLE 5.6

As shown in Figure E5.6, a solid rectangular bar of uniform relative density $\sigma_S$ (<1) has height $Z$, width $2L$, and length $S$. Show that the bar floats in water with its centre of gravity a distance $z_G$ below the surface given by

$$z_G = \left( \sigma_S - \frac{1}{2} \right) Z.$$

Show that the metacentre M is a distance MG above the centre of gravity G given by

$$MG = \frac{L^2}{3\sigma_S Z} - \frac{1}{2} (1 - \sigma_S) Z.$$

For given values of $L$ and $Z$, what is the minimum value of $\sigma$ for stability?

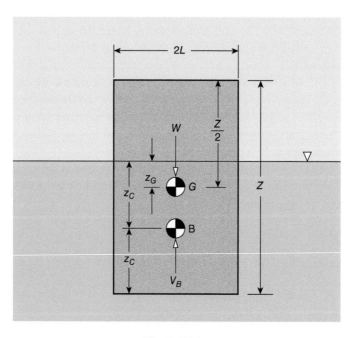

**Figure E5.6**

## Solution

The weight of the bar $W$ is given by

$$W = \rho_S 2LZSg$$

and the buoyancy force $V_B$ is

$$V_B = \rho_F 2L2z_C Sg$$

where the density of the bar $\rho_S = \sigma_S \rho_F$, and $\rho_F$ is the water density.
The condition for static equilibrium must be satisfied if the bar is floating freely, i.e.

$$W - V_B = 0 \quad \text{or} \quad \sigma_S \rho_F 2LZSg = \rho_F 4Lz_C Sg$$

from which $z_C = \sigma_S Z/2$. We can see from the geometry of Figure E5.6 that $Z/2 + z_G = 2z_C$ so that we get the following result after substituting for $z_C$:

$$z_G = \left( \sigma_S - \frac{1}{2} \right) Z.$$

From the ship example, we have

$$MG = \frac{2\rho_F L^3 Sg}{3W} - (z_C - z_G). \tag{5.28}$$

If we substitute for $W, z_C$, and $z_G$ we have

$$MG = \frac{L^2}{3\sigma_S Z} - \frac{1}{2}(1 - \sigma_S) Z.$$

Again from the ship example, we know that stability to an angular displacement requires $MG > 0$ so that from the equation for MG we have

$$\frac{L^2}{3\sigma_S Z} - \frac{1}{2}(1 - \sigma_S) Z > 0$$

which can be rearranged as

$$\sqrt{\frac{3}{2}\sigma_S (1 - \sigma_S)} < \frac{L}{Z}.$$

## Comment:

The final result shows that a bar of square cross section (i.e. $L/Z = 1/2$) will float upright if

$$\sqrt{\frac{3}{2}\sigma_S (1 - \sigma_S)} < \frac{1}{2}$$

or

$$6\sigma_S (1 - \sigma_S) < 1$$

which leads to $\sigma_S > 0.79$ or $\sigma_S < 0.21$, i.e. a square Styrofoam bar ($\sigma_S < 0.2$) will float upright but a square bar of wood ($\sigma_S \approx 0.6$) will not. The reader should think about the angular orientation adopted for stability by a square bar with $0.79 > \sigma_S > 0.21$.

More generally, a rectangular bar will be stable for any value of $\sigma_S$ (in the range 0 to 1) if

$$\frac{2L}{Z} > \sqrt{\frac{3}{2}} = 1.2250$$

i.e. a bar which has a rectangular cross section wider than it is deep is more stable than one which has a greater depth than width.

---

 ## 5.8  SUMMARY

In this chapter we have shown how to calculate the force which arises due to the hydrostatic pressure distributed over a surface or an object submerged in a fluid. For convenience we resolved the net force exerted on a surface into a vertical and a horizontal component. The vertical component was shown to be equal in magnitude to the weight of fluid which would occupy the volume directly above the surface and to act vertically downwards through the centroid of this volume. The buoyancy force exerted on a submerged or floating object was shown to equal the weight of the fluid displaced by the object and to act vertically upwards through the centroid of the displaced fluid. The relative positions of the centroid and the centre of gravity of the object were shown to determine the position of its metacentre and hence its stability.

We showed that, for a flat surface immersed vertically in a fluid, the magnitude of the net hydrostatic force is equal to the product of the area of the surface and the pressure at its centroid. Because the hydrostatic pressure increases with depth, the line of action of this force always lies below the centroid. For a curved surface, the magnitude of the horizontal component of the hydrostatic force was shown to equal the hydrostatic force on the projection of the curved surface onto a vertical plane.

The student should be able to

- calculate, both from first principles and from tabulated information for the properties of standard shapes, the magnitude and location of the line of action, and also specify the direction, of:
  - the vertical component of the hydrostatic force exerted on a submerged surface
  - the horizontal component of the hydrostatic force exerted on a submerged surface
  - the buoyancy force exerted on a submerged or floating object

The student should also be able to

- analyse the stability of floating objects

 ## 5.9  SELF-ASSESSMENT PROBLEMS

**5.1**  The centroid of a vertical surface of area 0.5 m$^2$ completely submerged in an oil of relative density 0.85 is 5 m below the surface of the oil. Calculate the hydrostatic force acting on the surface. Explain why this force always acts some distance below the centroid.
(Answer: 20.8 kN)

**5.2** An aperture in the vertical wall of a water tank is closed by a circular plate 600 mm in diameter. The plate is held in position by four stops, one at each end of the horizontal diameter, and one at each lower end of the two diameters at 60° to the horizontal. Determine the stop reactions when the water surface is 450 mm above the plate centre.
(Answers: 504.0 N, 120.1 N)

**5.3** (a) The spread of an oil slick of depth $Z$ and density $\rho_O$ is to be stopped by a floating boom, as shown in Figure P5.3. The boom, which is designed to float upright as shown in the figure, has a square cross section of side $t$ and weight per unit length $w$. Show that the maximum slick depth $Z_{MAX}$ which can be contained by the boom is given by

$$Z_{MAX} = \left[ \frac{1 - w/\rho_s g t^2}{1 - \rho_O/\rho_s} \right] t$$

where $\rho_s$ is the density of the sea water beneath the slick and $g$ is the acceleration due to gravity.

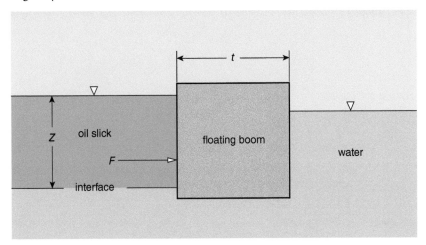

**Figure P5.3**

(b) Calculate the horizontal force per unit length acting on the water side of the boom if $t$ is 500 mm, $w$ is 2300 N/m, and the relative density of sea water is 1.025. Also calculate the depth below the top surface of the boom at which the horizontal force acts.

Would the horizontal force on the oil side of the boom be greater or smaller than that on the water side? Give a brief explanation for your answer.
(Answers: 1052 N/m, 0.348 m)

**5.4** (a) A square plate of side length $L$ is submerged in water at an angle $\theta$ to the vertical with its centroid a depth $Z$ below the surface and two sides parallel to the surface. Show that the net hydrostatic force on one face of the plate acts at a depth $Z_P$ given by

$$Z_P = \frac{z_2^3 - z_1^3}{3ZL \cos \theta}$$

where $z_2 = Z + L \cos \theta/2$, and $z_1 = Z - L \cos \theta/2$.

(b) The gate shown (side view) in Figure P5.4 is hinged at O, 3 m above the bed of a reservoir which contains water of depth 10 m. If the gate is a square of side length 5 m, calculate the force $\vec{R}$ applied vertically downwards at its centroid which is necessary to prevent the gate from opening. Neglect the weight of the gate, and any effects of leakage under or around the gate. There is no water on the right-hand side of the gate.

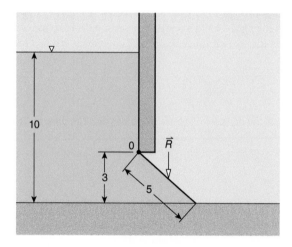

**Figure P5.4**

(Answer: 2.76 MN)

**5.5**    (a) A dam has the cross section shown in Figure P5.5 with $\tan\phi = 4$. Show that the resultant hydrostatic force $R$ acting on the dam is given by

$$R = \frac{\sqrt{13}}{6}\rho g Z^2 S$$

where $Z$ is the total water depth, as shown, $\rho$ is the water density, $g$ is the acceleration due to gravity, and $S$ is the width of the dam. Show also that the horizontal distance from O of the line of action of the vertical component of $R$ is given by $7Z/36$.

(b) Calculate the horizontal and vertical components of the hydrostatic force of a dam with the cross section shown in Figure P5.5 if the width is 600 m and the water depth is 30 m. What angle does the resultant force make with the vertical and at what depth does the horizontal component of the hydrostatic force act?

(Answers: 2.65 GN, 1.77 GN, 56.3°, 20 m)

**5.6**    (a) As shown in Figure P5.6, the water in a tank of depth $D$ and width $W$ is prevented from escaping by a gate of circular cross section hinged at O. If the radius of the gate is $D/2$, show that the net hydrostatic force $F$ acting on the gate surface is given by

$$F = \frac{1}{4}\rho g W D^2 \sqrt{\left(\frac{25}{4} - \pi + \frac{1}{16}\pi^2\right)}$$

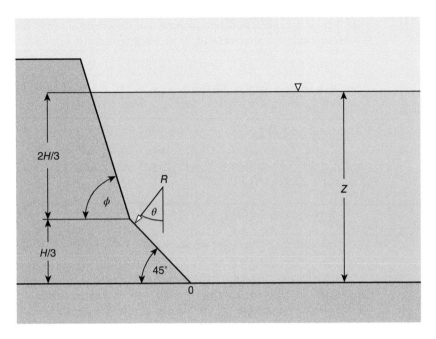

**Figure P5.5**

where $\rho$ is the water density and $g$ is the acceleration due to gravity. There is no water to the left of the gate. Show also that the vertical component of $F$ acts at a horizontal distance

$$X = \frac{D}{6\left(1 - \frac{1}{8}\pi\right)}$$

from the tank wall.

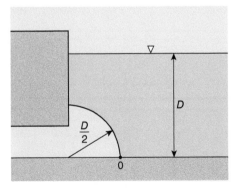

**Figure P5.6**

(b) If the depth $D$ is 10 m, and the width $W$ is 5 m, calculate the force which the gate exerts on the tank wall. The weight of the gate can be neglected.
(Answer: 1.29 MN)

**5.7**  (a) Due to increasing salt concentration and the presence of silt, the density $\rho$ of water in a reservoir increases with depth $z$ below the water surface according to

$$\rho = \rho_0 + Cz$$

where $\rho_0$ is the density at the surface ($z = 0$) and $C$ is a constant. Show from first principles that the hydrostatic pressure $p$ at depth $z$ is given by

$$p = \left( \rho_0 + \frac{1}{2}Cz \right) gz$$

and that the hydrostatic force $H$ exerted on a vertical rectangular wall of width $S$ due to water of depth $D$ is given by

$$H = \frac{1}{2}SD^2 g \left( \rho_0 + \frac{1}{3}CD \right).$$

Also calculate the depth below the water surface at which $H$ acts.

(b) If the reservoir depth is 50 m and the water density increases from 1000 kg/m$^3$ at the surface to 1100 kg/m$^3$ at the bottom, calculate the hydrostatic force on a horizontal circular plate of diameter 10 m at the bottom of the reservoir. Also calculate the horizontal component of the hydrostatic force exerted on a wall 10 m wide which is inclined at 60° to the horizontal.
(Answers: 40.4 MN, 126.7 MN)

**5.8**  (a) Figure P5.8 shows a flat plate which is immersed vertically in water to a depth $D$. The shape of the plate is given by

$$y = \frac{D^2 - z^2}{2D}$$

where $y$ is the half width of the plate a vertical distance $z$ below the surface. From first principles show the following:

 (i) The surface area of the plate $A = 2D^2/3$.
 (ii) The hydrostatic force acting on one side of the plate $H = 3\rho g A D/8$.
 (iii) The hydrostatic force acts at a distance $8D/15$ below the surface.

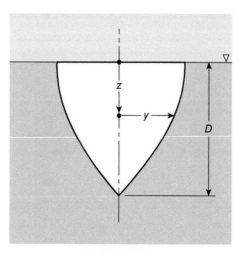

**Figure P5.8**

(b) A container of length 5 m and maximum width 2 m has the cross section shown in Figure P5.8. Calculate the maximum load which the container can carry (including its own weight) without sinking. Calculate the corresponding hydrostatic pressure acting at the centroid of one end of the container.
(Answers: 130.8 kN, 0.0736 bar)

5.9 (a) A rigid spherical balloon of diameter $D$ is filled with a light gas of density $r\rho_S$, where $\rho_S$ is the density of the atmospheric air at ground level. Show that the net upward force $F$ experienced by the balloon at ground level is given by

$$F = \frac{1}{6}\pi D^2 (1 - r) (p_L - p_H)$$

where $p$ is the atmospheric pressure and the subscripts $H$ and $L$ refer to the highest and lowest points on the balloon's surface, respectively. The weight and volume of the balloon's 'skin' may be neglected.

(b) A rigid balloon of diameter 3 m is filled with helium with a density of 0.2 kg/m$^3$. Calculate the value of $p_L - p_H$ at the altitude where the balloon just floats without rising or falling. Calculate the maximum mass the balloon could lift to an altitude at which the air density is 0.45 kg/m$^3$.
(Answers: 5.89 Pa, 3.53 kg)

5.10 (a) Figure P5.10 shows the cross section of a vertical rectangular barrier separating pure water of density $\rho$ on the right-hand side from a layer of pure water of depth $Z_1$ on the left-hand side above a layer of silt of depth $Z_2$. The silt may be treated as a liquid of density $\rho_2$. Determine the depth $Z$ of the water on the right-hand side if the net force on the barrier is to be zero. Is $Z$ greater or smaller than $Z_1 + Z_2$, and why?

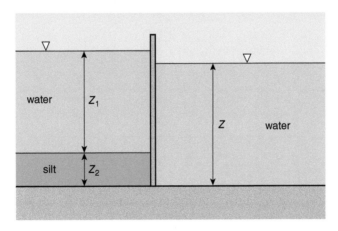

**Figure P5.10**

(b) If the depth of the water on the left-hand side is 1.5 m and that of the silt 0.5 m, calculate the total hydrostatic force exerted on that side of the barrier. Take the relative density of silt as 1.5 and the length of the wall as 2 m. Calculate the water depth on the right-hand side for zero net force on the barrier, and the net overall moment exerted on the barrier.
(Answers: 40.5 kN, 2.03 m, 1031 N · m anticlockwise)

**5.11**    Figure P5.11 shows the cross section of a yacht floating in sea water of density $\rho$. Excluding the keel, the weight of the yacht is $W$, and its centre of gravity a height $H$ above the water line. The submerged section of the hull is triangular in cross section, with apex angle $2\alpha$ and height $h$. The weight of the keel is $W_K$, and its centre of gravity is a depth $Z_K$ below the water line. Show that

$$h^2 = \frac{W + W_K}{\rho g L \tan \alpha}$$

where $L$ is the length of the yacht. The volume of the keel may be regarded as negligible. Show also that the metacentre is at a depth below the surface given by

$$\tfrac{1}{3} h \left( 1 - 2 \tan^2 \alpha \right).$$

Finally, show that for the yacht to float stably, the minimum weight of the keel is given by the equation

$$W_K Z_K - \frac{\left(W_K + W\right)^{3/2} \left(1 - 2 \tan^2 \alpha\right)}{3 \sqrt{\rho g L \tan \alpha}} - WH = 0.$$

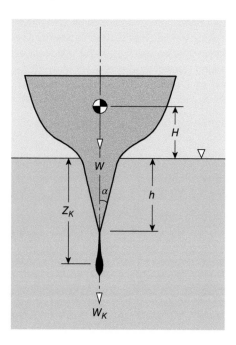

**Figure P5.11**

# 6 Kinematic description of fluids in motion and approximations

We start this brief chapter by introducing the concepts of fluid **particles**, **pathlines**, and **streamlines**, together with some of the other terms and ideas needed to describe fluid motion. We also point out some of the principal simplifications which can be made when analysing fluid flow to prevent the mathematics involved from becoming too demanding. With some minor exceptions, we restrict consideration to the **steady flow** of a **single-phase fluid**. As we pointed out in Section 2.11, all fluids are to some degree **compressible**, and the pressure variations which arise in high-speed gas flows are such that the consequential density variations have to be taken into account. So far as liquid flows are concerned, it is almost always adequate to treat them as **incompressible** (see Section 4.12). We introduce the concept of **one-dimensional internal flow**, whereby it is assumed that, over any cross section through which there is fluid flow, all flow and fluid properties are uniform. Application of the **principle of conservation of mass** is shown to result in a simple but important relationship between fluid density, flow velocity, and the cross-sectional area of the flow channel. The term **kinematic** in the chapter title indicates that at this stage we are concerned only with the description of flow velocity, not with the stresses and forces which cause fluid motion.

## 6.1 Fluid particles

In Section 2.5 we found that the average number of molecules contained in a cube of water of side length 0.1 $\mu$m (i.e. $10^{-7}$ m) is about thirty million while for a cube of air of the same volume, at a temperature of 20°C and pressure of 1 bar, the number is about thirty thousand. Such large numbers of molecules allows us to define average values for density, viscosity, and other fluid properties which are independent of the volume size (the **continuum hypothesis**). The flow of fluids through channels with submicron dimensions is becoming increasingly important (the study of such flows is termed **microfluidics**) but, in most situations of practical importance, 0.1 $\mu$m is several orders of magnitude smaller than any significant dimension of a flow channel. In normal circumstances, any changes in flow or fluid properties, such as pressure, velocity, density, or viscosity, would also be negligibly small from one side of the fluid volume to the other. The continuum hypothesis allows us to define a **fluid particle** as a tiny volume of fluid, which has fluid and flow properties independent of its size. A convenient way to think of a fluid particle is as a point-sized volume of fluid which has the temperature, pressure, velocity, etc., of its immediate surroundings. If we could mark and follow the movement of a number of fluid particles distributed throughout a flow, we would be able to form a visual impression of the flow. Although we cannot easily mark individual fluid particles in a real flow, there are a number of experimental **flow-visualisation** techniques which allow us to form such

*Introduction to Engineering Fluid Mechanics.* Marcel Escudier.
© Marcel Escudier 2017. Published 2017 by Oxford University Press.

an impression. For example, small amounts of dye or neutrally buoyant solid or liquid (for gas flows) particles can be introduced into a flow. In high-speed gas flows, the motion can be visualised using optical techniques (**interferometry, shadowgraphy,** and **Schlieren technique**) to detect the changes in refractive index which accompany density changes. The same techniques can also be used to visualise the flow of certain liquids for which the refractive index is sensitive to shearing of the liquid; the citrus oil **limonene** is an example.

## 6.2 Steady-flow assumption

With one or two exceptions, in this book we shall restrict consideration to **steady flows,** that is, to flows for which the velocity and pressure at any point in a flow do not change with time. In general there will be spatial variations in these quantities, often accompanied by changes in fluid properties, throughout the flowfield. Whenever a flow is created by the movement of an object, such as a car, a ship, or an aircraft, moving through an otherwise stationary fluid, it is possible to transform the resulting fluid motion into a steady flow relative to the moving object. This **Galilean transformation,** as it is called, in which the object is brought to rest and its velocity subtracted from that of the surroundings, is restricted to objects moving at constant velocity. For example, the airflow over the wings and fuselage of an aircraft would not appear to be steady when seen by an observer on the ground, but could be regarded as steady relative to the aircraft if it were flying at constant velocity.

## 6.3 Pathlines, streamlines, streamsurfaces, and streamtubes

The actual path followed by any fluid particle in a flow is called a **pathline**. A **streamline** is a line in a flow along which the flow direction at every point at any instant is tangential. In a steady flow, pathlines and streamlines are identical. An important consequence of the definition is that streamlines can never cross, since the flow at any point can have only one direction. A surface made up of streamlines is called a **streamsurface.** If the cross section of a streamsurface is a closed loop, the surface defines a **streamtube,** as shown in Figure 6.1. Since there can be no flow across a streamline, the same applies to a streamsurface, and a streamtube can therefore be thought of as representing the interior wall of a **duct** such as a **tube** or **pipe** through which there is flow.

Figure 6.2 shows the cross section of a stationary **aerofoil** with fluid flowing steadily over it from left to right. Each of the lines with arrowheads on them represents a streamline. Five of the streamlines shown pass over the upper (**suction**) **surface** of the aerofoil, and three over the lower (**pressure**) **surface.** The streamline which approaches the aerofoil and intercepts its surface at point P on its **leading edge** is said to be the **dividing streamline**. For a solid aerofoil, the velocity at P must be zero and such a point is called a **stagnation point**, the word stagnation meaning that the fluid concerned is at rest (i.e. it is stagnant). The location of P depends slightly upon the **angle of attack** $\alpha$ between the aerofoil and the approach flow. The surface pressure on the suction surface decreases with distance from P up to about one-third **chord** distance and then begins to increase. This increase in pressure opposes forward movement of the **boundary-layer fluid** (see below) and, once $\alpha$ exceeds a critical value (for the

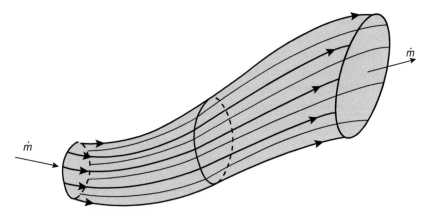

**Figure 6.1** Definition of a streamtube

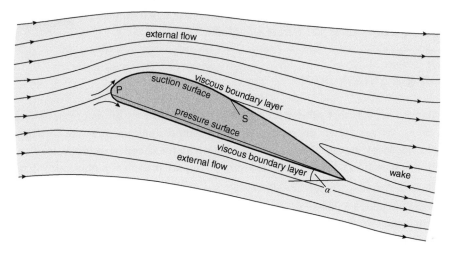

**Figure 6.2** Flow over a stationary aerofoil

particular aerofoil), can lead to **separation** of the boundary layer from the surface (at point S in Figure 6.2) and loss of lift (**stall**). A more detailed account of subsonic flow over an aerofoil and how lift is generated is given in Section 17.7

## 6.4 **No-slip condition and the boundary layer**

Although it is obvious that fluid cannot pass through a solid surface, and so the component of fluid velocity normal to any solid surface at rest must be zero, it is primarily a matter of experimental observation that the component of fluid velocity tangential to the surface is also zero. According to this **no-slip condition**, in the immediate vicinity of a solid surface a consequence of viscosity is that the fluid is brought to rest (or, more generally, if the surface is itself moving, to the same velocity as the surface so that the relative velocity is zero). In essence, the fluid adheres to the surface. In an external flow over a streamlined body such as an aerofoil, the change from zero to non-zero tangential velocity takes place across a thin layer of fluid called the viscous **boundary layer**. In an internal flow through a tube, as the flow develops from the inlet, velocity changes initially occur across a boundary layer but ultimately the entire cross section is influenced by viscosity and the flow becomes **fully developed** (i.e. unchanging with streamwise location). Chapter 16 is concerned primarily with the analysis of fully-developed internal flows, while boundary layers are the subject of Chapter 17. Further consideration is given to both types of flow in Chapter 18.

Since we stated in Chapter 2 that viscosity is the essential property that distinguishes a fluid from a solid, it may seem paradoxical that many flow problems can be analysed neglecting viscous effects entirely. This is the situation, for example, in an external flow beyond the near-wall viscous boundary layer. Such flows are said to be **inviscid** (the terms **frictionless** or **loss free** are also used) and the theory which has been developed to analyse them is termed **potential-flow theory**, a topic not included in this book. It is often the case that, in the absence of shockwaves, viscous effects can be neglected in both internal and external compressible-gas flows (see Chapter 11). An important consequence is that, in a frictionless flow, no mechanical energy is converted to heat (i.e. there is zero **dissipation**).

## 6.5 **Single-phase flow**

In Section 2.1 we saw that substances can exist in four different forms or phases (solid, liquid, vapour, or gas), often depending upon the temperature and pressure to which they are subjected. Many industrial processes, particularly in the chemical industry, involve flows, called **multiphase flows**, in which two or more phases are present simultaneously. Examples include

- liquids containing bubbles of vapour or gas, as would occur in boiling and **cavitation**, a phenomenon we explain in Section 8.11
- liquids containing droplets of another liquid with which it is immiscible, such as oil and water
- liquids containing solid particles, such as blood, the composition of which is about 54% **plasma**, an aqueous liquid, and 46% **blood cells** (**corpuscles**). Another example is lubricating oil contaminated with metal cuttings produced in machining operations.
- gases containing liquid droplets, such as the mixing of hot gas with atomised liquid fuel sprayed into a combustion chamber
- gases containing solid particles such as pollutants

Analysis of the flow of any liquid or gas where the second phase significantly alters the fluid properties is beyond the scope of this book, which is restricted to consideration of flow of **single-phase fluids**

## 6.6 **Isothermal, incompressible, and adiabatic flow**

A major simplification we can make in many flow situations is that the fluid properties which affect the flow (i.e. density, viscosity, and surface tension) are constant and uniform throughout the flowfield. Since all fluid properties depend to some extent on the fluid temperature, this essentially restricts consideration to constant-temperature (i.e. **isothermal**) flows. Flows for which density changes due to pressure variations are negligible are termed **incompressible**. Except for the analysis of **isothermal compressible flow** in Section 13.3 and **Rayleigh flow** in Section 13.4, we limit consideration to **adiabatic flows,** which means that there is no heat transfer to or from the fluid.

## 6.7 One-dimensional flow

There are many flows in which we can identify a main flow direction. The justification for this statement is evident from most of the figures in Chapter 1: the discharge from a centrifugal pump (Figure 1.2); flow through a convergent-divergent nozzle (Figure 1.7), a turbofan engine (Figure 1.8), a pipe bend (Figure 1.14), a rocket engine (Figure 1.15), a jet pump (Figure 1.16), and a cascade of guidevanes (Figure 1.17). All of these are examples of **internal flow**. The flow around a supersonic aerofoil, flow induced by a propeller, and the flow of water vapour emitted from a cooling tower, visible or invisible, as illustrated in Figure 6.3, are examples of **external flows** where the principal flow directions are readily identified.

Although most of the internal-flow examples in Chapter 1 are extremely complex when considered in detail, we can often make significant progress in their analysis by consideration of changes in the **spatial-average** (taken across a cross section) **conditions** between inlet and outlet while ignoring the interior details of the flow. There are many duct flows for which practically useful calculations can be made assuming that over any cross section of the duct the fluid velocity $V$, pressure $p$, and all fluid properties, such as density $\rho$ and viscosity $\mu$, are uniform and we account only for variations from location to location, i.e. $V, p, \rho, \mu$, etc., vary only with distance along the duct $s$. We shall make extensive use of this **one-dimensional approximation** in many of the following chapters. For the flow of a gas, in the absence of heat transfer, a decrease in pressure is inevitably accompanied by a decrease in density. As we have already indicated in Section 3.12, provided the **Mach number** does not exceed a value of about 0.3, the change in gas density is usually negligible (e.g. for air it is less than 5%) and the flow may be considered to be incompressible. For higher Mach numbers, compressibility effects, such as **shockwaves** and **choking**, where the Mach number reaches and is limited to unity, become important. Compressible-gas flows are the subject of Chapters 11 to 13, including internal one-dimensional flows with area change, frictionless pipe flow with wall heating (**Rayleigh flow**), and adiabatic pipe flow with wall friction (**Fanno flow**), and external flows with **shockwaves** and **expansion waves**. In Chapter 14 we analyse the compressibility effects which arise in the blading of gas compressors and gas turbines.

The fact that the flow of real fluids is affected by viscosity and the associated no-slip condition means that the uniform-velocity assumption is certainly invalid in the immediate vicinity of the inner surface of a duct and may well be of limited validity in interior regions of a flow. The uniform pressure assumption is quite different in character and is usually regarded as valid wherever streamline curvature is small but is inappropriate in situations where the streamlines are strongly curved.

As we shall show in Section 6.8, for steady flow of a constant-density fluid the streamwise-velocity variation $V(s)$ results directly from the shape of the streamtube through which flow

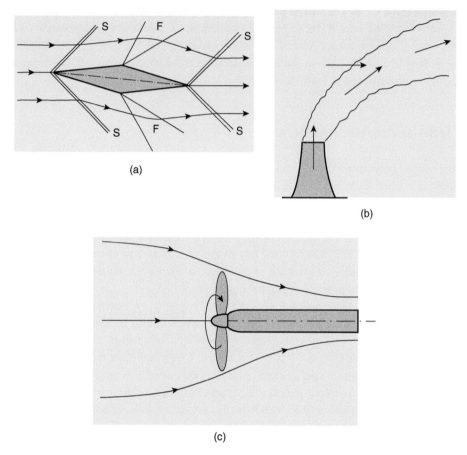

**Figure 6.3** Principal flow directions for various external flow situations: (a) double-wedge supersonic aerofoil with shockwaves (S) and expansion fans (F); (b) plume from chimney stack or cooling tower; (c) flow induced by a propeller

occurs. In Chapter 7 we shall show that these velocity changes are accompanied by pressure variations $p(s)$ and in Chapter 9 we shall derive a form of the **momentum equation** (essentially **Newton's second law of motion**) which will enable us to calculate the **hydrodynamic forces** which a moving fluid exerts on the surfaces with which it is in contact (Chapter 10).

External flows, which are usually more difficult to deal with than internal flows, will be discussed to a limited extent in Chapters 12, 16, 17, and 18.

## 6.8 One-dimensional continuity equation (mass-conservation equation)

We assume that the cross-sectional area $A$ of the streamtube shown in Figure 6.4 varies in some specified way with distance $s$ along the streamtube. As already stated, we also assume

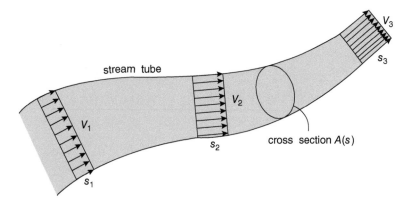

**Figure 6.4** Area and velocity variation along a streamtube

that the flow through the streamtube is steady and adopt the one-dimensional assumption that all fluid and flow properties are uniform across any given cross section but can vary from location to location. Our aim now is to find the variation in the flow velocity $V$ with location $s$ as a consequence of the area variation $A$ $(s)$.

The basis for our analysis is the principle of conservation of mass, according to which matter (in this case, the flowing fluid) is neither created nor destroyed. For a steady flow this principle requires that the same mass of fluid flows across every cross section in a given time.

We start by considering an infinitesimal slice of fluid of thickness $\delta s$ at some location $s$ along the streamtube, as shown in Figure 6.5. This slice has volume

$$\delta \mathcal{V} = A\delta s$$

and mass

$$\delta m = \rho \delta \mathcal{V} = \rho A \delta s$$

where $\rho$ is the density of the fluid within the slice (note that at this stage we do not need to assume that the density remains constant along the streamtube).

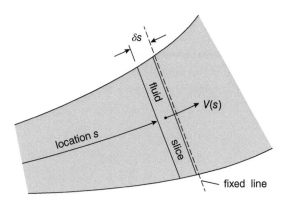

**Figure 6.5** Fluid slice moving through a streamtube

If the velocity of the fluid particles within the slice is $V$, then the slice will move a distance equal to its thickness $\delta s$ in time $\delta t$ given by

$$\delta t = \frac{\delta s}{V}.$$

If we use this equation to substitute for $\delta s$ in the expressions for $\delta \mathcal{V}$ and $\delta m$ above, we have

$$\delta \mathcal{V} = AV\delta t \quad \text{and} \quad \delta m = \rho AV\delta t$$

or

$$\frac{\delta \mathcal{V}}{\delta t} = AV \quad \text{and} \quad \frac{\delta m}{\delta t} = \rho AV.$$

We observe that $\delta m/\delta t$ represents the mass of fluid which crosses a section of the streamtube (in this case the section which instantaneously coincides with the elemental slice) per unit time. This quantity is constant for a steady flow and is called the **mass flowrate** $\dot{m}$, i.e.

$$\dot{m} = \rho AV = \text{constant}. \tag{6.1}$$

Equation (6.1) is referred to as either the **mass-conservation equation** or the **continuity equation**. The historical origin of the latter name is unclear, with some fluid dynamicists suggesting it reflects the continuum nature of a fluid, while others feel it refers to the continuous nature of fluid flow.

The quantity $\delta \mathcal{V}/\delta t$ represents the volume of fluid which crosses a section of the streamtube per unit time and is termed the **volumetric (or volume) flowrate** $\dot{Q}$, i.e.

$$\dot{Q} = AV. \tag{6.2}$$

We pointed out in Section 6.3 that a streamtube can be thought of as representing the interior wall of a duct such as a tube or pipe through which there is flow. In fact, equations (6.1) and (6.2) are important results, one or other of which is used in every one-dimensional, steady-flow analysis, including situations where the fluid density changes as a consequence of pressure changes, heating, or cooling.

---

### ILLUSTRATIVE EXAMPLE 6.1

As shown in Figure E6.1, water flows through a nozzle of circular cross section which contracts from an inlet area $A_1$ of 0.05 m$^2$ to an outlet area $A_2$ of 0.01 m$^2$. If the mass flowrate $\dot{m}$ is 110 kg/s, calculate the volumetric flowrate $\dot{Q}$ and the water velocity at inlet, $V_1$, and at outlet, $V_2$.

### Solution

$\dot{m} = 110$ kg/s; $A_1 = 0.05$ m$^2$; $A_2 = 0.01$ m$^2$; $\rho = 1000$ kg/m$^3$.
From the mass-conservation equation, we have $\dot{m} = \rho \dot{Q}$ so that

$$\dot{Q} = \frac{\dot{m}}{\rho} = \frac{110}{1000} = 0.11 \text{ m}^3/\text{s}.$$

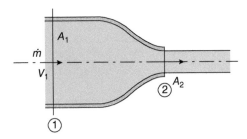

**Figure E6.1**

From $\dot{Q} = AV$, we have $\dot{Q} = A_1 V_1 = A_2 V_2$ so that

$$V_1 = \frac{\dot{Q}}{A_1} = \frac{0.11}{0.05} = 2.2 \text{ m/s}$$

and

$$V_2 = \frac{\dot{Q}}{A_2} = \frac{0.11}{0.01} = 11 \text{ m/s}.$$

**ILLUSTRATIVE EXAMPLE 6.2**

A cryogenic wind tunnel is being designed to develop the aerodynamic performance of a **Formula 1 car**. The wind tunnel will operate with an air density $\rho$ of 8 kg/m$^3$ at an airspeed $V_2$ of 90 m/s in the working section (see Figure E6.2). The working section is to have a rectangular cross section 1 m high and 2 m wide and will be just downstream of a **contraction** from a **plenum chamber**[39] with a cross-sectional area $A_1$ of 10 m$^2$. Calculate the air mass flowrate $\dot{m}$ through the wind tunnel and the airspeed $V_1$ in the plenum chamber.

Solution

$V_2 = 90$ m/s; $\rho = 8$ kg/m$^3$; $A_1 = 10$ m$^2$; $A_2 = 2$ m$^2$.
From the mass-conservation equation, $\dot{m} = \rho A_2 V_2$, we have

$$\dot{m} = 8 \times 2 \times 90 = 1440 \text{ kg/s}.$$

Also $\dot{Q} = A_1 V_1 = A_2 V_2$, so that

$$V_1 = \frac{A_2 V_2}{A_1} = \frac{2 \times 90}{10} = 18 \text{ m/s}.$$

[39] A brief outline of wind-tunnel design is given in Section 8.1.

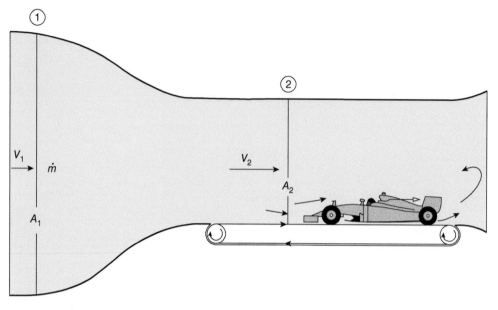

**Figure E6.2**

### ILLUSTRATIVE EXAMPLE 6.3

Helium flows through a nozzle of circular cross section with an outlet diameter $D$ of 50 mm. Calculate the maximum mass flowrate $\dot{m}$ for which the flow can be considered incompressible, i.e. for which the outlet Mach number $M$ is less than 0.3.

### Solution

$D = 0.05$ m; from Table A.6 in Appendix 2, for helium at STP, density $\rho = 0.166$ kg/m$^3$, and soundspeed $c = 1007$ m/s.

We require $M < 0.3$, and $M = V/c$, so the flow velocity $V < 0.3 \times 1007 = 302.1$ m/s.

The nozzle outlet area $A = \pi D^2/4 = 1.99 \times 10^{-4}$ m$^2$, and $\dot{m} = \rho A V$, so that

$$\dot{m} < 0.166 \times 1.99 \times 10^{-4} \times 302.1 = 0.01 \text{ kg/s}.$$

## 6.9 Average flow velocity $\overline{V}$

In any real flow through a duct, the fluid velocity varies from zero at the interior duct surface (the no-slip condition) to a maximum usually somewhere close to the duct axis. No matter

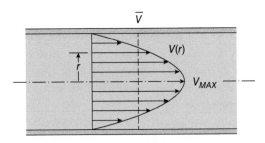

**Figure 6.6** Parabolic velocity distribution for fully-developed flow of a viscous fluid through a circular pipe

how complex the velocity variation (also called the **velocity distribution** or **profile**), it is often convenient to define a spatial-average[40] flow velocity $\overline{V}$ using equation (6.1) or (6.2), i.e.

$$\overline{V} = \frac{\dot{m}}{\rho A} = \frac{\dot{Q}}{A}. \tag{6.3}$$

The quantity $\overline{V}$ is also referred to as the **mean velocity** or **bulk-mean velocity**.

It is convenient and appropriate when applying the one-dimensional approximation to identify the velocity $V$ with the average velocity $\overline{V}$, but it has to be appreciated that other quantities, such as the momentum flowrate $\dot{m}\overline{V}$ and the kinetic-energy flowrate $\dot{m}\overline{V}^2/2$, do not accurately represent the average values of these quantities for a real flow. For example, for the flow of a viscous liquid through a long circular pipe at low flowrates, the velocity variation across the pipe is parabolic (**Poiseuille flow**, discussed in detail in Section 16.3), as depicted in Figure 6.6, and we find $\overline{V} = V_{MAX}/2$, where $V_{MAX}$ is the centreline velocity. The true momentum flowrate for Poiseuille flow is given by $4\dot{m}\overline{V}/3$, and the true kinetic-energy flowrate by $\dot{m}\overline{V}^2$. In both cases the correct values are considerably higher than would be the case for a flow with uniform velocity. Fortunately, the velocity distributions for many internal flows of engineering interest are much flatter (i.e. closer to uniform velocity) than the parabolic profile, and the one-dimensional approximation leads to results of acceptable accuracy.

## 6.10 Flow of a constant-density fluid

Equations (6.1) and (6.2) are related as follows

$$\dot{m} = \rho A \overline{V} = \rho \dot{Q} = \text{constant},$$

$\overline{V}$ being the spatial-average fluid velocity. If the fluid density $\rho$ is constant and uniform throughout the flow, then we have

$$\dot{Q} = A \overline{V} = \text{constant} \tag{6.4}$$

---

[40] For unsteady flows, especially turbulent flows (see Chapter 18), it is usual to introduce flow properties averaged with respect to time (temporal averages).

which shows that, for steady duct flow of an incompressible fluid, if the cross-sectional area of the duct $A(s)$ decreases with distance along the duct $s$, the fluid velocity $\overline{V}(s)$ must increase. If $A$ increases, $\overline{V}$ decreases. There is evidence that Leonardo da Vinci was aware of the continuity equation for a constant-density fluid and it may even have been known some 1400 years earlier to the Roman Sextus Julius Frontinus. As we shall see in Chapter 11, the situation can be very different for the flow of a compressible fluid.

This chapter has been limited to consideration of the kinematic description of fluid flow; in other words, we took no account of the forces and stresses which cause the flow. In Chapter 7 we derive **Bernoulli's equation**, which for steady flow of an incompressible, inviscid fluid allows us to relate pressure and velocity changes along a streamline. Engineering applications of Bernoulli's equation, particularly to the measurement of flowrate, are discussed in Chapter 8. Chapter 9 is concerned with the application of Newton's second law of motion to one-dimensional internal flow of an incompressible fluid, which leads to the **linear momentum equation** and allows us to calculate the **hydrodynamic forces** (Chapter 10) which arise from fluid flow through ducts with changes in area and/or direction.

 **6.11  SUMMARY**

In this chapter we have introduced some of the terminology and simplifications which enable us to begin to describe and analyse practical fluid-flow problems. The principle of conservation of mass applied to steady one-dimensional flow through a streamtube of varying cross-sectional area resulted in the continuity equation. This important equation relates mass flowrate $\dot{m}$, volumetric flowrate $\dot{Q}$, average fluid velocity $\overline{V}$, fluid density $\rho$, and cross-sectional area $A$

$$\dot{m} = \rho\dot{Q} = \rho A\overline{V} = \text{constant}.$$

For a constant-density fluid this result shows that fluid velocity increases if the cross-sectional area decreases, and vice versa.

The student should be able to

- explain what is meant by the following terms: fluid particle, steady flow, streamline, streamsurface, streamtube, no-slip condition, boundary layer, single-phase flow, incompressible, isothermal, adiabatic, one-dimensional flow, average velocity, mass flowrate, volumetric flowrate
- apply the continuity equation to one-dimensional duct flow

 **6.12  SELF-ASSESSMENT PROBLEMS**

**6.1**  Liquid medication of density 990 kg/m³ is injected from a hypodermic syringe with an internal barrel diameter of 10 mm through a needle with an internal diameter of 0.3 mm. If it takes 30 s to inject 2 ml of liquid, calculate the mass flowrate and the average liquid velocities within the syringe and the needle. Assume the plunger moves at constant speed.

(Answers: $6.6 \times 10^{-5}$ kg/s, $8.5 \times 10^{-4}$ m/s, 0.94 m/s)

**6.2**  Two pipes, one of internal diameter (I.D.) 0.5 m and the other of I.D. 1 m, are connected as shown in Figure P6.2 to a pipe of I.D. 1.2 m. Oil with a density of 880 kg/m$^3$ flows through the pipe system at a total flowrate of 15,000 t/h. If the liquid velocity in each of the two smaller-diameter pipes is the same, calculate this velocity, the corresponding volumetric and mass flowrates in the two pipes, and also the velocity in the large outlet pipe.

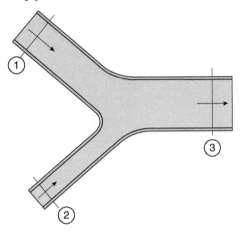

**Figure P6.2**

(Answers: 4.82 m/s, 0.95 m$^3$/s, 3.79 m$^3$/s, 3000 t/h, 12,000 t/h, 4.19 m/s)

**6.3**  Hot gas with a density of 0.4 kg/m$^3$ is exhausted from a rocket engine through a nozzle of exit diameter 1 m. If the mass flowrate through the nozzle is 370 kg/s, calculate the exhaust-gas velocity and the volumetric flowrate. If the soundspeed for the gas is 550 m/s, calculate the Mach number of the exhaust-gas flow. Can the exhaust flow be considered incompressible?
(Answers: 1178 m/s, 925 m$^3$/s, 2.14, no)

**6.4**  A water jet 50 mm in diameter impinges on a cone as shown in Figure P6.4. If the water velocity has the same magnitude at all points in the flow, calculate the thickness of the liquid layer at a location where the cone diameter is 0.5 m. If the mass flowrate of the water is 16 kg/s, calculate the flowspeed.

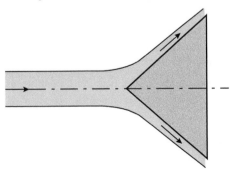

**Figure P6.4**

(Answers: 1.2 mm, 8.15 m/s)

# Bernoulli's equation

In this chapter we apply Newton's second law of motion to derive **Euler's equation**, which is a differential equation connecting the pressure, velocity, and height above a datum of a fluid particle moving steadily along a streamline in an **inviscid fluid**. By integrating Euler's equation for an incompressible fluid, we obtain **Bernoulli's equation**, which, in spite of the underlying restrictions, is arguably the most important and practically useful equation of fluid mechanics. It is shown that each of the terms in Bernoulli's equation can be interpreted as being a pressure, a form of energy, or the height of a fluid column.

## 7.1 Net force on an elemental slice of fluid flowing through a streamtube

In Chapter 6 we derived the one-dimensional continuity equation for steady flow through a streamtube

$$\dot{m} = \rho \dot{Q} = \rho A V \tag{6.1}$$

where $\dot{m}$ is the mass flowrate through the streamtube, $\dot{Q}$ is the volumetric flowrate, $\rho$ is the fluid density, $V$ is the magnitude of the fluid velocity, and $A$ is the cross-sectional area of the streamtube. To go further we need to introduce another of the basic laws of classical mechanics, the **principle of conservation of momentum**, usually referred to as **Newton's second law (of motion)**. Probably the most familiar form of Newton's second law is

$$F = ma \tag{7.1}$$

which states that a net force of magnitude $F$ exerted on a mass $m$ results in an acceleration of the mass of magnitude $a$ in the direction of $F$. As the words magnitude and direction suggest, $F$ and $a$ are both **vector** quantities.

We shall apply Newton's second law to the fluid slice of infinitesimal length $\delta s$ shown in Figure 7.1. The net force $\delta F$ which acts on the slice in the streamwise $s$-direction is made up of four components, as follows

- a pressure force $pA$ on the face at location $s$
- a pressure force $-(p + \delta p)(A + \delta A)$ on the opposite face at location $s + \delta s$
- a pressure force $(p + \delta p/2)\delta A$ on the curved face
- the component of the weight of the slice in the $s$-direction $= -\delta W \cos \theta$

*Introduction to Engineering Fluid Mechanics.* Marcel Escudier.
© Marcel Escudier 2017. Published 2017 by Oxford University Press.

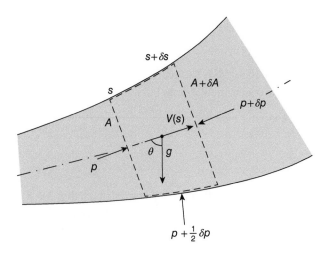

**Figure 7.1** Elemental fluid slice moving through a streamtube

so that

$$\delta F = pA - \left(p + \delta p\right)(A + \delta A) + \left(p + \frac{1}{2}\delta p\right)\delta A - \delta W \cos\theta.$$

In writing the equation for $\delta F$ we have assumed that the pressure $p$ is uniform across the cross section of the streamtube, consistent with the one-dimensional assumption, but varies with distance $s$ along it. The terms in the equation have been taken as positive in the direction of increasing $s$, which is also the flow direction. Although Figure 7.1 shows $A$ increasing with $s$, this in no way restricts the analysis, which is valid whether $A$ increases or decreases. The angle between the velocity vector at any location along the streamtube and the vertical is denoted by $\theta$. The third term on the right-hand side of the equation for $\delta F$ represents the component of force in the $s$-direction due to the pressure acting on the section of the surface of the streamtube, which coincides instantaneously with the moving slice. The average pressure acting on this strip of surface must have a value somewhere between $p$ and $p + \delta p$ and has been taken as the simple average $p + \delta p/2$, though, as we shall see shortly, the factor 1/2 is unimportant. Just as for the horizontal component of the elemental force due to the hydrostatic pressure acting on a curved surface (see Section 5.5), the net force in the $s$-direction due to the pressure $p + \delta p/2$ acting on the strip of streamtube surface is $(p + \delta p/2) \times$ projected area, where the projected area here is $\delta A$.

   If we now multiply out and simplify by cancellation the terms on the right-hand side of the equation for $\delta F$, we have

$$\delta F = -\delta p \left(A + \frac{1}{2}\delta A\right) - \delta W \cos\theta.$$

Since our elemental slice is infinitesimally thin, it is permissible to neglect the area change $\delta A$ in comparison with the area $A$ itself (which is why the factor 1/2 is unimportant), so that

$$\delta F = -\delta pA - \delta W \cos\theta.$$

The weight $\delta W$ of the infinitesimal slice is given by

$$\delta W = \delta m\, g = \rho \delta \mathcal{V} g = \rho A \delta s\, g$$

where $\delta m$ is the mass of the slice, $\delta \mathcal{V}$ is its volume, and $g$ is the acceleration due to gravity. If we substitute for $\delta W$ in the equation for $\delta F$, we have

$$\delta F = -\delta p\, A - \rho A \delta s\, g \cos \theta$$

or

$$\delta F = -\delta p\, A - \rho A \delta z' g \tag{7.2}$$

wherein we have made use of the fact that the vertical height change $\delta z'$, corresponding to the distance along the streamtube $\delta s$, is given by $\delta z' = \delta s \cos \theta$ (see Figure 7.1). Just as in Chapter 4, we use the symbol $z'$ to denote altitude, and $z$ to denote depth, so that $\delta z = -\delta z'$.

## 7.2 Acceleration of a fluid slice

Since we have restricted our attention to steady flow, it may seem a contradiction that we are now discussing acceleration of the fluid. However, as stated in Section 6.2, where this restriction was introduced, steady flow of a fluid implies that the fluid velocity in a flowfield is always the same at any fixed point but can vary from point to point. As shown in Figure 7.2, what this means is that the velocity $V$ of each fluid particle can change as it moves through the flowfield and so the particle experiences acceleration.

If the acceleration of a particle with instantaneous velocity $V$ at time $t$ is $a$, by definition we have

$$a = \frac{dV}{dt} = \frac{dV}{ds}\frac{ds}{dt}$$

wherein we have made use of the rule of differential calculus for differentiation of a function of a function. In this case $V$ is a function of $s$, which itself is a function of $t$. By definition the particle velocity along a streamline is given by

$$V = \frac{ds}{dt},$$

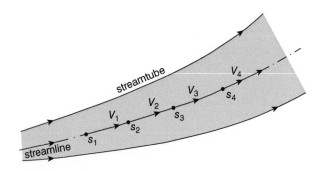

**Figure 7.2** Velocity variation for steady flow along a streamline

so the acceleration of a fluid particle can be written as

$$a = V\frac{dV}{ds}.$$

---

### ILLUSTRATIVE EXAMPLE 7.1

A liquid flows at a constant volumetric flowrate $\dot{Q}$ through a duct which decreases in area $A$ such that $A/A_0 = s_0/s$, where $s$ is the distance along the duct and $A_0$ is the area at $s = s_0$. Derive an expression for the fluid acceleration at any location along the duct. If $s_0 = 1$ m, $A_0 = 0.1$ m$^2$, and $\dot{Q} = 0.2$ m$^3$/s, calculate the fluid velocity and acceleration at $s = 5$ m.

### Solution

We have $\dot{Q} = AV$ so that

$$V = \frac{\dot{Q}}{A} = \dot{Q}\frac{s}{A_0 s_0}$$

from which

$$\frac{dV}{ds} = \frac{\dot{Q}}{A_0 s_0}.$$

Then, from $a = V(dV/ds)$, we have

$$a = \left(\frac{\dot{Q}}{A_0 s_0}\right)^2 s.$$

For the numerical part of the problem we have $s_0 = 1$ m, $A_0 = 0.1$ m$^2$, $\dot{Q} = 0.2$ m$^3$/s, and $s = 5$ m. At $s = 5$m,

$$A = \frac{A_0 s_0}{s} = \frac{0.1 \times 1}{5} = 0.02 \text{ m}^2.$$

Therefore,

$$V = \frac{\dot{Q}}{A} = \frac{0.2}{0.02} = 10 \text{ m/s}$$

and

$$a = \left(\frac{\dot{Q}}{A_0 s_0}\right)^2 s = \left(\frac{0.2}{0.1 \times 1}\right)^2 \times 5 = 20 \text{ m/s}^2.$$

### Comment:

Although the flow velocity here, 10 m/s, is quite modest, the particle acceleration produced by the area reduction is about $2g$.

---

## 7.3 **Euler's equation**

We are now in a position to apply Newton's second law to the infinitesimal fluid slice since we now have expressions for both the net force acting on the slice in the $s$-direction and for the acceleration of the fluid particles which constitute the slice. We start with the basic form of the second law

$$\delta F = \delta m \, a.$$

From Section 7.1 we have $\delta m = \rho A \delta s$ and, from Section 7.2, $a = V(dV/ds)$, so that

$$\delta F = \rho A \delta s \, V \frac{dV}{ds} \tag{7.3}$$

We can now substitute for $\delta F$ from equation (7.2) which, after dividing through by the area $A$, leads to

$$-\delta p - \rho \delta z' g = \rho \delta s \, V \frac{dV}{ds}.$$

We now divide through by $\delta s$ and rearrange to find

$$\frac{\delta p}{\delta s} + \rho g \frac{\delta z'}{\delta s} + \rho V \frac{dV}{ds} = 0$$

which, in the limit $\delta s \to 0$, gives

$$\frac{dp}{ds} + \rho g \frac{dz'}{ds} + \rho V \frac{dV}{ds} = 0. \tag{7.4}$$

Equation (7.4) is a first-order ordinary differential equation which connects the variation of pressure, velocity, and density along a streamline for the steady flow of an inviscid fluid and is known as **Euler's equation**. To be more precise, equation (7.4) is a restricted form of a much more general set of equations derived by Euler for the flow of an inviscid fluid.

As is the mass-conservation equation (6.1), Euler's equation is valid whether or not the fluid density is constant. We note too that equation (7.4) is independent of area and so applies to both internal and external flows. For a fluid at rest (i.e. $V = 0$), equation (7.4) reduces to the hydrostatic equation (4.5)

$$-\frac{dp}{dz'} = \frac{dp}{dz} = \rho g.$$

## 7.4 **Bernoulli's equation**

If the fluid density is taken as constant, then equation (7.4) is easily integrated and we have

$$p + \rho g z' + \frac{1}{2} V^2 = \text{constant} \tag{7.5}$$

which is known as **Bernoulli's equation** (or **theorem**) after Daniel Bernoulli, who included a form of it in his treatise *Hydrodynamica* (1738). It was not until 1755 that Bernoulli's close friend Leonhard Euler gave a complete derivation of equation (7.5). The constant of integration on the right-hand side is called the **Bernoulli constant**, although it is not an absolute constant but one that can vary from streamline to streamline in a given flow and is different for every flow. In a one-dimensional flow, however, the Bernoulli constant is uniform throughout the flowfield.

As we pointed out in Section 4.3, the combination of terms $p + \rho g z'$ which appears in Bernoulli's equation is termed the **piezometric pressure**, $P$.

Since we shall make extensive use of Bernoulli's equation in this and Chapters 8, 9, and 10, it is important to be aware of the assumptions on which its validity depends

- steady flow
- constant-density fluid
- inviscid fluid

For many flows, including compressible flow, which we discuss in some detail in Chapters 11, 12, and 13, the potential-energy (gravity) term in Euler's equation is generally negligible, and equation (7.4) then reduces to

$$\frac{dp}{ds} + \rho V \frac{dV}{ds} = 0 \qquad (7.6)$$

or, in integral form,

$$\frac{1}{2}V^2 + \int \frac{dp}{\rho} = \text{constant.} \qquad (7.7)$$

To evaluate the integral requires that the relationship between pressure and density be known (see Section 11.3).

---

**ILLUSTRATIVE EXAMPLE 7.2**

(a) Calculate the Bernoulli constant for water flowing through a pipe at zero altitude at a speed of 10 m/s if the water pressure is 1 bar. (b) If the elevation of the pipe falls by 20 m and the flowspeed decreases to 2 m/s, what is the new fluid pressure? The flow geometry is shown in Figure E7.2.

**Solution**

$z_1' = 0$; $V_1 = 10$ m/s; $p_1 = 10^5$ Pa; $z_2' = -20$ m; $V_2 = 2$ m/s; $\rho = 10^3$ kg/m³; $g = 9.81$ m/s².

(a) The terms in Bernoulli's equation are as follows

$$p_1 = 10^5 \text{ Pa}, \rho g z_1' = 0, \text{ and } 1/2\rho V_1^2 = 1/2 \times 10^3 \times 10^2 = 5 \times 10^4 \text{ Pa}$$

so the Bernoulli constant $= 10^5 + 0 + 5 \times 10^4 = 1.5 \times 10^5$ Pa or 1.5 bar.

(b) We have $z_2' = -20$ m, and $V_2 = 2$ m/s, so that

$$\rho g z_2' = 10^3 \times 9.81 \times (-20) = -1.96 \times 10^5 \text{ Pa and } 1/2 \, \rho V_2^2 = 1/2 \times 10^3 \times 2^2 = 2 \times 10^3 \text{ Pa.}$$

If the flow is steady, of constant density, and frictionless, the Bernoulli constant remains unchanged, so that

$$p_2 + (-1.96 \times 10^5) + 2 \times 10^3 = 1.5 \times 10^5 \text{ and } p_2 = 3.44 \times 10^5 \text{ Pa or 3.44 bar.}$$

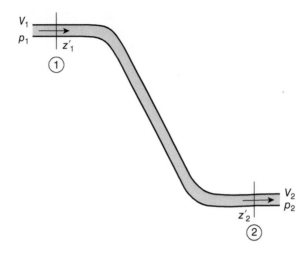

**Figure E7.2**

**Comment:**

It was important to take $z_2'$ as negative because the elevation $z_2'$ of the pipe at location ② was less than $z_1'$ at location ①.

## 7.5 Interpretations of Bernoulli's equation

The first term of Bernoulli's equation, $p$, is the pressure which would be sensed by an observer moving with the fluid (the stick man on the left in Figure 7.3) and is called the **static pressure**. As we discuss further in Section 15.1 this pressure is also called the **mechanical pressure** and is usually taken to be equal to the **thermodynamic pressure**. From the principle of dimensional homogeneity, which we discussed in Section 3.5, if one term in equation (7.4) is a pressure, then each of the other terms, including the constant of integration (i.e. the Bernoulli constant), must have the units and dimensions of pressure. In fact, it is common practice to refer to each of these terms individually, and certain combinations of them, as pressures

- $\rho g z = -\rho g z' =$ **hydrostatic pressure** $\hfill (7.8)$

- $p + \rho g z' = P =$ **piezometric pressure** $\hfill (7.9)$

- $p + \frac{1}{2}\rho V^2 = p_0 =$ **stagnation pressure** $\hfill (7.10)$

- $\frac{1}{2}\rho V^2 =$ **dynamic pressure** $\hfill (7.11)$

- $p + \rho g z' + \frac{1}{2}\rho V^2 = p_T =$ **total pressure.** $\hfill (7.12)$

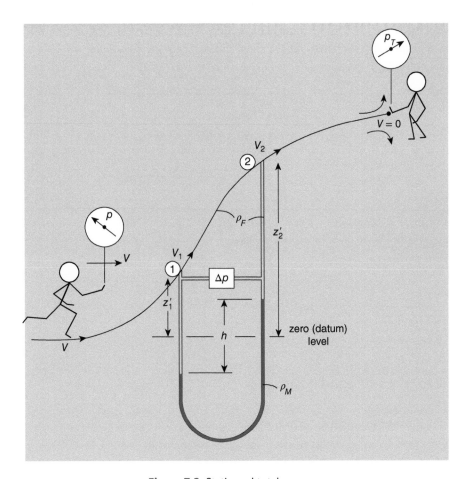

**Figure 7.3** Static and total pressures

The terms **hydrostatic pressure** and **piezometric pressure** were introduced in Section 4.3, where we considered fluids at rest. The **dynamic pressure** $\rho V^2/2$ is a new term which arises when a fluid is in motion. For a constant-density flow, the sum of the static and dynamic pressures is called the **stagnation pressure**. **Stagnation conditions** at any point in a flow are the conditions that <u>would be attained</u> if the flow there were brought to rest. A point where a flow is actually brought to rest, such as the point P (in reality a line) on the dividing streamline for flow around an aerofoil, as shown in Figure 6.1, is called a **stagnation point**. For flows where changes in the hydrostatic pressure $-\rho g z'$ are negligible compared with changes in the static and dynamic pressures, the stagnation pressure is essentially constant along a streamline. This condition obviously applies if the streamline lies in a horizontal plane, and is an excellent approximation for all gas flows. For high-speed gas flows where compressibility effects are significant, $\rho V^2/2$ is still called the dynamic pressure but, as will be seen in Section 11.3, is no longer simply the difference between the stagnation and static pressures as it depends upon both the static pressure and also the Mach number. The **total pressure** is identically equal to the Bernoulli constant and so constant along any streamline in steady inviscid flow. With reference to Figure 7.3, the 'stick man' on the left who is moving along the streamline at the same velocity as the fluid in his immediate vicinity would sense the local static pressure $p$ whereas the stationary stick man on the right and on the same streamline would sense the local total pressure $p_T$.

---

**ILLUSTRATIVE EXAMPLE 7.3**

The difference between the stagnation pressure $p_0$ and the static pressure $p$ of air of density $\rho = 4.4$ kg/m$^3$ and soundspeed $c = 323$ m/s is found to be 172 kPa. Calculate the gas velocity and determine whether the flow can be considered incompressible.

**Solution**

$\rho = 4.4$ kg/m$^3$; $c = 323$ m/s; $p_0 - p = 1.72 \times 10^5$ Pa.
Assuming the flow to be incompressible, we have

$$p_0 - p = \frac{1}{2}\rho V^2$$

so that

$$V = \sqrt{\frac{2\left(p_0 - p\right)}{\rho}}$$

and so

$$V = \sqrt{2 \times 1.72 \times 10^5/4.4} = 279.6\,\text{m/s}.$$

The Mach number $M = V/c = 279.6/323 = 0.87$, i.e. significantly greater than 0.3 so that compressibility effects cannot be regarded as negligible, and our calculation of the gas velocity must be in error. It can be shown, using the concepts presented in Chapter 11, that the error is about 8% and it depends upon the situation as to whether or not this is acceptable.

---

The dynamic pressure $\rho V^2/2$ can be thought of as a pressure which characterises the motion of a fluid, and its value at a particular location in a flow, usually the undisturbed flow **upstream** of an object, such as the uniform flow approaching the aerofoil of Figure 6.2, is frequently chosen to make other pressures, pressure differences, surface shear stress, etc., non-dimensional. The term **normalise** is often used to mean make non-dimensional. Several such non-dimensional quantities were introduced in Section 3.12, including

$$\frac{p_{REF} - p_V}{\frac{1}{2}\rho V^2} = \text{cavitation number,} \tag{7.13}$$

$p_{REF}$ being a reference pressure, typically the barometric pressure, and $p_V$ the saturated vapour pressure for a flowing liquid (see Sections 2.13 and 8.11), and

$$c_f = \frac{\tau_S}{\frac{1}{2}\rho V^2} = \text{friction factor} \tag{7.14}$$

where $\tau_S$ is the surface shear stress (see Chapters 16, 17, and 18).

The dynamic pressure is also used together with an appropriate area $A$ to non-dimensionalise forces such as the drag and lift forces $D$ and $L$ exerted on an object by a fluid flowing past it

$$C_D = \frac{D}{\frac{1}{2}\rho V^2 A} = \text{drag coefficient} \qquad (7.15)$$

$$C_L = \frac{L}{\frac{1}{2}\rho V^2 A} = \text{lift coefficient.} \qquad (7.16)$$

Either the projected frontal area (i.e. the area corresponding to a silhouette) or (for a wing) the planform area is frequently chosen for $A$.

To the above we can add

$$C_P = \frac{\Delta p}{\frac{1}{2}\rho V^2} = \text{pressure coefficient} \qquad (7.17)$$

where $\Delta p$ is a pressure loss or pressure difference with respect to a reference pressure such as the static pressure at the same location as that for $V$ (often equal to the barometric pressure $B$). The inclusion of the factor 1/2 in these definitions is conventional, and a consequence of its 'natural' occurrence in Bernoulli's equation. Its inclusion is not essential (without it all quantities are still non-dimensional) and it is sometimes omitted, as is the case for the **Euler number**

$$Eu = \frac{p - p_{REF}}{\rho V^2}. \qquad (7.18)$$

### 7.5.1 Energy

The dynamic pressure $\rho V^2/2$ represents the kinetic energy per unit volume of a flowing fluid. We can see that this is so by considering a mass $m$ of volume $\mathcal{V}$ moving at speed $V$. The kinetic energy of the mass is $mV^2/2$, and its kinetic energy per unit volume is therefore $mV^2/(2\mathcal{V})$ or $\rho V^2/2$ since the density $\rho \equiv m/\mathcal{V}$. Again on the basis of the principle of dimensional homogeneity, it must be the case that each of the terms in Bernoulli's equation can also be regarded as representing a form of energy

- $p$ = pressure energy per unit volume
- $\rho g z'$ = potential energy per unit volume
- $p_T$ = total energy per unit volume

and it follows that Bernoulli's equation itself can be thought of as an equation for the conservation of mechanical energy. In fact, Bernoulli's equation can be derived directly from the **first law of thermodynamics**, which is the basis for a general **energy-conservation equation** which we discuss in some detail in Chapter 11.

### 7.5.2 Head

If we divide through Bernoulli's equation by $\rho g$ we find

$$\frac{p}{\rho g} + z' + \frac{V^2}{2g} = \frac{p_T}{\rho g}. \qquad (7.19)$$

Once again the principle of dimensional homogeneity leads to the conclusion that, since $z'$ represents altitude or height, each term in equation (7.19) corresponds to a height or, as it is usually called (see Section 4.3), a **head**

- $\dfrac{p}{\rho g}$ = static head $\hspace{6cm}$ (7.20)

- $\dfrac{V^2}{2g}$ = dynamic head $\hspace{5.5cm}$ (7.21)

- $\dfrac{p_T}{\rho g}$ = total head. $\hspace{5.7cm}$ (7.22)

The head in each case corresponds to the vertical height of a column of fluid with the same density $\rho$ as that of the flowing fluid.

## 7.6 **Pressure loss versus pressure difference**

It is important to understand the distinction between pressure difference (or pressure change) and pressure loss. At points ① and ② on the streamline shown in Figure 7.3 the pressures, velocities, and heights are related by Bernoulli's equation as follows

$$p_1 + \rho g z_1' + \tfrac{1}{2}\rho V_1^2 = p_2 + \rho g z_2' + \tfrac{1}{2}\rho V_2^2 = p_T \hspace{3cm} (7.23)$$

or

$$\left(p_1 + \rho g z_1'\right) - \left(p_2 + \rho g z_2'\right) = \tfrac{1}{2}\rho\left(V_2^2 - V_1^2\right). \hspace{2.5cm} (7.24)$$

Equation (7.24) shows that a change in velocity between points ① and ② results in a change in the piezometric pressure $(p + \rho g z')$ and this is precisely the pressure difference that would be measured by a manometer or differential pressure transducer, as illustrated in the figure. If the velocities at points ① and ② were the same, both the manometer and the pressure transducer would indicate zero because the only change in pressure would be the hydrostatic pressure difference due to the height difference $z_2' - z_1'$ whereas the difference in the piezometric pressures is zero.

The difference in pressure associated with a velocity change becomes clearer if the hydrostatic pressure difference $\rho g\left(z_2' - z_1'\right)$ is negligible since we then have

$$p_1 - p_2 = \tfrac{1}{2}\rho\left(V_2^2 - V_1^2\right). \hspace{4cm} (7.25)$$

From equation (7.25) we see that an increase in velocity results in a decrease in static pressure and vice versa. If we couple this statement with the constant-density form of the continuity equation (6.4), then, for one-dimensional, steady flow of an inviscid, constant-density fluid through a streamtube, we can conclude that

- if $A_2 < A_1$, then $V_2 > V_1$ and $p_2 < p_1$
- if $A_2 > A_1$, then $V_2 < V_1$ and $p_2 > p_1$

It should be clear that, irrespective of whether the static and piezometric pressures increase or decrease, according to our assumptions the total pressure $p_T$ will remain constant. In practice the effect of fluid friction at a surface, due to viscosity, is for the total pressure to decrease in the absence of work or thermal-energy input to the fluid. This is what is meant by a **pressure loss**. According to the energy interpretation of Bernoulli's equation (Subsection 7.5.1), such a reduction in total pressure corresponds to a loss in mechanical energy. A more detailed analysis reveals that, for flow of a viscous fluid, mechanical energy is dissipated resulting in an increase in the internal energy of the fluid and hence an increase in fluid temperature. This **frictional heating** is usually negligible but can become a major factor at very high gas velocities as, for example, encountered in supersonic flight or re-entry of spacecraft into the earth's atmosphere.

 ## 7.7 SUMMARY

In this chapter we used Newton's second law of motion to derive Euler's equation for the flow of an inviscid fluid along a streamline. For a fluid of constant density $\rho$ Euler's equation can be integrated to yield Bernoulli's equation

$$p + \rho g z' + \frac{1}{2}\rho V^2 = p_T$$

which shows that the sum of the static pressure $p$, the hydrostatic pressure $\rho g z'$, and the dynamic pressure $\rho V^2/2$ is equal to the total pressure $p_T$. Each of the terms on the left-hand side of Bernoulli's equation can be regarded as representing different forms of mechanical energy and also equivalent to the hydrostatic pressure due to a vertical column of liquid. The dynamic pressure can be thought of as measuring the intensity or strength of a flow and is frequently combined with other fluid and flow properties to produce non-dimensional (or dimensionless) numbers which characterise various aspects of fluid motion.

The student should be able to

- state Bernoulli's equation in the forms

$$p + \rho g z' + \frac{1}{2}\rho V^2 = \text{constant} = p_T$$

  and

$$p_1 + \rho g z'_1 + \frac{1}{2}\rho V_1^2 = p_2 + \rho g z'_2 + \frac{1}{2}\rho V_2^2$$

- state the assumptions made in the derivation of Bernoulli's equation and the limitations on its applicability, i.e. to conditions along a streamline in the steady flow of an inviscid, constant-density fluid
- define the terms
  - Bernoulli constant
  - dynamic pressure and dynamic head
  - total pressure and total head
  - stagnation pressure

  in addition to the relevant terms introduced in Chapter 4

  - hydrostatic pressure and hydrostatic head
  - piezometric pressure and piezometric head

- interpret Bernoulli's equation in terms of pressure, mechanical energy, and head
- distinguish between pressure difference (or pressure change) and pressure loss

 ## 7.8  SELF-ASSESSMENT PROBLEMS

**7.1**  Water from a reservoir flows to the nozzles of a Pelton turbine through a pipe 2 m in diameter. The vertical height between the reservoir and the turbine is 400 m. Assuming steady, one-dimensional, frictionless flow, calculate the flow velocity in the pipe and at the nozzle outlet if the nozzle diameter is 200 mm. The static pressure at outlet is the same as that at the surface.
(Answers: 0.0886 m/s; 88.6 m/s)

**7.2**  The mass flowrate of methane gas through a pipeline 0.5 m in diameter is 10 kg/s. Calculate the gas velocity if the density of methane is taken as 0.66 kg/m$^3$. If the pipeline contracts linearly to a diameter of 0.35 m over a distance of 0.5 m, calculate the gas velocity and acceleration at the end of the contraction, the stagnation pressure if the upstream static pressure is 1 bar, and the drop in static pressure across the contraction. Assume steady, one-dimensional, incompressible, frictionless flow.
(Answers: 77.17 m/s; 157.5 m/s; $4.25 \times 10^4$ m/s$^2$; 1.0197 bar; 6219 Pa)

**7.3**  Calculate the Bernoulli constant and the stagnation pressure at a location in a pipeline where the water velocity is 25 m/s, the static pressure is 8 bar, and the elevation is 65 m. Also calculate the pressure head, the dynamic head, the total head, and the piezometric head. If the cross-sectional area of the pipeline is 1 m$^2$, what is the kinetic-energy flowrate?
(Answers: 17.5 bar; 11.125 bar; 81.55 m; 31.86 m; 178.4 m; 146.6 m; 7.81 MW)

**7.4**  The exhaust gas from a turbojet engine has a density of 0.18 kg/m$^3$ and a sound-speed of 600 m/s. If the exhaust-gas flowrate is 600 kg/s, and the exhaust has a cross-sectional area of 4 m$^2$, calculate the velocity of the gas and the Mach number. Calculate the stagnation pressure of the air entering the engine if its static pressure is 0.5 bar, its density is 0.7 kg/m$^3$, and the inlet area is 5 m$^2$. The mass flowrates of air and exhaust gas can be assumed to be the same.
(Answers: 833.3 m/s; 1.39; 0.603 bar)

# 8 Engineering applications of Bernoulli's equation

Bernoulli's equation is so valuable in analysing a wide range of fluid-flow problems that we now devote an entire chapter to illustrate how it is applied in practice, frequently together with the continuity equation. The basic design of a wind-tunnel contraction provides valuable insight into the interplay between pressure and kinetic energy. Instrumentation for flow measurement provides several application examples, including the Pitot-static tube for velocity measurement, and the Venturi-tube and orifice-plate meters for the measurement of total fluid flowrate. We show how Bernoulli's equation gives some insight into aerofoil lift and into the aerodynamics of modern racing cars. Liquid draining from a tank under the influence of gravity provides another example of the application of Bernoulli's equation. Another important application is the determination of the conditions for the onset of cavitation in a liquid flow.

## 8.1 Wind-tunnel contraction

A feature of most subsonic wind tunnels is a **contraction**, which is a smooth reduction in the cross section between the **settling chamber** (also called the **plenum chamber**) and the working section, as illustrated in Figure 8.1. The contraction is one of several **flow-conditioning** components designed to produce a flow of uniform velocity, low **swirl**, and low levels of fluctuation (known as **turbulence**, see Chapter 18). Swirl is minimised by installing a **honeycomb** (a composite structure consisting of an array of parallel cells of hexagonal cross section manufactured from thin aluminium sheets) and turbulence reduced using a sequence of fine wire-mesh screens. A **diffuser** is usually installed downstream of the working section to gradually increase the static pressure. Wind tunnels where the fan which produces the flow is upstream of the settling chamber, as in Figure 8.1, are called **blower tunnels**. Since the tunnel shown is not part of a closed loop, the configuration is referred to as **open-return**.

This section is concerned with applying Bernoulli's equation together with the continuity equation to demonstrate how a contraction produces a uniform flow.

We assume the flow is steady and incompressible with an upstream velocity $V_1$ and static pressure $p_1$, and corresponding values $V_2$ and $p_2$ downstream. According to the continuity,

$$A_1 V_1 = A_2 V_2 \tag{8.1}$$

where $A_1$ and $A_2$ are the upstream and downstream cross-sectional areas, respectively. What this equation shows is that the flow velocity increases as the flow passes through the contraction such that at exit we have

$$V_2 = \frac{A_1}{A_2} V_1.$$

*Introduction to Engineering Fluid Mechanics*. Marcel Escudier.
© Marcel Escudier 2017. Published 2017 by Oxford University Press.

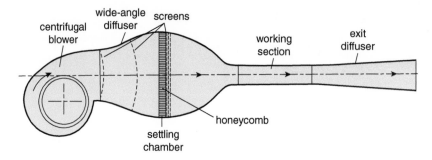

**Figure 8.1** Typical open-return wind-tunnel configuration

We now apply Bernoulli's in the form

$$p_1 + \frac{1}{2}\rho V_1^2 = p_2 + \frac{1}{2}\rho V_2^2 = p_0 \tag{7.10}$$

where $p_0$ is the constant stagnation pressure (assuming there is no appreciable change in elevation between locations ① and ② so that, in this case, no change in the total pressure $p_T$ implies no change in the stagnation pressure $p_0$). Substitution for $V_2$ then leads to

$$p_1 + \frac{1}{2}\rho V_1^2 = p_2 + \frac{1}{2}\rho \left(\frac{A_1}{A_2}\right)^2 V_1^2 = p_0.$$

This result shows that the contraction increases the kinetic energy of the flow by the factor $(A_1/A_2)^2$. Since we are assuming a frictionless flow, the mechanical energy of the flow must be conserved (as we pointed out in Section 7.5, Bernoulli's equation can be thought of as an energy-conservation equation). We can interpret this to mean that there has been a transformation of pressure energy into kinetic energy. We can draw another important conclusion: since the upstream static pressure $p_1$ must be practically uniform because of the low flow speed, and a large fraction of that pressure has been transformed into kinetic energy, the downstream flow speed $V_2$ must be practically uniform, which is precisely what is needed for the flow in the test section of a wind tunnel.

## 8.2 **Venturi-tube flowmeter**

The **Venturi tube**, developed by Clemens Herschel and named after Giovanni Battista Venturi, is one of a number of differential-pressure, **inline flowmeters**, designed on the basis of Bernoulli's equation (7.10), which are commonly used to measure the total volumetric rate $\dot{Q}$ at which a low-viscosity[41] gas or liquid flows through a pipe. A typical Venturi tube, which is a **convergent-divergent nozzle**, is illustrated in Figure 8.2. The essential features are a gradual conical contraction from the initial pipe diameter to a cylindrical **throat** section followed by an

---

[41]  As we saw in Chapter 3 the Reynolds number is the essential flow parameter involving viscosity. When we say 'low-viscosity fluid', we really mean 'high Reynolds number flow' and this, in turn, normally means turbulent flow, as will be seen in Chapter 15.

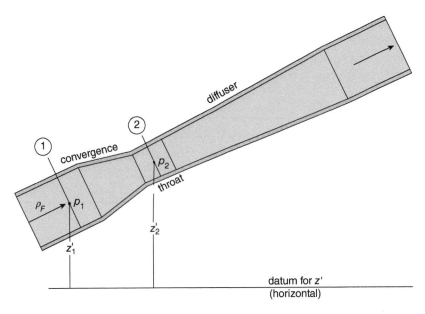

**Figure 8.2** Venturi-tube flowmeter

even more gradual area increase, usually back to the original pipe diameter[42]. The convergent section, between the inlet and throat, sometimes called a **confuser**, accelerates the fluid flowing through it, thereby reducing the fluid pressure. As we shall now show, the flowrate is derived from a measurement of the piezometric pressure drop across this upstream section of the Venturi tube. The absolute pressure is of significance only insofar as it influences the density of a gas or the tendency for a liquid to cavitate (see Section 8.11).

If we assume that the flow through the confuser is one dimensional, frictionless, and incompressible with density $\rho_F$, we can apply Bernoulli's equation between the sections marked ① and ② in Figure 8.2, as follows

$$p_1 + \rho_F g z_1' + \frac{1}{2}\rho_F V_1^2 = p_2 + \rho_F g z_2' + \frac{1}{2}\rho_F V_2^2$$

which we can rearrange as

$$\left(p_1 + \rho_F g z_1'\right) - \left(p_2 + \rho_F g z_2'\right) = \frac{1}{2}\rho_F(V_2^2 - V_1^2).$$

If the Venturi tube is installed with its axis horizontal, such that $z_1' = z_2'$, the terms on the left-hand side reduce to the static-pressure difference, $\Delta p = p_1 - p_2$. More generally, however, a **manometer** or **differential-pressure transducer** connected between sections ① and ② will measure the piezometric pressure difference $\Delta P$ given by

$$\Delta P = \Delta p + \rho_F g \Delta z' = \rho_F g \Delta H \tag{8.2}$$

where $\Delta z' = z_1' - z_2'$ is the height difference between ① and ②, such that $\Delta z' > 0$ if $z_2' < z_1'$, and $\Delta H$ is the piezometric head difference. In the event that either the static-pressure difference

---

[42] Further details are given in Section 8.3.

$\Delta p = p_1 - p_2$ or the individual static pressures $p_1$ and $p_2$ are measured directly, the piezometric pressure difference $\Delta P$ is determined by adding $\rho_F g \Delta z'$ to $\Delta p$, i.e. the absolute height of the Venturi tube is of no significance, only the height difference $\Delta z'$.

We now rewrite Bernoulli's equation in the convenient form

$$\Delta P = \frac{1}{2}\rho_F \left( V_2^2 - V_1^2 \right). \tag{8.3}$$

From the continuity equation for a constant-density flow, $\dot{Q} = AV$, the velocities $V_1$ and $V_2$ can be written in terms of the volumetric flowrate $\dot{Q}$ and the cross-sectional areas at sections ① and ②, $A_1$ and $A_2$, as

$$V_1 = \frac{\dot{Q}}{A_1} \quad \text{and} \quad V_2 = \frac{\dot{Q}}{A_2}.$$

Substitution for $V_1$ and $V_2$ in equation (8.3) then gives

$$\Delta P = \frac{1}{2}\rho_F \dot{Q}^2 \left( \frac{1}{A_2^2} - \frac{1}{A_1^2} \right)$$

so that, after rearrangement, we have

$$\dot{Q} = A_2 \sqrt{\frac{2\Delta P}{\rho_F \left[ 1 - (A_2/A_1)^2 \right]}} \tag{8.4}$$

from which $\dot{Q}$ can be calculated. The corresponding expression for the mass flowrate, $\dot{m} = \rho_F \dot{Q}$, is

$$\dot{m} = A_2 \sqrt{\frac{2\rho_F \Delta P}{\left[ 1 - (A_2/A_1)^2 \right]}}. \tag{8.5}$$

Since $\Delta P/\rho_F = g\Delta H$, equation (8.4) can be written in terms of the piezometric head difference $\Delta H$ as

$$\dot{Q} = A_2 \sqrt{\frac{2g\Delta H}{\left[ 1 - (A_2/A_1)^2 \right]}}. \tag{8.6}$$

## 8.3 Venturi-tube design and the coefficient of discharge $C_D$

Equations (8.4), (8.5), and (8.6) are all based directly on Bernoulli's equation and the continuity equation for one-dimensional, incompressible flow. The constant-density assumption is practically always valid for single-phase liquid flows and, as we showed in Chapter 7, remains very accurate for gas flows with Mach numbers up to about 0.3. Any error associated with the constant-density assumption is further minimised by the fact that the fluid density $\rho_F$ appears in the expression for the mass flowrate $\dot{m}$ within the square root.

The influence of the one-dimensional flow assumption on the accuracy of equations (8.4) and (8.5) is far less straightforward to quantify. The radial distribution of the fluid velocity upstream of the Venturi is likely to be far from uniform because any real fluid is affected by viscosity. As we shall see in Chapter 16, if the flow is laminar the velocity distribution will be parabolic; in Section 18.8, the velocity distribution is shown to be 'flatter' if the flow is turbulent, which is likely to be the case in most practical situations[43]. To compensate for the non-uniformity in the velocity distribution, it is usual to calibrate Venturi tubes against a standard of known, high accuracy to determine a performance factor called the **coefficient of discharge**, $C_D$, which is defined as the ratio of the actual (i.e. true) flowrate $\dot{Q}_A$ to the theoretical flowrate $\dot{Q}_{TH}$ based upon Bernoulli's equation for the measured $\Delta P$, i.e.

$$C_D \equiv \frac{\dot{Q}_A}{\dot{Q}_{TH}} = \frac{\dot{Q}_A}{A_2} \sqrt{\frac{\rho_F \left[1 - (A_2/A_1)^2\right]}{2\Delta P}}. \tag{8.7}$$

The coefficient of discharge can be regarded as a direct measure of the validity of the theory given in Section 8.2. It is quite remarkable, therefore, that values of $C_D$ for a low-viscosity liquid such as water flowing through a well-designed Venturi tube can be as high as 0.995, suggesting that this very simple theory is almost perfect in this application. Calibration is normally carried out over a wide range of flowrates, and the results presented in the form of a graph or table of $C_D$ versus the pipe-flow Reynolds number $\rho_F V_1 D_1 / \mu_F$, where $D_1$ is the upstream pipe diameter, $V_1 = 4\dot{Q}_A / \pi D_1^2$ is the upstream flowspeed, and $\mu_F$ is the dynamic viscosity of the fluid.

The section of a Venturi tube downstream of the throat, which is known as a **diffuser**, has a negligible influence on the characteristics of the Venturi tube and is designed to minimise the stagnation pressure loss between inlet and outlet.

Optimum values for the convergence angle of the confuser and the divergence angle of the diffuser are about 21° and 7° to 8°, respectively, while the throat-to-pipe diameter ratio is between 0.3 and 0.75. A key installation requirement is a run of straight undisturbed pipe (i.e. free of bends, valves, area changes, etc.) upstream of the Venturi at least 40 diameters in length and including a honeycomb flow conditioner to remove swirl. Detailed design information, including installation requirements, is given in various international and national standards, including the British Standard EN ISO 5167 and ASME MFC-3M[44]. As we discussed in Section 7.6, stagnation-pressure loss represents a loss (or, more correctly, conversion to heat) of mechanical energy. If the reduction in stagnation pressure is $\Delta p_0$ and the flowrate $\dot{Q}$, the power required to maintain the flow against this pressure difference is $\dot{Q}\Delta p_0$. The engineer who designs a pipework system which includes a Venturi-tube flowmeter has to trade off the long-term operating costs associated with this (and other) power requirements against the capital cost of the Venturi tube and the cost of correct installation.

---

**ILLUSTRATIVE EXAMPLE 8.1**

A Venturi tube installed in a horizontal pipe 80 mm in diameter has a throat diameter of 50 mm (see Figure E8.1). The flowing fluid is compressed air with a density of 5 kg/m³ and a dynamic viscosity of $1.8 \times 10^{-5}$ Pa · s. In a calibration test at a mass flowrate of 1.5 kg/s, the static pressure upstream of the Venturi tube was 4.20 bar, the throat pressure 3.69 bar, and

---

[43] See also comments on the installation requirements in the final paragraph of this section.

[44] Most standards are published in several parts and periodically updated.

the pressure at the Venturi-tube exit 4.19 bar. Calculate the coefficient of discharge, the pipe Reynolds number, the pressure-loss coefficient for the Venturi tube, and the rate of energy dissipation by the fluid flowing through the Venturi tube.

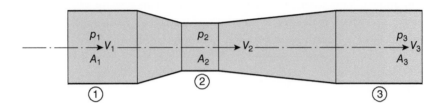

**Figure E8.1**

## Solution

$D_1 = 0.08$ m; $D_2 = 0.05$ m; $D_3 = 0.08$ m; $\rho_F = 5$ kg/m³; $\mu_F = 1.8 \times 10^{-5}$ Pa · s; $\dot{m}_A = 1.5$ kg/s; $p_1 = 4.20 \times 10^5$ Pa; $p_2 = 3.69 \times 10^5$ Pa; $p_3 = 4.19 \times 10^5$ Pa.

The coefficient of discharge $C_D \equiv \dot{m}_A/\dot{m}_{TH}$, where $\dot{m}_{TH}$ is the theoretical mass flowrate according to equation (8.5)[45]

$$\dot{m}_{TH} = A_2 \sqrt{\frac{2\rho_F\left(p_1 - p_2\right)}{\left[1 - (A_2/A_1)^2\right]}}$$

$$= \frac{\pi \times 0.05^2}{4} \sqrt{\frac{2 \times 5 \times (4.20 - 3.69) \times 10^5}{\left[1 - (0.05/0.08)^4\right]}}$$

$$= 1.523 \text{ kg/s}$$

so that

$$C_D = 1.5/1.523 = 0.985.$$

Since the pipe Reynolds number $Re$ is found from $Re = \rho_F V_1 D_1/\mu_F$, we need to calculate the average (or mean) flow velocity $V_1$ from the continuity equation (6.3), i.e.

$$V_1 = \frac{\dot{m}_A}{\rho_F A_1} = \frac{1.5}{5 \times \pi \times 0.08^2/4} = 59.7 \text{ m/s}$$

and so

$$Re = \frac{5 \times 59.7 \times 0.08}{1.8 \times 10^{-5}}.$$

The pressure-loss coefficient was defined in Section 7.5 as

$$C_P \equiv \frac{\Delta p}{\frac{1}{2}\rho_F V^2}.$$

In this case we take $\Delta p = p_1 - p_3 = 10^3$ Pa, i.e. the pressure difference between two sections where the areas are the same, and $V = V_1$ so that

---

[45] In an examination, equation (8.5) would either be given or have to be derived from Bernoulli's equation rather than remembered.

$$C_P = \frac{10^3}{0.5 \times 5 \times 59.7^2} = 0.112.$$

Finally, the rate of energy dissipation is given by

$$P = \frac{\dot{m}_A \Delta p}{\rho_F} = \frac{1.5 \times 10^3}{5} = 300.$$

### Comment:

While the closeness of the value of $C_D$ (i.e. 0.985) to unity indicates a well-designed upstream section, the $C_P$ value is rather high and this is reflected in the energy dissipation rate which would result in a small rise in fluid temperature (about 0.05 °C in this case).

## 8.4 Other Venturi-tube applications

The pressure reduction produced at the throat of a Venturi tube leads to its use as a suction device in a number of practical applications, including gas-fired water-heater control systems, carburettors, and firehose foam **injectors**. A typical application is illustrated in Illustrative Example 8.2.

### ILLUSTRATIVE EXAMPLE 8.2

The convergent-divergent nozzle arrangement shown in Figure E8.2 is used to inject liquid from a reservoir into a gas stream. If the stagnation pressure of the gas is $p_0$, show that the minimum volumetric flowrate $\dot{Q}_G$ of gas through the nozzle which will produce a liquid flow is given by

$$\dot{Q}_G = A\sqrt{2\left(p_0 - B + \rho_L g H\right)/\rho_G}$$

where $A$ is the cross-sectional area of the nozzle throat, $\rho_G$ is the gas density, $\rho_L$ is the liquid density, $B$ is the barometric pressure which acts on the liquid surface, $g$ is the acceleration due to gravity, and $H$ is the vertical height of the injection-tube tip above the liquid surface. Any effect of the injection tube on the gas flow can be neglected, the gas flow can be considered to be loss free, and the gas density neglected relative to that of the liquid.

If the relationship between the mass flowrate of liquid $\dot{m}_L$ and the frictional pressure drop $\Delta p_f$ between the injector-tube tip and the inlet to the injector tube is

$$\dot{m}_L = C\Delta p_f$$

where $C = 1.33 \times 10^{-6}$ m·s, calculate the volumetric flowrate of gas required to produce a liquid mass flowrate of $8 \times 10^{-3}$ kg/s. The gas density is 1.2 kg/m$^3$, and the stagnation pressure of the gas stream is 1.1 bar. The tip of the injector tube is 100 mm above the liquid surface, and the liquid density is 800 kg/m$^3$. The throat area of the nozzle is $10^{-3}$ m$^2$, and the barometric pressure is 1.01 bar.

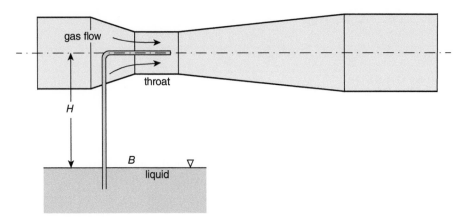

**Figure E8.2**

## Solution

According to Bernoulli's equation, as the gas flowrate $\dot{Q}_G$ through the nozzle is progressively increased, the static pressure $p_t$ at the throat drops such that

$$p_0 - p_t = \frac{1}{2}\rho_G V^2$$

where the velocity $V$ of the gas at the nozzle throat is obtained from the continuity equation, i.e. $V = \dot{Q}_G/A$. The two equations can be combined to give

$$p_0 - p_t = \frac{1}{2}\rho_G \left(\frac{\dot{Q}_G}{A}\right)^2.$$

Once the static pressure falls below the barometric pressure $B$, the injection tube behaves much like a piezometer tube, and the liquid will rise within the tube to a height $h$ above the level of the liquid surface, as given by the hydrostatic equation

$$p_t + \rho_L gh = B.$$

At a certain gas flowrate, the liquid rises to the top of the injection tube such that $h = H$. The corresponding value of $p_t$ is then given by

$$p_t = B - \rho_L gH$$

and the corresponding gas flowrate by

$$p_0 - B + \rho_L gH = \frac{1}{2}\rho_G \left(\frac{\dot{Q}_G}{A}\right)^2.$$

After rearrangement, this equation gives

$$\dot{Q}_G = A\sqrt{2\left(p_0 - B + \rho_L gH\right)/\rho_G}$$

which corresponds with the value of $\dot{Q}_G$, which must be exceeded to produce a flow of liquid into the gas stream.

Higher gas flowrates than this minimum will result in a liquid mass flowrate $\dot{m}_L$ according to

$$\dot{m}_L = C\Delta p_f$$

where $\Delta p_f$ is the frictional pressure drop over the total length of the injection tube corresponding to $\dot{m}_L$, the liquid density and viscosity, and the tube diameter, all of which are accounted for by the dimensional constant $C$. The overall pressure drop for the injection tube is the sum of $\Delta p_f$ and the hydrostatic pressure difference $\rho_L gH$, i.e.

$$B - p_t = \Delta p_f + \rho_L gH.$$

For the numerical part of the problem we have

$$C = 1.33 \times 10^{-6} \text{ m.s}; \; \dot{m} = 8 \times 10^{-3} \text{ kg/s}; \; \rho_G = 1.2 \text{ kg/m}^3; \; p_0 = 1.1 \times 10^5 \text{ Pa}; \; H = 0.1 \text{ m};$$
$$\rho_L = 800 \text{ kg/m}^3; \; A = 10^{-3} \text{ m}^2; \; B = 1.01 \times 10^5 \text{ Pa}.$$

To produce a liquid flowrate $\dot{m}_L = 8 \times 10^{-3}$ kg/s requires

$$\Delta p_f = \frac{\dot{m}_L}{C} = \frac{8 \times 10^{-3}}{1.33 \times 10^{-6}} = 6015 \text{ Pa}$$

so that

$$B - p_t = 6015 + 800 = 6815 \text{ Pa}.$$

Since we have $B = 1.01 \times 10^5$ Pa, we find $p_t = 9.42 \times 10^4$ Pa. The relationship between $p_0, p_t$, and $\dot{Q}_G$ is still valid, i.e.

$$p_0 - p_t = \frac{1}{2}\rho_G \left(\frac{\dot{Q}_G}{A}\right)^2.$$

Since the stagnation pressure $p_0$ is given as 1 bar, we have

$$\left(\frac{\dot{Q}_G}{A}\right)^2 = \frac{2\left(p_0 - p_t\right)}{\rho_G} = \frac{2 \times 5800}{1.2} = 9666 \text{ m}^2/\text{s}^2.$$

From this, the gas flowrate $\dot{Q}_G$ is

$$\dot{Q}_G = 10^{-3} \times \sqrt{9666} = 0.0983 \text{ m}^3/\text{s}.$$

The corresponding gas velocity in the throat is $0.0983/10^{-3} = 98.3$ m/s, which is well below the level at which compressibility effects become significant.

## 8.5 Orifice-plate flowmeter

A relatively simple, and therefore inexpensive, alternative to the Venturi-tube flowmeter is the **orifice-plate flowmeter**. An orifice plate is a thin disk with a hole (the orifice) in it which has an open area significantly smaller than that of the pipe cross section. As is the case for the Venturi tube, orifice-plate flowmeters can be used with both gases and liquids, and even a vapour such

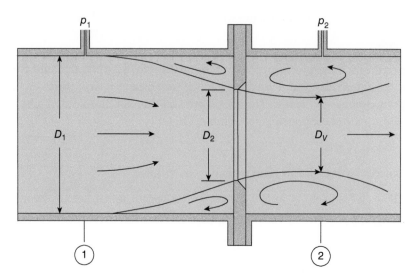

**Figure 8.3** Orifice-plate flowmeter

as steam. In principle, the orifice can be of any shape and located anywhere in the disk but is usually circular, concentric with the pipe bore, in the diameter range $0.75D_1 > D_2 > 0.2D_1$ (minimum value of $D_2$ is 12.5 mm), with a sharp bevelled edge. In a typical installation, as shown in Figure 8.3, the orifice plate is clamped between two flanges, and the pressure drop which results from the acceleration of the fluid passing through the orifice is measured between pressure tappings[46] located at distances $D_1$ upstream and $D_1/2$ downstream of the plate. A full specification for the design and installation of orifice-plate flowmeters is given in ISO 5167.

The basic analysis of flow through an orifice is identical to that for a Venturi tube, with the same final result for the theoretical volumetric flowrate $\dot{Q}_{TH}$ in terms of the pressure difference $p_1 - p_2$, the fluid density $\rho_F$, the cross-sectional area of the pipe $A_1$, and that of the orifice $A_2$, i.e.

$$\dot{Q}_{TH} = A_2 \sqrt{\frac{2(p_1 - p_2)}{\rho_F\left[1 - (A_2/A_1)^2\right]}}. \tag{8.8}$$

It turns out from calibration that the value of the coefficient of discharge,

$$C_D = \frac{\dot{Q}_A}{\dot{Q}_{TH}}$$

$\dot{Q}_A$ being the actual volumetric flowrate, is typically about 0.6. Such a low value, compared with a Venturi-tube flowmeter, suggests that Bernoulli's equation is a poor basis on which to analyse flow through an orifice plate. Reference to Figure 8.3 reveals that the fault is not so much with Bernoulli's equation but with the way it has been applied. In effect, the fluid passing through an orifice creates its own Venturi tube with a contraction starting at about location ① and a

---

[46] A pressure tapping is a small hole drilled into a surface such that the pressure of the fluid on the side exposed to the flow is communicated to a pressure sensor connected to the hole on the other side.

throat at location ②, the latter a distance of about $D_1/2$ downstream of the plate. This fluid throat is called the *vena contracta*, which means contracted vein, and if its cross-sectional area $A_V$ replaced $A_2$ in equation (8.8), the result would be a coefficient of discharge practically equal to unity. Unfortunately, the diameter $D_V$ is not defined by the geometry of the orifice plate nor is it easily measured. In practice, orifice plates are widely used with coefficients of discharge either based on a standard or determined from calibration tests. A low value for $C_D$ does not mean that an orifice flowmeter is inherently inaccurate: the accuracy is determined by that of $C_D$. As with the Venturi tube, for a given orifice plate, $C_D$ depends upon the Reynolds number.

Immediately upstream and downstream of the orifice plate the central stream of high-velocity fluid is surrounded by recirculating **eddies** of fluid within which the fluid velocity is relatively low (for this reason, sometimes called a **deadwater zone**) and the pressure roughly constant. Location ② corresponds with the position of minimum static pressure (at the *vena contracta*) so that the measured pressure difference $p_1 - p_2$ is as high as possible, thereby improving the accuracy of the flowrate measurement. The loss of stagnation pressure, i.e. the irrecoverable pressure loss, is much higher (for the same flowrate, fluid, and pipe diameter) for an orifice plate than for a Venturi tube. This is partly due to the contraction region but primarily the result of the rather violent way in which the flow recovers downstream of the orifice without the aid of a diffuser. As with the Venturi tube, there is an economic trade-off between the low capital cost of an orifice plate and the operating cost associated with the irrecoverable pressure loss.

---

**ILLUSTRATIVE EXAMPLE 8.3**

An orifice-plate flowmeter with an orifice diameter of 350 mm and a coefficient of discharge of 0.6 is used to monitor the flowrate of water in a pipe of 500 mm diameter. Calculate the volumetric flowrate if the pressure difference across the orifice plate is 1.26 bar and estimate the diameter of the *vena contracta*.

Solution

Using the symbols of Figure 8.3, $D_2 = 0.35$ m; $D_1 = 0.5$ m; $C_D = 0.6$; $\Delta p = p_1 - p_2 = 1.26 \times 10^5$ Pa; and $\rho_F = 1000$ kg/m$^3$ (water).
We have $A_1 = \pi D_1^2/4 = 0.196$ m$^2$; $A_2 = \pi D_2^2/4 = 0.0962$ m$^2$; and $A_2/A_1 = (0.35/0.5)^2 = 0.49$.
From equation (8.6),

$$\dot{Q}_{TH} = 0.0962 \sqrt{\frac{2 \times 1.26 \times 10^5}{10^3 \left(1 - 0.49^2\right)}} = 1.75 \text{ m}^3/\text{s}$$

and so

$$\dot{Q}_A = C_D \dot{Q}_{TH} = 1.05 \text{ m}^3/\text{s}.$$

As mentioned in the penultimate paragraph before this example, if the area $A_2$ in equation (8.6) is replaced by that of the *vena contracta* $A_V$, we have

$$\dot{Q}_A = A_V \sqrt{\frac{2(p_1 - p_2)}{\rho_F \left[1 - (A_V/A_1)^2\right]}}$$

wherein we have taken $C_D = 1$. This equation can be rearranged to give an explicit expression for $A_V$, as follows

$$\frac{1}{A_V^2} = \frac{1}{A_1^2} + \frac{2(p_1 - p_2)}{\rho_F \dot{Q}_A^2}.$$

Substitution of the values for $A_1$, $\rho_F$, $\dot{Q}_A$, and $p_1 - p_2$ leads to

$$\frac{1}{A_V^2} = \frac{1}{0.196^2} + \frac{2 \times 1.26 \times 10^5}{10^3 \times 1.05^2}$$

from which $A_V = \pi D_V^2/4 = 0.0627 \text{ m}^2$ and so $D_V = 0.283$ m or 283 mm, i.e. the diameter of the *vena contracta* is about 19% smaller than that of the orifice (350 mm).

## 8.6 Other differential-pressure inline flowmeters

The Venturi tube and the orifice plate are both examples of flowmeters designed on the basis of Bernoulli's equation where flowrate is determined from the measured static-pressure difference across an area reduction. Both devices are manufactured to high tolerances according to internationally agreed design specifications. There are numerous other differential-pressure inline flowmeters, such as flow nozzles, flow tubes, and the **Dall tube**, which is essentially a combination of a Venturi tube and an orifice plate. Which device is chosen and the material from which it is manufactured (bronze, mild steel, and stainless steel are common choices), depends upon the application, including such considerations as contamination of the flowing fluid by solid particles which increases wear, corrosivity, tolerable pressure loss, and required accuracy.

## 8.7 Formula One racing car

The modern **Formula One**, or Grand Prix, **racing car** is a complex package of mechanical and electronic components designed to a formula (hence the name) defined by the sport's governing body, the **Federation Internationale de l'Automobile (FIA)**, which prescribes a wide range of design parameters, including critical dimensions (length, width, wheelbase, tyre radius and width, etc.), weight, allowable fuel load, volumetric engine capacity, etc. Aerodynamic performance has long been a critical aspect of the design of high-performance racing cars, including Formula One, GP2, Indy Car, and LMP2. **Computational Fluid Dynamics** (CFD) and extensive wind-tunnel testing of large-scale models is an essential aspect of racing-car development. An essential feature of wind tunnels used to investigate the aerodynamic characteristics of cars is a **rolling road** to properly simulate the aerodynamic interaction between a car and the road surface over which it travels. Cryogenic wind tunnels, which

**Figure 8.4** Idealised Formula One car

operate at reduced temperature, are sometimes employed to achieve Reynolds numbers close to those which correspond to typical racing speeds (i.e. up to about 360 kph).

Figure 8.4 shows an idealised picture of a Formula One car. Multi-element front and rear aerofoils, designed to produce download (i.e. negative lift) on the front and rear wheels and thereby improve traction in corners, are the most obvious aspects of design motivated solely by aerodynamic considerations. Considerable download is also generated by the underside (floor) of the car which is designed to reduce the pressure of air flowing under the car. Just as for the wings and other **lifting surfaces** of aircraft, devices which generate aerodynamic download inevitably result in drag (known as **induced drag**) which adds to the drag associated with the exposed tyres, bodywork, radiators, oil coolers, engine inlet, and the driver. The complexity of the aerodynamic problem is made even worse by the interaction between these individual components and, in actual racing, other cars, particularly when one car is travelling in the wake of another car a short distance in front. It is the intense **trailing vortices**, often visible in humid or damp conditions swirling away from the endplates on either side of the rear wing, which are responsible for the asymmetric loss of download experienced by the following car. The same phenomenon affects one aircraft following another and can lead to catastrophic consequences for small aircraft following much larger aircraft.

On the basis of the material covered so far, we can make crude estimates of some aspects of the aerodynamic performance of a Formula One car. A more complete analysis would be immensely complicated, carried out on a powerful computer, and require knowledge of closely guarded design data.

We assume the following values which have been extracted or estimated from published information

| | |
|---|---|
| maximum speed 330 kph | $V = 91.7\,\text{m/s}$ |
| tractive power 900 hp | $P = 672\,\text{kW}$ |
| projected frontal area | $A_F = 1.5\,\text{m}^2$ |
| air density | $\rho = 1.2\,\text{kg/m}^3$ |
| area reduction for flow beneath car | 1.15:1 |
| mass of car including fuel and driver | $m = 800\,\text{kg}$ |
| projected plan area | $A_P = 7\,\text{m}^2$ |

At maximum speed $V$ we assume that the tractive power $P$ is used to overcome the aerodynamic drag force $D$, so that

$$P = DV$$

and we can therefore calculate the aerodynamic drag force to be

$$D = \frac{P}{V} = \frac{6.72 \times 10^5}{91.7} = 7328 \, \text{N}.$$

In the above estimate we have neglected the rolling resistance.

We can now calculate the overall drag coefficient $C_D$ from

$$C_D = \frac{D}{\frac{1}{2}\rho V^2 A_F} = \frac{7.328 \times 10^3}{0.5 \times 1.2 \times 91.7^2 \times 1.5} = 0.968.$$

This value of $C_D$ is about three times the value for a well-designed passenger car for which low drag is desirable in order to reduce fuel consumption and aerodynamic noise. Neither of these requirements is of paramount importance to the designer of a Formula One car, and the high value of $C_D$ is to a large extent a direct consequence of the induced drag associated with the high levels of download. Drag coefficients for various shapes of practical significance are listed in Section 18.15.

If we assume the airflow under the car is loss free, we can apply Bernoulli's equation to estimate the reduction in pressure below the ambient level $B$, i.e.

$$B + \frac{1}{2}\rho V^2 = p_2 + \frac{1}{2}\rho V_2^2$$

where $p_2$ is the pressure beneath the car and $V_2$ the corresponding airspeed. There are many things that can be criticised about this simple approach. For example, the area beneath the car is far from constant, especially towards the rear where a diffuser brings the flow back to ambient pressure. Also, the flow is likely to be three dimensional rather than one dimensional. Nevertheless, Bernoulli's equation incorporates much of the essential physics of many real flows and is unlikely to provide answers which are orders of magnitude different from reality.

For an area ratio of 1.15:1, the continuity equation (6.1) leads to $V_2 = 1.15V = 115$ m/s and so

$$B - p_2 = 0.5 \times 1.2 \times \left(115^2 - 100^2\right) = 7225 \, \text{Pa}.$$

If we now assume that this pressure difference is applied to the projected plan area $A_P$, the corresponding download is 50.6 kN compared with the weight of the car $mg = 7.85$ kN. The outcome of this calculation is critically dependent upon both the plan area and the cross-sectional area change. The overall download (including the wings) is certainly well in excess of the weight of the car and our crude calculation suggests that the contribution due to the underflow may be substantial.

We conclude this section by estimating the retardation due to aerodynamic drag. If all tractive force (whether due to the power unit or the brakes) is lost, according to Newton's second law of motion at maximum speed, we have a deceleration given by

$$ma = -D.$$

Our estimate for the drag force $D$ at maximum speed was 7328 N, which leads to

$$-a = \frac{D}{m} = \frac{7328}{800} = 9.16 \text{ m/s}^2 \quad \text{or} \quad 0.933\, g,$$

i.e. a deceleration about 7% less than $1g$, solely due to aerodynamic drag, a value well in excess of the braking capability of the majority of passenger cars.

## 8.8 Pitot tube

The simple L-shaped tube shown in Figure 8.5(a) is called a **Pitot tube** after the French engineer who devised it in the $18^{\text{th}}$ century. When immersed in a liquid flow to a depth $Z$ as shown, the liquid enters the tube and rises to a level $H$ above the free surface. If the liquid flow is steady, once the equilibrium situation is reached the liquid velocity within the tube falls to zero, and the point $P$ at its tip becomes a **stagnation point** at which the static pressure is equal to the stagnation pressure $p_0$. If the undisturbed liquid velocity upstream of the tip is $V$, and the corresponding static pressure is $p$, then from Bernoulli's equation we have

$$p_0 = p + \frac{1}{2}\rho_L V^2 \quad \text{or} \quad p_0 - p = \frac{1}{2}\rho_L V^2$$

where $\rho_L$ is the liquid density. From the hydrostatic equation (4.8) for a constant-density fluid, we have

$$p_0 = B + \rho_L g(H + Z) \quad \text{and} \quad p = B + \rho_L g Z$$

where $B$ is the static pressure (the **barometric pressure**) acting on the liquid surface. The validity of these two equations should be evident from Figure 8.5(a). If we eliminate $B$, we have

$$p_0 - p = \rho_L g H$$

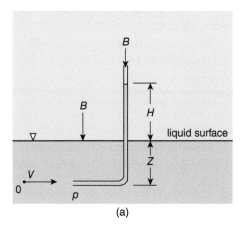
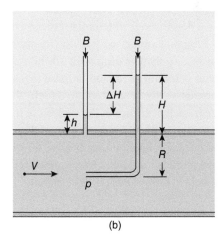

**Figure 8.5** Pitot tube: (a) free-surface flow; (b) pipe flow

so that

$$\frac{1}{2}\rho_L V^2 = \rho_L g H \quad \text{or} \quad V = \sqrt{2gH}. \tag{8.9}$$

The more usual arrangement for the measurement of the velocity of a flowing gas or liquid using a Pitot tube is in combination with a probe to measure the static pressure $p$ in the vicinity of the Pitot tube tip.

As shown in Figure 8.5(b) for liquid flow in a pipe the static pressure can be measured with an appropriately positioned piezometer tube (see Section 4.6). If the pipe radius is $R$ and the Pitot tube is aligned with the pipe axis, as shown, then

$$p_0 = B + \rho_L g (H + R) \quad \text{and} \quad p = B + \rho_L g (h + R).$$

As before, from Bernoulli's equation we have

$$p_0 - p = \frac{1}{2}\rho_L V^2$$

so that now

$$\frac{1}{2}\rho_L V^2 = \rho_L g (H - h) = \rho_L g \Delta H \tag{8.10}$$

where $\Delta H = H - h$ is the vertical height difference between the liquid levels in the Pitot and piezometer tubes. The velocity on the pipe centreline is then

$$V = \sqrt{2g\Delta H}.$$

For the flow through a pipe of an incompressible gas of density $\rho_G$, it would be necessary to measure the pressure difference $p_0 - p$ using a device such as a manometer or differential-pressure transducer and the velocity would then be determined from

$$V = \sqrt{2(p_0 - p)/\rho_G}. \tag{8.11}$$

There is an important difference between the application of Bernoulli's equation to the analysis of flow through differential-pressure inline flowmeters and to the analysis of the response of a Pitot tube immersed in a flowing fluid. For the flowmeters, the analysis in the early sections of this chapter dealt with all the fluid flowing through the pipe and so made use not only of Bernoulli's equation but also the continuity equation, with any departures from one-dimensional flow being accounted for by the coefficient of discharge. In the case of a Pitot tube, there is no dependence in the analysis of the one-dimensional assumption and it requires only that we know the difference between the stagnation and static pressures at the measurement location. It is often the case in pipe flow that there is negligible variation in the static pressure with radial distance from the centreline so that the flow velocity at any radius[47] can be determined using a Pitot tube at that radius.

---

[47] The variation of velocity with distance from a surface is termed the **velocity profile** or **distribution** (see Chapters 16, 17, and 18).

**ILLUSTRATIVE EXAMPLE 8.4**

The output from a differential-pressure transducer, with one side connected to a Pitot tube immersed in a gas flow and the other side to a static-pressure tapping in the near vicinity of the Pitot tube, is 3.9 kPa. If the gas density is 0.8 kg/m³ and its soundspeed 330 m/s, calculate the gas velocity and the corresponding Mach number to verify that the flow can be assumed incompressible. If the pressure transducer were to be replaced by a U-tube manometer, would kerosene (density 800 kg/m³) or mercury be the more suitable manometer liquid?

**Solution**

$p_0 - p = 3900$ Pa; $\rho_G = 0.8$ kg/m³; $c = 330$ m/s; $\rho_M = 800$ or $13.6 \times 10^3$ kg/m³.
We start with equation (8.11) derived above (in an examination the derivation would probably be part of the problem), so that

$$V = \sqrt{2\left(p_0 - p\right)/\rho_G} = \sqrt{2 \times 3900/0.8} = 98.7 \, \text{m/s}.$$

The corresponding Mach number $M = V/c = 98.7/330 = 0.299$, so the assumption of incompressible flow is only just valid (i.e. $M$ is below 0.3).

From Section 4.7, for a U-tube manometer we have $\Delta p = (\rho_M - \rho_F)g\Delta H$, where $\Delta H$ is the difference in the levels of the manometer liquid in the two arms of the U-tube. For mercury we thus find $\Delta H = 3900/[(13.6 \times 10^3 - 0.8) \times 9.81] = 0.292$ m or 29.2 mm, and for kerosene $\Delta H = 3900/[(800 - 0.8) \times 9.81] = 0.497$ m or 497 mm. On the basis of height difference, either liquid would be acceptable for the measurement although the inaccuracy of measuring 497 mm would be much less than for 29.2 mm. It is more likely that safety concerns might rule out both liquids: kerosene is inflammable, and mercury poisonous. In fact, in this instance, water would be preferable.

## 8.9 Pitot-static tube

The **Pitot-static tube** illustrated in Figure 8.6 consists of an inner tube to sense the stagnation pressure of a flow and a concentric outer tube, of outer diameter $D$, closed and rounded at its upstream end but perforated by a series of small holes to sense the static pressure $p$ of the flow. The static-pressure holes should be located sufficiently far downstream of the probe tip for any disturbance to the flow created by the probe to have died out: a distance of $6D$ is typical. Any bend in the two tubes should be a similar distance downstream of the static-pressure holes. To ensure an accurate measurement, the tube assembly must be aligned with the flow to within about 5°.

If the density of the flowing fluid is $\rho_F$, assumed constant, and the undisturbed flow velocity is $V$, then from Bernoulli's equation we have

$$p_0 = p + \tfrac{1}{2}\rho_F V^2 \quad \text{so that, as before, equation (8.11)} \quad V = \sqrt{\frac{2\left(p_0 - p\right)}{\rho_F}}.$$

In a typical application, as shown in Figure 8.5, the pressure difference $p_0 - p$ is measured using a differential-pressure transducer or a U-tube manometer. In the latter case, if the density of

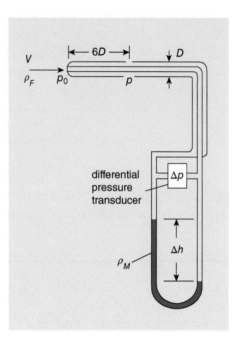

**Figure 8.6** Pitot-static tube

the manometer liquid is $\rho_M$ and the vertical height difference of the liquid in the two arms of the manometer is $\Delta H$, then we have

$$p_0 - p = (\rho_M - \rho_F) g \Delta H = \tfrac{1}{2} \rho_F V^2$$

from which

$$V = \sqrt{\frac{2\,(\rho_M - \rho_F)\,g \Delta H}{\rho_F}}.$$

As we shall see in Chapter 11, it is relatively straightforward to allow for the effects of compressibility on the behaviour of a Pitot tube.

## 8.10 Liquid draining from a tank

Figure 8.7 shows a cylindrical container of cross-sectional area $A_S$ open to atmospheric pressure $B$ and containing a liquid of density $\rho_L$. Under the influence of gravity, the liquid drains out of the tank through an orifice in the tank wall of cross-sectional area $A_O$. In practice, the orifice could be simply a hole, a nozzle such as a Venturi tube, or a control valve.

To determine the volumetric flowrate $\dot{Q}$ with which the liquid flows out of the tank, we start by writing Bernoulli's equation for a streamline connecting the liquid surface and the jet emerging from the orifice

$$p_T = B + \rho_L g h + \tfrac{1}{2} \rho_L V_S^2 = B + \tfrac{1}{2} \rho_L V_O^2$$

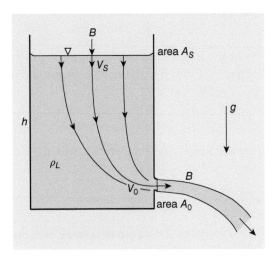

**Figure 8.7** Liquid draining from a tank

where $p_T$ is the total pressure of the liquid, $B$ is the barometric pressure, $h$ is the vertical height of the liquid surface above the orifice[48], $V_S$ is the surface velocity, and $V_O$ is the velocity of the liquid passing through the orifice. An unusual feature of this problem is that the static pressure of the liquid is equal to the ambient pressure $B$ both at the liquid surface and at the location of the orifice.

The continuity equation in this case is as follows

$$\dot{Q} = A_S V_S = A_O V_O.$$

In a typical situation, the surface area $A_S$ is far greater than the orifice area $A_O$ so that (from the continuity equation) $V_O \gg V_S$ and we may neglect the term $\rho_L V_S^2/2$ in comparison with $\rho_L V_O^2/2$ in Bernoulli's equation, which then simplifies to

$$V_O^2 = 2gh \quad \text{or} \quad V_O = \sqrt{2gh}$$

which is called **Torricelli's formula**.

The theoretical volumetric flowrate $\dot{Q}_{TH}$ can now be written as

$$\dot{Q}_{TH} = A_O V_O = A_O \sqrt{2gh}$$

and the actual flowrate $\dot{Q}_A$ as

$$\dot{Q}_A = C_D A_O \sqrt{2gh} \tag{8.12}$$

where we have introduced $C_D$, the coefficient of discharge (see Section 8.3) for flow through the orifice.

It is now straightforward to calculate the time $t$ required for the liquid level to fall to an intermediate value $h$ from an initial value $h_0$ at $t = 0$. We make use of the kinematic relation

[48] It is assumed that the orifice diameter is small compared with $h$ so that the vertical pressure variation across the orifice is negligible.

$$V_S = -\frac{dh}{dt}$$

i.e. the downward velocity of the liquid surface must equal the rate of change of the liquid level $h$. We can combine this result with the continuity equation and equation (8.12) for the actual volumetric flowrate $\dot{Q}_A$ as follows

$$\dot{Q}_A = A_S V_S = -A_S \frac{dh}{dt} = C_D A_O \sqrt{2gh}$$

which can be rearranged to yield a first-order ordinary differential equation for $h$ as a function of time $t$, i.e.

$$\frac{1}{\sqrt{h}} \frac{dh}{dt} = -C_D \frac{A_O}{A_S} \sqrt{2g}.$$

Since the right-hand side is a constant, this equation is easily integrated to give the desired relationship between liquid level $h$ and the time $t$

$$2\left(\sqrt{h} - \sqrt{h_0}\right) = -\left(C_D \frac{A_O}{A_S} \sqrt{2g}\right) t \qquad (8.13)$$

wherein the constant of integration has been determined from the initial condition $h = h_0$ at $t = 0$.

As a final step, we can combine equations (8.12) and (8.13) to give an equation for the flowrate $\dot{Q}_A$ as a function of time $t$

$$\dot{Q}_A = C_D A_O \sqrt{2gh_0} \left[ 1 - \left(C_D \frac{A_O}{A_S} \sqrt{\frac{g}{2h_0}}\right) t \right].$$

This equation shows that the flowrate decreases linearly with time from the initial value $C_D A_O \sqrt{2gh_0}$. The observant reader will have noticed that the problem we have just dealt with involves an unsteady flow (i.e. the flowrate $\dot{Q}_A$ varies with time) but has been analysed on the basis of steady-flow forms of Bernoulli's equation and the continuity equation. As we pointed out in Section 6.2, a steady flow is one in which the velocity and pressure at any given point in the flow do not change with time but there will usually be changes from one point to the next. In other words, even in a steady flow, a fluid particle moving along a streamline will experience acceleration and deceleration. The question to be answered in the draining-tank problem is whether the acceleration at a fixed point has a significant influence on the flow.

It is straightforward to estimate the fluid acceleration at the upper surface and at the orifice. For the upper surface we have

$$V_S = \frac{\dot{Q}_A}{A_S} = C_D \frac{A_O}{A_S} \sqrt{2gh_0} \left[ 1 - \left(C_D \frac{A_O}{A_S} \sqrt{\frac{g}{2h_0}}\right) t \right]$$

so that the corresponding acceleration is given by

$$\frac{dV_S}{dt} = -C_D^2 \left(\frac{A_O}{A_S}\right)^2 g.$$

At the orifice, the flow velocity is given by

$$V_O = \frac{\dot{Q}_A}{A_O} = C_D \sqrt{2gh_0} \left[ 1 - \left( C_D \frac{A_O}{A_S} \sqrt{\frac{g}{2h_0}} \right) t \right]$$

and the acceleration is

$$\frac{dV_O}{dt} = - C_D^2 \frac{A_O}{A_S} g.$$

Estimating the acceleration of a fluid particle at any point along a streamline is far more difficult. For a steady flow we can write

$$\frac{dV}{dt} = V \frac{dV}{ds}$$

where $V$ is the flow velocity a distance $s$ along the streamline. Although calculating $V(s)$ is beyond the scope of this book, we can at least make an order of magnitude estimate. We know that between the surface and the orifice the flow velocity increases from $V_S$ to $V_O$ over a distance comparable with $h_0$ so that we have, very roughly,

$$\frac{dV}{dt} \approx \frac{V_O^2}{h_0} = 2g C_D^2.$$

If we compare this result with that for $dV_O/dt$, we see that the latter is smaller by a factor $A_O/2A_S$ which is likely to be far less than unity since $A_O \ll A_S$ and we conclude that treating this unsteady flow as though it was steady is probably valid. Such flows are termed **quasi steady**.

As the following example illustrates, the foregoing analysis can be extended without great difficulty to the situation where the tank is closed at the top and a pressure $p_S$ is applied to the liquid surface.

---

**ILLUSTRATIVE EXAMPLE 8.5**

Gas at a pressure $p_S$ is used to force a liquid of density $\rho_L$ out of a container, as shown in Figure E8.5. The liquid leaves the container through a Venturi tube of exit area $A$. Show that the mass flowrate $\dot{m}$ of the liquid is given by

$$\dot{m} = A \sqrt{2\rho_L \left( p_S - B + \rho_L g h \right)}$$

where $h$ is the vertical height of the liquid surface above the Venturi-tube axis. The flow, including that through the Venturi tube, may be assumed to be steady, frictionless, and one dimensional, and the liquid pressure at exit from the Venturi tube equal to that of the surrounding atmosphere $B$. The downward velocity of the liquid surface may be neglected compared with that of the liquid jet.

If the liquid has a density of 800 kg/m$^3$ and its surface is 3 m above the Venturi tube, the applied pressure is 2 bar, the nozzle exit area is 0.01 m$^2$, and the atmospheric pressure is 1.01 bar, calculate the liquid mass flowrate. Calculate also the jet velocity and the downward velocity of the liquid surface if the cross-sectional area of the tank is 1 m$^2$.

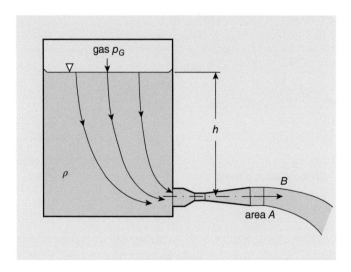

**Figure E8.5**

## Solution

Since the flow can be treated as steady, frictionless, and one dimensional, we can apply Bernoulli's equation between the liquid surface and the Venturi-tube exit, as follows

$$p_T = p_S + \rho_L gh + \tfrac{1}{2}\rho_L V_S^2 = B + \tfrac{1}{2}\rho_L V_J^2$$

where $p_T$ is the total pressure, which is constant along a streamline, $V_S$ is the downward velocity of the liquid surface, and $V_J$ is the jet velocity.

We can rearrange the equation as follows

$$V_J^2 - V_S^2 = 2\left(p_S + \rho_L gh - B\right)/\rho_L.$$

Since $V_S \ll V_J$, we have

$$V_J = \sqrt{2\left(p_S + \rho_L gh - B\right)/\rho_L}.$$

We now introduce the continuity equation, applied to the liquid jet, $\dot{m} = \rho_L A V_J$ and substitute for $V_J$ to obtain

$$\dot{m} = A\sqrt{2\rho_L\left(p_S - B + \rho_L gh\right)}.$$

For the numerical part of the problem we have

$$p_S = 2 \times 10^5 \text{ Pa}; \; B = 1.01 \times 10^5 \text{ Pa}; \; \rho_L = 800 \text{ kg/m}^3; \; h = 3 \text{ m}; \; A = 0.01 \text{ m}^2; \; A_S = 1 \text{ m}^2.$$

It is a matter of straightforward substitution to find $\dot{m} = 140$ kg/s.
From the continuity equation, $V_S = \dot{m}/(\rho_L A_S)$, and $V_J = \dot{m}/(\rho_L A)$, so that

$$V_S = \frac{140}{800 \times 1} = 0.175 \text{ m/s} \quad \text{and} \quad V_J = \frac{140}{800 \times 0.01} = 17.5 \text{ m/s}.$$

We shall return to this problem at the end of Section 8.11.

## 8.11 **Cavitation in liquid flows**

From Bernoulli's equation we can see that in a flowing fluid it is possible to develop very low pressure in regions of high flow velocity. If this pressure falls below the **saturated vapour pressure** for a liquid, tiny vapour bubbles begin to form, a process known as **cavitation** or **flow-induced boiling**. For a given temperature, the saturated vapour pressure is the pressure at which a liquid boils and is in equilibrium with its own vapour, i.e. it is the pressure which exists in pure vapour in contact with the liquid at a given temperature. The variation of the saturated vapour pressure with temperature for water, shown in Figure 2.7 and tabulated in Table A.3 in Appendix 2, is based upon the **saturation table** for water and steam (the vapour form of water). As we should expect, the saturated vapour pressure at 100 °C is 1.01 bar, i.e. at normal atmospheric pressure water boils at 100 °C. If the pressure is reduced to 0.1 bar, water boils at 45.8 °C and, at a pressure of 20 bar, the boiling point is raised to 212.4 °C. An application which takes advantage of the influence of pressure to raise the boiling point of water is the domestic pressure cooker.

Vapour bubbles formed due to the pressure reduction in a flowing liquid initially grow, are swept downstream, and then collapse implosively upon reaching a zone of sufficiently high pressure. Cavitation in pumps and hydraulic turbines is undesirable, first because it leads to a decrease in efficiency, and second because repeated impacts on blading and other components, due to the collapse of vapour bubbles, can be so intense as to cause serious wear (**surface pitting**). Much the same is true for ships' propellers where cavitation can occur at the tips. Further cavitation examples are provided by liquid flow through nozzles, valves, and pipes, where there are no moving parts but the liquid pressure is reduced by a sudden reduction in cross-sectional area. The examples given so far are for isothermal flows whereas in boilers and heating systems cavitation may result from a combination of increased temperature and reduced pressure. Cavitation is often detectable from the sound created by the implosive collapse of the vapour bubbles. In small-scale devices this is a harsh crackling sound whereas in very large structures, such as the **spillway tunnels** which carry water away from a dam, it can sound like rocks impacting the tunnel wall.

For a flow of a fluid of density $\rho_L$ with velocity $V$, the non-dimensional parameter used to characterise cavitation is the **cavitation number $Ca$**, defined as

$$Ca = \frac{p_{REF} - p_V}{\rho_L V^2/2}$$

where $p_{REF}$ is a reference pressure (often taken as the atmospheric or barometric pressure) and $p_V$ is the saturated vapour pressure. Cavitation within a given device occurs if the cavitation number falls below a critical value dependent upon the flow geometry.

---

### ILLUSTRATIVE EXAMPLE 8.6

Figure E8.6 shows water being drawn vertically upwards from a level $z' = 0$ at the bottom of a well to a level $z' = H$ at the inlet to a suction pump. Find the greatest depth of well $H$ from which water can be pumped if the water surface is at atmospheric pressure $B$, the saturated vapour pressure of the water is $p_V$, and the water density is $\rho_L$.

Calculate the water depth if the atmospheric pressure is 1.01 bar and the saturated vapour pressure is 1.23 kPa (corresponding to a water temperature of 10 °C).

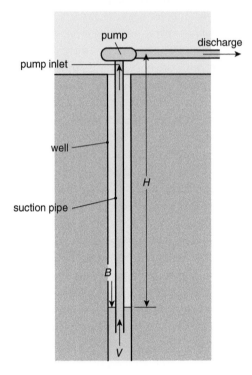

**Figure E8.6**

## Solution

We apply Bernoulli's equation to the flow in the suction pipe, between the pump inlet ($z' = H$) and the level of the water surface in the well ($z' = 0$)

$$p_T = p_I + \rho_L g H + \frac{1}{2}\rho_L V_I^2 = B + \frac{1}{2}\rho_L V_I^2.$$

Note that we have assumed that the water velocity $V_I$ is the same at entry to the suction pipe and the inlet to the pump. The terms involving $V_I$ cancel out and we have

$$H = \frac{B - p_I}{\rho_L g}$$

from which it is clear that $H$ is greatest when the inlet pressure $p_I$ is as low as possible, i.e. when $p_I = p_V$, so that

$$H_{MAX} = \frac{B - p_V}{\rho_L g}.$$

For the numerical part of the problem, we have

$$B = 1.01 \times 10^5 \text{ Pa}; \ p_V = 1.23 \times 10^3 \text{ Pa}; \ \rho_L = 10^3 \text{ kg/m}^3; \ g = 9.81 \text{ m/s}^2.$$

Thus,

$$H_{MAX} = \frac{1.01 \times 10^5 - 1.23 \times 10^3}{10^3 \times 9.81} = 10.2 \text{ m.}$$

---

**ILLUSTRATIVE EXAMPLE 8.7**

If the vapour pressure for the liquid in Illustrative Example 8.5 is $7 \times 10^4$ Pa, what is the smallest throat diameter of the Venturi tube if the liquid is not to cavitate?

**Solution**

If we take the throat velocity at $V_t$ and the corresponding pressure as $p_t$, then we can apply Bernoulli's equation between the throat and the Venturi-tube exit as

$$p_0 = p_t + \frac{1}{2}\rho_L V_t^2 = B + \frac{1}{2}\rho_L V_J^2$$

where $p_0$ is the stagnation pressure on the horizontal streamline coinciding with the nozzle centreline. We have $B = 1.01 \times 10^5$ Pa, and $\rho_L = 800 \text{ kg/m}^3$, and we calculated previously that $V_J = 17.5$ m/s, so that $p_0 = 2.54 \times 10^5$ Pa. Cavitation occurs when the lowest pressure in the Venturi tube $p_t$ falls to the value of the vapour pressure $p_V$, so that

$$p_0 = p_V + \frac{1}{2}\rho_L V_t^2$$

or

$$V_t = \sqrt{\frac{2\left(p_0 - p_V\right)}{\rho_L}} = \sqrt{\frac{2\left(2.54 \times 10^5 - 7 \times 10^4\right)}{800}} = 19.2 \text{ m/s.}$$

If the throat area is $A_t$, from the continuity equation,

$$\dot{m} = \rho_L A_t V_t$$

so that $A_t = \dot{m}/\rho_L V_t = 7.3 \times 10^{-3} \text{ m}^2$. Since $A_t = \pi D_t^2/4$, where $D_t$ is the throat diameter, we have finally $D_t = 0.096$ m or 96 mm.

---

 **8.12 SUMMARY**

In this chapter we have shown how Bernoulli's equation can be applied to practical fluid-flow problems. In the case of internal flows, such as that through a Venturi tube, we also needed the continuity equation to relate changes in cross-sectional area to changes in flow velocity. For liquid flows it was shown that for sufficiently high flowspeeds the static pressure could fall below the saturated vapour pressure and lead to cavitation.

The student should be able to

- identify flow problems where the application of Bernoulli's equation is appropriate
- identify problems where the continuity equation is also needed for their solution

- apply Bernoulli's equation to analyse such internal flow problems as the flow through a Venturi tube or an orifice plate
- understand the concept of a coefficient of discharge as a correction factor
- apply Bernoulli's equation to analyse the response of a Pitot-static tube to a fluid flow
- understand the significance of the saturated vapour pressure to liquid flow and its relevance to cavitation

## 8.13  SELF-ASSESSMENT PROBLEMS

**8.1**  (a) A fluid of density $\rho$ flows through a horizontal duct which contracts from a cross section of area $A_1$ to a minimum (throat) area $A_2$. Assume one-dimensional, incompressible, frictionless flow to show that the theoretical mass flowrate through the duct is given by

$$\dot{m} = A_1 A_2 \sqrt{\frac{2\rho \Delta p}{A_1^2 - A_2^2}}$$

where $\Delta p$ is the static-pressure difference between sections 1 and 2. How would the equation be modified using a coefficient of discharge to determine the actual mass flowrate?

(b) Water flows through a horizontal duct which changes from a circular pipe of diameter 100 mm to an annulus of outer diameter 100 mm and inner diameter 90 mm. Calibration tests show that the coefficient of discharge for this arrangement is 0.94. Calculate the pressure drop across the area change for a mass flowrate of 20 kg/s. Also calculate the velocity and static pressure in the annular section if the upstream stagnation pressure is 7 bar.
(Answers: 0.98 bar, 13.4 m/s, 5.99 bar)

**8.2**  (a) A pure liquid of density $\rho_L$ and saturated vapour pressure $p_V$ flows vertically upwards through a Venturi tube which contracts from a diameter $D_1$ to a throat diameter $D_2$. If the static pressure ahead of the Venturi tube is $p_1$, show that the maximum volumetric flowrate which can be measured before the onset of cavitation is

$$\dot{Q}_{CAV} = \pi D_1^2 D_2^2 \sqrt{\frac{p_1 - \rho_L g S - p_V}{8\rho_L \left(D_1^4 - D_2^4\right)}}$$

where $S$ is the distance from the throat to the location where $p_1$ is measured. Assume frictionless flow.

(b) If the liquid in the above situation is pure water at 90 °C, for which the vapour pressure is $7 \times 10^4$ Pa, calculate the mass flowrate corresponding to the onset of cavitation if the upstream static pressure is 2 bar, $D_1$ is 100 mm, $D_2$ is 50 mm, and $S$ is 5 m. Calculate the pressure differences for the same flowrate if the Venturi tube is operated (i) in a horizontal water line and (ii) in a vertical water line with downflow.
(Answers: 25.8 kg/s, 0.81 bar, 0.32 bar)

**8.3**  (a) A liquid of density $\rho_L$ and vapour pressure $p_V$ flows through a convergent-divergent nozzle which discharges to an ambient pressure $B$. If the exit area $A_E$ of the nozzle is a factor $r$ times the throat area, show that cavitation first occurs at a flowrate $\dot{Q}_{CAV}$, given by

$$\dot{Q}_{CAV} = A_E \sqrt{\frac{2(B - p_V)}{\rho_L (r^2 - 1)}}.$$

Assume one-dimensional, frictionless, incompressible flow.

(b) If the throat diameter is 60 mm and the exit diameter 90 mm, calculate the flowrate at which cavitation first occurs for a liquid of density 800 g/m³ and a vapour pressure of $5 \times 10^4$ Pa if the barometric pressure is 1 bar. Also calculate the stagnation pressure for the flow.
(Answers: 0.0353 m³/s, 1.123 bar)

8.4 (a) A Pitot-static tube in combination with an inclined-tube manometer is used to measure the speed $V$ of an incompressible fluid of density $\rho$, as shown in Figure P8.4. If the cross-sectional area of the inclined tube is $a$ and that of the reservoir is $A$, show that

$$V = \sqrt{2gL \left[ \left( \frac{\rho_M}{\rho} - 1 \right) \left( \sin\theta + \frac{a}{A} \right) \right]}$$

where $\rho_M$ is the density of the manometer liquid, $\theta$ is the inclination angle of the manometer tube, $g$ is the acceleration due to gravity, and $L$ is the change in level of the manometer reading (i.e. the level change measured along the tube).

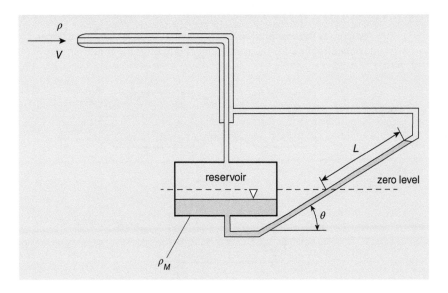

**Figure P8.4**

(b) A Pitot-static tube is used to measure the flowspeed of a gas of density 1.2 kg/m³. The manometer liquid is water. The internal diameter of the manometer tube is 5 mm and that of the reservoir is 100 mm; the inclination angle is 15°. Calculate the flowspeed and dynamic pressure if the manometer reading $L$ is 436 mm.
(Answers: 43.1 m/s, 1.115 kPa)

**8.5** (a) A jet of liquid of density $\rho_L$, surrounded by air, flows vertically downwards from a nozzle of cross-sectional area $A_N$. If the stagnation pressure of the jet at exit from the nozzle is $p_0$, and $B$ is the surrounding air pressure, show that the cross-sectional area of the jet $A_J$ changes with vertical distance below the nozzle exit $z$ according to

$$\left(\frac{A_N}{A_J}\right)^2 = 1 + \frac{\rho_L g z}{p_0 - B}$$

where $g$ is the acceleration due to gravity. Assume the flow is one dimensional and frictionless and regard the air density as negligible.

(b) If the nozzle area $A_N$ is $5 \times 10^{-4}\,\text{m}^2$ and the stagnation pressure $p_0$ is 2 bar, determine the jet velocity a distance 1000 m below the nozzle if the liquid is aviation fuel of density 700 kg/m$^3$ and the ambient pressure is 0.5 bar.
(Answer: 141.6 m/s)

**8.6** (a) The arrangement shown in Figure P8.6 is used to inject liquid detergent of density $\rho_W$ from a pool into the water, also of density $\rho_W$, flowing through a fire hose in order to create foam. The cross-sectional area of the contraction at section 1 is $A_T$ and the cross-sectional area of the outlet nozzle is $A_E$. The stagnation pressure of the water flow is $p_0$, the ambient pressure to which the nozzle discharges is $B$, and the contraction height above the surface of the detergent pool is $H$. Show that the stagnation pressure at which detergent just rises to the top of the vertical tube is given by

$$p_0 - B = \frac{\rho_W g H (A_T/A_E)^2}{1 - (A_T/A_E)^2}$$

where $g$ is the acceleration due to gravity. Assume one-dimensional, incompressible, frictionless flow for the water flow, and hydrostatic conditions for the detergent.

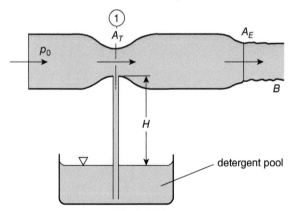

**Figure P8.6**

(b) If the outlet nozzle has a diameter of 70 mm and the internal diameter of the contraction is 60 mm, calculate the stagnation pressure for a water mass flowrate of 55 kg/s if the ambient pressure is 1.02 bar. Calculate also the static pressure at the location of the contraction. What is the maximum vertical height difference between the contraction and the surface of the detergent pool if the detergent is to rise to the top of the vertical tube? If the vapour pressure of water is taken as 2.3 kPa, calculate the water mass flowrate at which cavitation occurs.
(Answers: 2.04 bar, 0.15 bar, 8.9 m, 58.9 kg/s)

# 9 Linear momentum equation and hydrodynamic forces

This brief but important chapter is concerned with fluid flow through a duct which changes in direction (typically a bend) and/or cross-sectional area. Force must be exerted on the fluid to produce the changes in fluid momentum which are a consequence of such geometric changes. What is of interest from an engineering point of view is the external reaction force which has to be applied to a duct to counteract the force exerted by the fluid on its interior surface. We use Newton's second law of motion to derive the linear momentum[49] equation for a flowing fluid. We then identify the separate contributions to the net force acting on the fluid due to the fluid pressure at inlet and outlet to the duct, and the force exerted on the fluid by the duct's interior surface. We exclude from the analysis any body forces, including the weight of the fluid. The analysis is completed by applying the principle of static equilibrium to equate the internal and external forces acting on the duct. Emphasis is given to the vector nature of force and momentum flowrate.

## 9.1 Problem under consideration

We consider the flow of a fluid through a **duct**, such as that illustrated in Figure 9.1, which may be curved or straight and have a cross section which changes in shape and cross-sectional area $A$ with streamwise distance $s$. The term duct is used to mean any passage or channel through which there is fluid flow and includes, for example, pipes, bends, nozzles, Venturi tubes, engine intakes and exhausts, and rocket engines. We retain the assumption of steady, one-dimensional flow, but allow the interaction between the flowing fluid and the duct walls to involve not only the static pressure $p(s)$ but also the surface shear stress $\tau_S(s)$ due to the fluid viscosity, i.e. we no longer assume that the flow is frictionless. The restriction to constant density is also dropped for the basic analysis.

As indicated in Figure 9.1, we apply our analysis to a segment of the duct with an inlet (section ①) and an outlet (section ②). In many practical examples, sections ① and ② will correspond to an actual inlet or outlet, for example the intake to a jet engine or a nozzle exit. As we shall illustrate in Chapter 10, in other situations, such as flow through a complex duct system, an essential aspect of the analysis is to identify an appropriate duct segment for consideration. The volume between the inlet and outlet defined by the **wetted** interior surface of the duct segment is referred to as a **control volume**.

---

[49] We are concerned here with linear momentum because, for the flows under consideration, the effects of rotation of the fluid about an axis can be neglected. In Chapter 14, where we consider the flow within the blading of a turbomachine, the effects of rotation are important and it is essential then to consider the torque acting on the fluid and its angular momentum.

*Introduction to Engineering Fluid Mechanics*. Marcel Escudier.
© Marcel Escudier 2017. Published 2017 by Oxford University Press.

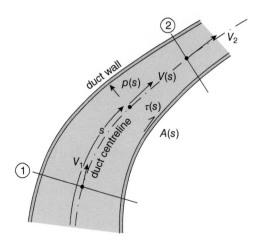

**Figure 9.1** Fluid flow through a duct

As we saw in Chapter 6, changes in the cross-sectional area of a duct result in changes in the velocity of fluid flowing through it. We shall show in Section 9.2 that, because a fluid stream has mass (or, more precisely, density), if its velocity and/or direction changes then so must a quantity we call the **momentum flowrate**, symbol $\dot{M}$[50], and this requires that a force is applied to the flow.

The forces acting on the fluid within the control volume are shown in Figure 9.2(a). The net force $F$ arises from the pressures at inlet and outlet, $p_1$ and $p_2$, and from the pressure and shear stress, $p(s)$ and shear stress $\tau_S(s)$, distributed over the wetted surface of the duct. The net force due to $p(s)$ and $\tau_S(s)$ we refer to as the **fluid-structure interaction force $S$**. As indicated in Figure 9.2(b), for the duct segment to be in **static equilibrium**, this force must be balanced by the external force (or forces) acting on the duct, usually made up of a force due to the external pressure $B$ distributed over its outer surface and an applied **restraining** (or **reaction**) **force**[51] $R$. Polygons illustrating the vector addition of these forces are given in Figure 9.2.

The aim of this chapter is to relate the forces $F$, $S$, and $R$, the external pressure $B$, the pressures $p_1$ and $p_2$, the mass flowrate of fluid through the duct $\dot{m}$, the fluid density $\rho$, and the cross-sectional areas at inlet and outlet, $A_1$ and $A_2$, respectively. As we shall see in Chapter 10, the results of this chapter can be applied directly even if there is more than inlet or outlet. Although for simplicity we restrict attention to flows for which the velocity and all other vector quantities have components only in two orthogonal directions, $x$ and $y$, the analysis is readily extended to include the third direction.

---

[50] In this chapter, vector quantities (i.e. those having both a magnitude and a direction), are shown in bold-face type.

[51] The fluid-structure interaction force and the external reaction force both arise as a consequence of Newton's third law of motion: to every action, there is an equal and opposite reaction.

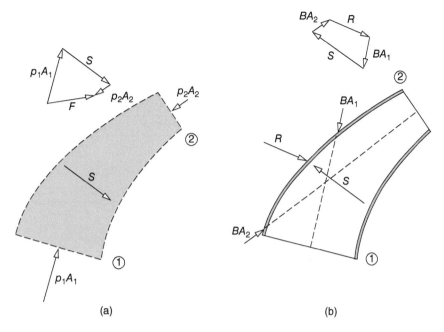

**Figure 9.2** Forces acting on (a) fluid control volume (b) duct segment

## 9.2 Basic linear momentum equation

In Chapter 7 we applied Newton's second law of motion to an elemental fluid slice flowing through a streamtube. In much the same way, as shown in Figure 9.3, we now apply Newton's second law to a fluid slice flowing through the control volume. As should already be evident, it is crucial that we take into account that force, velocity, acceleration, and momentum are all vector quantities. We do that here by considering separately the $x$- and $y$-directions.

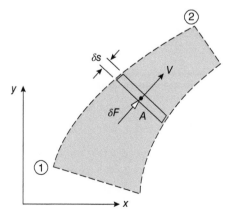

**Figure 9.3** Elemental fluid slice moving through a control volume

We consider first the $x$-direction. If the mass of the fluid slice is $\delta m$, then from Newton's second law of motion we have

$$\delta F_x = \delta m\, a_x \qquad (9.1)$$

where $\delta F_x$ is the component of force acting on the fluid slice in the $x$-direction and $a_x$ is the corresponding acceleration. Since the flow is steady, the acceleration of $\delta m$ is a consequence of changes in its velocity due to changes in the cross-sectional area of the duct, i.e. in a similar way to the treatment of the acceleration of fluid particles in Section 7.2 we may write

$$a_x = \frac{dV_x}{dt} = \frac{dV_x}{ds}\frac{ds}{dt} = V\frac{dV_x}{ds}$$

where $V$ is the resultant velocity of the slice a distance $s$ along the duct.

Substituting the final expression for $a_x$ into equation (9.1) gives

$$\delta F_x = \delta m\, V\frac{dV_x}{ds}$$

$$= \rho A\, \delta s V\frac{dV_x}{ds}$$

where we have replaced the mass of the slice by $\rho A\, \delta s$, $\rho$ being the fluid density, $A$ being the cross-sectional area of the duct at location $s$, and $\delta s$ being the thickness of the slice.

From the continuity equation (6.1) we have $\dot m = \rho A V$ so that the equation for $\delta F_x$ can be written as

$$\delta F_x = \dot m\frac{dV_x}{ds}\delta s.$$

In the limit of an infinitesimally thin fluid slice (i.e. $\delta s \to 0$),

$$\frac{dF_x}{ds} = \dot m\frac{dV_x}{ds}$$

or

$$dF_x = \dot m\, dV_x = d\dot M_x \qquad (9.2)$$

where $\dot M_x = \dot m V_x$ is the $x$-component of the momentum flowrate of the fluid flowing through the duct[52]. What equation (9.2) shows is that the $x$-components of the fluid velocity and the momentum flowrate will increase if the $x$-component of force acting on the fluid slice is positive.

Integration of equation (9.2) along the duct centreline between locations ① and ② produces the important result

$$F_x = \dot m\,(V_{2,x} - V_{1,x}) = \dot M_{2,x} - \dot M_{1,x} \qquad (9.3)$$

where $F_x$ is the $x$-component of the net force acting on the fluid within the control volume, $V_{2,x}$ is the $x$-component of the fluid velocity leaving the control volume, and $V_{1,x}$ is the $x$-component of the fluid velocity entering the control volume. Also in equation (9.3), $\dot M_{2,x}$

---

[52] For a solid of mass $m$ moving with velocity $V$, the product $mV$ is called the momentum. In a similar way, for a fluid flowing through a duct with mass flowrate $\dot m$, the momentum flowrate is $\dot M = \dot m V$ where $V$ is the fluid velocity.

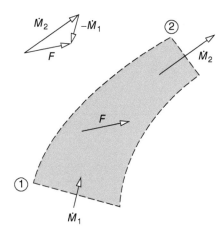

**Figure 9.4** Net force and momentum flowrates for control volume

is the $x$-component of the momentum flowrate at exit from the control volume, and $\dot{M}_{2,x}$ is the $x$-component of the momentum flowrate at the inlet to the control volume.

It should be apparent that the corresponding result to equation (9.3) for the $y$-direction is

$$F_y = \dot{m}\left(V_{2,y} - V_{1,y}\right) = \dot{M}_{2,y} - \dot{M}_{1,y}. \tag{9.4}$$

We can express equations (9.3) and (9.4) in words as follows: the net force acting on the fluid within the control volume, in a given direction, is equal to the change in the momentum flowrate (i.e. the rate of change of fluid momentum) in the same direction. The triangle in Figure 9.4 illustrates that the net force $F$ is equal to the vector difference between the momentum flowrates out of and into the control volume, $\dot{M}_2 - \dot{M}_1$.

---

**ILLUSTRATIVE EXAMPLE 9.1**

A fluid of density $\rho$ flows with mass flowrate $\dot{m}$ through a pipe which turns through 90° and at the same time halves in cross-sectional area. If the initial cross-sectional area is $A$, find the net force exerted on the fluid within the pipe bend and the direction in which it acts.

Solution

The problem under consideration is illustrated in Figure E9.1(a), and the corresponding fluid control volume in Figure E9.1(b). The inflow direction is taken as $x$, and the outflow direction as $y$. If the inflow velocity is $V_1$, the momentum equation for the $x$-direction gives

$$F_x = 0 - \dot{m}V_1 = -\frac{\dot{m}^2}{\rho A}$$

where we have made use of the continuity equation $\dot{m} = \rho A V_1$. The zero on the left-hand side of the equation for $F_x$ appears because the outflow has no velocity component in the $x$-direction.

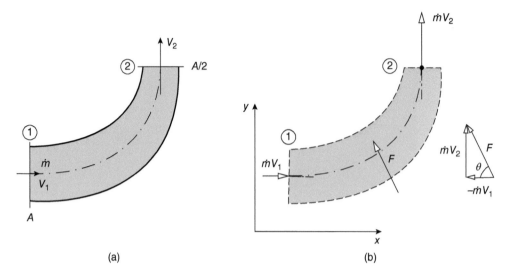

**Figure E9.1**

For the $y$-direction we have

$$F_y = \dot{m}V_2 - 0 = \frac{\dot{m}^2}{\rho A_2} = \frac{2\dot{m}^2}{\rho A}$$

where we have again made use of the continuity equation and also the area relation $A_2 = A/2$. The magnitude $F$ of the net force $\mathbf{F}$ acting on the fluid is then

$$F = \sqrt{F_x^2 + F_y^2} = \frac{\sqrt{5}\dot{m}^2}{\rho A}$$

and this force acts at an angle $\theta$ with respect to the $x$-axis (see Figure E9.1), given by

$$\tan\theta = \frac{F_y}{F_x} = -2 \text{ so that } \theta = -63.4°.$$

**Comments:**

(1) It is essential to draw a diagram showing the general arrangement of the flow situation under consideration, including symbols for any specified quantities, reference axes, etc. To further aid in the solution, draw a second diagram showing the control volume, again including all relevant information.

(2) As mentioned above, in the equation for $F_x$ the $x$-component of the outflow momentum is zero and, in consequence, $F_x$ is negative, i.e. opposed to the direction of the approach flow. The $y$-component of the outflow momentum, on the other hand, is positive while that of the inflow momentum is zero so that $F_y$ is positive. This is entirely what should be expected since the action of turning the flow through 90° requires that its initial momentum in the approach-flow direction is reduced to zero while the momentum in the outflow direction is increased from zero to the value $\dot{m}V_2$. In this case $V_2 = 2V_1$ so that the magnitude of the momentum outflow $\dot{m}V_2$ is double that of the momentum inflow $\dot{m}V_1$.

## 9.3 **Fluid-structure interaction force**

As pointed out in Section 9.1 and illustrated in Figure 9.2(a), the net force $F$ acting on the fluid within the control volume includes the pressure forces $p_1 A_1$ and $p_2 A_2$ together with the fluid-structure interaction force $S$.

For the $x$-direction we have

$$F_x = (p_1 A_1)_x - (p_2 A_2)_x + S_x = \dot{m} (V_{2,x} - V_{1,x}) \tag{9.5}$$

and, for the $y$-direction,

$$F_y = (p_1 A_1)_y - (p_2 A_2)_y + S_y = \dot{m} (V_{2,y} - V_{1,y}) \tag{9.6}$$

where the subscripts $x$ and $y$ again indicate the components of the vector quantities $F$, $pA$, $S$, and $V$ in the $x$- and $y$-directions, respectively.

In order to proceed further, the pressures $p_1$ and $p_2$ must be either specified or calculated. If, as is often the case, the flow can be regarded as frictionless, Bernoulli's equation can be used to relate $p_1, p_2, V_1,$ and $V_2$. If the flow discharges to atmosphere, then $p_2 = B$, the ambient pressure. In other situations it may be possible to calculate $p_1$ and $p_2$ from other information, or the values may be available from measurements.

---

**ILLUSTRATIVE EXAMPLE 9.2**

The flow through the pipe bend considered in Illustrative Example 9.1 discharges to atmospheric pressure $B$ and can be assumed frictionless and incompressible. Gravitational effects may also be ignored. Find the force exerted on the fluid by the internal surface of the pipe bend.

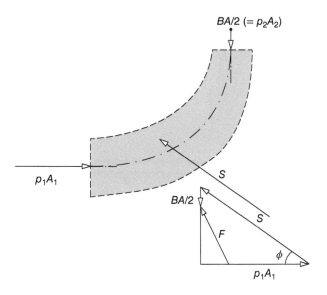

**Figure E9.2**

## Solution

Figure E9.2 again shows the fluid control volume but now includes the pressure forces $p_1 A_1$ and $p_2 A_2$ ($= BA/2$) and also the fluid-structure interaction force $S$ exerted on the fluid in the control volume by the wetted surface of the pipe.

For the $x$-direction we have

$$F_x = p_1 A_1 - S_x$$

and, for the $y$-direction,

$$F_y = -p_2 A_2 + S_y = -\frac{1}{2} BA + S_y$$

wherein we have made use of the fact that the flow discharges to atmospheric pressure, i.e. $p_2 = B$, and also that the outlet area is half the inlet area, i.e. $A_2 = A/2$.

Since the flow can be assumed frictionless, we introduce Bernoulli's equation to evaluate the unspecified inlet pressure $p_1$, i.e.

$$p_1 + \frac{1}{2} \rho V_1^2 = B + \frac{1}{2} \rho V_2^2.$$

As before, the velocities $V_1$ and $V_2$ can be determined from the continuity equation as

$$V_1 = \frac{\dot{m}}{\rho A} \quad \text{and} \quad V_2 = \frac{2\dot{m}}{\rho A}$$

so that

$$p_1 = B + \frac{3}{2\rho} \left( \frac{\dot{m}}{A} \right)^2.$$

We can now substitute in the equation for $F_x$ to find

$$F_x = BA + \frac{3\dot{m}^2}{2\rho A} - S_x.$$

From Illustrative Example 9.1 we have

$$F_x = -\frac{\dot{m}^2}{\rho A}$$

so that, by eliminating $F_x$ between the two equations, we find

$$S_x = BA + \frac{5}{2} \frac{\dot{m}^2}{\rho A}.$$

For $F_y$ we showed that $F_y = -BA/2 + S_y$ and from Illustrative Example 9.1,

$$F_y = \frac{2\dot{m}^2}{\rho A}$$

so that

$$S_y = \tfrac{1}{2}BA + \frac{2\dot{m}^2}{\rho A}.$$

From the two components of $S$ we can find its magnitude from $\sqrt{S_x^2 + S_y^2}$ and its direction $\phi$ relative to the $x$-axis (see Figure E9.2) from $\tan \phi = S_y/S_x$.

---

## 9.4 Hydrodynamic reaction force

So far in this chapter we have been concerned with the forces exerted on the fluid within a control volume and with the fluid-structure interaction force $S$. However, of principal interest in the majority of engineering applications is the **reaction force $R$** required to restrain the structure through which there is flow. At first sight it might appear that the reaction force would be equal in magnitude to the force exerted by the fluid on the duct surface, i.e. the fluid-structure interaction force $S$. That this is not so should be apparent from the **free-body diagram** in Figure 9.2(b), which shows that the forces due to the external pressure $B$ must also be taken into account. If the $x$- and $y$-components of the reaction force are $R_x$ and $R_y$, respectively, for the duct to be in static equilibrium we require

$$R_x + BA_{2,x} - BA_{1,x} - S_x = 0 \tag{9.7}$$

and

$$R_y + BA_{2,y} - BA_{1,y} - S_y = 0. \tag{9.8}$$

The form of the terms involving the external pressure $B$ can be explained as follows. As we showed in Section 5.1, the net force acting on a body subjected to uniform pressure over its entire outer surface is zero. In the present case, since there are openings in the outer surface at the inlet and outlet, unbalanced forces of magnitude $BA_1$ and $BA_2$ must arise on the opposite sides of the duct, as shown in Figure 9.2(b). It is the $x$- and $y$-components of these two forces which appear in equations (9.7) and (9.8).

If we now combine equations (9.5) and (9.7) to eliminate $S_x$ we find

$$R_x = \dot{m}(V_{2,x} - V_{1,x}) - \left(p_1 - B\right)A_{1,x} + \left(p_2 - B\right)A_{2,x} \tag{9.9}$$

and from equations (9.6) and (9.8) we have

$$R_y = \dot{m}\left(V_{2,y} - V_{1,y}\right) - \left(p_1 - B\right)A_{1,y} + \left(p_2 - B\right)A_{2,y}. \tag{9.10}$$

Equations (9.9) and (9.10) reveal that for the calculation of the reaction force it is the so-called **gauge pressures** $p_1 - B$ and $p_2 - B$, which we introduced in Section 4.1 which are important rather than the absolute static pressures $p_1$ and $p_2$. As in the following example, the reaction force $R$ frequently depends only upon the mass flowrate $\dot{m}$, the fluid density $\rho$, and the flow geometry and so is called the **hydrodynamic reaction force**.

**ILLUSTRATIVE EXAMPLE 9.3**

Calculate the external reaction force required to hold in place the pipe bend of Illustrative Example 9.1.

### Solution

Once again it is valuable to draw a diagram to aid the solution, this time showing the internal and external forces acting on the pipe bend: Figure E9.3(a) shows the internal fluid-structure interaction force $S$ (with components $S_x$ and $S_y$), the external reaction force $R$ (with components $R_x$ and $R_y$), and the pressure forces $BA$ (in the $-x$-direction and $BA/2$ in the $+y$-direction).

For the bend to be in static equilibrium we require

$$-R_x - BA + S_x = 0 \quad \text{and} \quad R_y + \frac{1}{2}BA - S_y = 0.$$

From Illustrative Example 9.2 we have

$$S_x = BA + \frac{5}{2}\frac{\dot{m}^2}{\rho A} \quad \text{and} \quad S_y = \frac{1}{2}BA + \frac{2\dot{m}^2}{\rho A}.$$

If we combine the two pairs of expressions we find

$$R_x = \frac{5}{2}\frac{\dot{m}^2}{\rho A} \quad \text{and} \quad R_y = \frac{2\dot{m}^2}{\rho A}$$

so that the magnitude of the external force $R$ is given by

$$R = \sqrt{R_x^2 + R_y^2} = \frac{\sqrt{41}\dot{m}^2}{2\rho A}.$$

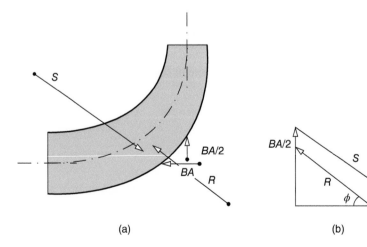

(a)                                        (b)

**Figure E9.3**

and the direction of $R$ is given by

$$\tan\phi = \frac{R_y}{R_x} = 0.8 \quad \text{so that} \quad \phi = 38.7^o.$$

The angle $\phi$ is shown in Figure E9.3.

## Comment:

This example underlines the point made earlier in this section that the external restraining force $R$ differs from the force $S$ exerted on the wetted surface of the pipe bend by the fluid flowing through it as a consequence of the forces arising from the pressure of the surroundings, and the final result is independent of the external pressure. That $R$ and $S$ are not of equal magnitude should be clear from the polygon of forces shown in Figure E9.3(b).

---

## ILLUSTRATIVE EXAMPLE 9.4

A pipe bend such as that shown in Figure E9.1 has an upstream diameter of 1 m. With water flowing through the bend, the external force required to hold it in place is 1200 N. Calculate the volumetric flowrate of the water and the pressure drop through the bend.

## Solution

$D_1 = 1$ m; $R = 1200$ N; $\rho = 1000$ kg/m$^3$.
We calculate first the cross-sectional area at inlet $A = \pi D_1^2/4 = 0.785$ m$^2$.
From Illustrative Example 9.3 we have $|R| = \sqrt{41}\,\dot{m}^2/2\rho A$ which gives

$$\dot{m} = \sqrt{1200 \times 2 \times 1000 \times 0.785/\sqrt{41}} = 543 \text{ kg/s}$$

and so the volumetric flowrate $\dot{Q} = \dot{m}/\rho = 0.543$ m$^3$/s.
To calculate the pressure drop through the bend, $\Delta p = p_1 - p_2$, we use Bernoulli's equation[53]

$$p_1 + \frac{1}{2}\rho V_1^2 = p_2 + \frac{1}{2}\rho V_2^2.$$

Since the cross-sectional area halves, according to the continuity equation the fluid velocity must double so that

$$\Delta p = p_1 - p_2 = \frac{3}{2}\rho V_1^2$$

$$= 1.5 \times 1000 \times \left(\frac{0.543}{0.785}\right)^2 = 716 \text{ Pa.}$$

## Comment:

The calculated pressure drop does not represent a pressure loss. It arises because in passing through the bend the fluid has been accelerated in a frictionless process (which is why

---

[53] Since we are given no information about any change in altitude between inlet and outlet, it is assumed that the axis of the duct lies in a horizontal plane.

Bernoulli's equation was applied). If a diffuser were attached to the bend outlet with an exit area equal to the duct inlet area, the initial static pressure would be recovered (in practice there would be frictional losses).

---

## 9.5  SUMMARY

The overall objective of this chapter was to show how we can calculate the external force which must be applied to a duct to counteract the hydrodynamic forces generated by a fluid flowing through it. After introducing the concept of a fluid control volume, we showed that the analysis involves three separate considerations. We started by applying Newton's second law of motion to fluid flow through a duct of arbitrary shape. The outcome was the linear momentum equation for a fluid flow which shows that the change in the momentum flowrate of the fluid is equal to the net force exerted on the fluid. The second step was to identify the individual forces which contribute to the net force: the pressure forces at inlet and outlet, and the forces which arise due to the static pressure and shear stress distributed over the wetted interior surface of the duct. Finally, we used the condition of static equilibrium for the duct to relate the external restraining force to the force exerted by the flowing fluid on the wetted surface, which we termed the fluid-structure interaction force. The vector nature of force and momentum was accounted for by considering components in orthogonal directions.

The student should be able to

- explain the concept of a fluid control volume
- apply the linear momentum equation

$$F_x = \dot{m}(V_{2,x} - V_{1,x}) \quad \text{and} \quad F_y = \dot{m}\left(V_{2,y} - V_{1,y}\right)$$

  to fluid flow through a specified control volume
- write down expressions for the separate forces which contribute to the net force acting on the fluid flowing through a control volume, i.e. the pressure forces acting on the fluid at inlet and outlet to the control volume and the fluid-structure interaction force
- use the condition of static equilibrium for a duct to write down equations relating the external reaction force, the fluid-structure interaction force, and the forces due to the pressure of the surroundings
- draw diagrams showing the forces acting on a fluid control volume, the net force and momentum flowrates for the control volume, and the forces acting on a duct through which there is fluid flow

## 9.6  SELF-ASSESSMENT PROBLEMS

**9.1**   A pipe of cross-sectional area $A$ turns through 90°. A fluid of density $\rho$ with static pressure equal to the external pressure flows through the pipe with mass flowrate $\dot{m}$. Draw two diagrams: (a) the fluid control volume including the forces and momentum flowrates, and (b) a free-body diagram showing the forces acting on the bend. Show that the net force required to hold the bend in place is $\sqrt{2}\dot{m}^2/\rho A$. Assume that the flow is one-dimensional and friction free, and that gravitational effects can be ignored.

**9.2**  Fluid of density $\rho$ flows through a turbomachine at a mass flowrate $\dot{m}$. If the inlet area is $A$ and the outlet area is $fA$, where $f$ is a numerical factor, derive a formula for the net force $F$ on the fluid within the turbomachine in terms of $\dot{m}, \rho, f$, and $A$ if the angle between the entry and outlet directions is 60°. Assume that the flow is one-dimensional and friction free, and that gravitational effects can be ignored.

**9.3**  A fluid of density $\rho$ flows at static pressure $p$ and velocity $V$ through a pipe of cross-sectional area $A$. If the pipe turns through 90°, find the force exerted on the fluid by the pipe wall. Assume that the flow is one-dimensional and friction free, and that gravitational effects can be ignored.

**9.4**  (a) A liquid of density $\rho$ flows through a circular pipe of diameter $D$ at a mass flowrate $\dot{m}$. The pipe turns through 90° in the horizontal plane and also reduces in diameter by 50%. The hydrodynamic loads exerted on the bend have to be supported externally. Assume that the flow is one-dimensional and friction free to show that the components of the external reaction, $R_x$ and $R_y$, are given by

$$R_x = \frac{\pi}{4}(p-B)D^2 + \frac{4}{\pi\rho}\left(\frac{\dot{m}}{D}\right)^2$$

and

$$R_y = \frac{\pi}{16}(p-B)D^2 + \frac{17}{2\pi\rho}\left(\frac{\dot{m}}{D}\right)^2$$

where $R_x$ is in the oppsite direction to the approach flow and $R_y$ is in the same direction as the leaving flow. $B$ is the external pressure and $p$ is the static pressure of the approach flow.

(b) The flow described in part (a) discharges at the exit from the bend to an external pressure of 1 bar, the liquid density is 2000 kg/m³, the mass flowrate is 100 kg/s, and the upstream pipe diameter is 150 mm. Calculate the resultant force required to support the bend and the direction of its line of action.
(Answers: 2658 N, 25.2°)

**9.5**  (a) A liquid flows through a pipe which turns through 180° in a horizontal plane and at the same time doubles in diameter. If the flow discharges at atmospheric pressure, show that the magnitude of the force $R$ needed to hold the pipe in place is given by

$$R = \frac{25\dot{m}^2}{32\rho A}$$

where $\dot{m}$ is the liquid mass flowrate, $\rho$ is the liquid density, and $A$ is the cross-sectional area of the pipe before the bend. Assume that the flow is one-dimensional and friction free, and that gravitational effects can be ignored.

(b) The external force required to hold in place the pipe bend described in part (a) is 1500 N. Calculate the volumetric flowrate of liquid of density 1000 kg/m³ and the pressure change through the bend if the pipe diameter is 0.5 m upstream of the bend.
(Answers: 0.614 m³/s, 0.046 bar)

# 10 Engineering applications of the linear momentum equation

In this chapter we use a number of specific problems to show how the analyses presented in Chapter 9 and other previous chapters can be applied to real engineering situations. Each problem, together with the numerical example which follows it, can be thought of as a case study. It is important not to think of the numerical examples in isolation, to be solved simply by substituting numbers into formulae. In an examination, the student would be expected to carry out the theoretical analysis first starting from the basic equations of continuity, momentum, and static equilibrium together with, where appropriate, Bernoulli's equation.

We consider flow through a convergent nozzle, a rocket engine, a turbojet engine, a turbofan engine, a sudden enlargement, and a jet pump—all examples where the generation of hydrodynamic force and changes in static pressure, velocity, and momentum flowrate are due entirely to changes in cross-sectional area. The remaining problems involve the additional complication of a change in flow direction, as in the internal flow through a pipe bend, a pipe junction, a set of guidevanes, and finally two situations where a free jet is deflected by impingement on a fixed or a moving object.

## 10.1 Force required to restrain a convergent nozzle

The conventional technique for dousing a major fire relies upon a **convergent nozzle** to create a high-speed jet of water directed towards the fire. Water at high pressure (typically 3 bar) flows at low speed through a long hose (typically 80 mm internal diameter) and is discharged at atmospheric pressure and high speed through a hand-held nozzle typically 50 mm in diameter at exit. As we shall see, the nozzle generates a considerable **reaction force** which has to be resisted by the firefighter.

As shown in Figure 10.1(a), we consider a convergent nozzle, of exit area $A_2$, connected to a circular-cylindrical hose of cross-sectional area $A_1$, through which an incompressible fluid of density $\rho$ flows at a mass flowrate $\dot{m}$. The fluid discharges to the surroundings, which are at static pressure $B$. Our aim is to determine the magnitude and direction of the restraining force $R$ required to hold the nozzle in place.

In this instance there is only one choice for the **fluid control volume**, as shown in Figure 10.1(b). Since there is no change in flow direction, we need consider only the axial direction. According to the **linear momentum equation**, the net force $F$ exerted on the fluid in the control volume is equal to the difference in the momentum flowrate between outlet and inlet, i.e.

$$F = \dot{M}_2 - \dot{M}_1 = \dot{m}(V_2 - V_1). \tag{10.1}$$

*Introduction to Engineering Fluid Mechanics.* Marcel Escudier.
© Marcel Escudier 2017. Published 2017 by Oxford University Press.

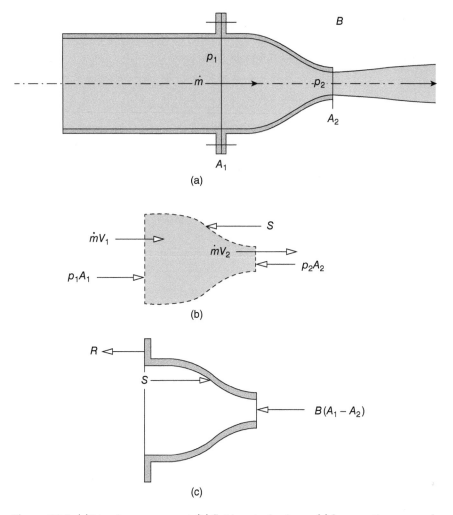

**Figure 10.1** (a) Nozzle arrangement; (b) fluid control volume; (c) forces acting on nozzle

The flow velocities at inlet and outlet, $V_1$ and $V_2$, respectively, can be found from the continuity equation (6.1) as $V_1 = \dot{m}/\rho A_1$, and $V_2 = \dot{m}/\rho A_2$, so that equation (10.1) leads to

$$F = \frac{\dot{m}^2}{\rho}\left(\frac{1}{A_2} - \frac{1}{A_1}\right). \qquad (10.2)$$

As indicated in Figure 10.1(b), there are three contributions[54] to the force $F$: the pressure forces $p_1A_1$ and $p_2A_2$, and the force exerted directly on the fluid by the internal surface of the nozzle, $S$ (the **fluid-structure interaction force**), i.e.

---

[54] The symmetry of the situation requires that all forces and momentum flowrates in this case are coaxial with the nozzle centreline.

$$F = p_1 A_1 - p_2 A_2 - S = \frac{\dot{m}^2}{\rho} \left( \frac{1}{A_2} - \frac{1}{A_1} \right).$$ (10.3)

Equation (10.3) can be rearranged to eliminate $F$ and give

$$S + \frac{\dot{m}^2}{\rho} \left( \frac{1}{A_2} - \frac{1}{A_1} \right) - p_1 A_1 + p_2 A_2 = 0.$$ (10.4)

Our aim is to calculate the external reaction force $R$ required to restrain the nozzle against the forces imposed on it. As shown in Figure 10.1(c), these forces comprise the internal force $S$ exerted directly on the fluid by the nozzle (equal in magnitude but opposite in direction to the force exerted on the fluid) and the forces due to the external pressure $B$.

From the condition for **static equilibrium** of the nozzle we have

$$R - S + BA_1 - BA_2 = 0$$ (10.5)

which we can combine with equation (10.4) to eliminate $S$ and find

$$R = -\frac{\dot{m}^2}{\rho} \left( \frac{1}{A_2} - \frac{1}{A_1} \right) + (p_1 - B) A_1 - (p_2 - B) A_2.$$ (10.6)

Since the nozzle discharges to the surroundings at ambient pressure $B$, we can take $p_2 = B$ so that

$$R = -\frac{\dot{m}^2}{\rho} \left( \frac{1}{A_2} - \frac{1}{A_1} \right) + (p_1 - B) A_1.$$ (10.7)

Equation (10.7) is as far as we can take our calculation without further information about the upstream static pressure $p_1$. If the flow within the nozzle can be regarded as frictionless, $p_1$ can be calculated from Bernoulli's equation, which was derived in Chapter 7, i.e.

$$p_0 = p_1 + \frac{1}{2}\rho V_1^2 = B + \frac{1}{2}\rho V_2^2$$ (7.10)

where $p_0$ is the stagnation pressure of the flow at inlet to the nozzle. As before, we can use the continuity equation (6.1) to substitute for $V_1$ and $V_2$, so that Bernoulli's equation can be written as

$$p_1 - B = \frac{\dot{m}^2}{2\rho} \left( \frac{1}{A_2^2} - \frac{1}{A_1^2} \right)$$ (10.8)

which we can substitute into equation (10.7) to obtain, after some simplification,

$$R = \frac{\dot{m}^2 A_1}{2\rho} \left( \frac{1}{A_2} - \frac{1}{A_1} \right)^2.$$ (10.9)

Although this result came about from an analysis in which frictional effects were excluded, it turns out that, as here, for many situations where the flow is highly turbulent and there are substantial frictional losses, hydrodynamic forces are proportional to the square of the flowrate or velocity, i.e. to the dynamic pressure.

---

**ILLUSTRATIVE EXAMPLE 10.1**

Water flows through a hose with an internal diameter of 80 mm and discharges to the atmosphere through a convergent nozzle with an exit diameter of 60 mm. If the water mass flowrate is 50 kg/s, calculate the force required to restrain the nozzle. If the ambient pressure is 1.03 bar, calculate the static and stagnation pressures of the flow within the hose assuming the flow to be frictionless.

**Solution**

$D_1 = 0.08$ m; $D_2 = 0.06$ m; $\dot{m} = 50$ kg/s; $p_2 = B = 1.03 \times 10^5$ Pa; and $\rho = 1000$ kg/m$^3$.
First, we calculate the areas $A_1 = \pi D_1^2/4 = 5.03 \times 10^{-3}$ m$^2$, and $A_2 = \pi D_2^2/4 = 2.83 \times 10^{-3}$ m$^2$.
Substitution in equation (10.9)[55] then gives

$$R = \frac{0.5 \times 50^2 \times 5.03 \times 10^{-3}}{1000}\left(\frac{1}{2.83} - \frac{1}{5.03}\right)^2 \times 10^6$$

$$= 150.4 \text{ N.}$$

Since the flow is frictionless, we can use Bernoulli's equation to relate the stagnation pressure of the water $(p_0)$, to the static pressures within the hose $(p_1)$ and at the nozzle exit $(p_2 = B)$ as follows

$$p_0 = p_1 + \frac{1}{2}\rho V_1^2 = B + \frac{1}{2}\rho V_2^2.$$

The corresponding velocities $V_1$ and $V_2$ are calculated from the continuity equation $\dot{m} = \rho A_1 V_1 = \rho A_2 V_2$, from which

$$V_1 = \frac{50}{1000 \times 5.03 \times 10^{-3}} = 9.95 \text{ m/s} \quad \text{and} \quad V_2 = \frac{50}{1000 \times 2.83 \times 10^{-3}} = 17.68 \text{ m/s.}$$

The stagnation pressure is then

$$p_0 = B + \frac{1}{2}\rho V_2^2 = 1.03 \times 10^5 + 0.5 \times 1000 \times 17.68^2 = 2.59 \times 10^5 \text{ Pa or 2.59 bar}$$

and the upstream pressure $p_1$ is

$$p_1 = p_0 - \frac{1}{2}\rho V_1^2 = 2.59 \times 10^5 - 0.5 \times 1000 \times 9.95^2 = 2.10 \times 10^5 \text{ Pa or 2.10 bar.}$$

---

## 10.2 Rocket-engine thrust

Figure 1.15, in Chapter 1, shows the general configuration (greatly simplified) of a liquid-propellant rocket engine such as those used for the space-shuttle main engines, and the Saturn V boosters in the Apollo programme. A typical propellant (i.e. fuel) would be liquid

---

[55] As we have made clear throughout this book, in an examination the student would be required to derive the necessary equations. A practicing engineer would be more likely to find them stated in a handbook or textbook (such as this!).

hydrogen, with liquid oxygen as the oxidiser. The fuel and oxidiser are pumped into the combustion chamber, where they mix and burn to produce a gas at high temperature and pressure, which is exhausted to the surrounding atmosphere through a convergent-divergent nozzle as a high-velocity jet. Although the nozzle resembles a **Venturi tube**, it functions very differently because the gas flow through it is highly compressible. The flow is subsonic in the combustion chamber itself, accelerates to **sonic conditions** (i.e. a Mach number of unity) at the nozzle throat, and becomes supersonic in the divergent section of the nozzle[56]. Unlike the situation for subsonic flow, the static pressure of the jet flow at exit from the nozzle is usually higher[57] than that of the surrounding atmosphere, which, at high altitude (above 30 km), is almost zero. The supersonic jet adjusts to the low pressure through a series of shock and expansion waves (see Section 11.9).

We now show that the thrust produced by a rocket engine is equal to the sum of the momentum flowrate of the exhaust jet and the pressure force due to the difference between the exhaust and ambient pressures. It is convenient to consider the gas flow relative to the rocket engine, as would be the case if it were on a test bed. As for the convergent nozzle of the previous section, the interior surface of the engine is a suitable choice to define a fluid control volume, as shown in Figure 10.2(a). The momentum flowrates associated with the inflow of

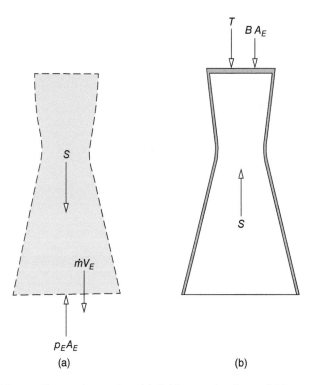

**Figure 10.2** Liquid-propellant rocket engine: (a) fluid control volume; (b) forces acting on rocket engine

---

[56] Compressible gas flow through a convergent-divergent nozzle is discussed in detail in Section 11.7.

[57] A supersonic jet exhausting into a low-pressure environment is said to be **underexpanded** (see Section 11.8).

fuel and oxidant are assumed to be negligible so that the momentum equation applied to the fluid control volume gives

$$F = \dot{m}V_E \tag{10.10}$$

where $\dot{m}$ is the mass flowrate of the exhaust gas, which must equal the combined mass flowrates of the fuel and oxidant. In the case of a solid-propellant rocket, the principal difference to the liquid-propellant case would be that the exhaust mass flowrate would equal the rate at which the propellant burned. In equation (10.10) the symbol $V_E$ denotes the velocity of the exhaust gas at exit from the nozzle, and $F$ represents the net force exerted on the fluid within the control volume.

If the static pressure of the exhaust gas at exit from the engine is $p_E$, and $S$ is the force exerted on the fluid within the control volume by the interior surfaces of the engine (the fluid-structure interaction force), including the combustion chamber and the exhaust nozzle, then

$$F = S - BA_E \tag{10.11}$$

where $A_E$ is the cross-sectional area of the nozzle at exit and $B$ is the ambient pressure. The condition for static equilibrium is

$$T - S + BA_E = 0 \tag{10.12}$$

where, as shown in Figure 10.2(b), the symbol $T$ represents the restraining force which would be required to keep the engine in place on a test bed, and this would be the **thrust** exerted on a moving rocket-powered vehicle such as the space shuttle.

We can now combine equations (10.10), (10.11), and (10.12) to yield the final result for the thrust

$$T = \dot{m}V_E + (p_E - B)A_E \tag{10.13}$$

which shows that thrust arises due to both the momentum flowrate of the exhaust gas $\dot{M}_E = \dot{m}V_E$ and also the exhaust pressure. Since the exhaust flow is supersonic and must be treated as a compressible flow, we cannot use Bernoulli's equation to connect $p_E$, $V_E$, $\dot{m}$, etc. Nevertheless, the continuity equation (6.1) does remain valid so that $\dot{m} = \rho_E A_E V_E$, where $\rho_E$, the gas density at exit from the nozzle, can be calculated from the ideal gas equation $p_E = \rho_E R_E T_E$, where $T_E$ is the exhaust-gas absolute temperature, and $R_E$ is the specific gas constant of the exhaust gas. The latter depends upon the molecular weight of the exhaust gas but as an approximation can be assumed to have the value 287 $m^2/s^2 \cdot K$, which is appropriate for air.

---

**ILLUSTRATIVE EXAMPLE 10.2**

The gas leaving the exhaust nozzle of a rocket engine has a temperature of 1400 °C, a pressure of 12 kPa, and a velocity of 4500 m/s. If the exit diameter of the exhaust nozzle is 2.3 m and the ambient pressure is 0.5 bar, calculate the thrust developed by the engine. The exhaust gas can be assumed to behave as an ideal gas with a specific gas constant of 280 $m^2/s^2 \cdot K$.

## Solution

$T_E = 1400 + 273 = 1673$ K; $p_E = 1.2 \times 10^4$ Pa; $R = 280$ m$^2$/s$^2 \cdot$ K; $V_E = 4500$ m/s; $D_E = 2.3$ m; and $B = 5 \times 10^4$ Pa.

We calculate first the exhaust-gas density using the ideal gas law

$$\rho_E = p_E/RT_E = 1.2 \times 10^4/(280 \times 1673) = 0.0256 \text{ kg/m}^3$$

and the exit cross-sectional area from $A_E = \pi D_E^2/4 = 4.15$ m$^2$.

The exhaust-gas momentum flowrate is then

$$\dot{M}_E = \rho_E V_E^2 A_E = 0.0256 \times 4500^2 \times 4.15 = 2.15 \times 10^6 \text{ N or 2.15 MN}$$

the pressure force is

$$(p_E - B)A_E = (1.2 \times 10^4 - 5 \times 10^4) \times 4.15 = -3.8 \times 10^4 \text{ N or } -0.38 \text{ MN}$$

and the total thrust is

$$T = \dot{M}_E + (p_E - B) A_E = 2.11 \text{ MN}.$$

## Comment:

The values used in this example are typical for a large liquid-propellant cryogenic rocket engine such as one of the three main engines which power the space shuttle[58]. The combined thrust provided by the two solid-propellant boosters required to launch the shuttle is about five times that of the three main engines.

---

## 10.3 Turbojet-engine thrust

A simplified cross section of a basic jet engine[59] is shown in Figure 10.3(a). The function of a jet engine for propulsion purposes[60] is to take in air from the surroundings and increase its momentum in a three-stage process: compression, combustion, and expansion. A kerosene-based liquid fuel is injected into combustion chambers, where it burns in the air which has been raised to high pressure by a multi-stage axial-flow compressor[61]. The power required to drive the compressor is produced by expansion of hot gas flowing from the combustion chambers through an axial-flow turbine. The inventor of the turbojet engine was the British Royal Air Force engineer Sir Frank Whittle. The supersonic airliner Concorde was powered

---

[58] The rated thrust of the Rocketdyne RS-25 Space Shuttle Main Engine (SSME) is given as about 1.86 MN at lift-off.

[59] The terms jet engine, turbojet engine, and gas-turbine engine are used interchangeably in the literature.

[60] For both commercial and combat aircraft, the turbojet engine has been largely superseded by the turbofan engine, which is the subject of Section 10.4. Other variants of the basic turbojet engine include the turboprop engine and turboshaft engines, in which all the useful power is transmitted by a shaft. The principal applications of turboshaft engines are to helicopters, ships, land-based power generation, the compression of natural gas at the point of extraction, and pumping oil through pipelines.

[61] Usually abbreviated to axial compressor.

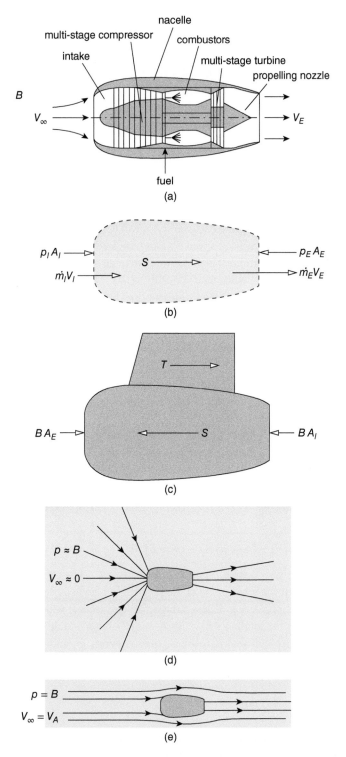

**Figure 10.3** Turbojet engine: (a) configuration; (b) fluid control volume; (c) forces acting on engine; (d) and (e) flow into and out of engine, respectively

by four **Rolls-Royce Olympus 593** turbojet engines: the final version of this engine developed 142 kN of thrust (169 kN with reheat).

As for the rocket engine, the fluid control volume corresponds to the interior of the engine casing. In the equations which follow, the subscript $I$ denotes airflow conditions at the engine intake, $E$ the exhaust-gas conditions at the engine outlet, and $F$ the fuel. If the engine is attached to a flying aircraft, it is convenient to consider flow relative to the engine, i.e. the engine is regarded as being stationary with the air flowing towards and around it at the cruising speed of the aircraft $V_A$ (Figure 10.3(e)).

Overall mass conservation requires that the mass flowrate of the exhaust gas $\dot{m}_E$ must equal the sum of the flowrates of air $\dot{m}_I$ and fuel $\dot{m}_F$ into the engine, i.e.

$$\dot{m}_E = \dot{m}_I + \dot{m}_F. \tag{10.14}$$

Since the velocity of the fuel at inlet to the engine $V_F$ will be low compared with that of the air $V_I$, which will be close to the cruising speed $V_A$, and the air-fuel ratio is considerably greater than unity, as with the rocket engine it is permissible to neglect the momentum flowrate of the inflowing fuel, $\dot{m}_F V_F$, compared with that of the air, $\dot{m}_I V_I$. The momentum equation applied to the fluid flowing through the control volume can then be written as

$$F = \dot{m}_E V_E - \dot{m}_I V_I \tag{10.15}$$

where the net force $F$ acting on the fluid in the control volume is given by

$$F = S - p_E A_E + p_I A_I. \tag{10.16}$$

In this case the fluid-structure interaction force $\mathbf{S}$ represents the net result of all the complex processes taking place within the engine.

The thrust $T$ is again equal to the reaction force applied to the airframe to which the engine is attached, so that, for static equilibrium of the engine (see Figure 10.3(c)),

$$T - S + BA_E - BA_I = 0. \tag{10.17}$$

We can now combine equations (10.15) to (10.17) to give the fundamental **thrust equation** for **jet propulsion**

$$T = (\dot{m}_I + \dot{m}_F)V_E - \dot{m}_I V_I + (p_E - B) A_E - (p_I - B) A_I \tag{10.18}$$

and we now need to assign values to the pressures $p_I$ and $p_E$. For <u>subsonic conditions</u>, it is reasonable to assume that the pressure $p_E$ at the propelling nozzle outlet is equal to the ambient pressure $B$. The inlet pressure $p_I$ is less straightforward to deal with. Since the continuity equation is still valid, for the intake we have $\dot{m}_I = \rho_I A_I V_I$, from which we can evaluate the air velocity at the engine inlet $V_I$ if we know the air mass flowrate $\dot{m}_I$ and the air density $\rho_I$. The airflow approaching the intake may be assumed to satisfy Bernoulli's equation (7.10), so that $p_I$ can be calculated from

$$p_0 = B + \frac{1}{2}\rho_I V_A^2 = p_I + \frac{1}{2}\rho_I V_I^2$$

where the airspeed of the aircraft $V_A$ is the velocity of the airflow relative to the engine far upstream of the aircraft, where the air pressure is equal to the ambient pressure $B$. Bernoulli's equation provides a more satisfactory way to specify the inlet static pressure $p_I$ than the more obvious assumption that $p_I = B$, which is only valid if the inlet velocity $V_I$ is close to $V_A$, as in Figure 10.3(e). In the case of an engine on a test bed or for an aircraft prior to take-off, as in Figure 10.3(d), the air far from the engine will be at rest so that $V_A = 0$ and it is clear that air must be drawn into the engine by reducing $p_I$ to a value much less than $B$. In some circumstances, the airspeed of the aircraft may exceed $V_I$, and if this occurs then $p_I$ is greater than $B$ and the engine benefits from what is called a **ram effect**.

If we combine Bernoulli's equation with the continuity equation, then

$$p_I - B = \rho_I \left[ V_A^2 - \left( \frac{\dot{m}_I}{\rho_I A_I} \right)^2 \right]. \qquad (10.19)$$

To complete the analysis, we introduce the continuity equation for the exhaust-gas flow together with equation (10.14)

$$\dot{m}_E = \rho_E A_E V_E = \dot{m}_I + \dot{m}_F$$

from which $V_E = (\dot{m}_I + \dot{m}_F)/\rho_E A_E$, which we can substitute into the thrust equation (10.18) together with equation (10.19) for $p_I - B$ and the exhaust-pressure condition $p_E = B$, to give

$$T = \frac{(\dot{m}_I + \dot{m}_F)^2}{\rho_E A_E} - \left[ 1 + \left( \frac{V_A}{V_I} \right)^2 \right] \frac{\dot{m}_I^2}{2\rho_I A_I}. \qquad (10.20)$$

The second of the two terms in equation (10.20) is called the **momentum drag** because it reduces the thrust below the value which would be obtained from the exhaust-gas flow alone. If the airspeed $V_A$ is equal to the airflow velocity into the engine $V_I$, such that $p_I = B$, the thrust equation reduces to

$$T = \frac{(\dot{m}_I + \dot{m}_F)^2}{\rho_E A_E} - \frac{\dot{m}_I^2}{\rho_I A_I} \qquad (10.21)$$

while for an engine at rest, with $V_A = 0$, we have

$$T = \frac{(\dot{m}_I + \dot{m}_F)^2}{\rho_E A_E} - \frac{\dot{m}_I^2}{2\rho_I A_I}. \qquad (10.22)$$

As can be seen from equation (10.18), the difference in the momentum drag between the two situations arises because, for $V_A = V_I$, the gauge pressure $p_I - B = 0$ while, when $V_A = 0$, $p_I - B = \rho_I [\dot{m}/(\rho_I A_I)]^2$.

---

**ILLUSTRATIVE EXAMPLE 10.3**

The cross-sectional area of the inlet to a turbojet engine is 2.9 m$^2$ and that of the exhaust nozzle is 2.6 m$^2$. The engine powers an aircraft flying at a speed of 250 m/s (900 km/h) at an altitude of 10 km, where the ambient pressure is 0.265 bar, the temperature is –50 °C, and the density is 0.41 kg/m$^3$. The engine consumes fuel at a rate of 13.5 kg/s and operates with an air:fuel ratio of 20:1. The exhaust-gas density is 0.19 kg/m$^3$. Calculate the air velocity and pressure at the engine intake, the velocity of the exhaust gas leaving the engine, and the thrust developed by the engine.

## Solution

$A_I = 2.9$ m$^2$; $A_E = 2.6$ m$^2$; $V_A = 250$ m/s; $\rho_I = 0.41$ kg/m$^3$; $T_I = 223$ K; $B = 2.65 \times 10^4$ Pa; $\dot{m}_F = 13.5$ kg/s; $\dot{m}_I/\dot{m}_F = 20$; and $\rho_E = 0.19$ kg/m$^3$.

From the value for $\dot{m}_F$ and the air:fuel ratio, we have $\dot{m}_I = 13.5 \times 20 = 270$ kg/s.

We can now calculate the airflow velocity at the engine intake $V_I$ from the continuity equation, i.e.

$$V_I = \frac{\dot{m}_I}{\rho_I A_I} = \frac{270}{0.41 \times 2.9} = 227.1 \text{ m/s.}$$

We see that $V_I$ is slightly lower than $V_A$, so that from Bernoulli's equation (see **Comment** below) the inlet pressure $p_I$ must be slightly higher than the ambient pressure $B$ according to

$$p_I = B + \frac{1}{2}\rho_I\left(\left(V_A^2 - V_I^2\right)\right)$$

$$= 2.65 \times 10^4 + 0.5 \times 0.41 \times \left(250^2 - 227.1^2\right)$$

$$= 2.87 \times 10^4 \text{ Pa or } 0.287 \text{ bar.}$$

The values of the two terms in equation (10.20) are as follows

$$\frac{(\dot{m}_I + \dot{m}_F)^2}{\rho_E A_I} = 1.627 \times 10^5 \text{ N or } 162.7 \text{ Kn}$$

and

$$-\left[1 + \left(\frac{V_A}{V_I}\right)^2\right]\frac{\dot{m}_I^2}{2\rho_I A_I} = -6.78 \times 10^4 \text{ N or } -67.8 \text{ Kn}$$

so that the overall thrust is 94.9 kN.

The gas velocity at outlet from the engine we obtain from the continuity equation as

$$V_E = \frac{\dot{m}_I + \dot{m}_F}{\rho_E A_E} = 574 \text{ m/s.}$$

## Comment:

The speed of sound $c_I$ corresponding to $-50$ °C is 299 m/s so that the flight Mach number $V_A/c_I = 0.835$, which is considerably above the incompressible threshold of about 0.3 (see Section 3.12). The use of Bernoulli's equation, which assumes incompressible flow, therefore introduces errors into the calculation. Since the exhaust-gas temperature $T_E$ will be much higher than the ambient temperature (ca 650 °C), the exhaust soundspeed will be much higher than $c_I$ and the exhaust velocity subsonic but, again, the assumption of incompressibility is unrealistic. A more detailed analysis accounting for the effects of compressibility would show whether the calculation provides reasonable estimates for the thrust and other quantities.

## 10.4 **Turbofan-engine thrust**

The majority of commercial and combat aircraft are now powered by turbofan engines such as that shown schematically in Figures 1.8 and 10.4(a). Most of the air flowing through a turbofan engine bypasses the **engine core** (i.e. the **compressor stages**, the **combustors**, and the **turbine stages**): a typical **bypass ratio**[62] for a large, modern, turbofan engine, such as the **Rolls-Royce Trent XWB**, which powers the **Airbus A350**, is 9:1. Much lower bypass ratios are typical of engines for combat aircraft, such as the **Pratt and Whitney F135**, which powers the **Lockheed Martin F-35 Lightning II** stealth fighter, which has a bypass ratio of 0.57.

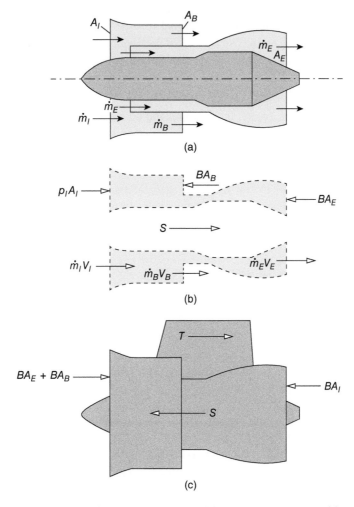

**Figure 10.4** Turbofan engine: (a) flow configuration; (b) fluid control volume; (c) forces acting on engine

[62]  The bypass ratio is the ratio of the air mass flowrate that bypasses the engine core to the air mass flowrate that passes through the core.

In Figure 10.4(a), the total mass flowrate of air into the engine is $\dot{m}_I$, the bypass flowrate is $\dot{m}_B$, and the fuel mass flowrate is $\dot{m}_F$. Overall mass conservation requires that

$$\dot{m}_I + \dot{m}_F = \dot{m}_E + \dot{m}_B \tag{10.23}$$

where $\dot{m}_E$ is the exhaust mass flowrate from the engine core. If the air density at inlet to the engine is $\rho_I$, the air density at outlet from the bypass ducting is $\rho_B$, and the exhaust-gas density is $\rho_E$, then from the continuity equation we have

$$\dot{m}_I = \rho_I A_I V_I, \quad \dot{m}_B = \rho_B A_B V_B, \text{ and } \dot{m}_E = \rho_E A_E V_E, \tag{10.24}$$

from which the fluid velocities at inlet $V_I$, bypass outlet $V_B$, and exhaust $V_E$, can all be calculated given the corresponding cross-sectional areas $A_I, A_B$, and $A_E$.

If $F$ is the net force acting on the two streams of fluid within the fluid control volume shown in Figure 10.4(b), then from the linear momentum equation we have

$$F = \dot{m}_B V_B + \dot{m}_E V_E - \dot{m}_I V_I. \tag{10.25}$$

The air pressure at the front face of the fan is $p_I$, and both the bypass air and exhaust gas are assumed to leave the engine at ambient pressure $B$, so that

$$F = S + p_I A_I - B(A_B + A_E) \tag{10.26}$$

where $S$ is the fluid-structure interaction force, again taking account of both fluid streams.

If $R$ is the reaction force required to hold the engine in place, then from Figure 10.4(c) we can see that the condition for static equilibrium is given by

$$R - S - BA_I + B(A_B + A_E) = 0 \tag{10.27}$$

so that the thrust $T$, which must be equal in magnitude to $R$ but opposite in direction, is given by

$$T = S + BA_I - B(A_B + A_E)$$
$$= \dot{m}_B V_B + \dot{m}_E V_E - \dot{m}_I V_I - \left(p_I - B\right) A_I. \tag{10.28}$$

For a stationary engine, we can use Bernoulli's equation to relate the static pressure at the inlet $p_I$ to the ambient pressure $B$ as follows

$$p_I + \frac{1}{2}\rho_I V_I^2 = B$$

where we have assumed that the air density is the same at the inlet and in the surroundings. If we now substitute $-\rho_I V_I^2/2$ for $p_I - B$ in equation (10.28) for the thrust, we have, finally,

$$T = \dot{m}_B V_B + \dot{m}_E V_E - \frac{1}{2}\dot{m}_I V_I. \tag{10.29}$$

In the following numerical example, the cross-sectional areas, flowrates, and bypass ratio are similar to those representative of a large turbofan engine such as the Rolls-Royce Trent XWB, for which the take-off thrust range is 330 to 430 kN.

**ILLUSTRATIVE EXAMPLE 10.4**

Air at a density of 0.9 kg/m$^3$ enters a turbofan engine at a flowrate of 1440 kg/s. The bypass ratio is 9.3:1 and the fuel flowrate is 4.4 kg/s. The bypass air leaves the engine with a density of 0.8 kg/m$^3$ and the exhaust-gas density is 0.15 kg/m$^3$. The bypass air and the exhaust gases leave the engine at ambient pressure, which is 1.01 bar. The cross-sectional areas are 6.5 m$^2$ at inlet, 5.2 m$^2$ at exit from the bypass ducting, and 1.9 m$^2$ at the exit of the exhaust nozzle. Calculate the thrust developed by the engine.

**Solution**

$\dot{m}_I$ = 1440 kg/s; $r$ = 9.3; $\dot{m}_F$ = 4.4 kg/s; $B$ = 1.01 × 10$^5$ Pa; $\rho_I$ = 0.9 kg/m$^3$; $\rho_B$ = 0.8 kg/m$^3$; $\rho_E$ = 0.10 kg/m$^3$; $p_B = p_E = B$; $A_I$ = 6.5 m$^2$; $A_B$ = 5.2 m$^2$; $A_E$ = 1.9 m$^2$;

$$\dot{m}_I = \dot{m}_B + \frac{\dot{m}_B}{r} = 1440$$

from which $\dot{m}_B$ =1330 kg/s, and the exhaust mass flowrate must be given by $\dot{m}_E = (\dot{m}_I - \dot{m}_B) + \dot{m}_F$ = 114 kg/s. We can now use the continuity equation to find the gas velocities as follows

$$V_I = \frac{\dot{m}_I}{\rho_I A_I} = 246 \text{ m/s} \quad V_B = \frac{\dot{m}_B}{\rho_B A_B} = 320 \text{ m/s} \quad V_E = \frac{\dot{m}_E}{\rho_E A_E} = 600 \text{ m/s}.$$

The thrust can then be calculated from

$$T = \dot{m}_B V_B + \dot{m}_E V_E - \frac{1}{2}\dot{m}_I V_I$$
$$= 1330 \times 320 + 114 \times 600 - 0.5 \times 1440 \times 246$$
$$= 3.17 \times 10^5 \text{ N or 317 kN.}$$

## 10.5 Flow through a sudden enlargement

From the continuity equation we know that, if there is an increase in the cross-sectional area of a duct through which fluid is flowing and if the fluid density remains constant, then the fluid velocity will decrease. If the area increase is gradual, as for the diffuser section of a Venturi tube, the assumption of frictionless flow is usually justified and Bernoulli's equation shows that the fluid static pressure will increase while the stagnation pressure remains constant. However, in many practical applications the area increase has to take place suddenly; for example, when there is inadequate space for a well-designed diffuser or, for other design considerations, a sudden area increase is advantageous. We now show how the static-pressure recovery and stagnation-pressure loss can be calculated for flow through such a sudden enlargement[63].

[63] The term **expansion** rather than enlargement is sometimes used but it should be understood that it is the area that is expanding, not the fluid.

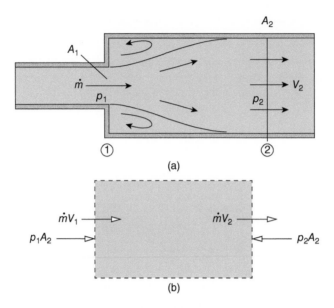

**Figure 10.5** Sudden enlargement: (a) flow geometry; (b) fluid control volume

The flow geometry under consideration is shown in Figure 10.5(a), and the fluid control volume in Figure 10.5(b). The duct downstream of the enlargement is taken to be cylindrical and circular[64]. A key assumption in the flow analysis is that the static pressure is uniform across section ①, where the flow enters the control volume, and equal to the static pressure in the pipe just upstream of the enlargement. This assumption, which in practice is well justified, implies that fluid enters the control volume with negligible streamline curvature. In the region immediately downstream of section ①, the flow forms a central jet, which is strongly affected by viscosity, becomes turbulent at all but very low Reynolds numbers, diverges until it reaches the interior wall of the duct at section ②, and then adjusts to the downstream cross section, where the static pressure can again assumed to be uniform. Although in reality the distribution of velocity both upstream of section ① and downstream of section ② would be non-uniform (see Chapter 15), for present purposes we shall retain the one-dimensional assumption of uniform velocity and negligible wall shear stress.

If we apply the momentum equation (9.3) to the flow through the control volume (between sections ① and ②), we have

$$F = \dot{M}_2 - \dot{M}_1 = \dot{m}(V_2 - V_1)$$

where $F$ is the force acting on the fluid within the control volume, $\dot{m}$ is the mass flowrate, $V_1$ and $V_2$ are the flow velocities at entrance to and exit from the control volume, respectively, and $\dot{M}_1$ and $\dot{M}_2$ are the corresponding momentum flowrates. Since the surface shear stress is assumed to be negligible, and the duct downstream of the enlargement is cylindrical, the

---

[64] The term cylindrical alone does not mean that the cross section is circular, but rather only that the duct is straight and has the same cross section at all axial locations.

fluid-structure interaction force in this case is negligible, and the force $F$ is due entirely to the pressures $p_1$ and $p_2$ so that (with reference to Figure 10.5(b))

$$F = p_1 A_2 - p_2 A_2 = (p_1 - p_2) A_2. \tag{10.30}$$

We note that the area associated with $p_1$ is $A_2$ not $A_1$ because this pressure is exerted over the entire cross section of the duct immediately downstream of the enlargement.

If we combine the two equations for $F$ we can derive an equation for the static-pressure difference

$$p_2 - p_1 = \frac{\dot{m}}{A_2} (V_1 - V_2). \tag{10.31}$$

We now introduce the continuity equation as $\dot{m} = \rho A_1 V_1 = \rho A_2 V_2$, which allows equation (10.31) to be written as

$$p_2 - p_1 = \frac{\dot{m}^2}{\rho A_2} \left( \frac{1}{A_1} - \frac{1}{A_2} \right). \tag{10.32}$$

The stagnation pressures upstream and downstream of the enlargement are given by

$$p_{0,1} = p_1 + \frac{1}{2}\rho V_1^2 \quad \text{and} \quad p_{0,2} = p_2 + \frac{1}{2}\rho V_2^2 \tag{7.10}$$

so that the loss in stagnation pressure is given by

$$p_{0,1} - p_{0,2} = \frac{1}{2}\rho \left( V_1^2 - V_2^2 \right) - (p_2 - p_1). \tag{10.33}$$

We can use equation (10.32) to substitute for $p_2 - p_1$, and the continuity equation to substitute for $V_1$ and $V_2$, so that after some rearrangement we have, finally,

$$p_{0,1} - p_{0,2} = \frac{\dot{m}^2}{2\rho} \left( \frac{1}{A_1} - \frac{1}{A_2} \right)^2 \tag{10.34}$$

which is a version of the **Borda-Carnot equation**. The reduction (or 'loss') in stagnation pressure, called the **Borda-Carnot** or **expansion pressure loss**, is often stated in terms of the reduction in stagnation-pressure head as

$$h_{0,1} - h_{0,2} = \frac{p_{0,1} - p_{0,2}}{\rho g} = \frac{V_1^2}{2g} \left( 1 - \frac{A_1}{A_2} \right)^2. \tag{10.35}$$

As we showed in Subsection 7.5.1, pressure can be thought of as mechanical energy per unit volume, so this loss of stagnation pressure represents mechanical energy which is dissipated by viscous effects (which includes turbulent dissipation), resulting in an increase in entropy and a small increase in fluid temperature. Equation (10.33) can also be written as

$$p_{0,1} - p_{0,2} = \frac{1}{2}\rho V_1^2 \left( 1 - \frac{A_1}{A_2} \right)^2 \tag{10.36}$$

i.e. the loss of stagnation pressure can also be viewed as a loss of kinetic energy. For the limiting case of a flow discharging from a duct of cross-sectional area $A_1$ into surroundings of effectively infinite extent (i.e. $A_2 \gg A_1$) we have the situation of a **free jet** and see that the loss in stagnation pressure is equal to $\rho V_1^2/2$, the dynamic pressure of the jet.

If a perfect diffuser were substituted for the sudden expansion, there would be no loss in stagnation pressure and, from Bernoulli's equation, the **static-pressure recovery** would be given by

$$p_2 - p_1 = \frac{1}{2}\rho\left(V_1^2 - V_2^2\right) = \frac{\dot{m}^2}{2\rho A_2^2}\left[\left(\frac{A_2}{A_1}\right)^2 - 1\right] \tag{10.37}$$

which is larger than the pressure recovery for a sudden expansion (given by equation (10.32)) by the factor $[(A_2/A_1) + 1]/2$. However, we emphasise that Bernoulli's equation does not apply to flow through the sudden enlargement, and the last result should be regarded as a basis for comparison.

---

### ILLUSTRATIVE EXAMPLE 10.5

Part of the exhaust system from an engine can be modelled as an axisymmetric gradual contraction followed by a sudden enlargement, as shown in Figure E10.5. The upstream and downstream pipes both have a cross-sectional area of $3 \times 10^{-3}$ m$^2$, and the contraction outlet area is $1.5 \times 10^{-3}$ m$^2$. Calculate the differences in static pressure between sections ① and ② and between sections ② and ③, and also the overall loss in stagnation pressure, for a gas of density 0.8 kg/m$^3$ with a mass flowrate of 0.2 kg/s. The entire flow may be considered incompressible and one dimensional, and the flow in the contraction as frictionless.

### Solution

$A_1 = 3 \times 10^{-3}$ m$^2$; $A_2 = 1.5 \times 10^{-3}$ m$^2$; $A_3 = 3 \times 10^{-3}$ m$^2$; $\rho = 0.8$ kg/m$^3$; and $\dot{m} = 0.2$ kg/s. We start by calculating the gas velocities at sections (1), (2), and (3), using the continuity equation

$$V_1 = \dot{m}/\rho A_1 = 0.2/\left(0.8 \times 3 \times 10^{-3}\right) = 83.3 \text{ m/s}; A_2 = A_1/2 \text{ so } V_2 = 2V_1 = 166.7 \text{ m/s; and}$$
$$A_3 = A_1 \text{ so } V_3 = V_1 = 83.3 \text{ m/s.}$$

Since the flow in the contraction, between sections ① and ②, is frictionless, Bernoulli's equation is applicable and we have

$$p_1 + \frac{1}{2}\rho V_1^2 = p_2 + \frac{1}{2}\rho V_2^2$$

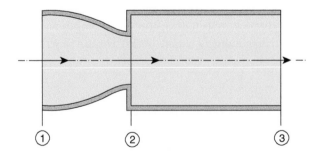

**Figure E10.5**

so that

$$p_1 - p_2 = \frac{1}{2}\rho \left( V_2^2 - V_1^2 \right)$$
$$= 0.5 \times 0.8 \times \left( 166.7^2 - 83.3^2 \right)$$
$$= 8.33 \times 10^3 \text{ Pa or } 8.33 \text{ kPa.}$$

For the sudden enlargement we can use the results of this section, equation (10.31), with appropriate changes to the subscripts, so that

$$p_3 - p_2 = \frac{\dot{m}}{A_3} \left( V_2 - V_3 \right)$$
$$= \frac{0.2 \times (166.7 - 83.3)}{3 \times 10^{-3}}$$
$$= 5.56 \times 10^3 \text{ Pa or } 5.56 \text{ kPa.}$$

Since there is no loss of stagnation pressure in the frictionless contraction, the entire stagnation-pressure loss occurs between sections ② and ③. Here again we can use the results of this section, equation (10.34), i.e.

$$p_{0,2} - p_{0,3} = \frac{\dot{m}^2}{2\rho} \left( \frac{1}{A_2} - \frac{1}{A_3} \right)^2$$
$$= \frac{0.5 \times 0.2^2}{0.8} \left( \frac{1}{1.5} - \frac{1}{3} \right)^2 \times 10^6$$
$$= 2.78 \times 10^3 \text{ Pa or } 2.78 \text{ kPa.}$$

**Comment:**

As we pointed out in Section 7.6, it is crucially important to understand the difference between a static-pressure loss and a static-pressure difference. The latter may be recoverable; the former is not.

## 10.6 Jet pump (or ejector or injector)

A basic jet pump consists of two concentric tubes, as shown in Figures 1.16 and 10.6(a). A secondary flow occurs in the annulus surrounding the inner tube due to fluid being drawn into the high-speed jet of fluid discharging from the central pipe. This process of low-speed or even stationary fluid being drawn into a fast-flowing stream is termed **entrainment**. Vigorous mixing takes place between the primary and secondary streams to produce a homogeneous exit flow. In practice, for liquids, the primary jet flow usually discharges into the body of the pump through a convergent-divergent nozzle to produce a high-speed, low-pressure jet flow. For gas flows a convergent-divergent nozzle producing a low-pressure supersonic jet is more common. The design of the outer tube depends upon the application, but is often divergent to act as a diffuser if a high outlet pressure is required. A convergent exit nozzle to reduce the pressure and increase the momentum of the outflow is more suitable in thrust applications.

Flow machines, such as the jet pump, which have no moving parts, often used in fluid control systems, are called **fluidic devices**. Jet pumps are cheap and robust, thereby requiring little maintenance, but are inefficient and can be noisy because of the high velocity of the primary flow. The low efficiency is often of little consequence because jet pumps are commonly used where the primary-flow energy would otherwise be wasted. The numerous applications of jet pumps include water-aeration systems, water-jet aspirators (commonly encountered in chemistry laboratories), feedwater pumps for boilers, air/gas mixers in domestic, laboratory, and industrial burners (the Bunsen burner is an example), thrust augmenters, and pumps for sand, gravel, foodstuffs, and other slurries.

For simplicity we shall consider an arrangement in which the outer body of the pump is circular and cylindrical with an inner injector tube which is concentric with the outer body. If we apply the momentum equation to the fluid control volume shown in Figure 10.6(b), we have

$$(p_N - p_2)A_2 = \dot{m}_2 V_2 - \dot{m}_N V_N - \dot{m}_S V_S \tag{10.38}$$

where $V$ is the flow velocity, $\dot{m}$ is the mass flowrate, $p$ is the static pressure, the subscript $S$ denotes the secondary flow (i.e. the flow in the annulus surrounding the injector tube), the subscript $N$ denotes conditions at the exit of the injector tube (section ①), and 2 refers to section ②, at which location it is assumed that the primary and secondary flow streams are fully mixed, and the flow velocity is uniform across the cross section, as is the pressure $p_2$. The static pressure in both streams at section ① is assumed to be the same and equal to the pressure $p_N$ of the jet flow, i.e. $p_I = p_N$. As was the case for the sudden enlargement (Section (10.5)), in writing equation (10.38), we have neglected any shear stresses acting on the fluid at the outer tube surface so that the only force exerted on the fluid in the control volume is due to the

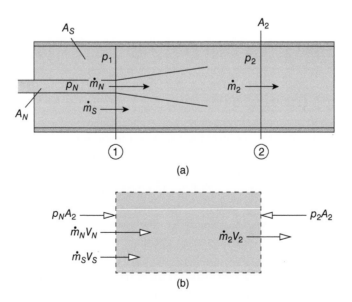

**Figure 10.6** Jet pump: (a) flow configuration; (b) fluid control volume

pressure difference $p_N - p_2$. It should be noted that an important difference between the analysis of the jet pump and the previous example is that we now have two fluid streams entering the control volume and it is necessary to account separately for the momentum flowrate of each.

Since the primary and secondary streams mix and leave the control volume as a single stream, overall mass conservation requires that

$$\dot{m}_2 = \dot{m}_N + \dot{m}_S \tag{10.39}$$

and since the geometry is cylindrical we have the area relationship

$$A_2 = A_N + A_S. \tag{10.40}$$

The continuity equation can be applied separately to the primary and secondary streams at section ① and to the mixed stream at section ②. Assuming the two streams at section ① have the same density $\rho$, we have

$$\dot{m}_N = \rho A_N V_N \quad \dot{m}_S = \rho A_S V_S \quad \text{and} \quad \dot{m}_2 = \rho A_2 V_2. \tag{10.41}$$

To go further we need to state which quantities can be regarded as specified and which are to be calculated, and this depends upon the application. For example, we may wish to calculate the mass flowrate of the induced secondary flow $\dot{m}_S$ for a pump of given dimensions (i.e. $A_N$ and $A_S$ specified), a fluid of specified density $\rho$, and given pressures $p_N$ and $p_2$. An alternative might be to calculate the final pressure $p_2$ for a given overall flowrate $\dot{m}_2$. In both cases we have sufficient information to carry out the calculations. In other circumstances we might be given the pressure of the secondary flow upstream of section ① and the outlet pressure downstream of section ② where the area is different from $A_2$. In such situations we need to introduce further assumptions. For example, for a pump completely submerged in a liquid, it might be appropriate to assume the flow up to section ① is frictionless and use Bernoulli's equation to relate $p_N$, the velocity of the secondary stream $V_S$, and the ambient fluid pressure $B$. A similar approach could be used to relate $p_2$ and $V_2$ to the conditions at outlet from a nozzle or diffuser downstream of section ②. It should be apparent that many other variations on this problem are possible.

### 10.6.1  Jet pump with specified mass flowrates

We return to the configuration shown in Figure 10.6(a), for which the momentum equation led to

$$\left(p_N - p_2\right)A_2 = \dot{m}_2 V_2 - \dot{m}_N V_N - \dot{m}_S V_S \tag{10.38}$$

and the continuity equation to

$$\dot{m}_N = \rho A_N V_N \quad \dot{m}_S = \rho A_S V_S \quad \text{and} \quad \dot{m}_2 = \rho A_2 V_2. \tag{10.41}$$

If we now substitute in equation (10.38) for $V_2$, $V_N$, and $V_S$ and rearrange, we have

$$p_2 - p_N = \frac{1}{\rho A_2} \left[ \frac{\dot{m}_N^2}{A_N} + \frac{\dot{m}_S^2}{A_2 - A_N} - \frac{(\dot{m}_N + \dot{m}_S)^2}{A_2} \right]. \tag{10.42}$$

This equation can be used to calculate the outlet pressure if the inlet pressure $p_N$ and the two mass flowrates are known. Alternatively if the pressure difference $p_N - p_2$ is known, then the secondary mass flowrate $\dot{m}_S$ can be calculated given $\dot{m}_N$.

---

**ILLUSTRATIVE EXAMPLE 10.6**

A jet pump with the configuration shown in Figure 10.6(a) has been designed to raise the pressure of a gas of density $1.1 \text{ kg/m}^3$. The inner tube has a cross-sectional area of $5 \times 10^{-4} \text{ m}^2$ while that of the outer tube is $8 \times 10^{-3} \text{ m}^2$. The primary flowrate is $0.15 \text{ kg/s}$ and the secondary flowrate is $0.5 \text{ kg/s}$. Calculate the pressure increase produced by the pump.

**Solution**

$\rho = 1.1 \text{ kg/m}^3$; $A_N = 5 \times 10^{-4} \text{ m}^2$; $A_2 = 8 \times 10^{-3} \text{ m}^2$; $\dot{m}_N = 0.15 \text{ kg/s}$; and $\dot{m}_S = 0.5 \text{ kg/s}$.
From the equation for $p_2 - p_N$ we have

$$p_2 - p_N = \frac{1}{1.1 \times 8 \times 10^{-3}} \left( \frac{0.15^2}{5 \times 10^{-4}} + \frac{0.5^2}{7.5 \times 10^{-3}} - \frac{0.65^2}{8 \times 10^{-3}} \right)$$
$$= 2900 \text{ Pa or } 2.9 \text{ kPa.}$$

As remarked already, if the pressure rise were specified, then the equation for $p_2 - p_N$ could be used to calculate $\dot{m}_S$ from

$$2900 \times 1.1 \times 8 \times 10^{-3} = \frac{0.15^2}{5 \times 10^{-4}} + \frac{\dot{m}_S^2}{7.5 \times 10^{-3}} - \frac{(\dot{m}_S + 0.15)^2}{8 \times 10^{-3}}$$

which can be simplified to give the quadratic equation for $\dot{m}_S$

$$\dot{m}_S^2 - 4.5\dot{m}_S + 2 = 0.$$

The two solutions are $\dot{m}_S = 0.5 \text{ kg/s}$, and $\dot{m}_S = 4.0 \text{ kg/s}$. The first solution corresponds to the secondary flowrate specified in our original problem while the second corresponds to a secondary flowspeed of 455 m/s and has to be ruled out since this would represent a supersonic flow condition (Mach number $\approx 1.34$) and we have taken no account of compressibility.
The situation becomes more difficult if we need to calculate $A_2$ given all other quantities because the equation for $p_2 - p_N$ results in a cubic equation for $A_2$.

---

## 10.6.2 Jet pump with specified external pressures

We consider the configuration shown in Figure 10.7(a) in which a submerged jet pump with an inlet contraction and a convergent exit nozzle is used to generate a high-speed flow of a liquid of density $\rho$. For the central section of the pump, between locations ① and ②, the momentum equation derived earlier is again applicable, i.e.

$$\left( p_N - p_2 \right) A_2 = \dot{m}_2 V_2 - \dot{m}_N V_N - \dot{m}_S V_S. \tag{10.38}$$

According to the continuity equation applied to the primary and secondary flows,

$$\dot{m}_N = \rho \dot{Q}_N = \rho A_N V_N \quad \dot{m}_S = \rho \dot{Q}_S = \rho A_S V_S \tag{10.41}$$

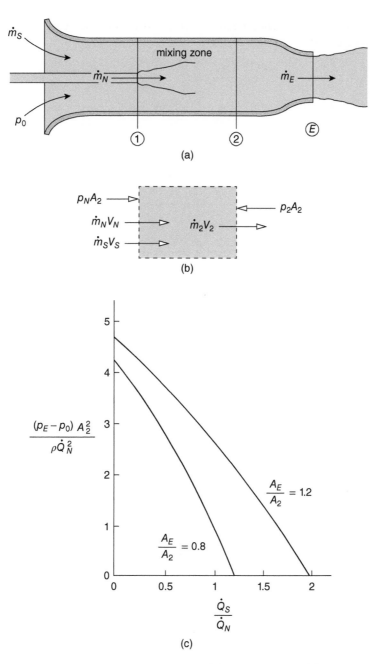

**Figure 10.7** Jet pump with convergent exit nozzle: (a) flow geometry; (b) fluid control volume; (c) performance curves

where $\dot{Q}_N$ and $\dot{Q}_S$ are the volumetric flowrates corresponding to the mass flowrates $\dot{m}_N$ and $\dot{m}_S$, $V_N$ and $V_S$ are the corresponding flow velocities, and $A_N$ and $A_S$ the appropriate cross-sectional areas. In addition we have the overall mass-conservation equation

$$\dot{m}_2 = \dot{m}_N + \dot{m}_S = \rho \left( \dot{Q}_N + \dot{Q}_S \right) = \rho A_2 V_2. \tag{10.39}$$

We can now recast the momentum equation in terms of the volumetric flowrates and the cross-sectional areas as

$$\frac{(p_2 - p_N) A_2}{\rho} = \frac{\dot{Q}_N^2}{A_N} + \frac{\dot{Q}_S^2}{A_S} - \frac{(\dot{Q}_N + \dot{Q}_S)^2}{A_2}. \tag{10.42}$$

We now need to relate the static pressure $p_N$ at the end of the injector pipe shall to the stagnation pressure in the reservoir $p_0$, and the pressure $p_2$ at the end of the **mixing zone** to the exit pressure $p_E$ which, finally, will give us an equation for the overall pressure rise across the pump $p_E - p_0$. We assume that the secondary flow is frictionless upstream of section ① and that the mixed flow is frictionless downstream of section ② so that in both cases we can use Bernoulli's equation as follows

$$p_0 = p_N + \frac{1}{2}\rho V_S^2,$$

where we have again made use of the pressure condition $p_S = p_N$ at section ①, and

$$p_2 + \frac{1}{2}\rho V_2^2 = p_E + \frac{1}{2}\rho V_E^2,$$

where the exit-flow condition corresponds to a jet discharging into the surroundings at a static pressure $p_E$. If we combine these two equations, we have

$$p_2 - p_N = p_E - p_0 + \frac{1}{2}\rho\left(V_E^2 - V_2^2 + V_S^2\right). \tag{10.43}$$

If we again make use of the continuity equation to eliminate the velocities, we have

$$p_2 - p_N = p_E - p_0 + \frac{1}{2}\left[\frac{(\dot{Q}_N + \dot{Q}_S)^2}{A_E} - \frac{(\dot{Q}_N + \dot{Q}_S)^2}{A_2} + \frac{\dot{Q}_S^2}{A_S}\right]. \tag{10.44}$$

If we substitute this equation for $p_2 - p_N$ in the momentum equation, after some algebra we find that the overall pressure rise is given by

$$\frac{(p_E - p_0) A_2^2}{\rho \dot{Q}_N^2} = \frac{A_2}{A_N} - \left(\frac{1}{2} + \frac{\dot{Q}_S}{\dot{Q}_N}\right)\left[1 + \left(\frac{A_2}{A_E}\right)^2\right]$$
$$+ \left(\frac{\dot{Q}_S}{\dot{Q}_N}\right)^2\left[\frac{A_2}{A_S} - \frac{1}{2}\left\{1 + \left(\frac{A_2}{A_S}\right)^2 + \left(\frac{A_2}{A_E}\right)^2\right\}\right]. \tag{10.45}$$

As can be seen, the final result is in non-dimensional form, all the quantities on the right-hand side appearing as ratios. As we argued in Chapter 3, it is preferable to present any result, whether theoretical or experimental, in non-dimensional form. In the present case one advantage is that the geometry of a range of geometrically similar pumps can be represented by the two parameters $A_N/A_2$ and $A_E/A_2$ ($A_S/A_2$ is not an independent parameter since $A_S = A_2 - A_N$). The non-dimensional pressure rise $(p_E - p_0) A_2^2/\rho \dot{Q}_N^2$ can be calculated for a range of values of the ratio $\dot{Q}_S/\dot{Q}_N$ to produce performance curves for any value of $A_N/A_2$. Figure 10.7(c) shows two such curves for $A_N/A_2 = 0.18$: one for $A_E/A_2 = 0.8$, corresponding to a convergent exit nozzle, and one for $A_E/A_2 = 1.2$, which corresponds to a diffuser. Although we carried out the analysis with the convergent arrangement in mind, it applies equally well to the situation of a well-designed diffuser[65]. As we should expect, for any given value of the flow ratio $\dot{Q}_S/\dot{Q}_N$, the results show that the diffuser produces a higher exit pressure than the nozzle does. We also see that the jet pump becomes increasingly effective as $\dot{Q}_S/\dot{Q}_N$ is reduced, i.e. as the primary flowrate is increased.

**ILLUSTRATIVE EXAMPLE 10.7**

A jet pump, similar to that shown in Figure 10.7(a), has dimensions corresponding to the following cross-sectional areas: $A_N = 5 \times 10^{-4}$ m$^2$, $A_2 = 8 \times 10^{-3}$ m$^2$, and $A_E = 6 \times 10^{-3}$ m$^2$. The pump operates completely submerged in water at a depth where the pressure is 1.51 bar, which can be taken as both the stagnation pressure of the secondary flow and also the pump outlet pressure. If the primary mass flowrate is 25 kg/s, calculate the secondary and overall mass flowrates and also the static pressures $p_N$ and $p_2$.

Solution

$A_N = 5 \times 10^{-4}$ m$^2$; $A_2 = 8 \times 10^{-3}$ m$^2$; $A_E = 6 \times 10^{-3}$ m$^2$; $p_0 = p_E = 1.51 \times 10^5$ Pa; $\dot{m}_N = 25$ kg/s; and $\rho = 1000$ kg/m$^3$.

From the specified cross-sectional areas, we can find the area ratios $A_N/A_2 = 0.0625$, $A_S/A_2 = (A_2 - A_N)/A_2 = 0.9375$, and $A_E/A_2 = 0.75$, so that the area terms in the final equation for $p_E - p_0$ are

$$\frac{A_2}{A_N} - \frac{1}{2}\left[1 + \left(\frac{A_2}{A_E}\right)^2\right] = 14.61$$

$$1 + \left(\frac{A_2}{A_E}\right)^2 = 2.78$$

and

$$\frac{A_2}{A_S} - \frac{1}{2}\left[1 + \left(\frac{A_2}{A_S}\right)^2 + \left(\frac{A_2}{A_E}\right)^2\right] = -0.89.$$

Since $p_0 = p_E$ in this case, we have

$$0.89\left(\frac{\dot{Q}_S}{\dot{Q}_N}\right)^2 + 2.78\left(\frac{\dot{Q}_S}{\dot{Q}_N}\right) - 14.61 = 0,$$

i.e. a quadratic equation for $\dot{Q}_S/\dot{Q}_N$ which we can solve to find $\dot{Q}_S/\dot{Q}_N = 2.78$. The second root is negative and is thus ruled out on physical grounds.

[65] We know from the continuity equation that an increasing cross-sectional results in a velocity decrease. From Bernoulli's equation this in turn leads to a pressure increase. As we shall see in Section 16.5, a pressure increase results in a reduction of surface shear stress and, in extreme circumstances, this may fall to zero, leading to **separation** of the boundary layer. The analysis presented here assumes the diffuser is running full (i.e. the boundary layer is not separated).

Since the flow is incompressible, the mass flowrates are in the same ratio as the volumetric flowrates, so that

$$\dot{m}_S = 2.78 \times 25 = 69.5 \text{ kg/s}$$

and the total mass flowrate is

$$\dot{m}_E = \dot{m}_N + \dot{m}_S = 94.5 \text{ kg/s.}$$

In the derivation of the equation for $p_E - p_0$ it was assumed that the flow upstream of section ① is frictionless. We can therefore use Bernoulli's equation to calculate the pressure $p_N$ from

$$p_0 = p_N + \frac{1}{2}\rho V_S^2$$

where the velocity of the secondary stream upstream of section ① $V_S$ is obtained from the continuity equation

$$\dot{m}_S = \rho A_S V_S$$

so that $V_S = \dot{m}_S/\rho A_S = 9.27$ m/s. From Bernoulli's equation we find

$$p_N = p_0 - \frac{1}{2}\rho V_S^2 = 1.51 \times 10^5 - 0.5 \times 1000 \times 9.27^2 = 1.08 \times 10^5 \text{ Pa or } 1.08 \text{ bar.}$$

To find the intermediate pressure $p_2$ we can use Bernoulli's equation between section ② and the outlet

$$p_2 + \frac{1}{2}\rho V_2^2 = p_E + \frac{1}{2}\rho V_E^2$$

so that

$$p_2 = p_E + \frac{1}{2}\rho\left(V_E^2 - V_2^2\right).$$

The outlet velocity $V_E$ and the intermediate velocity $V_2$ are again found from the continuity equation, $V_E = \dot{m}_E/\rho A_E = 15.75$ m/s, and $V_2 = \dot{m}_2/\rho A_2 = 11.81$ m/s, so that

$$p_2 = 1.51 \times 10^5 + 0.5 \times \left(15.75^2 - 11.81^2\right) = 2.05 \times 10^5 \text{ Pa or } 2.05 \text{ bar.}$$

---

## 10.7 Reaction force on a pipe bend

In the remaining sections of this chapter we consider situations where the flow geometry leads to a change in flow direction. We start with a pipe bend which turns through an angle $\theta$ and at the same time changes in cross-sectional area. This is a more general version of the 90° bend discussed in Chapter 9. Although we restrict consideration to a bend in the horizontal plane, the principles applied are general. The flow configuration is shown in Figures 1.14 and 10.8(a), the fluid control volume and the forces acting on it as well as the momentum flowrates in Figure 10.8(b), and the forces exerted on the bend itself in Figure 10.8(c). The inflow

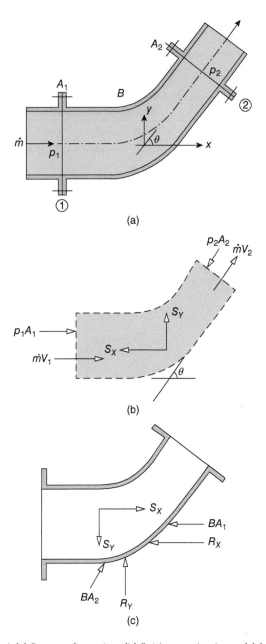

**Figure 10.8** Pipe bend: (a) flow configuration; (b) fluid control volume; (c) forces exerted on bend

cross-sectional area is $A_1$, the outflow area is $A_2$, the mass flowrate is $\dot{m}$, the flow velocities corresponding to $A_1$ and $A_2$ are $V_1$ and $V_2$, respectively, and the fluid density, assumed to be constant, is $\rho$. A uniform static pressure $B$ acts over the external surface of the bend. For convenience, the inflow at section ① is taken to be in the $x$-direction, and the outflow at section ② to be at an angle $\theta$ to the $x$-direction.

The momentum equation applied to the fluid in the control volume gives

$$p_1 A_1 - p_2 A_2 \cos\theta - S_X = \dot{m}(V_2 \cos\theta - V_1) \tag{10.46}$$

in the $x$-direction, $S_X$ being the $x$-component of the fluid-structure interaction force $\mathbf{S}$. The corresponding equation for the $y$-direction is

$$-p_2 A_2 \sin\theta + S_Y = \dot{m}V_2 \sin\theta. \tag{10.47}$$

With experience, it is usually straightforward to decide whether $S_X$ and $S_Y$ are positive or negative, but it is of no consequence if we choose incorrectly, as the directions of both $\mathbf{S}$ and the reaction force $\mathbf{R}$ are results of the analysis. In writing these two equations, we note that, because the inflow at section ① is in the $x$-direction, there is no contribution of either the pressure force $p_1 A_1$ or the momentum flowrate $\dot{m}V_1$ to the $y$-momentum equation. In contrast, unless the outflow at section ② is at 90° to the inflow, both $p_2 A_2$ and $\dot{m}V_2$ contribute to the $x$-momentum equation.

From Figure 10.8(c) we can see that the condition of static equilibrium leads to

$$-R_X + S_X - BA_1 + BA_2 \cos\theta = 0 \tag{10.48}$$

and

$$R_Y - S_Y + BA_2 \sin\theta = 0. \tag{10.49}$$

Between the $x$-momentum equation and the equation for static equilibrium in the $x$-direction, we can eliminate $S_X$ to give the equation for $R_X$

$$R_X = \dot{m}(V_1 - V_2 \cos\theta) + (p_1 - B)A_1 - (p_2 - B)A_2 \cos\theta. \tag{10.50}$$

A similar procedure for the $y$-direction leads to the equation for $R_Y$

$$R_Y = \dot{m}V_2 \sin\theta + (p_2 - B)A_2 \sin\theta. \tag{10.51}$$

The magnitude of the net reaction force $\mathbf{R}$ and its direction $\xi$ are then found from

$$R = \sqrt{R_X^2 + R_Y^2} \quad \text{and} \quad \xi = \tan^{-1}\left(\frac{R_Y}{R_X}\right).$$

As we remarked in Section 9.4, the reaction force depends upon the gauge pressures $p_1 - B$, and $p_2 - B$, rather than the absolute static pressures $p_1$ and $p_2$.

Up to this point the analysis has been quite general and to go further we need more information, for example about the pressures $p_1$ and $p_2$. A common outlet condition is that the flow discharges at ambient pressure $B$ so that $p_2 - B = 0$. If, in addition, the flow within the pipe bend can be assumed to be frictionless, then $p_1$ and $p_2$ (now equal to $B$) can be related using Bernoulli's equation, i.e.

$$p_1 + \frac{1}{2}\rho V_1^2 = B + \frac{1}{2}\rho V_2^2$$

which conveniently provides an equation for $p_1 - B$, i.e.

$$p_1 - B = \frac{1}{2}\rho\left(V_2^2 - V_1^2\right).$$

To combine this equation with the equation for $R_X$ requires the introduction of the continuity equation

$$\dot{m} = \rho A_1 V_1 = \rho A_2 V_2.$$

After some algebra we find

$$R_X = \frac{\dot{m}^2}{2\rho A_1}\left[\left(\frac{A_1}{A_2}\right)^2 - \frac{2A_1}{A_2}\cos\theta + 1\right] = \frac{1}{2}\rho A_1 V_1^2\left[\left(\frac{A_1}{A_2}\right)^2 - \frac{2A_1}{A_2}\cos\theta + 1\right] \qquad (10.52)$$

while the equation for $R_Y$ (with $p_2 - B = 0$) simplifies to

$$R_Y = \frac{\dot{m}^2}{\rho A_2}\sin\theta = \frac{\rho A_1^2 V_1^2}{A_2}\sin\theta. \qquad (10.53)$$

The equations for $R_X$ and $R_Y$ written in terms of $V_1$ rather than $\dot{m}$ have the form

reaction force = non-dimensional geometric factor × dynamic pressure × area

much like the equation for drag force

drag force = drag coefficient × dynamic pressure × area.

As we commented at the end of Section 10.1, it is typical of many practical flow situations to find that a hydrodynamic force is proportional to the dynamic pressure.

---

**ILLUSTRATIVE EXAMPLE 10.8**

Liquid of density $\rho$ flows through a pipe which turns through 180° in the horizontal plane and at the same time doubles in cross-sectional area (see Figure E10.8(a)). The static pressure before the bend is twice the external ambient pressure $B$, and the flow within the bend can be regarded as frictionless. Show that the external reaction force required to restrain the bend is given by

$$R = \frac{9\dot{m}^2}{4\rho A} + 3BA$$

where $\dot{m}$ is the mass flowrate and $A$ is the cross-sectional area upstream of the bend.

Solution

We start by substituting $p_1 = 2B$, $A_1 = A$, $A_2 = 2A$, and $\theta = 180°$ in equation (10.50) for $R_X$ (because $\theta = 180°$, $R_Y$ must be zero and so $R = R_X$), i.e.

$$R = \frac{3\dot{m}^2}{2\rho A} + BA + 2(p_2 - B)A$$

wherein we have also made use of the continuity equation, which gives

$$V_1 = 2V_2 = \frac{\dot{m}}{\rho A}.$$

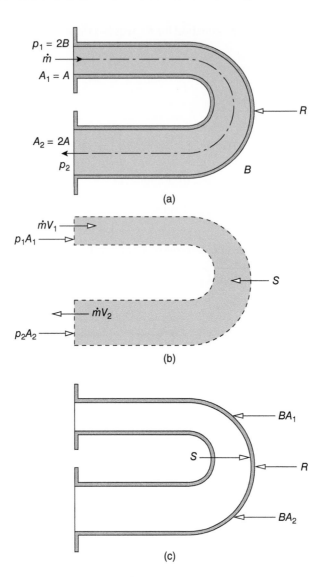

**Figure E10.8**

From Bernoulli's equation we have

$$p_2 = p_1 + \frac{1}{2}\rho \left( V_1^2 - V_2^2 \right)$$

$$= 2B + \frac{\dot{m}^2}{2\rho A^2} - \frac{\dot{m}^2}{8\rho A^2}$$

so that

$$p_2 - B = B + \frac{3\dot{m}^2}{8\rho A^2}.$$

If we substitute for $p_2 - B$ in the equation for $R$, we have

$$R = \frac{9\dot{m}^2}{4\rho A} + 3BA.$$

Although this is the required result, rather than starting by simply substituting in the equation for $R_X$, it would be better to carry out the analysis from first principles, as follows.

We draw a diagram, Figure E10.8(b), of the fluid control volume, including the forces acting on the fluid and the momentum flowrates. The momentum equation applied to the control volume gives

$$-S + p_1 A_1 + p_2 A_2 = -\dot{m} V_2 - \dot{m} V_1.$$

Note that care has to be taken over the sign given to the momentum flowrate $-\dot{m} V_2$ at the outlet to account properly for the fact that the bend has turned a full 180°.

From Figure E10.8(c) we see that the condition of static equilibrium applied to the pipe bend leads to

$$-R + S - BA_1 - BA_2 = 0$$

so that if we eliminate the fluid-structure interaction force $S$ between the two equations we find

$$R = \dot{m} (V_1 + V_2) + (p_1 - B) A_1 + (p_2 - B) A_2$$

which can be shown to lead to the same result as before.

## 10.8 Reaction force on a pipe junction

When we applied the momentum equation to the jet pump in Section 10.6 we pointed out that because there were two fluid streams entering the control volume it was essential to account separately for the momentum flowrate of each stream. The **pipe junction** shown in Figure 10.9(a) has two outlets, each at a different angle to the inlet flow direction, so we must now account for both the mass and momentum flowrates of each outlet stream and also for the pressure force acting on each outlet.

The momentum equation applied to the fluid control volume shown in Figure 10.9(b) leads to

$$-S_X + p_1 A_1 - p_2 A_2 \cos \theta_2 - p_3 A_3 \cos \theta_3 = \dot{m}_2 V_2 \cos \theta_2 + \dot{m}_3 V_3 \cos \theta_3 - \dot{m}_1 V_1 \tag{10.54}$$

for the $x$-direction, and

$$-S_Y - p_2 A_2 \sin \theta_2 + p_3 A_3 \sin \theta_3 = \dot{m}_2 V_2 \sin \theta_2 - \dot{m}_3 V_3 \sin \theta_3 \tag{10.55}$$

for the $y$-direction, where the symbols have their usual meanings.

The corresponding static-equilibrium conditions are

$$S_X - R_X + BA_2 \cos \theta_2 + BA_3 \cos \theta_3 = 0 \tag{10.56}$$

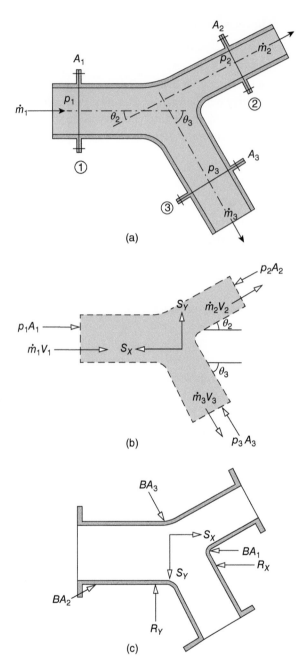

**Figure 10.9** Pipe junction: (a) flow configuration; (b) fluid control volume; (c) forces exerted on pipe junction

and

$$R_Y - S_Y + BA_2 \sin \theta_2 - BA_3 \sin \theta_3 = 0. \tag{10.57}$$

These equations can be combined to eliminate the components $S_X$ and $S_Y$ with the result

$$R_X = \dot{m}_1 V_1 - \dot{m}_2 V_2 \cos \theta_2 - \dot{m}_3 V_3 \cos \theta_3 + (p_1 - B) A_1 - (p_2 - B) A_2 \cos \theta_2$$
$$- (p_3 - B) A_3 \cos \theta_3 \tag{10.58}$$

and

$$R_Y = \dot{m}_2 V_2 \sin \theta_2 - \dot{m}_3 V_3 \sin \theta_3 + (p_2 - B) A_2 \sin \theta_2 - (p_3 - B) A_3 \sin \theta_3. \tag{10.59}$$

Overall mass conservation requires that

$$\dot{m}_1 = \dot{m}_2 + \dot{m}_3 \tag{10.60}$$

and the continuity equation applied to each of the three streams gives

$$\dot{m}_1 = \rho \dot{Q}_1 = \rho A_1 V_1 \quad \dot{m}_2 = \rho \dot{Q}_2 = \rho A_2 V_2 \quad \text{and} \quad \dot{m}_3 = \rho \dot{Q}_3 = \rho A_3 V_3. \tag{10.61}$$

To proceed further, we need information about the static pressures $p_1, p_2,$ and $p_3$. An interesting situation arises if the flow can be considered frictionless since this implies that the stagnation pressure of all three flow streams must be the same. From Bernoulli's equation (7.10) we then have

$$p_0 = p_1 + \frac{1}{2}\rho V_1^2 = p_2 + \frac{1}{2}\rho V_2^2 = p_3 + \frac{1}{2}\rho V_3^2.$$

It should be apparent that, by properly accounting for all the relevant pressure forces, momentum flowrates, flow angles, and cross-sectional areas, we could generalise the pipe-junction analysis to include any number of inlets and outlets.

## 10.9 Flow through a linear cascade of guidevanes

A linear cascade (or set) of curved guidevanes (or turning vanes), such as shown in Figures 1.17 and 10.10(a), is frequently used in wind tunnels, water channels, air-conditioning ducts, etc., to change the direction of a liquid or gas stream. In such applications, the guidevanes may well be formed from sheet metal. Cascades are also used with profiled guidevanes in the preliminary testing of compressor and turbine blades. The nozzle ring and stator stages of axial-flow turbomachines, which we consider in Chapter 14, are essentially cascades with a radial rather than a linear blade configuration. As we shall demonstrate in this section, consideration of the flow through a cascade gives some insight into aerofoil lift and drag.

We shall analyse incompressible flow through a single channel between adjacent guidevanes, as shown in Figure 10.10(a). If the guidevane pitch (i.e. the blade separation distance) is $w$, and the span is $s$, the cross-sectional area $A_1$ normal to the flow at inlet to the channel (section ①) is given by

$$A_1 = ws. \tag{10.62}$$

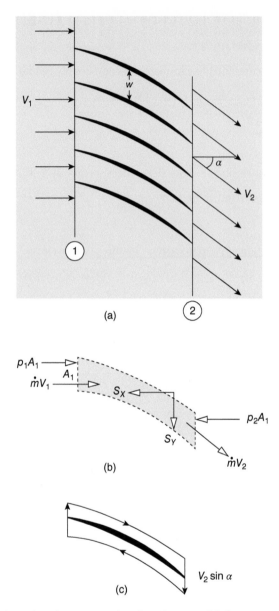

**Figure 10.10** Flow through a linear cascade of guidevanes: (a) flow geometry; (b) fluid control volume; (c) circulation loop

If the flow-deflection angle is $\alpha$, the effective width of the channel at outlet (section ②), taken normal to the flow, must be $w \cos \alpha$, and the effective outlet area $A_2$ is given by

$$A_2 = w \cos \alpha \, s = A_1 \cos \alpha. \tag{10.63}$$

From the continuity equation for the flow of an incompressible fluid of density $\rho$, the mass flowrate $\dot{m}$ through the channel is shown in Chapter 6 to be given by

$$\dot{m} = \rho \dot{Q} = \rho A_1 V_1 = \rho A_2 V_2 \tag{6.1}$$

so that the velocity at inlet $V_1$ is related to that at outlet $V_2$ according to

$$V_2 = \frac{V_1}{\cos \alpha}. \tag{10.64}$$

Since $\cos \alpha < 1$, it is apparent that one effect of the guidevanes is to accelerate the fluid flowing through the cascade (the other principal effect is to turn the flow).

If we assume the flow to be frictionless, we can use Bernoulli's equation to calculate the corresponding pressure drop, i.e.

$$p_0 = p_1 + \frac{1}{2}\rho V_1^2 = p_2 + \frac{1}{2}\rho V_2^2 \tag{7.10}$$

where $p_0$ is the stagnation pressure, $p_1$ is the static pressure at inlet to the cascade, and $p_2$ is the static pressure at outlet. After some algebra, we thus find

$$p_1 - p_2 = \frac{1}{2}\rho V_1^2 \tan^2 \alpha. \tag{10.65}$$

If we apply the momentum equation to the fluid control volume shown in Figure 10.10(b), we have

$$-S_X + p_1 A_1 - p_2 A_1 = \dot{m}V_2 \cos \alpha - \dot{m}V_1$$
$$= 0 \, (\text{since } V_2 \cos \alpha = V_1) \tag{10.66}$$

so that

$$S_X = (p_1 - p_2)A_1 = \frac{1}{2}\rho V_1^2 \tan^2 \alpha A_1 = \frac{1}{2}\dot{m}V_1 \tan^2 \alpha \tag{10.67}$$

and

$$S_Y = \dot{m}V_2 \sin \alpha = \dot{m}V_1 \tan \alpha. \tag{10.68}$$

Note that, in applying the momentum equation in the $x$-direction, the area acted on by the pressures $p_1$ and $p_2$ is $A_1$ in both cases and not $A_2$ for $p_2$. Also, the fluid-guidevane interaction force $S$ (with components $S_X$ and $S_Y$) takes into account the forces acting on both surfaces of the guidevane.

The components of the net force exerted by the fluid on a single guidevane are thus

$$L = S_Y = \dot{m}V_1 \tan \alpha \tag{10.69}$$

and

$$D = S_X = \frac{1}{2}\dot{m}V_1 \tan^2 \alpha. \tag{10.70}$$

We have introduced the symbols $L$ and $D$ here because the force exerted by the fluid on the guidevane normal to the approach-flow direction is the **lift force** $L$, and the corresponding force in the $x$-direction is the associated **drag force**[66] $D$. The drag associated with lift is termed the **induced drag**, and we see that

$$D = \frac{1}{2}L \tan \alpha \tag{10.71}$$

---

[66] These two forces are often referred to simply as **lift** and **drag**.

a result which holds good in much more general situations. We also note that another general aspect of aerodynamic lift is that it is invariably achieved through a change in momentum flowrate brought about by a change in flow direction. As we shall see in Chapter 14, there are similarities between the analysis of flow through a cascade of guidevanes and through the stator or rotor of an **axial-flow turbomachine.**

A rather sophisticated approach to the calculation of aerofoil lift is through the concept of **circulation.** If we draw a closed loop around an aerofoil and split the loop into infinitesimal segments each of length $\delta l$, then the circulation $\Gamma$ (upper case Greek letter gamma) is just the sum of the product of each $\delta l$ and the component $V$ of velocity tangential to the loop at the location of the segment, i.e.

$$\Gamma = \sum V \delta l = \oint V dl \tag{10.72}$$

where the line integral is calculated around the loop. For one of our guidevanes, we choose the loop shown in Figure 10.10(c), with identical curved segments in adjacent channels separated by a distance equal to the pitch $w$. In this case the circulation can be seen to be given by $\Gamma = V_2 w \sin \alpha$ because, over section ①, the channel inlet, the velocity $V_1$ is normal to the loop and so its contribution to $\Gamma$ is zero; the sum of all the contributions to $\Gamma$ along the upper curved section of the loop exactly cancel those along the lower curved section; and all that is left is $V_2 w \sin \alpha$ along the outlet section ②.

According to the Kutta-Joukowski theorem, the lift force per unit span on a body in a two-dimensional, inviscid flowfield is given by $V_1 \Gamma$, so that, for the guidevane,

$$L = \rho V_1 \Gamma s = \rho V_1 V_2 \sin \alpha w s = \dot{m} V_1 \tan \alpha$$

which is exactly the same result as we obtained from the momentum equation. Unfortunately, it is usually far more difficult to calculate the circulation for an isolated aerofoil than it was for the simple guidevane considered here. A more detailed discussion of subsonic aerofoil lift is given in Section 17.7.

---

**ILLUSTRATIVE EXAMPLE 10.9**

Hot gas with a density of 0.2 kg/m³ and velocity of 200 m/s leaves a combustion chamber and is deflected through an angle of 20° by a set of curved guidevanes. If the approach-flow cross-sectional area is 0.5 m², calculate the components of force exerted on the guidevanes in directions parallel to and perpendicular to the approach-flow direction. If the gas exhausts from the guidevanes at a pressure of 1.01 bar, calculate the stagnation pressure of the flow and the pressure drop across the guidevanes.

**Solution**

$\rho = 0.2$ kg/m³; $V_1 = 200$ m/s; $\alpha = 20°$; $A_1 = 0.5$ m²; and $p_2 = 1.01 \times 10^5$ Pa.
We start by calculating the mass flowrate

$$\dot{m} = \rho A_1 V_1 = 0.2 \times 0.5 \times 200 = 20 \text{ kg/s}.$$

The total lift force (i.e. the component of force perpendicular to the approach-flow direction) is thus

$$L = \dot{m}V_1 \tan \alpha = 20 \times 200 \times \tan 20° = 1456 \text{ N}$$

and the total drag force is

$$D = \frac{1}{2}L \tan \alpha = 0.5 \times 1456 \times \tan 20° = 265 \text{ N}.$$

The stagnation pressure $p_0$ can be calculated from the values of $p_2$ and $V_2$, so that we must first calculate the outlet velocity $V_2$ from the continuity equation, i.e.

$$V_2 = \frac{\dot{m}}{\rho A_1 \cos \alpha} = \frac{20}{0.2 \times 0.5 \times \cos 20°} = 213 \text{ m/s}$$

so that

$$p_0 = p_2 + \frac{1}{2}\rho V_2^2 = 1.01 \times 10^5 + 0.5 \times 0.2 \times 213^2 = 1.055 \times 10^5 \text{ Pa or } 1.055 \text{ bar}.$$

For the pressure drop across the guidevanes we have

$$p_1 - p_2 = \frac{1}{2}\rho \left(V_2^2 - V_1^2\right) = 0.5 \times 0.2 \times \left(213^2 - 200^2\right) = 530 \text{ Pa}.$$

## 10.10 Free jet impinging on an inclined flat surface

A **free jet** is one unaffected by solid boundaries. As in the example of a jet of water flowing through air, it is often the case that the surrounding fluid has a much lower viscosity and the two fluids are immiscible. Where the two fluids interact strongly, such as a jet of water flowing through water, the term **submerged jet** is used. In contrast to the internal flows we have considered so far, there are many situations in which the momentum flowrate of a free jet, usually liquid, is reduced in one direction and increased in another direction because the jet is deflected by impingement on a stationary or moving object. In this section, we analyse, first, the situation of a flat stationary plate held at an angle to the jet and, second, the situation where the plate is moving in the same direction as the jet. Also, we shall restrict ourselves to a liquid jet discharging into surroundings which have no influence on the jet other than to impose a uniform pressure $B$ on the free surface of the liquid. We shall also neglect gravitational effects and assume the flow to be frictionless throughout.

### 10.10.1 Stationary plate

If the jet shown in Figure 10.11(a) has a circular cross section and the plate is held normal to it (i.e. the angle $\alpha = 0°$), the deflected fluid will flow radially outwards over the plate. The net momentum outflow from the fluid control volume, shown in Figure 10.11(b), is then practically zero[67] and there is no difficulty in applying the momentum equation. The situation is more complicated for a round jet if the plate is held at an angle $\alpha$, as shown in

---

[67] The outflow will be axisymmetric so that the net momentum outflow normal to the jet is identically zero and any residual momentum flow in the original direction negligible.

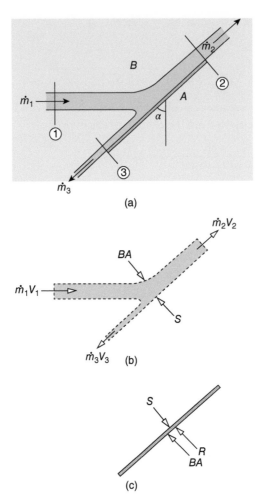

**Figure 10.11** Free jet impinging on a stationary flat plate: (a) flow geometry; (b) fluid control volume; (c) forces exerted on plate

Figure 10.11(a), because the flow over the plate is no longer axisymmetric and thus is unsuited to the simple analysis presented here. To simplify matters we consider a flat jet, i.e. what we see in Figure 10.11(a) is a section through a sheet of liquid leaving a rectangular nozzle of height $t$ and width $w$ where $w \gg t$. Following the impingement the flow splits into two separate streams as shown.

The frictionless-flow assumption has two important consequences: first, Bernoulli's equation can be used in the flow analysis and, second, the fluid-plate interaction force $S$ must act normal to the plate because the liquid pressure $B$ is the only stress exerted by the liquid on the plate surface.

As shown in Chapter 7, according to Bernoulli's equation, the stagnation pressure $p_0$ is given by

$$p_0 = B + \frac{1}{2}\rho V_1^2 = B + \frac{1}{2}\rho V_2^2 = B + \frac{1}{2}\rho V_3^2 \qquad (7.10)$$

from which we deduce that

$$V_1 = V_2 = V_3, \tag{7.11}$$

i.e. the liquid velocity is constant throughout the flowfield.

From the continuity equation we have

$$\dot{m}_1 = \rho A_1 V_1, \quad \dot{m}_2 = \rho A_2 V_2, \quad \text{and} \quad \dot{m}_3 = \rho A_3 V_3 \tag{6.1}$$

where $\dot{m}_2$ and $\dot{m}_3$ are the mass flowrates following impingement and $\dot{m}_1$ is the total mass flowrate of the liquid issuing from the nozzle, such that

$$\dot{m}_1 = \dot{m}_2 + \dot{m}_3 \tag{10.73}$$

from which we conclude that the cross-sectional areas are related by

$$A_1 = A_2 + A_3. \tag{10.74}$$

Because the fluid-plate interaction force acts normal to the plate, it is convenient to apply the momentum equation in directions normal to and parallel to the plate. Normal to the plate we have

$$-S + BA = -\dot{m}_1 V_1 \cos \alpha \tag{10.75}$$

and from the condition for static equilibrium of the plate (see Figure 10.11(c)), we have

$$S - R - BA = 0 \tag{10.76}$$

so that the magnitude of the hydrodynamic reaction force $R$ is given by

$$R = \dot{m}_1 V_1 \cos \alpha. \tag{10.77}$$

The momentum equation applied parallel to the plate gives

$$0 = \dot{m}_2 V_2 - \dot{m}_3 V_3 - \dot{m}_1 V_1 \sin \alpha. \tag{10.78}$$

If we substitute in this equation for the mass flowrates from the continuity equation and also use the results $V_1 = V_2 = V_3$, and $A_1 = A_2 + A_3$, we find

$$A_2 = \frac{1}{2} A_1 (1 + \sin \alpha) \quad \text{and} \quad A_3 = \frac{1}{2} A_1 (1 - \sin \alpha) \tag{10.79}$$

so that the impingement splits the initial flowrate $\dot{m}_1$ in the ratio

$$\frac{\dot{m}_2}{\dot{m}_3} = \frac{1 + \sin \alpha}{1 - \sin \alpha}. \tag{10.80}$$

The flowrates $\dot{m}_2$ and $\dot{m}_3$ are then

$$\dot{m}_2 = \frac{1}{2} (1 + \sin \alpha) \dot{m}_1 \quad \text{and} \quad \dot{m}_3 = \frac{1}{2} (1 - \sin \alpha) \dot{m}_1. \tag{10.81}$$

### 10.10.2 Moving plate

If the plate in Subsection 10.10.1 is moving at velocity $V_P$ in the same direction as the jet, relative to the surroundings, we now have an unsteady-flow problem. To convert this to a steady-flow problem, we apply a Galilean transformation (see Section 6.2) by imposing a velocity $-V_P$ on both the plate and the jet. The relative velocity between the jet and the plate is now $V_{REL} = V_1 - V_P$ and if, in the analysis of Subsection 10.10.1, $V_1$ is replaced by $V_{REL}$, all results remain valid.

---

**ILLUSTRATIVE EXAMPLE 10.10**

A water jet with a velocity of 10 m/s has a rectangular cross section of 25 mm by 100 mm. Calculate the hydrodynamic reaction force exerted on a flat plate held at 60° (measured from the jet centreplane) to the jet (a) if the plate is stationary and (b) if the plate is moving in the same direction as the jet at a velocity of 5 m/s. Calculate also the ratio in which the approach flow is split by the plate.

**Solution**

(a) $V_P = 0$ m/s; $\rho = 1000$ kg/s; $V_1 = 10$ m/s; $A_1 = 0.025 \times 0.1 = 2.5 \times 10^{-3}$ m$^2$; and $\alpha = 30°$.
We calculate first the mass flowrate $\dot{m}_1$ of the jet

$$\dot{m}_1 = \rho A_1 V_1 = 10^3 \times 2.5 \times 10^{-3} \times 10 = 25 \text{ kg/s}.$$

From the analysis of Subsection 10.10.1, we have the magnitude of the reaction force,

$$R = \dot{m}_1 V_1 \cos \alpha = 25 \times 10 \times \cos 30° = 217 \text{ N}$$

and the flow is split in the ratio

$$\frac{1 + \sin \alpha}{1 - \sin \alpha} = 3$$

i.e. the mass flowrate of the upper stream (see Figure 10.11(a)) is three times that of the lower stream.

(b) $V_P = 5$ m/s.
As we pointed out in Subsection 10.10.2, the essential change is that $V_1$ is now replaced by $V_{REL} = V_1 - V_P$, i.e. by 5 m/s. Then,

$$\dot{m}_{REL} = \rho A_1 V_{REL} = 10^3 \times 2.5 \times 10^{-3} \times 5 = 12.5 \text{ kg/s}$$

and

$$R = \dot{m}_{REL} V_{REL} \cos \alpha = 12.5 \times 5 \times \cos 30° = 54.1 \text{ N}.$$

---

## 10.11 Pelton impulse hydraulic turbine

There are two principal types of hydraulic turbine: **reaction turbines** and **impulse turbines**. In the more common reaction turbine, power is generated by water losing pressure and momentum as it passes through fully encased fixed and moving blades (the **stator** and **runner**). In an impulse turbine, water at high pressure flows through nozzles to create high-speed jets which impinge on the blades (often called **buckets**) of a rotor. The rotor does not need to be encased and the water pressure remains essentially constant downstream of the nozzles. The classic bucket shape of the **Pelton turbine** shown in Figure 1.9 is designed to split the impinging water jet into two, turn each half through about 165° and eject the water to either side of the runner. The symmetrical design of the bucket ensures that there is no axial load on the runner, and the high deflection angle produces a momentum change within a few per cent of the maximum possible. A large Pelton turbine generating about 350 MW will have a runner diameter of about 4 m diameter with about 20 buckets equally spaced around its rim. Water is supplied to the turbine through a number of nozzles (2, 4, and 6 are common choices) with **spear** (or **needle**) valves and deflectors to control the flowrate. The high water pressure required by a Pelton turbine is achieved by supplying water to it through a pipe (called a **penstock**) fed from a reservoir 200 to 1800 m above the turbine.

In the simple analysis which follows we neglect the influence of rotation on the water jet-bucket interaction and assume that a water jet impinges on a typical bucket moving at a linear velocity $V_B$ (see Figure 10.12(a)) equal to the peripheral speed of a bucket at pitch radius $R$ on a runner rotating at $N$ rps, i.e.

$$V_B = 2\pi NR. \tag{10.82}$$

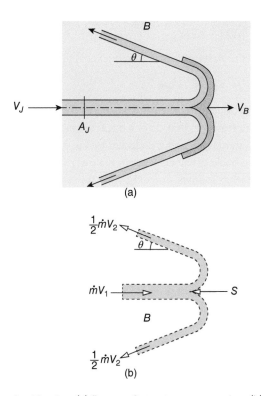

**Figure 10.12** Pelton-wheel bucket: (a) flow configuration cross section; (b) fluid control volume

As we did for a free jet impinging on a moving plate in Subsection 10.10.2, we consider the water flow relative to the Pelton bucket. If we apply the linear momentum equation to the fluid control volume shown in Figure 10.12(b), we have

$$-S + BA_B = -\dot{m}_{EFF} V_{REL} \cos \theta - \dot{m}_{EFF} V_{REL} \tag{10.83}$$

where $S$ is the fluid-bucket interaction force, $A_B$ is the projected area of the bucket (normal to the incoming water jet), $B$ is the ambient pressure, $\theta$ is the angle of the exiting water stream as shown in Figure 10.12(b), and $V_{REL}$ is the velocity of the inflowing water jet relative to the bucket, i.e.

$$V_{REL} = V_J - V_B, \tag{10.84}$$

$V_J$ being the actual velocity of the jet produced by the stationary nozzle. The magnitude of the relative water velocity is the same at the inlet and outlet of the control volume because we have assumed that the static pressure on the surface of the water, and hence within the water, is constant and equal to the ambient pressure $B$, and also that there is negligible friction between the bucket surface and the water flowing over it. Although the actual mass flowrate of the water jet leaving the nozzle $\dot{m}_J$ is given by

$$\dot{m}_J = \rho A_J V_J, \tag{10.85}$$

$A_J$ being the cross-sectional area of the nozzle exit, the effective mass flowrate entering the control volume $\dot{m}_{EFF}$ must correspond to the relative velocity $V_{REL}$ so that

$$\dot{m}_{EFF} = \rho A_J V_{REL} = \rho A_J (V_J - V_B). \tag{10.86}$$

The net force exerted by the water on the Pelton bucket is then given by

$$F_B = S - BA_B = \rho A_J (V_J - V_B)^2 (1 + \cos \theta). \tag{10.87}$$

Some care is necessary in calculating the power generated by the turbine due to the force $F_B$ acting on the runner. The reason is that, as one bucket sweeps into the path of the water jet, a substantial quantity of water is still being deflected by the previous bucket so that for a short period of time at least two buckets will be generating power simultaneously. We take this effect into account for a single jet as follows. Since the time for one complete rotation of the runner is $1/N$, if there are $n_B$ buckets in total, the time $t_I$ for which the jet is intercepted by any one bucket during one rotation must be given by

$$t_I = \frac{1}{n_B N}. \tag{10.88}$$

During this time, the total mass of liquid leaving the nozzle $m = \dot{m}_J t_I = \rho A_J V_J t_I$. Since the mass flowrate through the fluid control volume is only $\dot{m}_{EFF}$, the period of time the mass of fluid $m$ is in contact with the bucket $t_C$ and so generating power, must be given by

$$t_C = \frac{m}{\dot{m}_{EFF}} = \frac{V_J}{V_J - V_B} t_I. \tag{10.89}$$

The angular displacement of the runner during this time is $2\pi N t_C$, and the work done by the torque $F_B R$ due to the force $F_B$ acting at radius $R$ is $2\pi F_B R N t_C$. Since there are $n_B$ buckets in total, the total work for one rotation of the runner is $2\pi n_B F_B R N t_C$, and the corresponding power $P$ generated by the turbine must be given by

$$P = \frac{2\pi n_B F_B R N t_C}{1/N} = 2\pi n_B F_B R N^2 t_C. \tag{10.90}$$

If we substitute for $F_B$, $t_I$, and $t_C$, after some algebra, we find

$$P = \dot{m}_J V_B \left( V_J - V_B \right) \left( 1 + \cos \theta \right).$$

This result applies to a Pelton turbine supplied by a single nozzle. For a turbine with $n_J$ nozzles, the power output is

$$P = n_J \dot{m}_J V_B \left( V_J - V_B \right) \left( 1 + \cos \theta \right). \tag{10.91}$$

As a final point, we can see from the equation (10.91) that the power is a maximum for a given jet velocity $V_J$ when the peripheral bucket velocity $V_B$ is $V_J/2$.

---

**ILLUSTRATIVE EXAMPLE 10.11**

A Pelton turbine with a runner diameter (i.e. a bucket pitch circle) of 2.85 m rotates at a speed of 8.3 rps. Water is supplied through six nozzles, each producing a jet 270 mm in diameter with a speed of 139 m/s. If the Pelton buckets deflect the water through 165°, calculate the power generated by the turbine.

**Solution**

$R = 1.425$ m; $\rho = 1000$ kg/m$^3$; $n_J = 6$; $D_J = 0.17$ m; $V_J = 139$ m/s; $N = 8.3$ rps; and $\theta = 15°$. We start by calculating $A_J$, $\dot{m}_J$, and $V_B$ as follows

$$A_J = \pi D_J^2/4 = 0.0573 \text{ m/s}^2$$

$$\dot{m}_J = \rho A_J V_J = 7958 \text{ kg/s}$$

$$V_B = 2\pi N R = 74.3 \text{ m/s}.$$

The total power generated from all six nozzles is then

$$P = n_J \dot{m}_J V_B \left( V_J - V_B \right) \left( 1 + \cos \theta \right) = 4.51 \times 10^8 \text{ W or 451 MW}.$$

**Comment:**

The data used for this example are close to those for two vertical Pelton-turbine sets, among the most powerful ever built, installed in Sellrain-Silz, Austria.

---

 10.12 SUMMARY

In this chapter we have shown how to apply the linear momentum equation, together with the continuity equation and either Bernoulli's equation or some other information about static pressure, to the analysis of a diverse range of practical problems. A key aim was to demonstrate that we have established a relatively simple theoretical basis which can give quite accurate and useful information about the performance of such complex machines as jet and rocket engines, the jet pump, and the Pelton turbine. Other examples include flow through pipe bends, pipe junctions, and a cascade of guidevanes.

For a given problem, the student should be able to

- identify a fluid control volume appropriate for the application of the linear momentum equation
- apply the linear momentum equation to the flow through the control volume, either in one direction or two orthogonal directions
- apply the continuity equation to the flow through the control volume and also, for flow geometries with multiple inlets and/or outlets, the overall mass-conservation equation
- use either Bernoulli's equation or other information to determine the static pressure throughout the flow
- where appropriate, calculate the hydrodynamic forces created by fluid passing through or around a component or machine
- sketch and label appropriate diagrams to show (a) the geometric arrangement for a given problem, (b) the fluid control volume, and (c) the forces exerted on the device concerned

## 10.13  SELF-ASSESSMENT PROBLEMS

**10.1**  The jet engine shown on the left-hand side of Figure P10.1 takes in air at standard atmospheric conditions (i.e. 20 °C and 1.01 bar) at section ①, where the flow velocity $V_1$ is 200 m/s and the inlet area $A_1$ is 0.3 m². The fuel:air mass ratio is 1:40. The exhaust gases leave the engine at atmospheric pressure at section ②, where the gas velocity $V_2$ is 1000 m/s and the cross-sectional area $A_2$ is 0.25 m². Compute the reaction force required to balance the thrust of the engine.

What will be the reaction force if a deflector, as shown on the right-hand side of the figure, is attached to the engine? How does this reaction force relate to the braking effectiveness of the deflector?

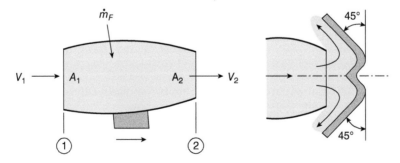

**Figure P10.1**

(Answers; 59.4 kN, –66.6 kN)

**10.2**  (a) A liquid atomiser has the configuration shown in Figure P10.2. The liquid to be atomised is accelerated through the circular nozzle and impinges on a cone attached to the nozzle by a thin rod. The entire arrangement is axisymmetric. Show that the magnitude $R$ of the net hydrodynamic force to hold the nozzle-cone arrangement in place is given by

$$R = \frac{\rho \dot{Q}^2}{2A_2} \left( \frac{A_1}{A_2} + \frac{A_2}{A_1} - 2\cos\theta \right)$$

where $\rho$ is the liquid density, $\dot{Q}$ is the liquid volumetric flowrate, $A_1$ is the cross-sectional area of the nozzle upstream of the contraction, $A_2$ is the cross-sectional area of the nozzle exit, and $\theta$ is the half angle of the cone. Assume that, external to the nozzle, the liquid static pressure is equal to that of the surroundings, frictional and gravitational effects are negligible, and the thin rod has no effect on the flow.

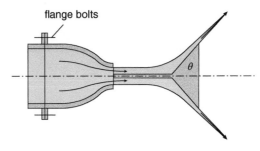

**Figure P10.2**

(b) A light oil of density 900 kg/m$^3$ flows through a nozzle-cone atomiser as shown in Figure P10.2. The upstream nozzle diameter is 50 mm and the exit diameter is 15 mm. If the liquid mass flowrate is 1.35 kg/s and the ambient pressure is 1.01 bar, calculate the stagnation pressure of the liquid and the separate forces acting on the nozzle and on the cone, which has a half angle of 60°.
(Answers: 1.33 bar, 52.7 N, 5.73 N)

10.3  (a) Water flows into a hydraulic turbine through a horizontal pipe of cross-sectional area $A$ and then over a hub, concentric with the pipe, of cross-sectional area $CA$, where $C$ is a numerical constant less than unity. The arrangement is shown in Figure P10.3. If $V$ is the velocity of the water in the pipe, and $\rho$ its density, show that the water exerts a force on the hub of magnitude $R$ given by

$$R = \frac{1}{2}\rho A \left( \frac{CV}{1-C} \right)^2$$

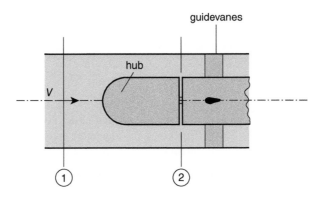

**Figure P10.3**

and find the direction (upstream or downstream) of the force $R$. Assume that the flow is steady and frictionless, that gravitational effects are negligible, and that the pressure and velocity are uniformly distributed across sections ① and ②. Note that the hub is connected rigidly to the shaft supporting the guidevanes by a spindle of negligible cross section in such a way that the static pressure at section ② is constant across the back of the hub and equal to the value in the water stream at section ②.

(b) Calculate $R$ for a turbine with an inlet pipe 10 m in diameter and a hub 8.5 m in diameter for a water flowrate of $2 \times 10^5$ kg/s. Calculate also the static-pressure difference between the forward stagnation point on the nose of the hub and the back of the hub.
(Answers: 1.73 MN, 0.42 bar)

**10.4**   (a) The flowrate of a liquid through a duct of square cross section is controlled by a hinged plate, as shown in Figure P10.4. Show that when the plate is at an angle $\theta$ as shown, the magnitude $F$ of the force acting on the hinge is given by

$$F = (p_0 - p_E)\, A \cos \theta$$

where $p_0$ is the upstream stagnation pressure, $p_E$ is the liquid pressure immediately downstream of the plate, and $A$ is the area of the duct cross section. Assume that the flow is steady, one dimensional, frictionless, and incompressible.

**Figure P10.4**

(b) For a duct of cross-sectional area 4 m$^2$ the mass flowrate is 1000 kg/s when the opening angle of the plate is 30°. Calculate the upstream flow velocity and the force $F$ if the liquid density is 800 kg/m$^3$.
(Answers: 0.313 m/s, 7539 N)

**10.5**   (a) A liquid stream is split and deflected by the pipe junction shown in Figure P10.5 and discharged to surroundings at ambient pressure. Gravitational effects may be neglected and the flow may be assumed to be steady, one dimensional, incompressible, and frictionless. Each of the two exits has an area one quarter that of the inlet area $A$. Show that the magnitude $R$ of the reaction force required to hold the pipe junction in place is given by

$$R = \frac{5\dot{m}^2}{2\rho A}$$

where $\dot{m}$ is the mass flowrate of the liquid and $\rho$ its density. In which direction does the reaction force $\mathbf{R}$ act? **Hint:** Show that the outflow conditions are the same for each exit and use this information to guide the solution.

(b) In a particular case, the liquid is crude oil of relative density 0.85, and the inlet is a circular pipe of diameter 1 m. The static pressure of the oil at inlet to the junction

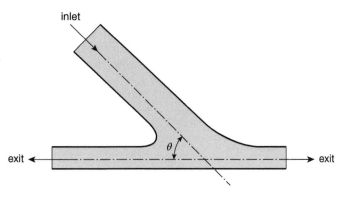

**Figure P10.5**

is 1.04 bar, and the ambient pressure is 1.01 bar. Calculate the oil mass flowrate, the stagnation pressure, and the magnitude of the reaction force.
(Answers: 1024 kg/s, 1.05 bar, 3.93 kN)

**10.6** (a) Figure P10.6 shows the design of a simple jet pump in which a secondary airflow is sucked through a tube of large diameter $D$ by air flowing at high speed through a concentric central tube of diameter $d$. If the mass flowrate of air through the central tube is $\dot{m}_P$, show that the secondary mass flowrate $\dot{m}_S$ is given by the quadratic equation

$$(\dot{m}_P + \dot{m}_S)^2 - \frac{\dot{m}_P^2}{x} - \frac{\dot{m}_S^2}{1-x} + \frac{\dot{m}_S^2}{2(1-x)^2} = 0$$

where $x = (d/D)^2$. Assume that the flow is one dimensional and incompressible throughout, that up to plane ① both flows are frictionless, and that the static pressure is the same in both flows at plane ①. The secondary flow is drawn from the surrounding atmosphere, and the total discharge is to the atmosphere. Note that Bernoulli's equation does not apply downstream of plane ①. **Hint**: Write equations for both mass flowrates and for the stagnation pressure of the secondary flow, and apply the momentum equation to the flow downstream of plane ①.

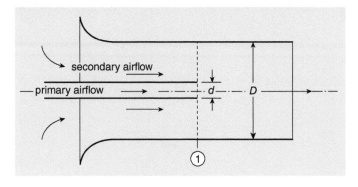

**Figure P10.6**

(b) A jet pump is designed with $D = 100$ mm and $d = 10$ mm. Calculate the ratio between the two mass flowrates $\dot{m}_S/\dot{m}_P$. If the air velocity in the central tube is 150 m/s, calculate the velocity of the secondary flow, the static pressure at plane (1), and the stagnation pressure of the primary airflow. The air density may be taken as 1.2 kg/m$^3$ and the atmospheric pressure as 1.01 bar.

(Answers: 12.2, 18.5 m/s, 1.0079 bar, 1.143 bar)

# 11 Compressible fluid flow

In this chapter we shall introduce the thermodynamic principles which, together with the mass- and momentum-conservation equations, underlie the analysis of the flow of a compressible fluid. For the most part, we shall limit consideration to one-dimensional, internal gas flows, first through a convergent nozzle or duct, then through a convergent-divergent nozzle. We show the important role played by the Mach number $M$ and how subsonic fluid flows ($M < 1$) are fundamentally different from supersonic fluid flows ($M > 1$). We show that the existence of a minimum cross section, termed the **throat** of a convergent-divergent nozzle, leads to a limiting condition called **choking** and is required for transition from subsonic to supersonic flow for an internal compressible flow. While the transition from subsonic to supersonic flow occurs gradually, the change from supersonic to subsonic conditions may be practically discontinuous across a thin zone called a **shockwave**.

## 11.1 Introductory remarks

In Section 2.11 we introduced a material property called the bulk modulus of elasticity and defined by $K = -\mathcal{V}\delta p/\delta \mathcal{V}$, where $\mathcal{V}$ represents volume and $p$ represents thermodynamic pressure. What this definition implies is that the higher the value of $K$ for a material, the less **compressible** it is since a large change in pressure produces only a small change in volume. The inverse of $K$ is called the **compressibility**. From Table A.5 in Appendix 2 we see that typical values of $K$ for liquids are in the range $10^4$ to $10^5$ bar. Also in Section 2.11 we showed that, for a perfect gas, the **isentropic bulk modulus** $K_S = \gamma p$, $\gamma$ being the ratio of specific heats, which allows us to quantify the statement that gases are far more compressible than liquids. We conclude that the subject of compressible fluid flow primarily concerns gas flow[68] and, in consequence, is frequently referred to as **gas dynamics**.

## 11.2 Thermodynamics

### 11.2.1 Specific-entropy change

Throughout this chapter we shall assume that flow processes are **adiabatic**, i.e. no heat is transferred to or from the flow, and **reversible**. A fluid process is reversible if the fluid can return to its initial state with no effect on the surroundings. According to the **second law of**

---

[68] In principle we should also include vapour flow as dry vapour is usually dealt with as a perfect gas.

*Introduction to Engineering Fluid Mechanics*. Marcel Escudier.
© Marcel Escudier 2017. Published 2017 by Oxford University Press.

thermodynamics, the entropy remains constant for a process which is adiabatic and reversible. Such a process is said to be **isentropic**. Although in reality viscous effects cause all flow processes to be **irreversible**, the isentropic assumption is common in compressible-flow theory and leads to practically useful results for both internal and external flows. Flow through a shockwave, which we discuss in Section 11.8 and Chapter 12, is not isentropic but the flow on either side of the shock[69] may be treated as isentropic.

From the **first and second laws of thermodynamics** for a pure substance, the entropy change in any process is given by

$$Tds = dh - \frac{1}{\rho}dp \tag{11.1}$$

where $T$ is the absolute temperature, $s$ is the **specific entropy**, $h$ is the **specific enthalpy**, $p$ is the static pressure, and $\rho$ is the density. Equation (11.1) is called the **second $Tds$ equation**. For a **perfect gas**[70], $p = \rho RT$ (equation (2.9)), where $R$ is the **specific gas constant** for the gas under consideration and is related to the **universal (or molar) gas constant** $\mathcal{R}$, which has the value $\mathcal{R} = 8.3144621$ kJ/kmol $\cdot$ K, through the equation $R = \mathcal{R}/\mathcal{M}$, $\mathcal{M}$ being the molecular weight of the gas with units kg/kmol. For the specific enthalpy we have $dh = C_P dT$ so that, if $C_P$, the **specific heat at constant pressure**, is constant, the change in specific entropy from state 1 to state 2 is given by

$$s_2 - s_1 = C_P \ln\left(\frac{T_2}{T_1}\right) - R \ln\left(\frac{p_2}{p_1}\right). \tag{11.2}$$

### 11.2.2 Isentropic flow

For isentropic flow (or an isentropic process) of a perfect gas, $s_2 - s_1 = 0$, and from equation (11.2) we thus find

$$\frac{p_2}{p_1} = \left(\frac{T_2}{T_1}\right)^{\gamma/\gamma-1} = \left(\frac{\rho_2}{\rho_1}\right)^{\gamma} \tag{11.3}$$

where $\gamma = C_P/C_V$ is the ratio of the specific heats, $C_V$ being the specific heat at constant volume. The term involving density ratio $\rho_2/\rho_1$ arises from the perfect-gas law, equation (2.9). Since $R = C_P - C_V$ (see Section 2.4), it is easily shown that $C_P = \gamma R/(\gamma - 1)$ and $C_V = R/(\gamma - 1)$. Values for $\mathcal{M}, R$, and $\gamma$ for a range of common gases are listed in Table A.6 in Appendix 2. For dry air[71] we have $\gamma = 1.402$ (usually rounded to 1.4 for engineering calculations), $\mathcal{M} = 28.96$ kg/kmol, $R = 287.0$ m$^2$/s$^2 \cdot$ K, and $C_P = 1000.9$ m$^2$/s$^2 \cdot$ K.

A convenient general form of the relationship between $p$ and $\rho$ for an isentropic process is

$$\frac{p}{\rho^{\gamma}} = \text{constant}. \tag{11.4}$$

---

[69] It is common practice to drop the word 'wave' and to refer to a shockwave simply as a shock.
[70] The definition of a perfect gas is given in Section 2.4.
[71] It is easily shown that the units of $R$ and $C_P$ can also be stated as J/kg $\cdot$ K.

### 11.2.3 Steady-flow energy equation

The **steady-flow energy equation** represents a form of the first law of thermodynamics which, together with the continuity and momentum equations, is the basis for analysing compressible duct flow. As shown in Figure 11.1, we consider a duct through which fluid flows with mass flowrate $\dot{m}$. The dashed lines define a control volume between sections ① and ②. It is assumed that the mechanical power (i.e. work) input or output is zero while, in general, the rate of heat transfer $\dot{q}$ into the control volume is non-zero. For present purposes, the steady-flow energy equation can be written in terms of the change in specific[72] **stagnation** (or **total**) **enthalpy**, $\Delta h_0 = h_{0,2} - h_{0,1}$ as follows

$$\dot{q} = \dot{m}\Delta h_0 \tag{11.5}$$

where $h_0$ is defined as

$$h_0 = h + \frac{1}{2}V^2 \tag{11.6}$$

$h$ being the thermodynamic property **specific enthalpy**. Since we shall be limiting consideration to gas flow, no account has been taken of minimal potential-energy variations due to altitude changes. The term $V^2/2$ will be recognised as **kinetic energy** per unit mass. The specific enthalpy $h$ is defined by

$$h = u + pv = u + \frac{p}{\rho} \tag{11.7}$$

where $v$ is the **specific volume** ($v = 1/\rho$) of the fluid and $u$ is its **specific internal energy**, a measure of the kinetic energy of the molecules comprising the fluid. Both $h$ and $u$ are intensive

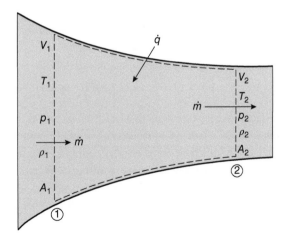

**Figure 11.1** Control volume for compressible fluid flow through a duct

---

[72] The word specific, indicating per unit mass, is normally omitted for the stagnation enthalpy.

thermodynamic properties[73] determined by pressure and temperature. The quantity $pv = p/\rho$ is termed the **flow energy** or **flow work**. If we substitute equations (11.6) and (11.7) into the steady-flow energy equation (11.5), we have

$$\dot{q} = \dot{m}\Delta\left(u + \frac{p}{\rho} + \frac{1}{2}V^2\right) = \Delta\left(\dot{m}u + \frac{\dot{m}p}{\rho} + \frac{1}{2}\dot{m}V^2\right). \qquad (11.8)$$

We note that, for one-dimensional flow, the continuity equation is $\dot{m} = \rho AV$, so that the combination $\dot{m}p/\rho = pAV$, i.e. a force $pA$ multiplied by a velocity $V$, therefore represents work per unit time or power. This interpretation of the term $\dot{m}p/\rho$ justifies calling the combination $p/\rho$ flow work.

From the foregoing, we see that the steady-flow energy equation may be written as

$$\dot{q} = \dot{m}(h_{0,2} - h_{0,1}) \qquad (11.9)$$

$$= \dot{m}\left[\left(h_2 + \frac{1}{2}V_2^2\right) - \left(h_1 + \frac{1}{2}V_1^2\right)\right]. \qquad (11.10)$$

For a perfect gas we have $h = C_P T$ so that

$$\dot{q} = \dot{m}\left[\left(C_P T_2 + \frac{1}{2}V_2^2\right) - \left(C_P T_1 + \frac{1}{2}V_1^2\right)\right]. \qquad (11.11)$$

For a gas with constant specific heat, i.e. a calorically perfect gas, equation (11.11) can thus be written as

$$\dot{q} = \dot{m}C_P\Delta T_0 \qquad (11.12)$$

where $T_0$ is the **stagnation** (or **total temperature**) defined by

$$T_0 = T + \frac{V^2}{2C_P}. \qquad (11.13)$$

As stated earlier, in this chapter we shall limit consideration to **adiabatic flows**[74], i.e. flows for which $\dot{q} = 0$, so that our steady-flow energy equation reduces to

$$T_0 = T + \frac{V^2}{2C_P} = T\left(1 + \frac{V^2}{2C_P T}\right) = T\left[1 + \left(\frac{\gamma - 1}{2}\right)M^2\right] = \text{constant} \qquad (11.14)$$

where $M = V/c$ is the Mach number (see Section 11.4). In deriving equation (11.14) we have made use of the relationships $C_P = \gamma R/(\gamma - 1)$ and $c = \sqrt{\gamma RT}$.

While the equations including the heat-transfer rate $\dot{q}$ apply to one-dimensional duct flow, those for adiabatic flow ($\dot{q} = 0$) also apply along a streamline in an external compressible-gas flow. It is important to keep in mind that the static temperature $T$, and so the speed of sound $c$, will usually vary along a duct or streamline so that the Mach number $M$ is the ratio of two varying quantities, $V$ and $c$.

---

[73] An intensive thermodynamic property is one that is independent of the size of a thermodynamic system. Examples are temperature, pressure, and density.

[74] In Chapter 13 we discuss **Rayleigh flow**, which is flow through a pipe of constant cross section with heat transfer, i.e. non-adiabatic or **diabatic**.

## 11.3 Bernoulli's equation and other relations for compressible-gas flow

In Section 7.3 we derived **Euler's** (momentum) **equation** for flow along a streamline in an inviscid flow

$$\frac{dp}{ds} + \rho g \frac{dz'}{ds} + \rho V \frac{dV}{ds} = 0, \tag{7.4}$$

$s$ being distance along the streamline.

As we saw in Section 7.4, for a constant-density fluid, equation (7.4) can be integrated to yield **Bernoulli's equation**. As we pointed out in Section 11.2, for gas flow the potential-energy term (involving changes in the altitude $z'$) in Euler's equation can be neglected and the equation can then be written as

$$\frac{1}{\rho}\frac{dp}{ds} + V\frac{dV}{ds} = 0. \tag{11.15}$$

If we restrict consideration to **isentropic flow** of a perfect gas, we have

$$\frac{p}{\rho^\gamma} = \text{constant}. \tag{11.4}$$

We can then combine this with equation (11.15) to show that

$$\left(\frac{\gamma}{\gamma-1}\right)\frac{d}{ds}\left(\frac{p}{\rho}\right) + V\frac{dV}{ds} = 0. \tag{11.16}$$

Since the specific-heat ratio $\gamma$ is a constant, equation (11.16) can be integrated to yield

$$\left(\frac{\gamma}{\gamma-1}\right)\frac{p}{\rho} + \frac{1}{2}V^2 = \text{constant} = \left(\frac{\gamma}{\gamma-1}\right)\frac{p_0}{\rho_0}, \tag{11.17}$$

which can be regarded as Bernoulli's equation for steady compressible flow of a perfect gas. Unlike the total temperature $T_0$, which remains constant throughout an adiabatic flow, the local total pressure $p_0$ and the local total density $\rho_0$ remain constant throughout an adiabatic flow only if it is also reversible, i.e. **isentropic**. As we shall see in Section 11.8, while there is no change in $T_0$ across a shockwave, there is an entropy increase and an associated decrease in both $p_0$ and $\rho_0$.

In compressible flow, the **dynamic pressure** $\rho V^2/2$, which appears in Bernoulli's equation, is no longer the difference between the stagnation and static pressures, as in the incompressible-flow case. For a perfect gas, we have

$$\frac{1}{2}\rho V^2 = \frac{1}{2}\rho c^2 M^2 = \frac{1}{2}\gamma \rho R T M^2 = \frac{1}{2}\gamma p M^2 \tag{11.18}$$

so that the dynamic pressure depends upon the static pressure $p$ and the Mach number $M$.

If we now introduce into equation (11.17) the perfect-gas equation, $p = \rho R T$, together with $c = \sqrt{\gamma R T}$ and the Mach-number definition, $M = V/c$, we find

$$T\left[1 + \left(\frac{\gamma - 1}{2}\right)M^2\right] = \text{constant} = T_0 \tag{11.19}$$

i.e. the same result as equation (11.14), which we obtained from the steady-flow energy equation. This equation is more usually written as

$$\frac{T_0}{T} = 1 + \left(\frac{\gamma - 1}{2}\right)M^2. \tag{11.20}$$

From equation (11.3) for an isentropic flow,

$$\frac{\rho_0}{\rho} = \left(\frac{T_0}{T}\right)^{1/(\gamma - 1)}$$

so, from equation (11.20),

$$\frac{\rho_0}{\rho} = \left[1 + \left(\frac{\gamma - 1}{2}\right)M^2\right]^{1/(\gamma - 1)}. \tag{11.21}$$

Again using equation (11.3) for an isentropic flow we have

$$\frac{p_0}{p} = \left[1 + \left(\frac{\gamma - 1}{2}\right)M^2\right]^{\gamma/(\gamma - 1)}. \tag{11.22}$$

For subsonic flow ($M < 1$), a Pitot tube can be used to measure the stagnation pressure $p_0$, and the Mach number can be determined if an independent measurement is made of the static pressure $p$. For supersonic flow ($M > 1$), a **detached shock** forms ahead of a Pitot tube, a situation we consider in Subsection 11.8.3.

Values for the ratios $p_0/p$, $\rho_0/\rho$, and $T_0/T$, or their reciprocals, together with other quantities of interest in compressible flow, including normal and oblique shock-wave properties, have been tabulated for different values of the specific-heat ratio $\gamma$ in numerous textbooks. The source of many of these tabulations is NACA[75] Report 1135, *Equations, Tables, and Charts for Compressible Flow*, published by NACA AMES Research Staff in 1953. Also included are many of the equations for supersonic flow discussed in this chapter and Chapter 12. Numerical values for many compressible-flow relations are now available free in the form of a *Compressible Aerodynamics Calculator* at <http://www.dept.aoe.vt.edu/~devenpor/aoe3114/calc.html>[76]. This **Calculator**, as we shall refer to it, consists of six sections: Isentropic flow, Normal shocks, Oblique shocks, Conical shocks, Fanno flow, and Rayleigh flow. Throughout this chapter and Chapters 12 and 13, we shall make extensive use of the Calculator in the numerical solution of compressible-flow problems. The advantage of the Calculator over tables is that very accurate calculations can be made for any values of $M$ and $\gamma$ without the necessity of interpolation.

---

[75] NACA was the *National Advisory Committee for Aeronautics* and the predecessor of NASA, the *National Aeronautics and Space Administration*, which came into being on 1 October 1958.

[76] The basic calculator was written by Professor William Devenport of the Department of Aerospace and Ocean Engineering at Virginia Tech. The last update, at the time of writing, was 5 January 2014.

## 11.4  Subsonic flow and supersonic flow

In Subsection 3.12.6 we defined the **Mach number** $M$ as the ratio of a flowspeed[77], $V$, to the local **soundspeed** of the fluid $c$, i.e. $M = V/c$. The adjective 'local' is important because $c$ is not constant but a variable in most flow situations (in the case of a perfect gas, $c = \sqrt{\gamma RT}$ so that the soundspeed varies if the static temperature changes). As will become apparent, the Mach number is the key parameter in the description of compressible flows with the **sonic condition**, $M = 1$, taking on special significance.

Flows may be classified according to the Mach number as follows

**subsonic flow**, $M < 1$, with compressibility effects becoming increasingly significant as $M$ increases although no new physical phenomena arise until $M$ approaches unity.

**transonic flow**, $M \approx 1$, with some regions remaining subsonic while others are supersonic and shockwaves begin to appear.

**supersonic flow**, $M > 1$, with shock and **expansion waves** arising (see Chapter 12).

**hypersonic flow**, $M \gg 1$, with new phenomena (beyond the scope of this text), such as gas ionisation, arising.

In this text we shall limit attention to one-dimensional **internal flows**, including the effects of heat transfer and surface friction on compressible duct flow (Chapter 13), and to two-dimensional **external flows** (Chapter 12).

## 11.5  Mach wave and Mach angle

An infinitesimal pressure pulse (i.e. sound) originating from a point source within any medium with uniform properties propagates spherically outwards, with the centre of each spherical wavefront being **advected**[78] at the flow velocity $V$. A wavefront moves away from the sphere centre at the speed of sound $c$ so that after a time $t$ the front will have moved a distance $ct$ relative to the instantaneous centre, which itself will have moved a distance $Vt$. Figure 11.2 shows four situations for a series of such a pulses emitted at fixed time intervals $\Delta t$, corresponding to a constant frequency $f$, from the source and propagating through a gas flowing from left to right at steady velocity $V$: (a) $M = 0$, (b) $M < 1$, (c) $M = 1$, and (d) $M > 1$. This figure, based upon one first given in a paper presented by **Ernst Mach** in 1887, is known as **Mach's construction**. We observe the following

Figure 11.2(a), $M = 0$. The wavefronts are concentric with the outermost front a distance $ct$ from the origin. A stationary observer anywhere in the wavefield would hear the sound at frequency $f$.

Figure 11.2(b), $M < 1$. Each wavefront and its centre has been advected to the right, with the instantaneous centre of the outermost wavefront a horizontal distance $Vt$ from the origin, and the wavefront a distance $ct$ ($> Vt$) from its instantaneous centre. On the centreline,

---

[77] We can apply a **Galilean transformation** (see Section 6.2) to transform the situation of an object moving at steady velocity through a stationary fluid to that of steady flow past a stationary object.

[78] The term advection should not be confused with **convection**, which is the combination of advection and diffusion.

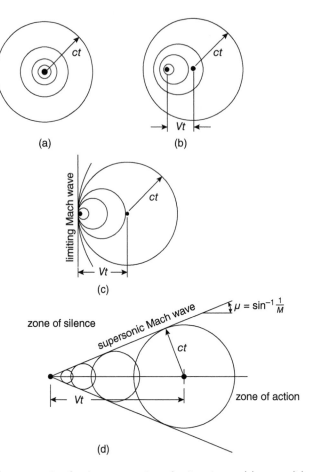

**Figure 11.2** Mach's construction for the propagation of a disturbance (a) $M = 0$, (b) $M < 1$, (c) $M = 1$, (d) $M > 1$

left-moving wavefronts propagate upstream at velocity $(c - V)t$ while those moving downstream to the right move at velocity $(c + V)t$. A stationary observer would now hear the sound at different frequencies, depending upon his or her position in the wavefield: higher frequencies to the right, lower frequencies to the left. This dependence of frequency upon location is an example of the **Doppler effect**.

Figure 11.2(c), $M = 1$. Since we now have $V = c$, wavefronts on the centreline coalesce to the left, forming a **Mach wave**, while those moving to the right do so at velocity $2c$.

Figure 11.2(d), $M > 1$. The situation is now completely different from $M < 1$. The wavefronts are confined within a cone, termed the **Mach cone**[79], of semi-angle $\mu = \sin^{-1} (1/M)$, called the **Mach angle**. The region within the cone is referred to as the **zone of action** (also called the **region of influence**), while the undisturbed region outside is the **zone of silence**.

Mach's construction may also be applied to a solid surface over which there is a supersonic flow with Mach number $M$, as shown in Figure 11.3. Any slight irregularity in the surface

---

[79] In two dimensions the wavefronts are cylindrical and are confined within a **Mach wedge**.

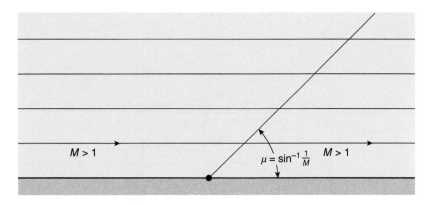

**Figure 11.3** Mach wave at a surface

results in a small disturbance communicated to the flow along a Mach wave of infinitesimal strength at the Mach angle $\mu$ to the surface.

## 11.6 Steady, one-dimensional, isentropic, perfect-gas flow through a gradually convergent duct

We consider isentropic flow through the gradually convergent duct (or nozzle) shown in Figure 11.4(a), for which the cross-sectional area $A$ decreases monotonically to the exit area $A_E$, and the flow exhausts to an external (or **back**) **pressure** $p_B$, which will be less than or equal to the static pressure $p_E$ in the exit plane of the duct. The limitation to a gradually convergent duct is to rule out sudden changes in cross section, such as we considered in Chapter 10 for incompressible flow. In the absence of viscous effects, flow in a gradually convergent (or divergent) nozzle will be very nearly one dimensional. It may help the reader to imagine that the duct is axisymmetric (i.e. circular) and straight, although the analysis is not limited to straight circular ducts.

The continuity equation for one-dimensional flow through the duct may be written, as derived in Chapter 6, as follows

$$\dot{m} = \rho A V. \tag{6.1}$$

If we substitute for $\rho$ from the perfect-gas equation, $p = \rho RT$, for $V$ using the Mach-number definition, $M = V/c$, and, for the soundspeed, as shown in Chapter 2,

$$c = \sqrt{\gamma RT}, \tag{2.50}$$

we can rewrite the continuity equation as

$$\dot{m} = p A M \sqrt{\frac{\gamma}{RT}}. \tag{11.23}$$

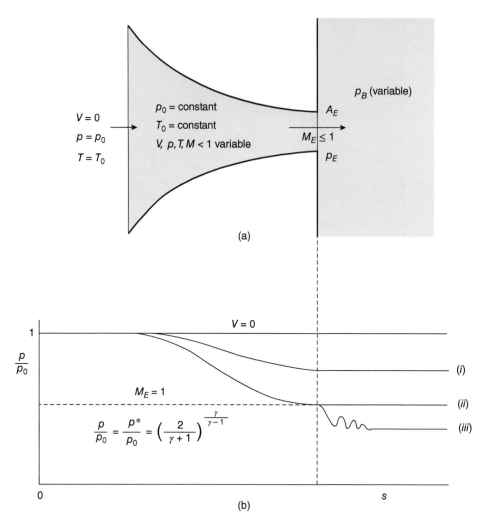

**Figure 11.4** (a) Flow through a convergent duct (b) Static pressure variation along a convergent duct for (i) $p_B > p^*$, (ii) $p_B = p^*$, (iii) $p_B < p^*$

We now introduce the stagnation (or reservoir) conditions $p_0$ and $T_0$, both of which remain constant throughout the flow since we are assuming isentropic conditions, to find

$$\frac{\dot{m}\sqrt{RT_0}}{Ap_0} = \sqrt{\gamma}M\frac{p}{p_0}\sqrt{\frac{T_0}{T}}. \tag{11.24}$$

We can substitute in equation (11.24) for $T_0/T$ using equation (11.20) and for $p/p_0$, using equation (11.22) so that, finally, we have

$$\frac{\dot{m}\sqrt{RT_0}}{Ap_0} = \sqrt{\frac{\gamma M^2}{\left[1 + \left(\frac{\gamma-1}{2}\right)M^2\right]^{(\gamma+1)/(\gamma-1)}}}. \tag{11.25}$$

The combination of terms $\dot{m}\sqrt{RT_0}/\left(Ap_0\right)$ is referred to as the **mass-flow function**. At the exit of the duct we have

$$\frac{\dot{m}\sqrt{RT_0}}{A_E p_0} = \sqrt{\frac{\gamma M_E^2}{\left[1 + \left(\frac{\gamma-1}{2}\right) M_E^2\right]^{(\gamma+1)/(\gamma-1)}}} \tag{11.26}$$

where $M_E$ is the Mach number at the duct exit. It is sometimes useful to write equation (11.25) in terms of the pressure ratio $p/p_0$ rather than $M$

$$\frac{\dot{m}\sqrt{RT_0}}{A p_0} = \sqrt{\frac{2\gamma}{\gamma-1}\left[\left(\frac{p}{p_0}\right)^{2/\gamma} - \left(\frac{p}{p_0}\right)^{(\gamma+1)/\gamma}\right]} \tag{11.27}$$

so that, at the duct exit,

$$\frac{\dot{m}\sqrt{RT_0}}{A_E p_0} = \sqrt{\frac{2\gamma}{\gamma-1}\left[\left(\frac{p_E}{p_0}\right)^{2/\gamma} - \left(\frac{p_E}{p_0}\right)^{(\gamma+1)/\gamma}\right]}. \tag{11.28}$$

At the beginning of this section, it was stated that the flow exhausts to a back pressure $p_B$, which must be less than or equal to $p_E$ so that equation (11.28) shows how the mass flowrate, for given values of the stagnation conditions, $p_0$ and $T_0$, and the exit area $A_E$, is determined by the back pressure $p_B (= p_E)$. When $p_B$ and $p_0$ are equal, there is no flow but, as $p_B$ is progressively reduced below $p_0$, $\dot{m}$ increases until it reaches a maximum value $\dot{m}^*$ for

$$\frac{p_E}{p_0} = \left(\frac{2}{\gamma+1}\right)^{\gamma/(\gamma-1)} = 0.528 \text{ with } \gamma = 1.4. \tag{11.29}$$

Substitution of this value for $p_E/p_0$ into equation (11.28) leads to

$$\frac{\dot{m}^*\sqrt{RT_0}}{A_E p_0} = \sqrt{\gamma\left(\frac{2}{\gamma+1}\right)^{(\gamma+1)/(\gamma-1)}} = 0.685 \tag{11.30}$$

for the maximum possible flowrate for given values of the stagnation conditions and nozzle exit area. The corresponding exit Mach number $M_E$ is then equal to unity. For any other pressure ratio $p_E/p_0$, above that corresponding to equation (11.29), the flowrate can be calculated from equation (11.28), the pressure variation along the duct from equation (11.27), and the Mach-number variation from equation (11.25).

Since it is possible to derive from equation (11.28) values of $\dot{m}\sqrt{RT_0}/\left(A_E p_0\right)$ for values of $p_E/p_0$ less than that given by equation (11.29), it might be thought that the mass flowrate would continue to increase. In fact, not only are the mass flowrates lower but, as can be seen from equation (11.22), the corresponding values of the exit Mach number exceed the maximum allowable value of unity. From a practical point of view, we can reduce the back pressure $p_B$ to any value we wish, down to absolute zero (i.e. vacuum), but the lowest possible value for the exit pressure $p_E$ cannot be lower than that given by equation (11.29), and the highest possible mass flowrate $\dot{m}^*$ is that given by equation (11.30) when the duct is said to be **choked**.

The corresponding flow conditions at the exit are said to be **critical**[80] and are identified by a superscript asterisk[81], as follows

$$\frac{T^*}{T_0} = \frac{2}{\gamma + 1} = 0.8333 \tag{11.31}$$

$$\frac{p^*}{p_0} = \left(\frac{2}{\gamma + 1}\right)^{\gamma/(\gamma-1)} = 0.5283 \tag{11.32}$$

$$\frac{\rho^*}{\rho_0} = \left(\frac{2}{\gamma + 1}\right)^{1/(\gamma-1)} = 0.6339. \tag{11.33}$$

The conclusions drawn here for compressible, subsonic flow through a convergent duct are generalised in the following section for flow through a convergent-divergent duct.

Figure 11.4(b) shows the static pressure variation along a convergent duct, calculated from equation (11.27), and Figure 11.5 the variation of non-dimensional mass flowrate as a function of $p_B/p_0$, calculated from equation (11.28).

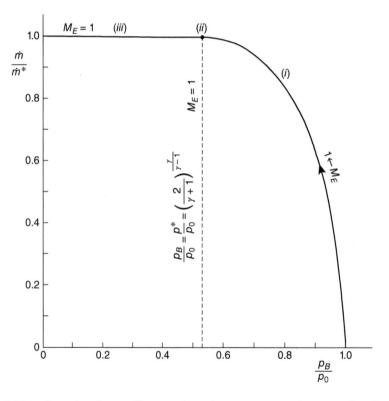

**Figure 11.5** Non-dimensional mass flowrate through a convergent duct as a function of back pressure

[80] The term 'critical' in gas dynamics should not be confused with the thermodynamics use of the term with regard to phase changes.

[81] In some texts a subscript $c$ is used instead of a superscript asterisk to indicate critical conditions.

The three conditions $(i)$, $(ii)$, and $(iii)$, identified in Figure 11.4(b), are also shown in Figure 11.5, which shows the variation of the ratio $\dot{m}/\dot{m}^*$ with overall pressure ratio $p_B/p_0$.

---

**ILLUSTRATIVE EXAMPLE 11.1**

An industrial furnace is supplied with propane gas through a convergent nozzle with an exit diameter of 25 mm. The stagnation pressure of the gas is 3 bar, and the stagnation temperature is 300 K. Calculate the exit conditions and mass flowrate if (a) the nozzle exit is choked and (b) the furnace pressure is 2.4 bar.

**Solution**

For propane $R = 188.6 \text{ m}^2/\text{s}^2 \cdot \text{K}$, $\gamma = 1.136$, $p_0 = 3 \times 10^5$ Pa, $T_0 = 300$ K, $D_E = 0.25$ m, and $A_E = \pi D_E^2/4 = 4.909 \text{ m}^2$.

(a) If the nozzle is choked, $M_E = 1$, $p_E = p^*$, $T_E = T^*$, and $V_E = c^*$. From the Isentropic-flow Calculator, $p^*/p_0 = 0.528$ so $p^* = 1.585 \times 10^5$ Pa, and $T^*/T_0 = 0.833$ so $T^* = 250$ K.
The soundspeed corresponding with $T^*$ is $c^* = \sqrt{\gamma R T^*} = 231.5$ m/s.
From the perfect-gas equation, $\rho^* = p^*/RT^* = 3.361 \text{ kg/m}^3$.
From the continuity equation, $\dot{m} = \rho^* A_E V^* = 0.382 \text{ kg/m}^3$.

(b) If $p_E = 2.4$ bar, $p_E/p_0 = 0.8$ and from the Isentropic-flow Calculator, $M_E = 0.574$, and $T_E/T_0 = 0.938$, so that $T_E = 281.5$ K.
The soundspeed corresponding with $T_E$ is $c_E = \sqrt{\gamma R T_E} = 245.6$ m/s so that $V_E = M_E c_E = 140.9$ m/s.
From the perfect-gas equation, $\rho_E = p_E/RT_E = 4.521 \text{ kg/m}^3$.
From the continuity equation, $\dot{m} = \rho_E A_E V_E = 0.313$ kg/s.

---

## 11.7 Steady, one-dimensional, isentropic, perfect-gas flow through a convergent-divergent nozzle

### 11.7.1 General considerations

We begin with general considerations of steady, one-dimensional, isentropic flow of a perfect gas through a convergent-divergent nozzle, as shown schematically in Figure 11.6. A nozzle of this form is called a **Laval nozzle** or an **effusor**. As before, the continuity equation is

$$\dot{m} = \rho A V \qquad (6.1)$$

where the nozzle cross-sectional area $A$ decreases to a minimum throat area $A_T$ and then increases again. If we differentiate equation (6.1) with respect to the distance along the nozzle $s$ and divide through by $\dot{m}$, we find

$$0 = \frac{1}{\rho}\frac{d\rho}{ds} + \frac{1}{A}\frac{dA}{ds} + \frac{1}{V}\frac{dV}{ds}. \qquad (11.34)$$

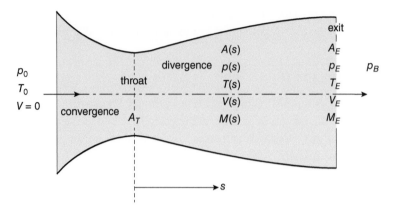

**Figure 11.6** Geometry of a convergent-divergent nozzle

From equation (11.15), Euler's (momentum) equation, we have

$$\frac{1}{\rho}\frac{dp}{ds} + V\frac{dV}{ds} = 0$$

or

$$\frac{1}{\rho V^2}\frac{dp}{ds} = -\frac{1}{V}\frac{dV}{ds}. \tag{11.35}$$

From Section 2.12 we can write, for isentropic flow,

$$\frac{dp}{d\rho} = c^2 \tag{2.49}$$

so that

$$\frac{dp}{ds} = c^2\frac{d\rho}{ds} \tag{11.36}$$

or

$$\frac{1}{\rho}\frac{d\rho}{ds} = \frac{1}{\rho c^2}\frac{dp}{ds} = \frac{M^2}{\rho V^2}\frac{dp}{ds} \tag{11.37}$$

where $p$ is the static pressure, $c$ is the soundspeed, and $M \equiv V/c$ is the local Mach number. We can combine equations (11.35) and (11.37) to show that

$$\frac{1}{\rho}\frac{d\rho}{ds} = -\frac{M^2}{V}\frac{dV}{ds}. \tag{11.38}$$

If we substitute $d\rho/ds$ from equation (11.38) into equation (11.34), we find

$$\frac{1}{A}\frac{dA}{ds} = \left(M^2 - 1\right)\frac{1}{V}\frac{dV}{ds} \tag{11.39}$$

or

$$\frac{1}{V}\frac{dV}{ds} = \frac{1}{\left(M^2 - 1\right)}\frac{1}{A}\frac{dA}{ds} \tag{11.40}$$

and equation (11.38) then gives

$$\frac{1}{\rho}\frac{d\rho}{ds} = \frac{-M^2}{(M^2 - 1)}\frac{1}{A}\frac{dA}{ds}.$$  (11.41)

Equation (11.40) is known as the Hugoniot equation.

Since

$$\frac{dp}{ds} = c^2\frac{d\rho}{ds},$$  (11.36)

with $c^2 = \gamma RT$ (equation (2.50) and $p = \rho RT$ (equation (2.9)) we have

$$\frac{1}{p}\frac{dp}{ds} = \frac{\gamma}{\rho}\frac{d\rho}{ds}$$  (11.42)

so that, from equation (11.41)

$$\frac{1}{p}\frac{dp}{ds} = \frac{-\gamma M^2}{(M^2 - 1)}\frac{1}{A}\frac{dA}{ds}.$$  (11.43)

From equation (11.20) for $T_0/T$ we can show that, for an adiabatic flow

$$-\left[1 + \left(\frac{\gamma - 1}{2}\right)M^2\right]\frac{1}{T}\frac{dT}{ds} = (\gamma - 1)M\frac{dM}{ds}$$  (11.44)

and, from the definition of $M = V/c$, we have the general result

$$\frac{1}{c}\frac{dc}{ds} = \frac{1}{V}\frac{dV}{ds} - \frac{1}{M}\frac{dM}{ds}.$$  (11.45)

Since $c = \sqrt{\gamma RT}$, we have

$$\frac{2}{c}\frac{dc}{ds} = \frac{1}{T}\frac{dT}{ds}$$  (11.46)

so that, if we combine the last three equations with equation (11.40), we can show that

$$\frac{1}{M}\frac{dM}{ds} = \frac{\left[1 + \left(\frac{\gamma - 1}{2}\right)M^2\right]}{(M^2 - 1)}\frac{1}{A}\frac{dA}{ds}$$  (11.47)

and, from equation (11.44)

$$\frac{1}{T}\frac{dT}{ds} = \frac{-(\gamma - 1)M^2}{(M^2 - 1)}\frac{1}{A}\frac{dA}{ds}.$$  (11.48)

The equations above reveal important fundamental differences in the behaviour of subsonic and supersonic flows in response to an area change. These differences may be summarised as follows

|  | Subsonic $M < 1$ | Supersonic $M > 1$ |
|---|---|---|
| If $dA/ds < 0$ (convergent) | $dV/ds > 0$ $d\rho/ds < 0$ $dp/ds < 0$ $dM/ds > 0$ $dT/ds < 0$ | $dV/ds < 0$ $d\rho/ds > 0$ $dp/ds > 0$ $dM/ds < 0$ $dT/ds > 0$ |
| If $dA/ds > 0$ (divergent) | $dV/ds < 0$ $d\rho/ds > 0$ $dp/ds > 0$ $dM/ds < 0$ $dT/ds > 0$ | $dV/ds > 0$ $d\rho/ds < 0$ $dp/ds < 0$ $dM/ds > 0$ $dT/ds < 0$ |

The differences between subsonic and supersonic duct flow are remarkable

- For subsonic flow, an area decrease causes a velocity increase and a pressure decrease, in accordance with Bernoulli's equation for incompressible flow, though now the density also decreases, i.e. for subsonic flow, a convergent duct acts as a **subsonic nozzle**.
- For subsonic flow, an area increase causes a velocity decrease and a pressure increase, again in accordance with Bernoulli's equation for incompressible flow, though now the density also increases, i.e. for subsonic flow, a divergent duct acts as a **subsonic diffuser**.
- For supersonic flow, an area decrease causes a velocity decrease, a pressure increase, and a density increase, i.e. for supersonic flow, a convergent duct acts as a **supersonic diffuser**.
- For supersonic flow, an area increase causes a velocity increase, a pressure decrease, and a density decrease, i.e. for supersonic flow, a divergent duct acts as a **supersonic nozzle**.

We can draw other conclusions from the equations derived in this section

- Equation (11.41) shows why the Mach number is a measure of compressibility in a flow: as $M \to 0$, $d\rho/ds \to 0$, i.e. for $M \ll 1$ the gas density is effectively constant.
- Equation (11.47) shows that for subsonic flow, an area decrease (i.e. $dA/ds < 0$) causes a Mach-number increase and for supersonic flow a Mach-number decrease, i.e. the Mach number always tends to unity if the area decreases.
- Since infinite acceleration (i.e. $dV/ds = \infty$) or deceleration is a physical impossibility, equation (11.40) shows that isentropic duct flow can only pass through **sonic velocity** ($M = 1$) to change from being subsonic to supersonic, or from supersonic to subsonic, if $dA/ds = 0$, i.e. if the cross-sectional area of the duct is a minimum, at what is termed a **throat**.
- Equation (11.40) also shows that the velocity of a subsonic flow reaches a maximum at a throat while a supersonic flow reaches a minimum. Equation (11.43) then shows that the pressure is a minimum at the throat for subsonic flow and a maximum for supersonic flow.

## 11.7.2 Pressure and Mach number variations through a Laval nozzle

Many of the equations derived in Section 11.6 for flow through a convergent duct are still valid here. Of particular interest is equation (11.27) for the variation of the static pressure $p$ with respect to the cross-sectional area $A$

$$\frac{\dot{m}\sqrt{RT_0}}{Ap_0} = \sqrt{\frac{2\gamma}{\gamma-1}\left[\left(\frac{p}{p_0}\right)^{2/\gamma} - \left(\frac{p}{p_0}\right)^{(\gamma+1)/\gamma}\right]} \tag{11.27}$$

and equation (11.28) for the relationship between the exit area $A_E$ and the exit pressure $p_E$ for the same flowrate

$$\frac{\dot{m}\sqrt{RT_0}}{A_E p_0} = \sqrt{\frac{2\gamma}{\gamma-1}\left[\left(\frac{p_E}{p_0}\right)^{2/\gamma} - \left(\frac{p_E}{p_0}\right)^{(\gamma+1)/\gamma}\right]}. \tag{11.28}$$

We have also equation (11.25) for the Mach-number variation through the duct

$$\frac{\dot{m}\sqrt{RT_0}}{Ap_0} = \sqrt{\frac{\gamma M^2}{\left[1 + \left(\frac{\gamma-1}{2}\right)M^2\right]^{(\gamma+1)/(\gamma-1)}}}. \tag{11.25}$$

We have shown that, for isentropic gas flow, sonic conditions ($M = 1$) are only possible at a throat. If we set $M = 1$ in equation (11.25), we find that the cross-sectional area $A^*$ for a given flowrate $\dot{m}$ and stagnation conditions, $p_0$ and $T_0$, is given by

$$\frac{\dot{m}\sqrt{RT_0}}{A^* p_0} = \sqrt{\gamma\left(\frac{2}{\gamma+1}\right)^{(\gamma+1)/(\gamma-1)}} \tag{11.49}$$

or

$$A^* = \frac{\dot{m}\sqrt{RT_0}}{p_0}\sqrt{\frac{1}{\gamma}\left(\frac{\gamma+1}{2}\right)^{(\gamma+1)/(\gamma-1)}} \tag{11.50}$$

which can be used as a reference area, called the **critical area**. It is perhaps worth noting that $A^*$ may not correspond to a physical area for a given duct but is the cross-sectional area that would result in sonic conditions for given values of $\dot{m}, p_0$, and $T_0$. If the throat area $A_T > A^*$ for a given convergent-divergent duct, an initially subsonic flow through that duct will remain subsonic throughout.

Equations (11.25) and (11.50) can be combined to give

$$\frac{A}{A^*} = \frac{1}{M}\sqrt{\left[\left(\frac{2}{\gamma+1}\right)\left[1 + \left(\frac{\gamma-1}{2}\right)M^2\right]\right]^{(\gamma+1)/(\gamma-1)}} \tag{11.51}$$

while combining equations (11.27) and (11.50) leads to

$$\frac{A^*}{A} = \sqrt{\left(\frac{2}{\gamma-1}\right)\left(\frac{\gamma+1}{2}\right)^{(\gamma+1)/(\gamma-1)}\left[\left(\frac{p}{p_0}\right)^{2/\gamma} - \left(\frac{p}{p_0}\right)^{(\gamma+1)/\gamma}\right]}. \tag{11.52}$$

The static pressure and temperature can also be made non-dimensional using the throat conditions, $p^*$ and $T^*$. If we combine equations (11.20) and (11.31) we have

$$\frac{T^*}{T} = \left(\frac{2}{\gamma+1}\right)\left[1 + \left(\frac{\gamma-1}{2}\right)M^2\right]. \tag{11.53}$$

and, from equations (11.22) and (11.32),

$$\frac{p^*}{p} = \left\{ \left(\frac{2}{\gamma+1}\right)\left[1 + \left(\frac{\gamma-1}{2}\right)M^2\right] \right\}^{\gamma/(\gamma-1)}. \tag{11.54}$$

In addition, instead of defining a Mach number in terms of the local (i.e. varying) soundspeed $c$ we can define a Mach number (i.e. a normalised flow velocity) in terms of the soundspeed at the throat for choked flow $c^*$, i.e. $V/c^{*82}$. Since $V/c^* = Mc/c^* = M\sqrt{T/T^*}$, we have

$$\frac{V}{c^*} = M\sqrt{\left[\frac{\gamma+1}{2\left[1 + \left(\frac{\gamma-1}{2}\right)M^2\right]}\right]}. \tag{11.55}$$

The variation with Mach number of $T/T_0$, $T/T^*$, $p/p_0$, $p/p^*$, $A/A^*$, and $V/c^*$, according to equations (11.20), (11.53), (11.22), (11.54), (11.51), and (11.55), respectively, is shown graphically in Figure 11.7.

As we did for a convergent duct, it is instructive to investigate flow conditions within a convergent-divergent duct as the back pressure $p_B$ is reduced below $p_0$. For $p_B \approx p_0$ (but $p_B < p_0$) an essentially incompressible subsonic flow will occur throughout the duct, with

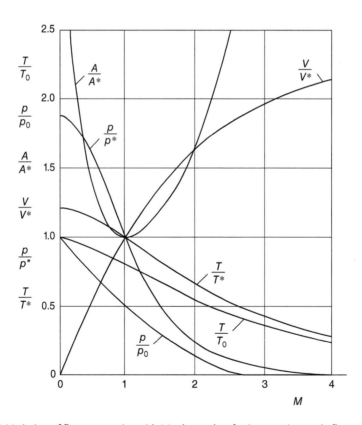

**Figure 11.7** Variation of flow properties with Mach number for isentropic nozzle flow of air ($\gamma = 1.4$)

---

[82] The symbol $V^*$ is also used to represent the soundspeed at the throat for choked flow.

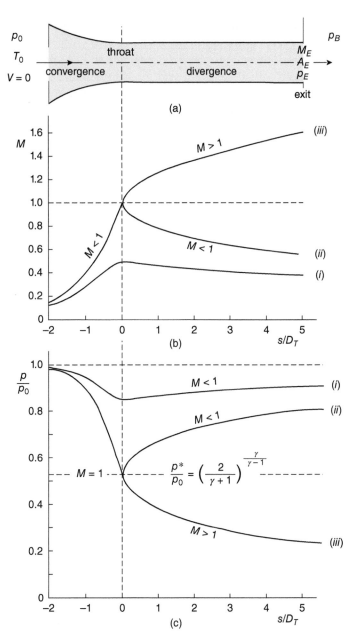

**Figure 11.8** Isentropic flow of a perfect gas through a convergent-divergent nozzle: (a) flow geometry, (b) Mach-number variation, (c) static pressure variation; curves labelled $(i)\, p_B/p_0 = 0.91$, $(ii)\, p_B/p_0 = 0.81$, $(iii)\, p_B/p_0 = 0.24$

an increase in velocity to a maximum at the throat, and a corresponding minimum pressure followed by pressure recovery in the divergent section (curve (*i*) in Figure 11.8). As $p_B$ is further reduced, the mass flowrate progressively increases and the flow remains subsonic throughout the duct, again with a maximum velocity and minimum pressure at the throat, until sonic conditions are reached at the throat, i.e. the duct is choked and the mass flowrate is a maximum (curve (*ii*) in Figure 11.8). For the corresponding static pressure in the nozzle exit plane we introduce the symbol $p_{E,SUB}$. No matter by how much more $p_B$ is reduced, the flow in the convergent section up to the throat remains unchanged, and isentropic flow in the divergent section is impossible until supersonic flow is established from the throat to the exit (curve (*iii*) in Figure 11.8). This situation corresponds to the **design pressure ratio**, and the nozzle is said to be **perfectly expanded**. The static pressure in the exit plane is now $p_{E,SUP}$. The static pressure variation throughout the duct for the three pressure ratios was calculated from equation (11.52) and the Mach number variation from equation (11.51), using the Isentropic-flow Calculator. The nozzle length is $5D_T$, and the area ratio, $A_E/A_T = 1.25$, where the throat area $A_T = \pi D_T^2/4$, $D_T$ being the throat diameter. A conical nozzle with this geometry would have a total included angle of $1.35°$[83].

---

### ILLUSTRATIVE EXAMPLE 11.2

Nitrogen gas with a stagnation pressure of 5 bar and a stagnation temperature of 400 K flows through a convergent-divergent nozzle. The throat area is $0.01$ m$^2$ and the exit area is $0.015$ m$^2$. Assuming isentropic flow, calculate the throat and exit conditions if (a) the exit static pressure is 4.5 bar, (b) the flow is choked but the flow in the divergence is subsonic, and (c) the flow in the divergence is supersonic.

### Solution

For nitrogen $R = 296.8$ m$^2$/s$^2 \cdot$ K, $\gamma = 1.401$, $p_0 = 5 \times 10^5$ Pa, $T_0 = 400$ K, $A_T = 0.01$ m$^2$, and $A_E = 0.015$ m$^2$.
(a) At the nozzle exit, $p_E = 4.5 \times 10^5$ Pa so $p_E/p_0 = 0.9$.
From the Isentropic-flow Calculator, the exit Mach number $M_E = 0.391$, and $T_E/T_0 = 0.970$, so that $T_E = 388.1$ K. Also, from the same Calculator, $A_E/A^* = 1.620$.
The exit soundspeed $c_E = \sqrt{\gamma R T_E} = 401.7$ m/s so that $V_E = M_E c_E = 157.0$ m/s.
From the perfect-gas equation, $\rho_E = p_E/R T_E = 3.906$ kg/m$^3$.
From the continuity equation, $\dot{m} = \rho_E A_E V_E = 9.202$ kg/s.
We have already $A_E/A^* = 1.620$ so that $A_T/A^* = A_T/A_E \times A_E/A^* = 1.620 \times 0.01/0.015 = 1.080$.
From the Isentropic-flow Calculator, $M_T = 0.721$ so that the flow at the throat is subsonic and the nozzle is not choked for an exit static pressure $p_E = 4.5$ bar. Also, from the same Calculator, for $M_T = 0.721$ we have $p_T/p_0 = 0.708$, and $T_T/T_0 = 0.906$, so that $p_T = 3.538$ bar and $T_T = 362.4$ K.
The throat soundspeed $c_T = \sqrt{\gamma R T_T} = 388.2$ m/s so that $V_T = M_T c_T = 279.7$ m/s.
From the perfect-gas equation, $\rho_T = p_T/R T_T = 3.290$ kg/m$^3$.
From the continuity equation, $\dot{m} = \rho_T A_T V_T = 9.202$ kg/s, the same value as we calculated at the nozzle exit, as it should be.

---

[83] Figure 11.8 is not to scale and shows a divergence angle larger than $1.35°$.

(b) Since the flow is choked, $A^* = A_T = 0.01$ m$^2$, $p^* = p_T$, and $p^*/p_0 = [2/(\gamma + 1)]^{\gamma/(\gamma-1)} = 0.528$, a result which can also be obtained from the Isentropic-flow Calculator, as can $T^*/T_0 = 2/(\gamma + 1) = 0.833$. We thus find $p_T = 2.641$ bar and $T_T = T^* = 333.3$ K. In addition, with $A_E/A_T = 1.5$, from the Calculator we find $M_E = 0.430$ if the flow remains subsonic from the throat to the exit, and $M_E = 1.854$ if the flow becomes supersonic downstream of the throat. The throat soundspeed is thus $c_T = \sqrt{\gamma R T_T} = 372.3$ m/s and this is also the gas velocity at the throat $V_T$.

From the perfect-gas equation, $\rho_T = p_T/RT_T = 2.670$ kg/m$^3$.

From the continuity equation, $\dot{m}^* = \rho_T A_T V_T = 9.940$ kg/s, where we have again used the symbol $\dot{m}^*$ to emphasise that the flowrate is a maximum for choked flow.

If the flow remains subsonic from the throat to the exit, we have already $M_E = 0.430$ so that, again from the Isentropic-flow Calculator, $p_E/p^* = 1.667$, and $T_E/T^* = 1.157$, so that $p_{E,SUB} = 4.403$ bar and $T_E = 385.7$ K.

The exit soundspeed is thus $c_E = \sqrt{\gamma R T_E} = 400.5$ m/s, and the exit velocity $V_E = M_E c_E = 172.3$ m/s.

From the perfect-gas equation, $\rho_E = p_E/RT_E = 3.846$ kg/m$^3$.

From the continuity equation, $\dot{m}_E = \rho_E A_E V_E = 9.940$ kg/s, which equals the throat value $\dot{m}^*$, as it should.

(c) We again have $A_E/A_T = 1.5$ but the flow downstream of the throat is now supersonic with $M_E = 1.854$ and from the Isentropic-flow Calculator $p_E/p^* = 0.303$ and $T_E/T^* = 0.711$, so that $p_{E,SUP} = 0.801$ bar and $T_E = 237.0$ K.

The exit soundspeed is thus $c_E = \sqrt{\gamma R T_E} = 313.9$ m/s and the exit velocity $V_E = M_E c_E = 582.0$ m/s.

From the perfect-gas equation, $\rho_E = p_E/RT_E = 1.138$ kg/m$^3$.

From the continuity equation, $\dot{m}_E = \rho_E A_E V_E = 9.940$ kg/s, the same value we calculated for the subsonic flow. This is as it should be, because the throat conditions for the two flows are identical.

Comments:

(i) The exit-flow conditions for the subsonic and supersonic flows downstream of the choked throat are significantly different in magnitude even though the mass flowrates are the same. The flow velocities, for example, differ by a factor of 3.4.

(ii) The exit pressure is $p_{E,SUB} = 4.403$ bar for the subsonic flow and $p_{E,SUP} = 0.801$ bar for the supersonic flow. It is clearly possible to impose a back pressure $p_B$ on the nozzle such that $p_{E,SUB} > p_E > p_{E,SUP}$ and we should ask how this affects the flow downstream of the throat. We could select, for example, $p_E/p_0 = 0.528$ (i.e. $p_E = 2.642$ bar), for which $M_E = 1$. Following the steps detailed above for other values of $p_E$ we find $\dot{m}_E = 14.9$ kg/s, i.e. 50% higher than $\dot{m}^*$, the value for the choked throat. Other values for exit pressures in the range $p_{E,SUB} - p_{E,SUP}$ also lead to higher values for $\dot{m}_E$ than for $\dot{m}^*$. This is clearly inadmissible, since the choking value is the maximum possible. The discrepancy is explained by the incorrect assumption that the flow in the divergence is isentropic: there has to be an entropy increase and a stagnation pressure decrease such that the mass flowrate remains equal to $\dot{m}^*$. The entropy increase is a consequence of an important new phenomenon, which we shall now discuss in detail: the occurrence of a **shockwave**.

## 11.8 **Normal shockwaves**

An object moving through a compressible fluid at a speed that exceeds the speed of sound produces wavefronts in the fluid within which the fluid pressure, temperature, and density appear to change discontinuously. These wavefronts are called shockwaves. Depending upon the Mach number of the moving object, the shockwaves may be curved or straight, much like the infinitesimally weak waves considered in Section 11.5. The simplest form of shockwave, which can be generated by a close-fitting piston moving at constant speed into a long cylindrical tube filled with stationary gas, is called a **normal**[84] **shockwave**. The shock moves away from the face of the piston and propagates into the stationary gas at a speed higher than that of the piston. Shockwaves can also occur in the divergent section of a convergent-divergent nozzle (discussed in Section 11.8.4) or in supersonic flow in a cylindrical pipe with wall friction or heat transfer (Chapter 13). As we shall show, the static pressure, density, and static temperature of a gas all increase across a shock, and the stagnation pressure decreases, while the stagnation temperature remains unchanged.

### 11.8.1 **Shock analysis**

To analyse the change in flow properties across a normal shock, it is convenient to consider the flow relative to the shock, as shown in Figure 11.9. Location ① is immediately upstream of the shock, and location ② is immediately downstream. Although we shall treat the shock as a discontinuity (the internal structure of a shock is discussed in Subsection 11.8.2), we follow Liepmann and Roshko (1957) in representing it with two closely separated lines.

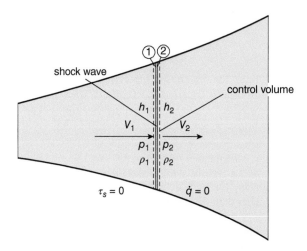

**Figure 11.9** Control volume for steady flow through a normal shock

---

[84] The adjective 'normal' is used because it is only the velocity component normal to the shock which undergoes change, as we shall see when we consider oblique shocks.

Since we are treating the shock as a discontinuity, there is no area change between locations ① and ②, the continuity equation (6.1) reduces to $\rho V$ = constant, and we have

$$\rho_1 V_1 = \rho_2 V_2 \tag{11.56}$$

Equation (11.15) represents the momentum equation for frictionless flow of a perfect gas

$$\frac{1}{\rho}\frac{dp}{ds} + V\frac{dV}{ds} = 0. \tag{11.15}$$

Since $\rho V$ = constant, we can integrate equation (11.15) to give

$$p_1 + \rho_1 V_1^2 = p_2 + \rho_2 V_2^2 \tag{11.57}$$

Equation (11.57) should not be confused with Bernoulli's equation for an incompressible flow where pressure and velocity changes are a consequence of area changes (there is no area change across a shock).

Finally, for adiabatic flow ($\dot{q} = 0$)

$$h_1 + \frac{1}{2}V_1^2 = h_2 + \frac{1}{2}V_2^2. \tag{11.58}$$

From equation (11.57) combined with equation (11.56) we see that

$$p_1 - p_2 = (\rho V)^2 \left(\frac{1}{\rho_2} - \frac{1}{\rho_1}\right) \tag{11.59}$$

and, from equation (11.58) again combined with equation (11.56),

$$2(h_1 - h_2) = (\rho V)^2 \left(\frac{1}{\rho_2^2} - \frac{1}{\rho_1^2}\right) = (\rho V)^2 \left(\frac{1}{\rho_2} - \frac{1}{\rho_1}\right)\left(\frac{1}{\rho_2} + \frac{1}{\rho_1}\right). \tag{11.60}$$

If equations (11.59) and (11.60) are combined to eliminate $\rho V$ we have

$$2(h_1 - h_2) = (p_1 - p_2)\left(\frac{1}{\rho_2} + \frac{1}{\rho_1}\right) \tag{11.61}$$

which is a form of the **Rankine-Hugoniot relation** for the change of thermodynamic variables across a shockwave. If we substitute $C_P(T_1 - T_2)$ for $h_1 - h_2$ and use the perfect-gas equation $p = \rho RT$ to substitute for $T_1$ and $T_2$, after some rearrangement, we have

$$\frac{p_2}{p_1} = \frac{(\gamma + 1) - (\gamma - 1)(\rho_1/\rho_2)}{(\gamma + 1)(\rho_1/\rho_2) - (\gamma - 1)}. \tag{11.62}$$

In deriving equation (11.62) we have assumed that the flow through the shock is adiabatic but not reversible, i.e. the flow is not isentropic. Although solutions of equation (11.62) can be obtained for $p_2/p_1 < 1$, these can be shown to violate the second law of thermodynamics and so are inadmissible.

If we again introduce the perfect-gas equation $p = \rho RT$ together with the equation for the soundspeed $c = \sqrt{\gamma RT}$, the equation $C_P = \gamma R/(\gamma - 1)$, the enthalpy-temperature relation $h = C_P T$, and the Mach-number definition $M = V/c$, it is straightforward to rewrite equations (11.56), (11.57), and (11.58) in terms of the Mach numbers $M_1$ and $M_2$

$$\frac{p_2}{p_1} = \frac{1 + \gamma M_1^2}{1 + \gamma M_2^2} \tag{11.63}$$

$$\frac{T_2}{T_1} = \frac{1 + \left(\frac{\gamma - 1}{2}\right) M_1^2}{1 + \left(\frac{\gamma - 1}{2}\right) M_2^2} \tag{11.64}$$

and

$$\frac{\rho_2}{\rho_1} = \left(\frac{1 + \gamma M_1^2}{1 + \gamma M_2^2}\right) \left[\frac{1 + \left(\frac{\gamma - 1}{2}\right) M_2^2}{1 + \left(\frac{\gamma - 1}{2}\right) M_1^2}\right]. \tag{11.65}$$

Since $p = \rho RT$, we can combine these three equations to show that

$$\frac{M_1}{M_2} = \left(\frac{1 + \gamma M_1^2}{1 + \gamma M_2^2}\right) \sqrt{\frac{1 + \left(\frac{\gamma - 1}{2}\right) M_2^2}{1 + \left(\frac{\gamma - 1}{2}\right) M_1^2}} \tag{11.66}$$

which can be rearranged as a quadratic equation for $M_2$ in terms of $M_1$ with the solution[85]

$$M_2 = \sqrt{\frac{\frac{2}{\gamma - 1} + M_1^2}{\left(\frac{2\gamma}{\gamma - 1}\right) M_1^2 - 1}}. \tag{11.67}$$

If we now substitute for $M_2$ in equation (11.63) for $p_2/p_1$ we find

$$\frac{p_2}{p_1} = \left(\frac{2\gamma}{\gamma + 1}\right) M_1^2 - \left(\frac{\gamma - 1}{\gamma + 1}\right). \tag{11.68}$$

The quantity $\Pi$, defined by $\Pi \equiv \Delta p/p_1$, where $\Delta p \equiv p_2 - p_1$, is a convenient definition of **shock strength**[86]. From equation (11.68) we see that $\Pi$ is given by

$$\Pi \equiv \frac{\Delta p}{p_1} = \left(\frac{2\gamma}{\gamma + 1}\right) \left(M_1^2 - 1\right). \tag{11.69}$$

A **weak shock** is one for which $\Pi \ll 1$ while for a **strong shock** $\Pi \gg 1$.
Equation (11.64) for $T_2/T_1$ can also be written in terms of $M_1$

$$\frac{T_2}{T_1} = \frac{2(\gamma - 1)}{(\gamma + 1)^2 M_1^2} \left[\left(\frac{\gamma - 1}{2}\right) M_1^2 + 1\right] \left[\left(\frac{2\gamma}{\gamma - 1}\right) M_1^2 - 1\right] \tag{11.70}$$

as can equation (11.65) for $\rho_2/\rho_1$

$$\frac{\rho_2}{\rho_1} = \frac{V_1}{V_2} = \frac{\left(\frac{\gamma + 1}{2}\right) M_1^2}{1 + \left(\frac{\gamma - 1}{2}\right) M_1^2} \tag{11.71}$$

wherein use has been made of equation (11.56).

---

[85] A second (trivial) solution is simply $M_2 = M_1$.
[86] Shock strength is sometimes defined as $\Delta p/\gamma p_1$.

The quantity $(\rho_2 - \rho_1)/\rho_1$, referred to as the **condensation**, is a measure of the density increase across the shock and can also be regarded as an indicator of shock strength.

Since the flow is adiabatic, to the equations derived so far, we can add

$$\frac{T_{02}}{T_{01}} = 1. \tag{11.72}$$

Although the flow is not isentropic so far as the shock itself is concerned, we may still use the isentropic relationship for $p_0/p$, equation (11.22), on either side of the shock, so that

$$\frac{p_{02}}{p_{01}} = \frac{p_{02}}{p_2}\frac{p_1}{p_{01}}\frac{p_2}{p_1} = \left[\frac{\left(\frac{\gamma+1}{2}\right)M_1^2}{1+\left(\frac{\gamma-1}{2}\right)M_1^2}\right]^{\gamma/(\gamma-1)} \left[\left(\frac{2\gamma}{\gamma+1}\right)M_1^2 - \frac{(\gamma-1)}{(\gamma+1)}\right]^{-1/(\gamma-1)}. \tag{11.73}$$

According to equation (11.2) the entropy change for a perfect gas in a thermodynamic process is given by

$$s_2 - s_1 = C_P \ln\left(\frac{T_2}{T_1}\right) - R\ln\left(\frac{p_2}{p_1}\right). \tag{11.2}$$

Since we are assuming that the flow on either side of the shock is isentropic (i.e. the specific entropy is constant), it must be that

$$s_{01} = s_1 \quad \text{and} \quad s_{02} = s_2$$

where $s_{01}$ and $s_{02}$ are the specific entropies for stagnation conditions on either side of the shock. We can then write

$$s_2 - s_1 = s_{02} - s_{01} = C_P \ln\left(\frac{T_{02}}{T_{01}}\right) - R\ln\left(\frac{p_{02}}{p_{01}}\right) \tag{11.74}$$

so that for adiabatic flow of a perfect gas, for which $T_{02} = T_{01}$, for the entropy change across a shock we find

$$s_2 - s_1 = -R\ln\left(\frac{p_{02}}{p_{01}}\right). \tag{11.75}$$

Equation (11.73) shows that $p_{02}/p_{01} < 1$ for $M_1 > 1$, so that from equation (11.75) there must be an entropy increase across a shock. Since $M_1 > 1$, equations (11.68), (11.70), and (11.71) show that pressure, temperature, and density all increase across a shock. The ratios $p_2/p_1$, $T_2/T_1$, $\rho_2/\rho_1$, and $p_{02}/p_{01}$ are also included in NACA Report 1135, as are various forms of the equations derived here. Numerical values for these and other ratios can be found from the Calculator. The ratios are shown in graphical form plotted versus $M_1$ in Figure 11.10.

In Subsection 11.7.2 it was shown that the mass flowrate $\dot{m}$ for flow of a perfect gas through a choked convergent-divergent nozzle can be determined from

$$A^* = \frac{\dot{m}\sqrt{RT_0}}{p_0}\sqrt{\frac{1}{\gamma}\left(\frac{\gamma+1}{2}\right)^{(\gamma+1)/(\gamma-1)}}. \tag{11.50}$$

Since we have shown that the stagnation pressure decreases across a shock, it is evident from equation (11.50) that the critical or choking area $A^*$ must increase according to

$$p_{01}A_1^* = p_{02}A_2^* \tag{11.76}$$

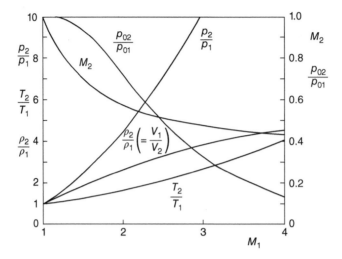

**Figure 11.10** Shock-wave properties as a function of upstream Mach number

which plays an important role in calculations of flow through a convergent-divergent nozzle with a shockwave downstream of the throat, as will become apparent in Illustrative Example 11.4.

We can summarise as follows the principal conclusions reached in this subsection regarding the flow changes which occur across a shockwave

- the changes are effectively discontinuous
- the flow changes from supersonic to subsonic so that $M_1 > 1$ and $M_2 < 1$
- the gas static pressure $p$ increases
- the gas static temperature $T$ increases
- the gas density $\rho$ increases, i.e. the gas is compressed
- the stagnation temperature $T_0$ is unchanged
- the stagnation pressure $p_0$ decreases
- the specific entropy $s$ increases

## 11.8.2  Shock thickness

It should be evident that an infinitesimally thin shockwave across which occur discontinuous changes in fluid and flow properties is a theoretical idealisation. Photographs of normal shocks in supersonic-flow wind tunnels show that shocks are indeed very thin but often not quite straight, especially at the ends where they interact with the boundary layers on the wind-tunnel sidewalls. Theoretical calculations and experimental observations show that the thickness of a normal shock is of the order of 1 $\mu$m, which is about an order of magnitude greater than the mean free path $\Lambda$ of the gas molecules: for air at ambient pressure $\Lambda \approx 0.07$ $\mu$m. In reality the velocity and all fluid properties vary continuously through the shockwave in the direction of motion and involve complex viscous and heat-conduction phenomena. Nevertheless, calculations based upon the simplified model, which assumes that discontinuous changes occur across infinitesimally thin shocks, compare well with observations.

An approximate one-dimensional theory for the **internal structure of a weak shockwave** was published by Taylor in 1910[87]. Without going into the details, the theory, which is based upon the conservation equations for mass, momentum, and energy, includes the influence of bulk viscosity $\mu_B$ and heat conduction. The end result may be stated as follows

$$\frac{(\gamma + 1)}{2\alpha'} (V_1 - V_2) x = \ln\left(\frac{V_1 - V}{V - V_2}\right) = \ln\left(\frac{\frac{\rho}{\rho_1} - 1}{1 - \frac{\rho}{\rho_1}\frac{\rho_1}{\rho_2}}\right) \tag{11.77}$$

where $V_1$ and $\rho_1$ are the flow velocity and density far upstream of the shock, respectively, $V_2$ and $\rho_2$ are the velocity and density far downstream, respectively, $x$ is the distance measured from the centreplane of the shock, $\gamma$ is the ratio of specific heats, and $\alpha'$ is a constant involving $\gamma, C_P$, the thermal conductivity $k, \rho$, and the full viscosity coefficient $\mu + 3\mu_B/4$

$$\rho\alpha' = \frac{4}{3}\mu + \mu_B + (\gamma - 1)\frac{k}{C_P} \tag{11.78}$$

so that $\rho\alpha'$ has the dimensions of dynamic viscosity $M/LT$. From equation (11.77) we find, by differentiation, that the velocity gradient at the midplane ($x = 0$) of the shock is given by

$$\left.\frac{dV}{dx}\right|_0 = -\frac{(\gamma + 1)(V_1 - V_2)^2}{8\alpha'} \tag{11.79}$$

and we can define a representative shock thickness $\Delta_S$ as

$$\Delta_S = \frac{V_1 - V_2}{-dV/dx|_0} = \frac{8\alpha'}{(\gamma + 1)(V_1 - V_2)}. \tag{11.80}$$

For the continuum hypothesis to be valid, we require that $\Delta_S/\Lambda \gg 1$. Based upon the **kinetic theory of gases**, a rough estimate for the viscosity is given by

$$\mu = \sqrt{\frac{2}{\pi\gamma}}\rho c\Lambda \tag{11.81}$$

where $\Lambda$ is the molecular mean free path (see Section 2.8), and $c$ is the soundspeed. Equations (11.80) and (11.81), together with equation (11.71) for $V_1/V_2$, can be combined to give an equation for the ratio $\Delta_S/\Lambda$

$$\frac{\Delta_S}{\Lambda} = 16\sqrt{\frac{2}{\pi\gamma}}\left[\frac{\left(\frac{4}{3} + \frac{\mu_B}{\mu} + \frac{\gamma - 1}{Pr}\right)M_1}{M_1^2 - 1}\right]. \tag{11.82}$$

The quantity $Pr = C_P\mu/k$ is the non-dimensional Prandtl number, which plays a key role in heat transfer. It is apparent immediately that as $M_1 \to 1, \Delta_S/\Lambda \to \infty$, and the weak-shock theory is valid. However, the validity of the continuum approximation deteriorates for $M_1 \gg 1$. Note that, in deriving equation (11.82) it has been assumed that fluid properties remain constant through the shock, an assumption that is only valid for very weak shocks. Finally, the ratio $\Delta_S/\Lambda$ can be regarded as a **Knudsen number**, based upon the thickness of the shockwave.

[87] More complete details of Taylor's analysis are given by Thompson (1972).

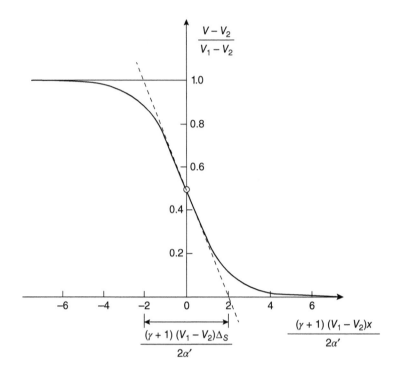

**Figure 11.11**  Velocity distribution within a shockwave

The density distribution, calculated by Thompson from equation (11.77), agrees well with experimental data for a shock in helium at $M_1 \approx 1.59$. Figure 11.11 shows the velocity variation within a shockwave according to equation (11.77), with the dashed line corresponding to the midplane velocity gradient upon which $\Delta_S$ is based (equation (11.80)).

---

**ILLUSTRATIVE EXAMPLE 11.3**

Calculate the thickness of a shockwave $\Delta_S$ in a flow of air at a temperature of 27 °C, a static pressure of 2 bar, and a Mach number of 1.1. The dynamic viscosity $\mu$ of air can be taken as $1.86 \times 10^{-5}$ Pa·s, the ratio of the bulk viscosity $\mu_B$ to the dynamic viscosity is 0.8, and the thermal conductivity of air $k$ has the value $2.6 \times 10^{-2}$ W/m K. Calculate the mean free path $\Lambda$ for air and hence determine the ratio $\Delta_S/\Lambda$.

**Solution**

We have $T_1 = 273 + 27 = 300$ K, $p_1 = 2 \times 10^5$ Pa, and $\mu_B = 0.8\mu = 1.488 \times 10^{-5}$ Pa·s.
For air, $\gamma = 1.4$, and $R = 287$ m$^2$/s$^2$K, so that $C_P = \gamma R/(\gamma - 1) = 1004.5$ m$^2$/s$^2$·K.
From the perfect-gas equation, $\rho_1 = p_1/RT_1 = 2.323$ kg/m$^3$.
The soundspeed $c_1 = \sqrt{\gamma RT_1} = 347.2$ m/s so that $V_1 = M_1 c_1 = 381.9$ m/s.
From the Calculator, with $M_1 = 1.1$, $\rho_2/\rho_1 = 1.169 = V_1/V_2$ so that $V_2 = V_1/1.169 = 326.7$ m/s, and $V_1 - V_2 = 55.2$ m/s.

From equation (11.78),

$$\rho_1 \alpha' = \tfrac{4}{3} \times 1.86 \times 10^{-5} + 0.8 \times 1.86 \times 10^{-5} + 0.4 \times 2.6 \times 10^{-2}/1004.5$$
$$= 5.003 \times 10^{-5}\,\text{Pa}\cdot\text{s so that } \alpha' = 2.154 \times 10^{-5}\,\text{m}^2/\text{s}.$$

From equation (11.80), $\Delta_S = 8 \times 2.154 \times 10^{-5}/(2.4 \times 55.2) = 1.30 \times 10^{-6}$ m or 1.30 $\mu$m.
From equation (11.81), $\Lambda = \sqrt{\pi\gamma}\mu\big/\left(\sqrt{2}\rho_1 c_1\right) = 3.42 \times 10^{-8}$ m or 34.2 nm. The ratio $\Delta_S/\Lambda$
is thus 38.0, which suggests that the continuum hypothesis is just valid in this case.

### 11.8.3 Pitot tubes in supersonic flow

As illustrated in Figure 11.12, a detached bow shock forms a short distance upstream of the
mouth of a Pitot tube facing a supersonic gas stream. The segment of the shock on the stagnation streamline may be regarded as a normal shock so that the pressure registered by the Pitot
tube, $p_{02}$, is given by equation (11.73)

$$\frac{p_{02}}{p_{01}} = \left[ \frac{\left(\frac{\gamma+1}{2}\right)M_1^2}{1 + \left(\frac{\gamma-1}{2}\right)M_1^2} \right]^{\gamma/(\gamma-1)} \left[ \left(\frac{2\gamma}{\gamma+1}\right)M_1^2 - \frac{(\gamma-1)}{(\gamma+1)} \right]^{-1/(\gamma-1)} \tag{11.73}$$

and the stagnation pressure of the flow ahead of the shock $p_{01}$ can be computed if the Mach
number ahead of the shock $M_1$ is known. Otherwise, an additional measurement is needed.

The static pressure $p_1$ and stagnation pressure $p_{01}$ for the flow are related to the Mach
number through equation (11.22)

$$\frac{p_{01}}{p_1} = \left[ 1 + \left(\frac{\gamma-1}{2}\right)M_1^2 \right]^{\gamma/(\gamma-1)} \tag{11.22}$$

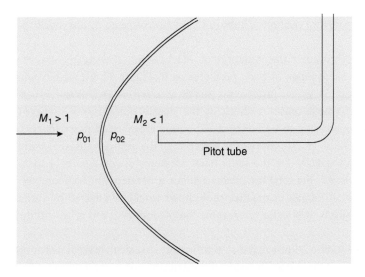

**Figure 11.12** A Pitot tube in supersonic flow

which allows $p_{01}$ to be eliminated from equation (11.73) to provide an equation for $M_1$ in terms of $p_1$ and $p_{02}$

$$\frac{p_1}{p_{02}} = \frac{\left[\left(\frac{2\gamma}{\gamma+1}\right)M_1^2 - \frac{(\gamma-1)}{(\gamma+1)}\right]^{1/(\gamma-1)}}{\left[\left(\frac{\gamma+1}{2}\right)M_1^2\right]^{\gamma/(\gamma-1)}} \tag{11.83}$$

which is called **Rayleigh's supersonic Pitot formula**. This formula is useful when $p_1$ and $p_{02}$ are both known, e.g. from independent measurements, while $p_{01}$ is not known.

### 11.8.4 Normal shockwave in a convergent-divergent duct

In Section 11.7, which is concerned with isentropic flow of a perfect gas through a convergent-divergent duct, we concluded that as the exit pressure $p_E$ was reduced below the stagnation pressure $p_0$ a level $p_{E,SUB}$ was reached at which the throat of the duct became choked (i.e. the throat Mach number $M_T = 1$), the mass flowrate was a maximum, $\dot{m}^*$, and the flow downstream of the throat remained subsonic. For a second, much lower exit pressure, $p_{E,SUP}$, the flow within the divergence became supersonic while the mass flowrate remained unchanged. For $p_{E,SUB} > p_E > p_{E,SUP}$ the assumption of isentropic flow led to the contradictory conclusion $\dot{m}_E > \dot{m}^*$. This contradiction is resolved by the realisation that a flow process must occur within the divergence, resulting in a reduction of stagnation pressure downstream of the throat: at some location between the throat and the exit a **normal shockwave** occurs across which there is a sudden increase in static pressure accompanied by an entropy increase and a decrease in stagnation pressure. Flow in the region between the throat and the shockwave is supersonic and can be treated as isentropic, while downstream of the shock the flow is now subsonic and again isentropic. The sudden increase in static pressure across the shock is followed by a gradual increase to the exit static pressure, $p_E$. As the back pressure is decreased further, the shock moves towards the nozzle exit, while the Mach number ahead of the shock increases, as does the **shock strength**. The shock location within the nozzle is determined by the exit pressure: the shock 'finds' the location where the pressure increase across it is just sufficient for the pressure rise in the subsonic flow downstream of the shock to match the exit pressure.

Figure 11.13(b) shows the variation of Mach number along the convergent-divergent nozzle for the same range of back pressures as in Figure 11.8, together with a back pressure $(p_B/p_{01} = 0.73)$ which produces a shock at $x/D_T = 3$ (curve (iv)), a back pressure $(p_B/p_{01} = 0.66)$ which causes a shock at the nozzle exit ($x/D_T = 5$, curve (v)), and back pressures corresponding to underexpanded and overexpanded flows (curves (vi) and (vii), respectively). Figure 11.13(c) shows the variation of the non-dimensional static pressure $p/p_{01}$ with axial location $x/D_T$ for the same back pressures as for Figure 11.13(b). As indicated in Subsection 11.8.3, the total included angle for a nozzle with the geometry under consideration is 1.35°. Although in practice these flows would be affected by surface friction and other viscous effects, it is quite remarkable that shockwaves can arise within such a modest divergence.

If the shock location is known, it is straightforward to calculate the flow conditions upstream and downstream of the shock. However, if the exit pressure is specified, an iterative calculation is required to find the shock location and the corresponding flow properties on either side.

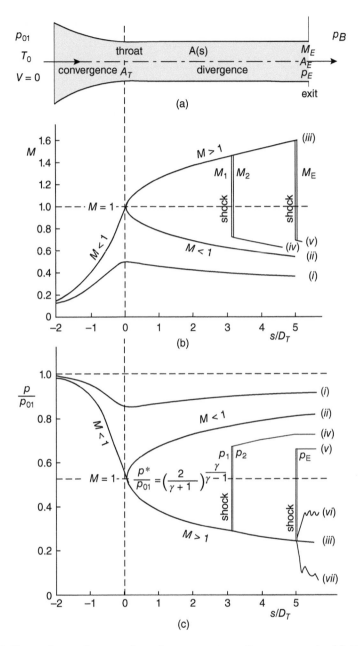

**Figure 11.13** Flow of a perfect gas through a convergent-divergent nozzle: (a) flow geometry, (b) Mach-number variation, (c) static pressure variation; curves labelled $(i)\ p_B/p_{01} = 0.91$, $(ii)\ p_B/p_{01} = 0.81$, $(iii)\ p_B/p_{01} = 0.24$, $(iv)\ p_B/p_{01} = 0.73$, $(v)\ p_B/p_{01} = 0.66$, $(vi)\ p_B/p_{01}$ corresponding to underexpanded flow, $(vii)\ p_B/p_{01}$ corresponding to overexpanded flow

Illustrative Example 11.4 shows how a typical calculation is carried out for the two situations, again making use of the Calculator.

---

**ILLUSTRATIVE EXAMPLE 11.4**

As in Illustrative Example 11.2, we consider the situation where nitrogen gas with a stagnation pressure of 5 bar and a stagnation temperature of 400 K flows through a convergent-divergent nozzle. The throat area is 0.01 m$^2$, and the exit area is 0.015 m$^2$. A shockwave arises in the divergence (a) at a location where the cross-sectional area is 0.0125 m$^2$, (b) in the exit plane, and (b) at a location determined by an exit pressure of 3.35 bar. In all cases the flow upstream and downstream of the shock can be assumed to be isentropic. Calculate the flow conditions upstream and downstream of the shock and also the exit conditions. The flow can be assumed to be isentropic upstream of the shock in all three cases, and downstream of the shock in cases (a) and (c).

Solution

For nitrogen $R = 296.8$ m$^2$/s$^2 \cdot$ K, $\gamma = 1.401$, $p_{01} = 5 \times 10^5$ Pa, $T_0 = 400$ K, $A_T = 0.01$ m$^2$, and $A_E = 0.015$ m$^2$.

(a) The shock is located where the cross-sectional area $A_S = 0.0125$ m$^2$ so $A_S/A_T = 1.25$. Since there is a shock downstream of the throat, the flow in this region must be supersonic. From the Isentropic-flow Calculator, for $A_S/A_1^* = 1.25$ we find $M_1 = 1.600$ for the Mach number just ahead of the shock. We have also $T_1/T_0 = 0.661$, and $p_1/p_{01} = 0.235$, from which $T_1 = 264.6$ K, and $p_1 = 1.177$ bar. The symbol $A_1^*$ ($= A_T$) has been introduced for the choking area upstream of the shock in anticipation of the increased value $A_2^*$, which will correspond with the Mach number $M_2$ ($< 1$) and stagnation pressure $p_{02}$ ($< p_{01}$) just behind the shock.

The soundspeed $c_1 = \sqrt{\gamma RT_1} = 331.7$ m/s so that $V_1 = M_1 c_1 = 530.6$ m/s.

From the perfect-gas equation, $\rho_1 = p_1/RT_1 = 1.499$ kg/m$^3$.

From the continuity equation, $\dot{m} = \rho_1 A_S V_1 = 9.940$ kg/s, which is precisely the same as the throat value found in Illustrative Example 2 and so provides a check on the calculated values for $\rho_1$ and $V_1$.

From the Normal-shock Calculator, for conditions just behind the shock we have $M_2 = 0.669$, $p_{02}/p_{01} = 0.895$, $p_2/p_1 = 2.819$, and $T_2/T_1 = 1.388$, from which $p_{02} = 4.477$ bar, $p_2 = 3.317$ bar, and $T_2 = 367.2$ K.

The soundspeed $c_2 = \sqrt{\gamma RT_2} = 390.7$ m/s so that $V_2 = M_2 c_2 = 261.2$ m/s.

From the perfect-gas equation, $\rho_2 = p_2/RT_2 = 3.044$ kg/m$^3$.

From the continuity equation, $\dot{m} = \rho_2 A_S V_2 = 9.940$ kg/s, which is again precisely the same as the throat value found in Illustrative Example 2 and so provides a check on the calculated values for $\rho_2$ and $V_2$.

To determine the exit conditions we first calculate the increased choking area $A_2^*$ corresponding to $M_2$ and the reduced stagnation pressure. From the Isentropic-flow Calculator we have $A_2/A_2^* = 1.119$, from which $A_2^* = 0.0112$ m$^2$. The same result can be obtained from equation (11.51), $p_{01} A_1^* = p_{02} A_2^*$.

We thus have $A_E/A_2^* = 1.343$ and, from the Isentropic-flow Calculator $M_E = 0.498$, $T_E/T_0 = 0.953$, and $p_E/p_{02} = 0.844$, so that $T_E = 381.1$ K and $p_E = 3.778$ bar.

The soundspeed at exit $c_E = \sqrt{\gamma R T_E} = 398.1$ m/s so that $V_E = M_E c_E = 198.4$ m/s.
From the perfect-gas equation, $\rho_E = p_E / R T_E = 3.340$ kg/m$^3$.
From the continuity equation, $\dot{m} = \rho_E A_E V_E = 9.940$ kg/s, which provides a check on the calculated values for $\rho_E$ and $V_E$.

(b) The flow upstream of the exit plane is identical with that for case (c) of Illustrative Example 11.2. Now, however, a shock is located in the exit plane, indicating that the back pressure must be higher.

In Illustrative Example 11.2 we found $M_E = 1.854, p_E = 0.801$ bar, and $T_E = 237.0$ K so these correspond to the conditions just ahead of the shock, i.e. $A_S = A_E = 0.015$ m$^2$, $M_1 = 1.854, p_1 = 0.801$ bar, and $T_1 = 237.0$ K.

From the Normal-shock Calculator we find $M_2 = 0.605 (= M_E), p_2/p_1 = 3.844, p_{02}/p_{01} = 0.788$, and $T_2/T_1 = 1.572$, from which $p_2 (= p_E) = 3.079$ bar, $p_{02} (= p_{0E}) = 3.942$ bar, and $T_2 (= T_E) = 372.7$ K.

The soundspeed at exit $c_E = \sqrt{\gamma R T_E} = 393.7$ m/s so that $V_E = M_E c_E = 238.1$ m/s.
From the perfect-gas equation, $\rho_E = p_E / R T_E = 2.783$ kg/m$^3$.
From the continuity equation, $\dot{m} = \rho_E A_E V_E = 9.940$ kg/s, which again provides a check on the calculated values for $\rho_E$ and $V_E$.

(c) The calculation proceeds as in part (a) but now we do not know the location (i.e. the cross-sectional area $A_S$) of the shockwave and this has to be determined by trial and error until the calculated mass flowrate matches the throat value of 9.940 kg/s.

The exit-plane pressure (behind the shock) is now 3.35 bar, which is higher than the value $p_E = 3.079$ bar for a shock in the exit plane (part (b) above) but lower than the value 3.778 bar for part (a) above. We conclude that the shock must lie between $A_S/A_T = 1.125$ and the exit plane.

As a first guess we try $A_S = 0.0145$ m$^2$, which leads to $M_1 = 1.810, M_2 = 0.614, p_{02} = 4.040$ bar, $T_E = 374.8$ K, and $p_E = 3.216$ bar. This indicates that $A_S < 0.0145$ m$^2$, i.e. the shock is located closer to the throat.

A second guess with $A_S = 0.0140$ m$^2$ leads to $M_1 = 1.763, M_2 = 0.625, p_{02} = 4.145$ bar, $T_E = 376.4$ K, and $p_E = 3.353$ bar, which may be regarded as close enough to the specified value. The step-by-step procedure for calculating these values is identical to that for part (a).

---

## 11.9 Perfectly expanded, underexpanded, and overexpanded nozzle flow

When a supersonic flow leaves a convergent-divergent nozzle at precisely the same pressure as that of the surroundings, i.e. $p_E = p_B$, there is no shock, and the nozzle is said to be **perfectly** (or **correctly**) **expanded** (curve (*iii*) in Figure 11.13). If the back pressure is higher than that of the flow in the exit plane, but still lower than that which causes a normal shock in the exit plane, the nozzle is **underexpanded**. Finally, if the back pressure is lower than that for the nozzle to be correctly expanded, the nozzle is said to be **overexpanded**.

An overexpanded flow leaves the nozzle as a supersonic jet within which there is a series of oblique shockwaves and expansion fans[88] which raise the jet's static pressure from the exit-plane pressure $p_E$ to that of the surroundings $p_B$. A flow pattern typical of an overexpanded jet is shown in Figure 11.14(a). In the situation of an underexpanded jet, the role of shockwaves and expansion fans is to reduce the static pressure from $p_E$ to $p_B$, as shown in Figure 11.14(b).

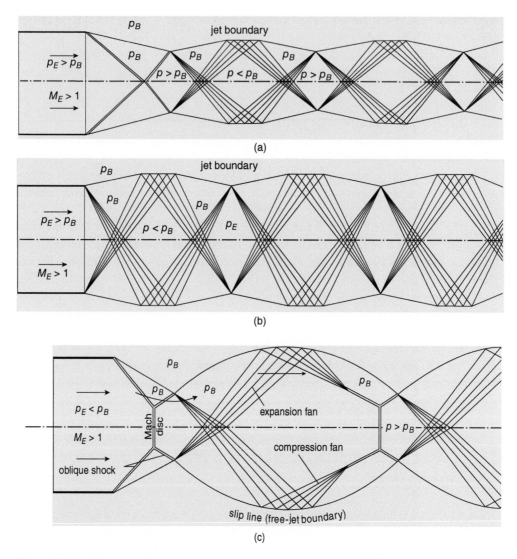

**Figure 11.14** Schematic diagrams showing the internal structures of (a),(c) overexpanded and (b) underexpanded supersonic nozzle flows

---

[88] Oblique shocks and expansion fans are the subject of Chapter 12.

The patterns appearing within an overexpanded or underexpanded jet are termed **Mach (or shock) diamonds**. Regions of high pressure, such as those apparent in the overexpanded jet, occur in aircraft and rocket exhausts. Because such regions are also hot, they emit light and so are visible.

 11.10 SUMMARY

In this chapter we analysed compressible-gas flow through convergent and convergent-divergent nozzles based upon the conservation laws for mass, momentum, and energy. We showed that in both cases the key parameter in describing the flow is the Mach number, which is used to distinguish between subsonic and supersonic flow. So that significant results could be achieved, the flowing fluid was treated as a perfect gas, and the flow as one dimensional. We also discussed flow through a normal shockwave, which is an important feature of supersonic flow. No account was taken of surface friction or heat transfer, and the flow upstream and downstream of a shockwave was treated as isentropic.

The student should be able to

- define the terms Mach number and choking
- understand the trends for convergent-divergent nozzle flows based upon the differential equations derived
- perform calculations using the using the Virginia Tech Compressible Aerodynamics Calculator for flow through a convergent nozzle or a convergent-divergent nozzle depending upon the back pressure
- where appropriate, in carrying out the calculations, allow for the presence of a normal shock within the divergent section of a convergent-divergent nozzle

 11.11 SELF-ASSESSMENT PROBLEMS

It is recommended that, where appropriate, the Virginia Tech Compressible Aerodynamics Calculator be used in the solution of all numerical calculations.

**11.1** Nitrogen with a stagnation pressure of 8 bar and stagnation temperature of 500 K flows through a convergent nozzle with an exit area of 0.1 m². Calculate the mass flowrate if the back pressure is (a) 5 bar and (b) 2 bar. Assume that the flow is isentropic.
(Answers: (a) 129.2 kg/s (b) 142.2 kg/s)

**11.2** Sulphur hexafluoride ($SF_6$; $R = 56.93$ m²/s²·K, and $\gamma = 1.098$) with an initial stagnation pressure of 10 bar and a stagnation temperature of 350 K flows through a convergent-divergent nozzle with a throat area of 0.04 m² and an exit area of 0.08 m². Calculate the flow conditions (static pressure, static temperature, stagnation pressure, density, velocity, and Mach number) in the exit plane if (a) the back pressure is 8.5 bar, (b) the flow is just choked at the throat, and the flow in the divergence is subsonic, (c) the flow is just choked at the throat, and the flow in the divergence is supersonic, (d) there is a normal shockwave in the exit plane, and (e) there is a normal shockwave where the cross-sectional area is 0.055 m². (f) For case

(e) calculate the flow conditions on either side of the shock. The flow may be assumed to be isentropic throughout for (a), (b), and (c); isentropic upstream of the shock for (d); and isentropic upstream and downstream of the shock for (e).
(Answers: (a) 8.5 bar, 344.9 K, 10 bar, 43.3 kg/m³, 80.2 m/s, 0.546; (b) 9.47 bar, 348.3 K, 10 bar, 47.8 kg/m³, 46.6 m/s, 0.316; (c) 1.38 bar, 293.3 K, 10 bar, 8.27 kg/m³, 268.9 m/s, 1.99; (d) 5.64 bar, 345.3 K, 6.56 bar, 28.2 kg/m³, 77.6 m/s, 0.53; (e) 7.91 bar, 347.6 K, 8.55 bar, 40.0 kg/m³, 55.7 m/s, 0.378; (f) upstream of shock 2.50 bar, 309.2 K, 10 bar, 14.2 kg/m³, 228.0 m/s, 1.64; downstream of shock 6.92 bar, 343.4 K, 8.55 bar, 35.4 kg/m³, 91.5 m/s, 0.624. Mass flowrate for (a) 277.7 kg/s, and 177.9 for all other cases.)

**11.3**  Starting with equations (11.56), (11.57), and (11.58), derive expressions for the ratios $T_2/T_1$, $p_2/p_1$, and $V_2/V_1$ in terms of $M_1$ and $M_2$ for the flow of a perfect gas through a normal shockwave.

**11.4**  Calculate the thickness of a shockwave $\Delta_S$ in a flow of $SF_6$ at a temperature of 20 °C, a static pressure of 2 bar, and a Mach number of 1.2. The dynamic viscosity $\mu$ of $SF_6$ can be taken as $1.43 \times 10^{-5}$ Pa·s, the ratio of the bulk viscosity $\mu_B$ to the dynamic viscosity is 0.8, and the thermal conductivity $k$ of $SF_6$ has the value $1.3 \times 10^{-2}$ W/m K. Calculate the mean free path $\Lambda$, the quantity $\alpha'$ for $SF_6$, and the shock thickness $\Delta_S$ for the conditions given, and hence determine the ratio $\Delta_S/\Lambda$.
(Answers: 11.57 nm, $2.711 \times 10^{-6}$ m²/s, 0.219 μm, 18.9)

**11.5**  A blast wave from a nuclear explosion is moving through the lower atmosphere at a speed of 6000 m/s. Estimate the static and stagnation pressure, static temperature, velocity, and Mach number behind the blast wave. Assume the atmospheric conditions are 1 atm and 20°C and treat the blast wave as a normal shock.
(Answers: 357 bar, 394 bar, $1.77 \times 10^4$ K, 1016 m/s, 0.381)

**11.6**  Air at 25°C and 0.7 bar flows through a pipe at a speed of 120 m/s. The flow is stopped by a valve which closes suddenly, causing a shockwave to propagate back into the pipe. Calculate the speed of the shockwave and the pressure and temperature of the air which has been brought to rest.
(Answers: 305.4 m/s, 1.119 bar, 68.7°C)

# 12 Oblique shockwaves and expansion fans

This chapter concerns the changes in fluid and flow properties which occur when a supersonic flow is forced to make a sudden change of direction, for example by a sharp corner. For a 'concave' corner the direction change occurs through an isentropic expansion fan while for a 'convex' corner the flow passes through an oblique (i.e. inclined) shockwave. For the analysis of these two flow phenomena, no further equations are needed beyond those employed in Chapter 11. We use these equations to derive the working equations which are the basis for practical calculations performed, most conveniently, using the **Virginia Tech Compressible Aerodynamics Calculator**[89]. We consider in detail supersonic flow over an inclined flat-plate and a diamond-profile aerofoil to illustrate the application of oblique shock-wave and expansion-fan theory, usually referred to as shock-expansion theory.

## 12.1 Oblique shockwaves

In Chapter 11 we considered the situation of a **normal shockwave**, where there is a practically discontinuous change in flow conditions from supersonic to subsonic across a region of negligible thickness normal to the flow. We showed that, for certain back pressures, a normal shock is essential to the operation of supersonic flow through a divergent nozzle. We now consider an external supersonic flow approaching a two-dimensional wedge-shaped object, as illustrated in Figure 12.1(a). If the included half angle of the wedge is $\theta$, the flow above the wedge[90] has to turn through the same angle so that it is parallel to the wedge surface. The angle $\theta$ is thus the flow **turning angle**[91]. Turning is accomplished through an **oblique shockwave**[92] at an angle $\beta$, called the **wave angle**, measured with respect to the approach flow. An identical flowfield results for supersonic flow approaching a concave corner, as shown in Figure 12.1(b). As shown in Figure 12.1(c), we can analyse flow through an oblique shock simply by superimposing on the flowfield of a normal shock, with velocity components $V_{1N}$ and $V_{2N}$, a uniform velocity $V_T$ parallel to the shock. The resultant velocity components on the two sides of the shock are then $V_1 = \sqrt{V_{1N}^2 + V_T^2}$, and $V_2 = \sqrt{V_{2N}^2 + V_T^2}$, and the inclination of the shock to the approach-flow velocity $V_1$ is given by $\beta = \tan^{-1}(V_{1N}/V_T)$. As in the case of a **Galilean transformation**, superimposition of the uniform velocity $V_T$ does not affect the static pressure or any other of

---

[89] Details of the **Calculator** were given in Section 11.3.

[90] For convenience, we assume here that the centreline of the wedge is aligned with the approach flow so that the flow divides symmetrically above and below the wedge.

[91] The term **turn angle** is used in the Calculator.

[92] The term **inclined shockwave** is also used.

*Introduction to Engineering Fluid Mechanics.* Marcel Escudier.
© Marcel Escudier 2017. Published 2017 by Oxford University Press.

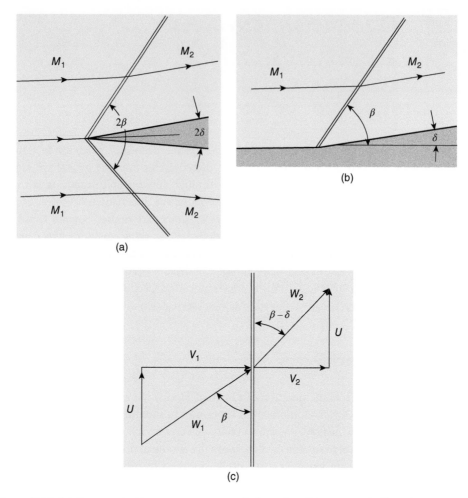

**Figure 12.1** (a) Supersonic flow over a symmetrical wedge (b) supersonic flow into a corner (c) oblique shock-wave notation

the static parameters of the flow. As shown in Figure 12.1(c), the flow angle downstream of the shock is $\beta - \theta$ with respect to the shock and must be less (because $V_{2N} < V_{1N}$) than the angle $\beta$ upstream of the shock, i.e. the flow has turned through an angle $\theta$ towards the shock.

We can transform the equations derived in Chapter 11, for property changes across a normal shock, to the corresponding equations for an oblique shock by observing that these changes are all associated with the components of velocity normal to the shock. Thus wherever $M_1$ occurs in the normal-shock relations it must be replaced by $V_{1N}/c_1 = M_1 \sin \beta$, where the approach-flow Mach number $M_1 = V_1/c_1$, and $M_2$ must be replaced by $V_{2N}/c_2 = M_2 \sin (\beta - \theta)$, where $M_2 = V_2/c_2$. We then have, from equation (11.67)

$$M_2 \sin (\beta - \theta) = \sqrt{\frac{\frac{2}{\gamma - 1} + M_1^2 \sin^2 \beta}{\left(\frac{2\gamma}{\gamma - 1}\right) M_1^2 \sin^2 \beta - 1}}.$$

(12.1)

From equation (11.68),

$$\frac{p_2}{p_1} = \left(\frac{2\gamma}{\gamma + 1}\right) M_1^2 \sin^2 \beta - \left(\frac{\gamma - 1}{\gamma + 1}\right). \tag{12.2}$$

From equation (11.70),

$$\frac{T_2}{T_1} = \frac{2(\gamma - 1)}{(\gamma + 1)^2 M_1^2 \sin^2 \beta} \left[\left(\frac{\gamma - 1}{2}\right) M_1^2 \sin^2 \beta + 1\right]\left[\left(\frac{2\gamma}{\gamma - 1}\right) M_1^2 \sin^2 \beta - 1\right]. \tag{12.3}$$

From equation (11.71)

$$\frac{\rho_2}{\rho_1} = \frac{\left(\frac{\gamma + 1}{2}\right) M_1^2 \sin^2 \beta}{1 + \left(\frac{\gamma - 1}{2}\right) M_1^2 \sin^2 \beta}. \tag{12.4}$$

From equation (11.73)

$$\frac{p_{02}}{p_{01}} = \left[\frac{\left(\frac{\gamma + 1}{2}\right) M_1^2 \sin^2 \beta}{1 + \left(\frac{\gamma - 1}{2}\right) M_1^2 \sin^2 \beta}\right]^{\gamma/(\gamma-1)} \left[\left(\frac{2\gamma}{\gamma + 1}\right) M_1^2 \sin^2 \beta - \frac{(\gamma - 1)}{(\gamma + 1)}\right]^{-1/(\gamma-1)}. \tag{12.5}$$

We can also derive an equation relating the flow directions on either side of an oblique shock. From Figure 12.1(c), ahead of the shock we have

$$\tan \beta = \frac{V_{1N}}{V_T} \tag{12.6}$$

and, behind the shock,

$$\tan (\beta - \theta) = \frac{V_{2N}}{V_T}. \tag{12.7}$$

According to the continuity equation (6.1)

$$\rho_1 V_{1N} = \rho_2 V_{2N}, \tag{12.8}$$

so that

$$\frac{\tan (\beta - \theta)}{\tan \beta} = \frac{V_{2N}}{V_{1N}} = \frac{\rho_1}{\rho_2} = \frac{1 + \left(\frac{\gamma - 1}{2}\right) M_1^2 \sin^2 \beta}{\left(\frac{\gamma + 1}{2}\right) M_1^2 \sin^2 \beta}. \tag{12.9}$$

Equation (12.9) can be rearranged as follows

$$\tan \theta = 2 \cot \beta \frac{M_1^2 \sin^2 \beta - 1}{M_1^2 (\gamma + \cos 2\beta) + 2} \tag{12.10}$$

from which, if $M_1$ and the wave angle $\beta$ are known, it is straightforward to calculate the turning angle $\theta$. Equation (12.10) is sometimes referred to as the $\theta$ – $\beta$ – **M relation**. Emanuel (2000) has provided a direct relationship which enables calculation of the wave angle if, as is more usual in a practical flow problem, $M_1$ and $\delta$ are given:

$$\tan \beta = \frac{M_1^2 - 1 + 2\lambda \cos \left( \frac{4\pi \xi + \cos^{-1} \chi}{3} \right)}{3 \left[ 1 + \left( \frac{\gamma - 1}{2} \right) M_1^2 \right] \tan \theta} \qquad (12.11)$$

where

$$\lambda = \sqrt{\left( M_1^2 - 1 \right)^2 - 3 \left[ 1 + \left( \frac{\gamma - 1}{2} \right) M_1^2 \right] \left[ 1 + \left( \frac{\gamma + 1}{2} \right) M_1^2 \right] \tan^2 \theta}$$

$$\qquad (12.12)$$

$$\chi = \frac{1}{\lambda^3} \left\{ \left( M_1^2 - 1 \right)^3 - 9 \left[ 1 + \left( \frac{\gamma - 1}{2} \right) M_1^2 \right] \left[ 1 + \left( \frac{\gamma - 1}{2} \right) M_1^2 + \left( \frac{\gamma + 1}{4} \right) M_1^4 \right] \tan^2 \theta \right\}$$

and

$\xi = 1$ for the **weak-shock solution**, $\xi = 0$ for the **strong-shock solution**.

The strong-shock solution corresponds to the larger of the two values for the wave angle $\beta$. Equation (12.11) is called the **$\beta$ – $\theta$ – $M$ relation**.

Numerical calculations can be made using the equations derived here directly but it is more likely (also more convenient and less prone to error) that calculations will be carried out using either the Virginia Tech Compressible Aerodynamics Calculator or a computer program.

Figure 12.2 shows the relation between the wave angle $\beta$ and the turning angle $\theta$ for various values of the upstream Mach number $M_1$. We can make several observations

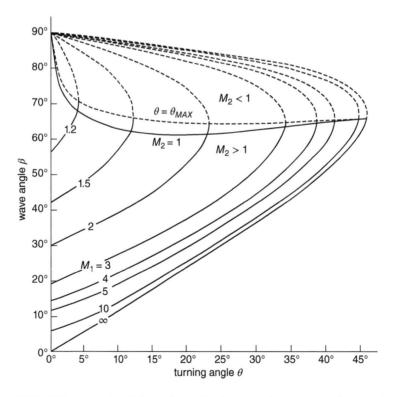

**Figure 12.2** Oblique shock solutions: strong shocks above $\theta = \theta_{\text{MAX}}$; weak shocks below

- For each value of $M_1$ there are two values of $\beta$ for which there is zero flow deflection ($\theta = 0$): $\mu = \sin^{-1}(1/M_1)$ corresponding to a **Mach line** (or wave of infinitesimal strength), and $\pi/2$ for a normal shockwave.
- For each value of $M_1$ there is a maximum value of the turning angle $\theta = \theta_{MAX}$, and a corresponding value for the wave angle $\beta_{MAX}$. An explicit expression for the latter is

$$\sin^2 \beta_{MAX} = \frac{1}{\gamma M_1^2}\left[\frac{(\gamma+1)}{4}M_1^2 - 1 + \sqrt{(\gamma+1)\left\{1 + \left(\frac{\gamma-1}{2}\right)M_1^2 + \left(\frac{\gamma+1}{16}\right)M_1^4\right\}}\right].$$

$$(12.13)$$

The maximum turning angle $\theta_{MAX}$ can be found by substituting into equation (12.10) the value of $\beta_{MAX}$ found from equation (12.13).

- The question arises as to what happens when the turning angle $\theta$ imposed on a supersonic flow by a wedge exceeds the maximum possible value $\theta_{MAX}$, corresponding to the flow Mach number $M_1$. Observations show that the flow is then compressed as it passes through a curved **bow shock** detached from the wedge and located some distance ahead of it. The same holds for supersonic flow over a blunt object and this is what we assumed when analysing the behaviour of a Pitot tube immersed in a supersonic flow in Subsection 11.8.3. The segment of the bow shock on the body centreline can be treated as a normal shock but a more general analytical treatment of the flowfield is beyond the scope of this text.
- From equation (12.11) for any value of $M_1$, for $\theta < \theta_{MAX}$, there are two possible values for the wave angle $\beta$. The larger value of $\beta$ corresponds to the so-called **strong-shock solution** (shown in Figure 12.2 by the broken curves above the line representing $\theta = \theta_{MAX}$), and the smaller value to the **weak-shock solution**.
- Behind a strong shock, the flow becomes subsonic (i.e. $M_2 < 1$) and the associated pressure ratio $p_2/p_1$ is higher than for a weak shock.
- Behind a weak shock the flow remains supersonic ($M_2 > 1$) except for turning angles just less than $\theta_{MAX}$.
- Since there are two possible solutions for $\theta < \theta_{MAX}$, the question arises as to which will occur in any given situation. The answer is that much depends on the downstream pressure. For external flow past a wedge, for example, the downstream pressure must eventually return to the level upstream of the wedge, suggesting that the weak solution is likely in this case, and this is consistent with observations. For compressive turning in internal duct flow, a strong shock occurs when the back pressure is high.
- Just below the curve in Figure 12.2 for $\theta = \theta_{MAX}$, is a narrow region for weak shocks in which the flow becomes subsonic behind the shock.
- The solid line just below the curve labelled $\theta = \theta_{MAX}$ represents the locus of solutions to equation (12.1) for which $M_2 = 1$. The wave angle $\beta^*$ corresponding to $M_2 = 1$ is given by

$$\sin^2 \beta^* = \frac{1}{\gamma M_1^2}\left[\frac{(\gamma+1)}{4}M_1^2 - \frac{(3-\gamma)}{4}\right.$$
$$\left. + \sqrt{(\gamma+1)\left\{\frac{9+\gamma}{16} - \left(\frac{3-\gamma}{8}\right)M_1^2 + \left(\frac{\gamma+1}{16}\right)M_1^4\right\}}\right]$$

$$(12.14)$$

and the corresponding deflection angle $\theta^*$ can be found by substituting into equation (12.10) the value of $\beta^*$ found from equation (12.14)[93].

[93] The equations for $\beta_{MAX}$ and $\beta^*$ are taken from Chapter 3 of Ferri (1949).

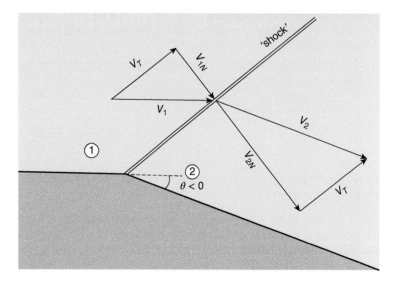

**Figure 12.3** Rarefaction shockwave

### 12.1.1 **Rarefaction shock**

In the paragraphs above we discussed positive turning angles between zero and $\theta_{MAX}$ and what happens when $\theta > \theta_{MAX}$. At first sight there would appear to be no reason why an oblique shock should not occur for a negative turning angle, as depicted in Figure 12.3, which shows a supersonic flow being turned around a convex corner by a single oblique wave. Since the velocity component $V_T$ parallel to the wave must remain unchanged, the flow geometry shows that for this situation to that the normal velocity would have to increase, i.e. $V_{2N} > V_{1N}$. As we now show, such a change violates the second law of thermodynamics and so has to be ruled out.

The combination of terms within the first square bracket of equation (12.5) can be recognised from equation (12.9) as corresponding to $V_{1N}/V_{2N}$ while, from equation (12.2), the terms within the second square bracket correspond to $p_2/p_1$. Equation (12.5) can thus be rewritten as

$$\frac{p_{02}}{p_{01}} = \left[\frac{V_{1N}}{V_{2N}}\right]^{\gamma/(\gamma-1)} \left[\frac{p_1}{p_2}\right]^{1/(\gamma-1)} . \tag{12.15}$$

Equations (12.2) and (12.9) can be combined to eliminate $M_1 \sin \beta$ to give

$$\frac{p_1}{p_2} = \frac{\left(\frac{\gamma-1}{\gamma+1}\right) \frac{V_{1N}}{V_{2N}} - \left(\frac{\gamma+1}{\gamma-1}\right)}{\left(\frac{\gamma-1}{\gamma+1}\right) - \left(\frac{\gamma+1}{\gamma-1}\right) \frac{V_{1N}}{V_{2N}}} \tag{12.16}$$

so that, finally, we can write

$$\frac{p_{02}}{p_{01}} = \left[\frac{V_{1N}}{V_{2N}}\right]^{\gamma/(\gamma-1)} \left[\frac{\left(\frac{\gamma-1}{\gamma+1}\right) \frac{V_{1N}}{V_{2N}} - \left(\frac{\gamma+1}{\gamma-1}\right)}{\left(\frac{\gamma-1}{\gamma+1}\right) - \left(\frac{\gamma+1}{\gamma-1}\right) \frac{V_{1N}}{V_{2N}}}\right]^{1/(\gamma-1)} . \tag{12.17}$$

Equation (12.17) shows that, if $V_{2N}/V_{1N} > 1$, $p_{02}/p_{01} > 1$ so that, from equation (11.75), $s_2 < s_1$, i.e. a decrease in specific entropy. For an adiabatic flow this would represent a violation of the second law of thermodynamics, leading to the conclusion that an increase in the normal velocity component across a shock is impossible. Since $V_{2N}/V_{1N} = \rho_1/\rho_2$ we have also demonstrated the impossibility of a **rarefaction shock** (i.e. one across which density decreases).

## 12.2 Prandtl-Meyer expansion fan (centred expansion fan)

We have just shown that, while a compressive oblique shockwave is produced by a supersonic flow flowing into a concave corner, a rarefaction shock cannot result from supersonic flow over a convex corner. Instead, expansion of a supersonic flow around a convex corner occurs progressively through an isentropic **centred wave** known as a **Prandtl-Meyer expansion fan**. The fan is defined by radial Mach lines centred on the corner, as shown in Figure 12.4, with the leading Mach line at the Mach angle $\mu_1 = \sin^{-1}(1/M_1)$ measured from the upstream flow direction and the terminating Mach line at the Mach angle $\mu_2 = \sin^{-1}(1/M_2)$ measured from the downstream flow direction. If the turning angle between the upstream and downstream flow directions is $\theta_{12}$, the angle between the two Mach lines is given by $\mu_1 - \mu_2 + \theta_{12}$ and is termed the **fan angle**. The fundamental difference between compression through a shock and expansion through a fan is that that non-linear mechanisms tend to steepen a compression whereas the opposite occurs in an expansion. Since the individual Mach waves which make up an expansion fan are infinitesimally weak, flow through such a fan is isentropic throughout. This is important as it means that the stagnation pressure upstream $p_0$ remains constant through an expansion fan, in contrast to the reduction in stagnation pressure across a shockwave. In the absence of heat addition, the stagnation temperature $T_0$ remains constant for both an expansion fan and a shockwave.

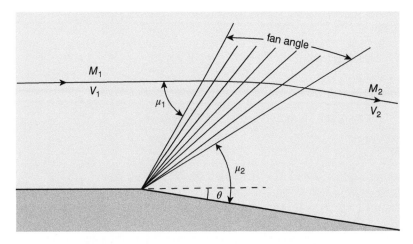

**Figure 12.4** Prandtl-Meyer centred expansion fan

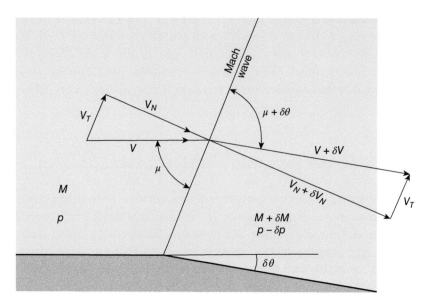

**Figure 12.5** Infinitesimal flow deflection through a Mach wave

We consider an elemental segment of the centred expansion wave, as illustrated in Figure 12.5, in which a supersonic flow with initial velocity $V$ and Mach number $M$ expands by turning through an infinitesimally small angle $\delta\theta$. After the expansion the flow velocity is $V + \delta V$, and the Mach number is $M + \delta M$. The velocity component normal to the Mach wave increases from $V_N$ to $V_N + \delta V_N$ while the component parallel to the wave, which remains unchanged, is $V_T$. We see from the geometry of the figure that

$$V_T = V \cos\mu = (V + \delta V) \cos(\mu + \delta\theta)$$

from which, after expanding the terms on the right-hand side and neglecting second-order terms, we obtain

$$\frac{\delta V}{V} = \delta\theta \tan\mu.$$

Since $\sin\mu = 1/M$, this leads to

$$\delta\theta = \sqrt{M^2 - 1}\frac{\delta V}{V}.$$

From the definition of the Mach number, $M = V/c$, we have

$$\frac{\delta V}{V} = \frac{\delta M}{M} + \frac{\delta c}{c}$$

so that

$$\delta\theta = \sqrt{M^2 - 1}\left(\frac{\delta M}{M} + \frac{\delta c}{c}\right).$$

Since the flow is adiabatic, the soundspeed $c_0$ corresponding to the stagnation temperature $T_0$ is constant and we have $c_0^2 = c^2\left[1 + (\gamma - 1)M^2/2\right]$ = constant, so that

$$\frac{\delta c}{c} = \frac{-\left(\frac{\gamma-1}{2}\right)M^2\frac{\delta M}{M}}{\left[1+\left(\frac{\gamma-1}{2}\right)M^2\right]}$$

which leads to

$$\delta\theta = \frac{\sqrt{M^2-1}}{\left[1+\left(\frac{\gamma-1}{2}\right)M^2\right]}\frac{\delta M}{M}.$$

In the limit $\delta M \to 0$, we then have

$$d\theta = \frac{\sqrt{M^2-1}}{\left[1+\left(\frac{\gamma-1}{2}\right)M^2\right]}\frac{dM}{M}. \tag{12.18}$$

Integration of this equation from an initial Mach number $M = 1$ and $\theta = 0$ to an arbitrary Mach number $M$ leads to an important quantity $v(M)$ is called the **Prandtl-Meyer function**, which is another quantity tabulated in NACA 1135 and incorporated into the isentropic-flow section of the Calculator. The angle $v(M)$, in radians, is thus the angle through which an initially sonic flow (i.e. $M = 1$) turns and expands isentropically to supersonic conditions. We have

$$v(M) = \theta(M) - \theta(1) = \int_1^M \frac{\sqrt{M^2-1}}{\left[1+\left(\frac{\gamma-1}{2}\right)M^2\right]}\frac{dM}{M}$$

$$= \sqrt{\frac{\gamma+1}{\gamma-1}}\arctan\left[\sqrt{\frac{\gamma-1}{\gamma+1}(M^2-1)}\right] - \arctan\left(\sqrt{M^2-1}\right). \tag{12.19}$$

According to equation (12.19), as the Mach number increases without limit, $v$ asymptotes to the value

$$v_{MAX} = \left[\sqrt{\left(\frac{\gamma+1}{\gamma-1}\right)}-1\right]\frac{\pi}{2} = 2.277^c\ (\approx 130.45^o)\ \text{for}\ \gamma = 1.4. \tag{12.20}$$

This theoretical limit to the turning angle corresponds to infinite Mach number, zero absolute pressure, and zero absolute temperature, none of which is physically realisable.

The Prandtl-Meyer function can be used to calculate the Mach number $M_2$ following supersonic expansion through a finite turning angle $\theta_{12}$ from an initial Mach number $M_1$ using the relation

$$\theta_{12} = v(M_2) - v(M_1)$$

or

$$v(M_2) = v(M_1) + \theta_{12}. \tag{12.21}$$

Once $M_2$ is known, other flow properties can be calculated using the Isentropic-flow Calculator or the relations derived in Section 11.3.

## ILLUSTRATIVE EXAMPLE 12.1

As shown in Figure E12.1, a supersonic airflow with Mach number $M_1$ = 2 turns through an angle $\theta_{12}$ = 30° through an expansion fan and then back $\theta_{23}$ = 30° through an oblique shockwave. If the initial static pressure $p_1$ is 0.5 bar and the static temperature $T_1$ is 280 K, with the aid of the Calculator, calculate the wave angle $\mu_1$, the Mach number $M_2$, the wave angle $\mu_2$, the static temperature $T_2$, the static pressure $p_2$, and the density $\rho_1$. Determine the Mach number $M_3$, the static temperature $T_3$, the static pressure $p_3$, the density $\rho_3$, and the shock angle $\beta_{23}$ for both a weak shock and a strong shock. Calculate the stagnation pressure $p_{03}$ for both cases.

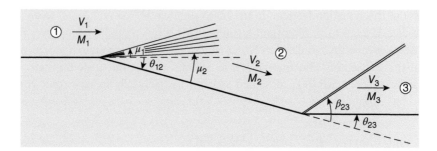

**Figure E12.1**

## Solution

We have $M_1$ = 2, $p_1$ = 5 × 10$^4$ Pa, $T_1$ = 280 K, $\theta_{12}$ = 30°, and $\theta_{23}$ = 30°.
We consider first region ①.
From the perfect-gas equation we have $\rho_1 = p_1/RT_1$ = 0.622 kg/m$^3$.
From the Isentropic-flow Calculator with $M_1$ = 2, we find $p_1/p_{01}$ = 0.128, $T_1/T_0$ = 0.556, the Mach angle $\mu_1$ = 30°, and the Prandtl-Meyer angle $\nu_1$ = 26.38°, from which $p_{01}$ = 3.91 × 10$^5$ Pa, and $T_0$ = 504.0 K. The soundspeed $c_1 = \sqrt{\gamma RT_1}$ = 335.4 m/s so that $V_1 = M_1 c_1$ = 670.8 m/s.
We consider next the flow between regions ① and ②.
The flow is turned through an angle $\theta_{12}$ = 30° by an expansion fan so that $\nu_2 = \nu_1 + \theta_{12}$ = 56.38°. From the Isentropic-flow Calculator, with $\nu_2$ = 56.38° we find $M_2$ = 3.368, $p_2/p_{02}$ = 0.0158, $T_2/T_0$ = 0.306, and $\mu_2$ = 17.27. Since flow through an expansion fan is isentropic, we have $p_{02} = p_{01}$ = 3.91 × 10$^5$ Pa so that $p_2$ = 6.194 × 10$^3$ Pa, and $T_2$ = 154.2 K. The soundspeed $c_2 = \sqrt{\gamma RT_2}$ = 248.9 m/s so that $V_2 = M_2 c_2$ = 838.3 m/s.
We consider next the flow from region ② to region ③ through a <u>weak</u> oblique shockwave.
The flow is turned through an angle $\theta_{23}$ = 30°. From the Oblique-shock Calculator, for a weak shock with $M_2$ = 3.368 and a turning angle $\theta_{23}$ = 30°, we find $M_3$ = 1.596, $p_3/p_2$ = 7.292, $T_3/T_2$ = 2.166, $p_{03}/p_{02}$ = 0.488, and the shock angle $\beta_{23}$ = 48.65°. Thus $p_3$ = 4.516 × 10$^4$ Pa, $T_3$ = 333.9 m/s, and $p_{03}$ = 1.908 × 10$^5$ Pa. The soundspeed $c_3 = \sqrt{\gamma RT_3}$ = 366.3 m/s so that $V_3 = M_3 c_3$ = 584.6 m/s.
For flow from region ② to region ③ through a <u>strong</u> oblique shockwave, the flow is turned through an angle $\theta_{23}$ = 30°. From the Oblique-shock Calculator, for a strong shock with

$M_2 = 3.368$ and a turning angle $\theta_{23} = 30°$, we find $M_3 = 0.626$, $p_3/p_2 = 12.43$, $T_3/T_2 = 3.032$, $p_{03}/p_{02} = 0.256$, and the shock angle $\beta_{23} = 77.33°$. Thus $p_3 = 7.701 \times 10^4$ Pa, $T_3 = 467.4$ m/s, and $p_{03} = 1.003 \times 10^5$ Pa. The soundspeed $c_3 = \sqrt{\gamma R T_3} = 433.4$ m/s so that $V_3 = M_3 c_3 = 271.2$ m/s.

---

## 12.3  Supersonic aerofoils and shock-expansion theory

The classic streamlined aerofoil profile used in low subsonic-flow applications is entirely unsuited to supersonic flight as it would result in a strong attached or detached shockwave resulting in excessively high drag forces on the aerofoil. A typical supersonic aerofoil has a thin profile with a sharp leading edge, such as the **diamond profile** shown in Figure 12.6(a). Even the flat-plate aerofoil shown in Figure 12.6(b) is well suited to supersonic flight. The latter is simpler to analyse than the diamond profile but structurally impractical.

In the flat-plate case, flow over the upper surface is turned through a Prandtl-Meyer expansion fan, centred on the **leading edge**, while the flow over the lower surface is compressed through a weak oblique shock attached to the leading edge. The situation is reversed at the trailing edge where the turning angles ensure the same pressure and direction in each stream. The two streams downstream of the plate have different velocities and so are separated by a surface of discontinuity termed a **contact surface**[94]. The direction of the flow downstream of the **trailing edge** will, usually, be slightly different from that of the flow upstream of the leading edge, as indicated by the angle $\varepsilon$ in Figure 12.6(a). **Upwash** corresponds with the situation where $\varepsilon > 0$, and **downwash** where $\varepsilon < 0$. For the diamond aerofoil $\varepsilon = \theta_{12} - \theta_{23} + \theta_{34}$ and for the flat-plate aerofoil $\varepsilon = -\theta_{12} + \theta_{23}$, the latter being too small to show in Figure 12.6(b).

If the pressure on the upper surface of the flat plate is $p_U$ and that on the lower surface is $p_L$, which will be greater than $p_U$ since the flow over the upper surface has undergone expansion while that over the lower surface has undergone compression, it is easily seen that the lift $L$ and drag $D$ exerted on the plate are given by

$$L = (p_L - p_U)\, cS\cos\alpha \tag{12.22}$$

and

$$D = (p_L - p_U)\, cS\sin\alpha \tag{12.23}$$

where $c$ is the chord length, $S$ is the span, and $\alpha$ is the angle of attack. These two results can be written in terms of lift and drag coefficients as

$$C_L = \frac{L}{\frac{1}{2}\rho_1 V_1^2 cS} = \frac{(p_L - p_U)\cos\alpha}{\frac{1}{2}\gamma p_1 M_1^2} \tag{12.24}$$

[94] Among other names for the contact surface are **slip surface**, which assumes that the velocity difference between the two streams does not result in a shear stress, and **vortex sheet**, which confines shearing effects to a region of infinitesimal thickness.

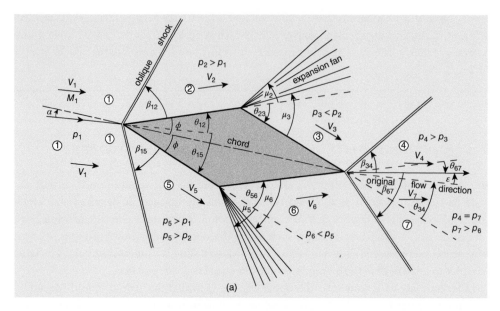

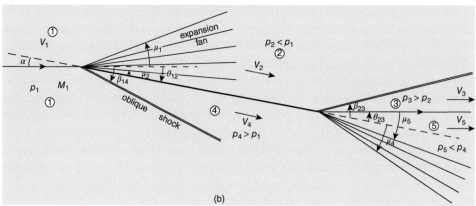

**Figure 12.6** Supersonic aerofoils: (a) diamond profile (b) inclined flat plate

and

$$C_D = \frac{D}{\frac{1}{2}\rho_1 V_1^2 cS} = \frac{(p_L - p_U)\sin\alpha}{\frac{1}{2}\gamma p_1 M_1^2} \tag{12.25}$$

where $p_1$ is the static pressure upstream of the plate, $\rho_1$ is the corresponding density, $M_1$ is the upstream Mach number, and $\gamma$ is the ratio of specific heats. Equation (11.18) has been used in equations (12.24) and (12.25) to replace $\rho_1 V_1^2/2$ by $\gamma p_1 M_1^2/2$.

These simple results reveal another feature which distinguishes supersonic flow from subsonic flow: the plate experiences drag, termed **wave drag**, even though no account has been taken of viscous effects on its surfaces. The leading-edge shockwave is the cause of the so-called **sonic boom** experienced at ground level when a supersonic aircraft flies overhead.

For supersonic flow of a symmetrical diamond aerofoil (see Figure 12.6(a)), weak shocks occur at the leading edge[95], followed by supersonic expansion at the shoulders, and oblique shocks at the trailing edge. Clearly, the details of the flow over a diamond aerofoil depend upon the total included **wedge angle** $2\phi$, the angle of attack $\alpha$, and the upstream Mach number. Since shock (and expansion) waves will occur on a symmetric diamond aerofoil in a supersonic flow, even at zero angle of attack, such an aerofoil will experience wave drag without generating lift. The foregoing comments apply to flow close to the aerofoil. Further away, in what is sometimes called the **far field**, the expansion waves attenuate, curve, and ultimately dissipate the shockwaves.

The simple analysis used to determine the lift and drag coefficients for an inclined flat plate can be extended to the diamond aerofoil, as follows. The lift $L$ and drag $D$ resulting from the static pressures $p_2$, $p_3$, $p_5$, and $p_6$ are given in terms of the turning angles by

$$L = \left[-p_2 \cos\theta_{12} - p_3 \cos(\theta_{23} - \theta_{12}) + p_5 \cos\theta_{15} + p_6 \cos(\theta_{56} - \theta_{15})\right]cS \tag{12.26}$$

and

$$D = \left[p_2 \sin\theta_{12} - p_3 \sin(\theta_{23} - \theta_{12}) + p_5 \sin\theta_{15} - p_6 \sin(\theta_{56} - \theta_{15})\right]cS. \tag{12.27}$$

It is straightforward to rewrite these two equations in terms of the angle of attack $\alpha$ and the half wedge angle $\phi$ as

$$L = \left[-p_2 \cos(\phi - \alpha) - p_3 \cos(\phi + \alpha) + p_5 \cos(\phi + \alpha) + p_6 \cos(\phi - \alpha)\right]cS \tag{12.28}$$

and

$$D = \left[p_2 \sin(\phi - \alpha) - p_3 \sin(\phi + \alpha) + p_5 \sin(\phi + \alpha) - p_6 \sin(\phi - \alpha)\right]cS. \tag{12.29}$$

After some simplification, in terms of lift and drag coefficients we then have

$$C_L = \left[(p_5 - p_3) \cos(\phi + \alpha) - (p_2 - p_6) \cos(\phi - \alpha)\right] \Big/ \left(\tfrac{1}{2}\gamma p_1 M_1^2\right) \tag{12.30}$$

and

$$C_D = \left[(p_5 - p_3) \sin(\phi + \alpha) + (p_2 - p_6) \sin(\phi - \alpha)\right] \Big/ \left(\tfrac{1}{2}\gamma p_1 M_1^2\right). \tag{12.31}$$

The theory outlined here in which fully supersonic flow over simple two-dimensional shapes is analysed by patching together flow turned through oblique shockwaves and centred expansion fans is termed **shock-expansion theory**.

For both the flat plate and the diamond shape, calculations of the pressures, velocities, temperatures, and Mach numbers are straightforward with the aid of the Calculator, as illustrated by the following examples.

---

[95] For an angle of attack $\alpha$ greater than the half wedge angle $\phi$, an expansion fan, rather than an oblique shock, would occur at the leading edge between regions ① and ③.

**ILLUSTRATIVE EXAMPLE 12.2**

A flat-plate aerofoil flies at Mach 3 at an angle of attack of $10°$ at an altitude where the static pressure is 5000 Pa and the static temperature is 217 K. Calculate the static pressure and static temperature, the stagnation pressure, the Mach number, and the flow velocity in each of the five flow regions identified in Figure E12.2. Calculate also the wave angles for the shock and expansion waves shown in the figure and the flow direction downstream of the trailing edge. Assume that the shockwaves are weak.

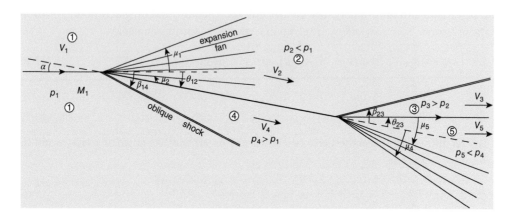

**Figure E12.2**

## Solution

We have $M_1 = 3$, $p_1 = 5 \times 10^3$ Pa, $T_1 = 217$ K, and $\alpha = 10°$.

We consider first region ① upstream of the plate.

From the Isentropic-flow Calculator[96] with $M_1 = 3$, we find $p_1/p_{01} = 0.0272$, $T_1/T_0 = 0.357$, the Prandtl-Meyer angle $\nu_1 = 49.76°$, and the wave angle $\mu_1 = 19.47°$, from which $p_{01} = 1.837 \times 10^5$ Pa, and $T_0 = 607.6$ K. The soundspeed $c_1 = \sqrt{\gamma R T_1} = 295.3$ m/s so that $V_1 = M_1 c_1 = 885.8$ m/s.

We consider next the flow between regions ① and ②.

The flow is turned through an angle $\theta_{12} = \alpha = 10°$ by an isentropic expansion fan so that $\nu_2 = \nu_1 + \theta_{12} = 59.76°$. From the Isentropic-flow Calculator, with $\nu_2 = 59.76°$ we find $M_2 = 3.578$, $p_2/p_{01} = 0.0117$, $T_2/T_0 = 0.281$, and $\mu_2 = 16.23°$. Thus $p_2 = 2156$ Pa, and $T_2 = 170.6$ K. The soundspeed $c_2 = \sqrt{\gamma R T_2} = 261.8$ m/s so that $V_2 = M_2 c_2 = 936.9$ m/s.

We consider next the flow between regions ① and ④.

The flow is turned through an angle $\theta_{14} = \alpha = 10°$ by an oblique shockwave. From the Oblique-shock Calculator, for a weak shock with $M_1 = 3$ and turning angle $\theta_{14} = 10°$, we find $M_4 = 2.505$, $p_4/p_1 = 2.054$, $p_{04}/p_{01} = 0.963$, $T_4/T_1 = 1.242$, $\nu_4 = 39.24°$, $\mu_4 = 23.53°$,

---

[96] Throughout the calculation, the subscripts correspond to those in Figure E12.2 whereas in the Calculator subscript 1 indicates conditions upstream of a shock or expansion wave, and subscript 2 conditions downstream.

and $\beta_{14} = 27.4°$. Thus $p_4 = 1.027 \times 10^4$ Pa, $p_{04} = 1.769 \times 10^5$ Pa, and $T_4 = 269.4$ K while $T_{04} = T_{01}$. The soundspeed $c_4 = \sqrt{\gamma R T_4} = 329.0$ m/s so that $V_4 = M_4 c_4 = 824.2$ m/s.

So far the calculation has been a straightforward, step-by-step process. Calculating the conditions downstream of the trailing edge is less straightforward because the flow direction itself is unknown and its determination entails an iterative procedure. The conditions to be satisfied are that the flow directions in regions ③ and ⑤ are the same, i.e. the turning angles satisfy $\theta_{23} = \theta_{45}$, and the static pressures in these two regions, $p_3$ and $p_5$, are also the same.

We start with a first guess that the flow from region ② turns through a weak oblique shock with a turning angle $\theta_{23} = 10°$, which would bring the flow back to its original direction. From the Oblique-shock Calculator we find $p_3/p_2 = 2.305$ so that $p_3 = 4969$ Pa. The flow from region ④ turns through an expansion fan with a turning angle $\theta_{45} = 10°$ so that $v_5 = 49.24°$. From the Isentropic-flow Calculator we find $p_5/p_{04} = 0.0283$ so that $p_5 = 5012$ Pa which is very close to $p_3$ indicating that the turning angle is indeed about $10°$.

The mean of 4969 and 5012 Pa is 4991 Pa, which we can use to determine a new estimate for the turning angle. If $p_3/p_2 = 4991/2156 = 2.315$, from the Oblique-shock Calculator we find $\theta_{23} = 10.06°$. With $\theta_{45} = 10.06°$ we have $v_5 = v_4 + \theta_{45} = 49.30°$ so that from the Isentropic-flow Calculator $p_5/p_{04} = 0.0282$, and $p_5 = 4990$ Pa, in almost perfect agreement with the estimate of 4991 Pa.

Finally, with $M_2 = 3.578$, and $\theta_{23} = 10.06°$, from the Oblique-shock Calculator we have $M_3 = 2.962$, $T_3/T_2 = 1.293$, $\beta_{23} = 24.05°$, and $p_{03}/p_{02} = 0.942$, from which $T_3 = 220.6$, K and $p_{03} = 1.731 \times 10^5$ Pa. The soundspeed $c_3 = \sqrt{\gamma R T_3} = 297.7$ m/s so that $V_3 = M_3 c_3 = 881.8$ m/s. With $v_5 = 49.30°$, from the Isentropic-flow Calculator we have $M_5 = 2.976$, $T_5/T_0 = 0.361$, and $\mu_5 = 19.63°$, from which $T_5 = 219.2$ K. The soundspeed $c_5 = \sqrt{\gamma R T_5} = 296.8$ m/s so that $V_5 = M_5 c_5 = 883.4$ m/s. The upwash angle $\varepsilon$, between the initial and final flow directions is $0.06°$, which is evidently negligible.

Figure E12.2 has been constructed using the values for the turning, shock, and wave angles calculated here.

---

Although the inclined flat plate provides a useful example for the application of shock-expansion theory, the symmetric diamond aerofoil of the following example is practically more realistic.

---

**ILLUSTRATIVE EXAMPLE 12.3**

A diamond-profile aerofoil, symmetrical about its chordline, flies at Mach 3 at an altitude where the static pressure is 5000 Pa and the static temperature is 217 K. If the semi-included aerofoil wedge angle is $20°$ and the angle of attack is $10°$, calculate the static pressure, the static temperature, the stagnation pressure, the Mach number, and the flow velocity in each of the seven flow regions identified in Figure E12.2. Calculate also the wave angles for the shock and expansion waves and the flow direction downstream of the trailing edge. Assume that the shockwaves are weak.

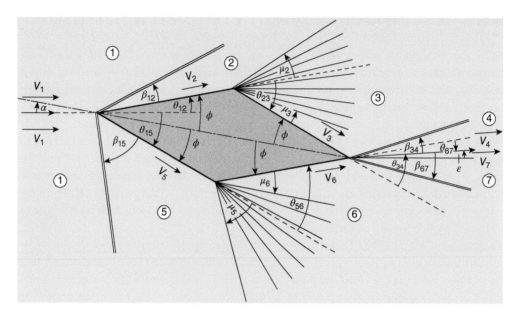

**Figure E12.3**

### Solution

We have $M_1 = 3$, $p_1 = 5 \times 10^3$ Pa, $T_1 = 217$ K, $\alpha = 10°$, and $\phi = 20°$.

The upstream conditions in region ① are identical to those in Illustrative Example 12.2, i.e. $p_{01} = 1.837 \times 10^5$ Pa, $T_0 = 607.6$ K, and $V_1 = 885.8$ m/s.

Between regions ① and ② the flow is turned through an angle $\theta_{12} = \phi - \alpha = 10°$ by an oblique shockwave, which is identical to the situation in Illustrative Example 12.2 between regions ① and ②. The conditions in region ② are therefore as follows: $M_2 = 2.505$, $p_2 = 1.027 \times 10^4$ Pa, $p_{02} = 1.769 \times 10^5$ Pa, $T_2 = 269.4$ K, the wave angle $\mu_2 = 23.53°$, the Prandtl-Meyer angle $\nu_2 = 39.24°$, and the flow velocity $V_2 = 824.2$ m/s. The shock angle $\beta_{12} = 27.4°$.

Between regions ② and ③ the flow is turned through an angle $\theta_{23} = 2\phi = 40°$ so that $\nu_3 = \nu_2 + \theta_{23} = 79.24°$. Since flow through an expansion fan is isentropic, $p_{03} = p_{02} = 1.769 \times 10^5$ Pa.

In region ③, from the Isentropic-flow Calculator with $\nu_3 = 79.24°$, we find $M_3 = 5.258$, $p_3/p_{03} = 1.405 \times 10^{-3}$, $T_3/T_0 = 0.153$, and $\mu_3 = 10.96°$. Thus $p_3 = 248.6$ Pa, and $T_3 = 93.0$ K. The soundspeed $c_3 = \sqrt{\gamma R T_3} = 193.4$ m/s so that $V_3 = M_3 c_3 = 1017$ m/s.

We consider next regions ⑤, and ⑥ on the lower surface.

Between regions ① and ⑤ the flow is turned through an angle $\theta_{12} = \phi + \alpha = 30°$ by an oblique shockwave. From the Oblique-shock Calculator, for a weak shock with $M_1 = 3$ and turning angle $\theta_{15} = 30°$, we find $M_5 = 1.406$, $p_5/p_1 = 6.356$, $p_{05}/p_{01} = 0.555$, $T_5/T_1 = 2.007$, and $\beta_{15} = 52.01°$. Thus $p_5 = 3.178 \times 10^4$ Pa, $p_{05} = 1.020 \times 10^5$ Pa, and $T_5 = 435.5$ K. The soundspeed $c_5 = \sqrt{\gamma R T_5} = 418.3$ m/s so that $V_5 = M_5 c_5 = 588.1$ m/s.

In region ⑤, from the Isentropic-flow Calculator with $M_5 = 1.406$, $\mu_5 = 45.34°$, and $\nu_5 = 9.16°$.

Between regions ⑤ and ⑥ the flow is turned through an angle $\theta_{56} = 2\phi = 40°$ so that $v_6 = v_5 + \theta_{56} = 49.16°$. Since flow through an expansion fan is isentropic, $p_{06} = p_{05} = 1.020 \times 10^5$. In region ⑥, from the Isentropic-flow Calculator with $v_6 = 49.16°$, we find $M_6 = 2.969$, $p_6/p_{06} = 2.852 \times 10^{-2}$, $T_6/T_0 = 0.362$, and $\mu_6 = 19.68°$. Thus $p_6 = 2.908 \times 10^3$ Pa, and $T_6 = 219.9$ K. The soundspeed $c_6 = \sqrt{\gamma R T_6} = 297.2$ m/s so that $V_6 = M_6 c_6 = 822.6$ m/s.

As in Illustrative Example 12.2, so far the calculation has been a straightforward, step-by-step process. Calculating the conditions downstream of the leading edge is again less straightforward because the flow direction itself is unknown and its determination entails an iterative procedure. The conditions to be satisfied are that the flow directions in regions ④ and ⑦ are the same, i.e. the turning angles satisfy $\theta_{34} + \theta_{67} = 40°$, and also that the static pressures in these two regions, $p_4$ and $p_7$, are the same.

We start with a first guess that the flow from region ③ turns through a weak oblique shock with a turning angle $\theta_{34} = 40°$. From the Oblique-shock Calculator we find $p_4/p_3 = 23.6$ so that $p_4 = 5867$ Pa. Since the corresponding turning angle $\theta_{67}$ for the flow from region ⑥ is zero, $p_7 = p_6 = 2908$ Pa, which is much lower than $p_4$.

To increase $p_7$ and decrease $p_4$ we reduce $\theta_{34}$ to $35°$ and increase $\theta_{67}$ to $5°$, where the flow again turns through a weak oblique shock. We then find $p_4/p_3 = 18.4$, and $p_7/p_6 = 1.449$, so that $p_4 = 4574$ Pa, and $p_7 = 4214$ Pa, i.e. the two values are much closer.

Ultimately, we find $\theta_{34} = 34.3°$, $\theta_{67} = 5.7°$, and $p_4 \approx p_7 = 4.420 \times 10^3$ Pa. From the Oblique-shock Calculator we also find $M_4 = 1.822$, $T_4/T_3 = 3.924$, $p_{04}/p_{03} = 0.148$, $\beta_{34} = 48.2°$, $M_7 = 2.689$, $T_7/T_6 = 1.129$, $p_{07}/p_{06} = 0.993$, and $\beta_{67} = 23.9°$, Thus $T_4 = 365.2$ K, $V_4 = 697.9$ m/s, $p_{04} = 2.625 \times 10^4$ Pa, $T_7 = 248.4$ K, $V_7 = 849.5$ m/s, and $p_{07} = 1.012 \times 10^4$ Pa. From the Isentropic-flow Calculator, $\mu_4 = 33.23°$, and $\mu_7 = 21.84°$. The upwash angle is obtained from $\varepsilon = \theta_{12} - \theta_{23} + \theta_{34} = -\theta_{15} + \theta_{56} - \theta_{67} = 4.3°$, i.e. significantly greater than was the case for the inclined flat plate of Illustrative Example 12.2.

Figure E12.3 has been constructed using the values for the turning, shock, and wave angles calculated here.

 ### 12.4 SUMMARY

In this chapter we analysed external supersonic gas flow in which changes in the fluid and flow properties were brought about by direction change. We showed that flow over a corner between two flat surfaces resulted in an oblique shockwave if the angle between the two surfaces is less than 180° (a concave corner). The analysis of flow through an oblique shockwave was based upon the superposition of the flowfield for a normal shock onto a uniform flow parallel to the shock. For an angle in excess of 180° (a convex corner), the flow is turned through an isentropic Prandtl-Meyer expansion fan. Analysis of a Prandtl-Meyer expansion fan started from consideration of an infinitesimal flow deflection through a Mach wave.

The student should be able to

- explain the terms turning angle, shock angle, Mach angle, wave angle, and Prandtl-Meyer angle

- understand the changes in oblique-shock behaviour as the turning angle is increased from zero to the maximum possible
- understand the difference between a weak oblique shock and a strong oblique shock
- understand the changes in flow behaviour for a Prandtl-Meyer expansion fan as the turning angle is increased from zero to the maximum possible
- perform calculations using the equations derived in this chapter for an oblique shock and for a Prandtl-Meyer expansion fan
- for an oblique shock, perform calculations using the Virginia Tech Compressible Aerodynamics Calculator
- for a Prandtl-Meyer expansion fan, perform calculations using the Virginia Tech Compressible Aerodynamics Calculator
- carry out a shock-expansion calculation for an inclined flat-plate aerofoil or a diamond-section aerofoil

## ? 12.5  SELF-ASSESSMENT PROBLEMS

**12.1**  Air flows at a Mach number of 3.0 with static pressure 2.0 bar and static temperature 50 °C. Calculate the Mach number, the static and stagnation temperatures, the static and stagnation pressures, the flow velocity, and the shock angle following a 20° turn through (a) a weak oblique shock and (b) a strong oblique shock.

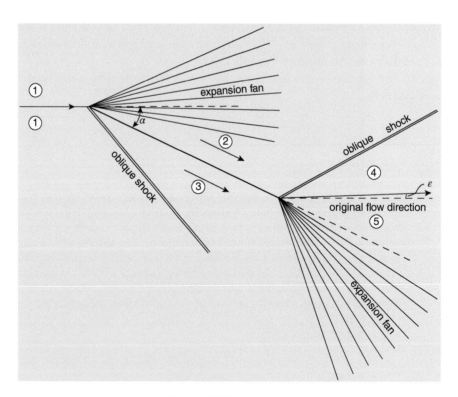

**Figure P12.3**

(Answers: (a) 1.994, 503.8 K, 904.4 K, 7.54 bar, 58.48 bar, 897.2 m/s, 37.8°; (b) 0.539, 854.7 K, 904.4 K, 20.27 bar, 24.71 bar, 316.1 m/s, 82.15°)

**12.2** For the same upstream conditions as in Problem 12.1, calculate the Mach number, Mach angle, static temperature, static and stagnation pressure, and flow velocity following a 20° expansion turn.
(Answers: 4.318, 13.39°, 191.2 K, 0.319 bar, 73.47 bar, 1197 m/s)

**12.3** As shown schematically in Figure P12.3, air flows over a flat-plate aerofoil at an angle of attack $\alpha$ of 25°. If the upstream Mach number is 2.5, the static pressure is 0.4 bar, and the static temperature is 260 K, calculate the Mach number, static pressure, static temperature, and flow velocity in regions ②, ③, ④, and ⑤, shown in the figure, and also the final flow direction $\varepsilon$. Assume both shockwaves are weak.
(Answers: ② 3.877, 0.0531 bar, 146.0 K, 939.1 m/s; ③ 1.386, 1.657 bar, 422.5 K, 571.3 m/s; ④ 2.047, 0.391 bar, 318.2 K, 732.1 m/s; ⑤ 2.335, 0.391 bar, 279.8 K, 783.1 m/s, 1.53° upwash)

# 13 Compressible pipe flow

This chapter concerns the flow of a perfect gas through a pipe of constant cross section. The analysis is simplified, so that a number of complete analytical solutions are possible, by again assuming one-dimensional flow. Three separate situations are discussed: adiabatic flow with wall friction (**Fanno flow**), **isothermal pipe flow** with wall friction, and frictionless flow with heating or cooling through the pipe wall (**Rayleigh flow**). If the initial flow is subsonic, it is shown that the flow becomes **choked** after a certain length. An initially supersonic flow also chokes after a certain length but a shockwave may be required to match the exit condition.

## 13.1 **Basic equations**

In Chapters 11 and 12 we considered compressible gas flows in which property changes were brought about by area or direction change in the absence of surface friction or heat transfer. In reality, all flows are affected by surface friction and in this chapter we present the one-dimensional analysis of the steady flow of a perfect gas through a straight pipe of constant cross-sectional area, including the influence of friction and/or heat transfer. The analysis is based upon the three conservation equations

- mass-conservation equation (continuity equation; derived in Chapter 6)

$$\dot{m} = \rho A V = A G = \text{constant} \qquad (6.1)$$

where $\dot{m}$ is the mass flowrate, $\rho$ is the fluid density, $A$ is the (constant) cross-sectional area of the pipe, $V$ is the flow velocity, and $G$ is (constant) the flow per unit area defined as $G = \rho V = \dot{m}/A$, often termed the **mass velocity** (or **mass flux**)

- linear momentum-conservation equation

The flow situation under consideration is shown in Figure 13.1(a), in which we have defined an elemental control volume of infinitesimal length $\delta x$ at some axial location $x$. The forces acting on the control volume and the momentum flowrates into and out of it are shown in Figure 13.1(b). If we apply the linear momentum equation (see Section 9.2) to the flow through the control volume, for steady flow we have

$$pA - (p + \delta p) A - \tau_S \quad \delta A_F = \dot{m} (V + \delta V) - \dot{m} V$$

which, after cancellation, simplifies to

$$-A \delta p - \tau_S \delta A_F = \dot{m} \delta V.$$

In this equation, $p$ represents the static pressure, $\tau_S$ the pipe-wall shear stress, and $\delta A_F$ the surface area over which $\tau_S$ acts. For a circular duct[97] of diameter $D$, we have $A = \pi D^2/4$,

---

[97] The analysis can be generalised for non-circular ducts by replacing $D$ with the **hydraulic diameter** $D_H$ (discussed in Section 16.2).

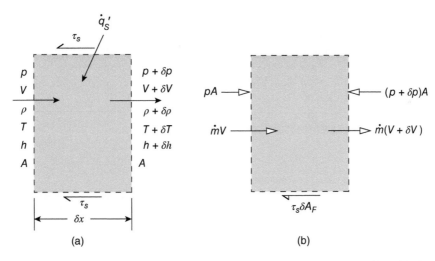

**Figure 13.1** (a) Elemental control volume for the analysis of compressible pipe flow (b) The forces acting on the control volume and the momentum flowrates into and out of it

and $\delta A_F = \pi D \delta x$, so that, if we divide through by $A\delta x$ and take the limit $\delta x \to 0$, the momentum equation results in

$$-\frac{dp}{dx} - \frac{4\tau_S}{D} = \rho V \frac{dV}{dx} \tag{13.1}$$

wherein we have substituted for $\dot{m}$ using equation (6.1). The wall shear stress can be written in terms of the Fanning friction factor $f_F \equiv 2\tau_S/\rho V^2$, leading to

$$-\frac{dp}{dx} - \frac{2\rho V^2 f_F}{D} = \rho V \frac{dV}{dx}. \tag{13.2}$$

- energy-conservation equation

In the absence of external work, the steady-flow energy equation derived in Chapter 11, equation (11.11), for the elemental control volume, reduces to

$$\delta \dot{q} = \dot{m}\delta h_0 = \dot{m}\left[\delta h + \frac{\delta\left(V^2\right)}{2}\right]$$

where $\delta\dot{q}$ is the rate of thermal-energy addition to the fluid within the control volume, $h$ is the specific enthalpy, and $h_0$ is the specific stagnation enthalpy. For a perfect gas, $\delta h = C_P \delta T$ where $C_P$ is the specific heat at constant pressure and $T$ is the absolute temperature, so that the energy equation becomes

$$\delta\dot{q} = \dot{m}C_P\delta T_0 = \dot{m}\left[C_P\delta T + \frac{\delta\left(V^2\right)}{2}\right].$$

If $\dot{q}'$ represents the rate of heat transfer into the control volume per unit length, such that $\delta\dot{q} = \dot{q}'\delta x$, the differential form of the energy equation may be written as

$$\dot{q}' = \dot{m}C_P\frac{dT_0}{dx} = \dot{m}\left(C_P\frac{dT}{dx} + V\frac{dV}{dx}\right), \tag{13.3}$$

$T_0$ being the stagnation temperature (see Section 13.2). In the three conservation equations (13.1), (13.2), and (13.3), $\dot{m}, f_F, D$, and $\dot{q}'$ can all be regarded as specified, although in principle $f_F$ depends upon the pipe Reynolds number which, for compressible flow, varies with $x$, and, as will be seen, we are not completely free to specify either $\dot{m}$ or $\dot{q}'$. Since we have four variables, $\rho, V, p$, and $T$, we need one more equation: an equation of state that relates $p, \rho$, and $T$

- Equation of state

The equation of state for a perfect gas, given in Chapter 2, is

$$p = \rho R T \tag{2.9}$$

where $R$ is the specific gas constant.

Although this completes the set of equations which incorporate the underlying physics of compressible pipe flow, to proceed further it is convenient to introduce the Mach number $M$, which was defined in Chapter 3 as

$$M = \frac{V}{c} \tag{3.42}$$

where $c = \sqrt{\gamma R T}$ is the local speed of sound and $\gamma$ is the ratio of specific heats.

Useful analytical results can be obtained by considering three specific situations

- adiabatic flow with wall friction, known as **Fanno flow**
- isothermal flow with wall friction
- frictionless flow with heating or cooling, known as **Rayleigh flow**[98]

## 13.2 Adiabatic pipe flow with wall friction: Fanno flow

From the energy-conservation equation (13.3),

$$T_0 = T + \frac{V^2}{2C_P} = T\left(1 + \frac{V^2}{2C_P T}\right) = T\left[1 + \left(\frac{\gamma - 1}{2}\right)M^2\right] = \text{constant} \tag{13.4}$$

so that

$$\frac{T}{T^*} = \frac{c^2}{c^{*2}} = \frac{\frac{\gamma + 1}{2}}{1 + \left(\frac{\gamma - 1}{2}\right)M^2}. \tag{13.5}$$

As will be seen later, for both subsonic and supersonic flow of a perfect gas through a pipe under adiabatic conditions, the Mach number tends to unity. The flow properties corresponding to $M = 1$ are indicated by an asterisk, i.e. $c^*(= V^*), p^*, p_0^*, s^*, T^*$, and $\rho^*$, and used as reference conditions. Since $A = $ constant, from the mass-conservation equation (13.1)

$$\frac{\rho^*}{\rho} = \frac{V}{c^*} = M\frac{c}{c^*}. \tag{13.6}$$

---

[98] Instead of Fanno flow and Rayleigh flow, the terms Fanno-line flow and Rayleigh-line flow are also used, for reasons that will become apparent shortly.

which can be combined with equation (13.5) to give

$$\frac{V}{c^*} = \frac{V}{V^*} = M \sqrt{\frac{\gamma + 1}{2\left[1 + \left(\frac{\gamma - 1}{2}\right) M^2\right]}} \tag{13.7}$$

and

$$\frac{\rho}{\rho^*} = \frac{1}{M} \sqrt{\frac{2\left[1 + \left(\frac{\gamma - 1}{2}\right) M^2\right]}{\gamma + 1}}. \tag{13.8}$$

From the perfect-gas equation (2.9),

$$\frac{p}{p^*} = \frac{\rho}{\rho^*} \frac{T}{T^*}, \tag{13.9}$$

which can be combined with equations (13.5) and (13.8), for $T/T^*$ and $\rho/\rho^*$, respectively, to give

$$\frac{p}{p^*} = \frac{1}{M} \sqrt{\frac{\gamma + 1}{2\left\{1 + \left(\frac{\gamma - 1}{2}\right) M^2\right\}}}. \tag{13.10}$$

The stagnation pressure $p_0$ is defined by equation (11.22)

$$p_0 = p \left[1 + \left(\frac{\gamma - 1}{2}\right) M^2\right]^{\frac{\gamma}{\gamma-1}}$$

so that, from equation (13.10)

$$\frac{p_0}{p_0^*} = \frac{1}{M} \left[\frac{2\left\{1 + \left(\frac{\gamma - 1}{2}\right) M^2\right\}}{\gamma + 1}\right]^{\frac{\gamma+1}{2(\gamma-1)}}. \tag{13.11}$$

Finally, from equation (11.2) for the change in specific entropy $s$ up to the choking location, as derived in Chapter 11

$$s - s^* = C_P \ln\left(\frac{T}{T^*}\right) - R \ln\left(\frac{p}{p^*}\right), \tag{11.2}$$

from which, since $C_P = \gamma R/(\gamma - 1)$,

$$\frac{s - s^*}{R} = \ln\left[\left(\frac{T}{T^*}\right)^{\gamma/(\gamma-1)} \left(\frac{p^*}{p}\right)\right]. \tag{13.12}$$

Substitution in equation (13.12) for $T/T^*$ from equation (13.5) and for $p/p^*$ from equation (13.10) leads to

$$\frac{s - s^*}{R} = \ln\left\{M \left[\frac{(\gamma + 1)}{2\left\{1 + \left(\frac{\gamma - 1}{2}\right) M^2\right\}}\right]^{\frac{\gamma+1}{2(\gamma-1)}}\right\}. \tag{13.13}$$

As already mentioned, the asterisked quantities in equations (13.5) to (13.13), i.e. $c^*$, $p^*$, $p_0^*$, $s^*$, $T^*$, and $\rho^*$, indicate that all these quantities correspond to the reference **choking state**, $M = 1$ for adiabatic pipe flow.

We now introduce the momentum-conservation equation

$$-\frac{dp}{dx} - \frac{2\rho V^2 f_F}{D} = \rho V \frac{dV}{dx} \qquad (13.2)$$

where $f_F$ is the Fanning friction factor. Substitution for $V$, $\rho$, and $p$ in equation (13.2), using equations (13.7), (13.8), and (13.10), leads to

$$\frac{1}{M^2} \frac{dM^2}{dx} = \frac{\gamma M^2 \left[ 1 + \left( \frac{\gamma - 1}{2} \right) M^2 \right]}{1 - M^2} \frac{4 f_F}{D}. \qquad (13.14)$$

Equation (13.14) shows that $dM/dx > 0$ if the initial flow is subsonic, i.e. $M < 1$, and $M$ increases along the pipe, while, if the initial flow is supersonic, i.e. $M > 1$, we see that $dM/dx < 0$, and $M$ decreases along the pipe. We can thus draw an important conclusion from equation (13.14): in adiabatic pipe flow of a perfect gas, the influence of wall friction causes the Mach number, whether subsonic or supersonic, to approach unity. If the pipe is sufficiently long, at a certain length $L^*$, depending upon the value of the friction factor $f_F$, the pipe diameter $D$, and the initial Mach number $M$ at $x = 0$, the flow becomes **frictionally choked** ($M = 1$).

For subsonic flow, if $L > L^*$, the mass flow and initial Mach number have to be reduced until the exit flow is sonic, with a new value of $L^*$ corresponding to the reduced inlet Mach number. If the pipe is preceded by an appropriately designed convergent-divergent nozzle (see Section 11.7), the flow entering the pipe may be supersonic. In this situation, slight increase in the pipe length beyond $L^*$ causes a normal shock to form within the pipe just upstream of the exit such that the subsonic flow downstream of the shock is just choked at the exit. As the pipe length is increased further, the shock moves further upstream until it reaches the inlet. As we shall see in Illustrative Example 13.2, if the shock location is known the calculation is straightforward. If instead the pipe length is specified, an iterative trial-and-error procedure is needed to determine the shock location.

To enable practical calculations, a value for the Fanning friction factor averaged over the length $L^*$ is usually defined by

$$\bar{f}_F = \frac{1}{L^*} \int_0^{L^*} f_F dx \qquad (13.15)$$

and the weak dependence of $f_F$ on the pipe Reynolds number is neglected. With an average value for the friction factor, $\bar{f}_F$, equation (13.14) can be integrated to give

$$\frac{4 \bar{f}_F L^*}{D} = \frac{1 - M^2}{\gamma M^2} + \frac{(\gamma + 1)}{2\gamma} \ln \left[ \frac{(\gamma + 1) M^2}{2 \left( 1 + \frac{\gamma - 1}{2} M^2 \right)} \right]. \qquad (13.16)$$

Values for $4\bar{f}_F L^*/D$ for any initial Mach number $M$, together with the ratios $T/T^*$, $p/p^*$, $V/V^*$, $\rho/\rho^*$, $p_0/p_0^*$, and $(s^* - s)/R$, can be calculated from the equations above, or obtained from the **Fanno-flow Calculator**. It is worth noting that the friction factor used in the Calculator is the Fanning friction factor[99].

Figure 13.2 shows the variation with Mach number of selected flow properties for Fanno flow according to the equations derived above. The curve of $\bar{f}_F L^*/D$ versus $M$ is particularly interesting: (a) for $M < 1$, the curve steepens considerably as $M \to 0$, so that small changes in the inlet Mach number have a significant effect on the maximum possible pipe length, and (b) for $M > 1$, small changes in the pipe length have a significant effect on the shock location. We note too that, as $M \to \infty$, $4\bar{f}_F L^*/D$ approaches the asymptotic value

$$\frac{4}{\gamma}\left[\frac{(\gamma+1)}{2}\ln\left(\frac{\gamma+1}{\gamma-1}\right) - 1\right] \approx 3.286 \tag{13.17}$$

so that the flow chokes after a relatively short length of pipe, e.g. with $\bar{f}_F = 0.005$, $D = 50$ mm, and $L^* = 2.5$ m.

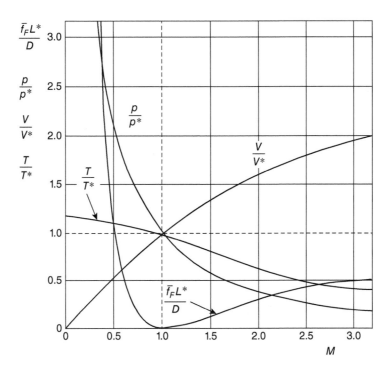

**Figure 13.2** Variation of flow properties with Mach number for Fanno flow of a perfect gas with $\gamma = 1.4$

[99] The factor 4 on the left-hand side of equation (13.16) can be regarded as a reminder that the friction factor used in this textbook is the Fanning friction factor $f_F$ and not the Darcy friction factor $f_D = 4f_F$. The Darcy friction factor is used in the tabulated values presented in many textbooks, and great care is necessary to avoid confusing the two friction factors.

### 13.2.1  Fanno-flow calculation procedure

When confronted with a Fanno-flow problem, the reader may find it somewhat confusing to be dealing with so many quantities: the stagnation pressure and temperature, $p_0$ and $T_0$; the static pressure and temperature, $p$ and $T$; the soundspeed, $c$; the density, $\rho$; the flow velocity, $V$; the Mach number $M$; the mass flowrate $\dot{m}$; the pipe length; the cross-sectional area $A$; the friction factor, $\bar{f}_F$; and the sonic conditions, indicated by the superscript $*$. In addition, apart from $T_0$, $\dot{m}$, and the sonic conditions, all the flow variables vary with distance along the pipe $x$. The situation is greatly simplified if it is realised that the key quantities involved in solving many problems are the stagnation pressure and temperature at the pipe inlet, and one other quantity, often the inlet velocity or Mach number. It is then straightforward to calculate all other quantities sequentially as follows

- specify $T_0$, which is a constant for adiabatic pipe flow
- specify $p_{01}$, the stagnation pressure at inlet
- specify $D$ (or $A$)
- specify $\bar{f}_F$, which is a constant for the flow
- specify $V_1$, the flow velocity at inlet (or $M_1, p_1, T_1, \rho_1$, or $c_1$)
- calculate $T_1$ from $T_1 = T_0 - V_1^2/2C_P$
- calculate $c_1$ from $c_1 = \sqrt{\gamma R T_1}$
- calculate $M_1$ from $M_1 = V_1/c_1$
- calculate $p_1$ from $p_1 = p_{01}/\left[1 + (\gamma - 1)M_1^2/2\right]$
- calculate $\rho_1$ from $\rho_1 = p_1/RT_1$
- calculate $\dot{m}$ from $\dot{m} = \rho_1 A V_1$, which is a constant for the flow
- calculate $4\bar{f}_F L_1^*/D$ from equation (13.16), or the Fanno-flow Calculator, corresponding to the Mach number $M_1$, which may be less than or greater than unity

Although in this sequence it is assumed that the calculation starts with specified conditions at the pipe inlet, numerous other possibilities can be envisaged, for example, specified stagnation conditions at the inlet, $p_{01}$ and $T_0$, and a specified exit pressure, $p_E$. An iterative procedure is now required to carry out the calculation: assuming the flow is choked at the pipe exit, such that $p_E = p_1^*$, then, for subsonic flow, a first guess for $M_1$ would lead to $p_1/p_{01}$ from the Isentropic-flow Calculator and to $p_1/p_1^*$ from the Fanno-flow Calculator. We can thus obtain $p_1^*/p_{01} = (p_1/p_{01})/(p_1/p_1^*)$ corresponding to the guessed value for $M_1$. This value can be compared with $p_E/p_{01}$ and the process repeated with a new guess for $M_1$ until the calculated value of $p_1^*/p_{01}$ is sufficiently close to $p_E/p_{01}$.

We now consider separately how to proceed for pipes where the length $L \neq L_1^*$, first for subsonic flow then for initially supersonic flow.

### 13.2.1.1 Subsonic Fanno flow

- If the actual pipe length $L_1 < L_1^*$, calculate the exit Mach number $M_E$ from equation (13.16), or the Fanno-flow Calculator, with $L_E^* = L_1^* - L_1, L_E^*$ being the choking length corresponding with an initial Mach number $M_E$; the procedure is illustrated in Figure 13.3 with $M_1 = 0.35$, and $L_1 = 0.7\, L_1^*$.

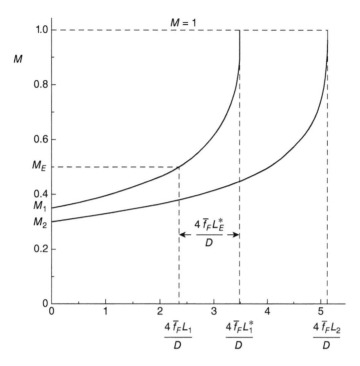

**Figure 13.3** Calculation of flow conditions for subsonic Fanno flow with $L_1 = 0.7 L_1^*$, and $L_2 = 1.5 L_1^*$.

- If the actual pipe length $L_2 > L_1^*$, calculate a new (lower) initial Mach number $M_2$ from equation (13.16) with $L_2^* = L_2$; the procedure is illustrated in Figure 13.3, with $L_2 = 1.5 L_1^*$.

---

### ILLUSTRATIVE EXAMPLE 13.1

Air at a stagnation pressure of 2.5 bar and a stagnation temperature of 600 K flows into a well-insulated 50 mm diameter pipe. The average Fanning friction factor is 0.005. For the specified stagnation pressure and temperature at inlet, calculate (a) the maximum possible pipe length, the mass flowrate, and the flow properties at inlet and exit if the inlet velocity is 120 m/s; (b) the mass flowrate and the flow and fluid properties at inlet and exit, if the pipe length compared with part (a) is reduced by 30% but the inlet flowspeed is still 120 m/s; (c) the mass flowrate and the flow and fluid properties at inlet and exit, if the pipe length is doubled.

### Solution

Since the pipe is well insulated, we assume the flow is adiabatic and so can be treated as a Fanno flow. The subscript $E$ will be used to indicate exit conditions: 1 for the inlet of the full-length pipe and the half-length pipe, 2 for the inlet of the double-length pipe, and 3 for the outlet of the reduced-length pipe. The three flow situations are shown schematically in Figure E13.1, assuming subsonic flow.

(a) $p_{01} = 2.5 \times 10^5$ Pa, $T_0 = 600$ K, $D = 0.05$ m, $V_1 = 120$ m/s, $\bar{f}_F = 0.005$, $\gamma = 1.4$, $R = 287$ m$^2$/s$^2$K, and $C_P = 1004.5$ m$^2$/s$^2$K.

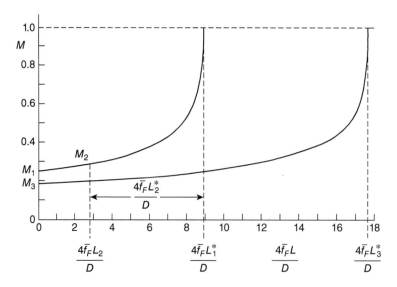

**Figure E13.1** Mach-number variation for subsonic Fanno flow through a full-length, reduced-length, and double-length pipe

For specified inlet conditions, the maximum possible pipe length in subsonic Fanno flow is the choking length $L_1^*$, and $4\bar{f}_F L_1^*/D$ can be evaluated from equation (13.16) once the inlet Mach number $M_1$ is known.

To calculate $M_1$ we start with $C_P T_0 = C_P T_1 + V_1^2/2$, from which the inlet static temperature $T_1$ is given by

$$T_1 = T_0 - V_1^2/2C_P = 592.8\,\text{K}.$$

The corresponding soundspeed $c_1$ is obtained from

$$c_1 = \sqrt{\gamma R T_1} = 488.0\,\text{m/s}$$

and so the inlet Mach number $M_1 = V_1/c_1 = 0.246$, confirming that the flow is subsonic.

The value of $4\bar{f}_F L_1^*/D$ can now be found by calculation from equation (13.16), by interpolation in tables, or from the Fanno-flow Calculator. The simplest method is to use the Calculator, which gives $4\bar{f}_F L_1^*/D = 8.83$ so that $L_1^* = 22.1$ m.

From equation (11.22), or the Isentropic-flow Calculator, $p_1/p_{01} = 0.959$ so that $p_1 = 2.398$ bar. The inlet gas density $\rho_1$ now follows from the perfect-gas equation, $\rho_1 = p_1/RT_1 = 1.409$ kg/m³.

The mass flowrate is then $\dot{m}_1 = \rho_1 A V_1 = 0.332$ kg/s, wherein we have used $A = \pi D^2/4 = 1.963 \times 10^{-3}$ m² for the cross-sectional area.

Since the flow is adiabatic, the stagnation temperature at exit is the same as at inlet, i.e. $T_{0E} = T_0 = 600$ K.

The pipe length $L_1^*$ corresponds to choked flow, i.e. the exit conditions are sonic with $M_E = 1$. So far as the temperature is concerned we thus have $T_{E1}^* = T_0/\left[1 + (\gamma - 1)M_E^2/2\right] = 2T_0/(\gamma + 1) = 500$ K.

The exit soundspeed and flow velocity are then $V_{E1}^* = c_{E1}^* = \sqrt{\gamma R T_{E1}^*} = 448.2$ m/s.

The mass flowrate is unchanged from the inlet so the exit gas density $\rho_{E1}^* = \dot{m}_1/AV_{E1}^* = 0.377$ kg/m$^3$.

The static pressure at exit $p_{E1}^* = \rho_{E1}^* RT_{E1}^* = 5.415 \times 10^4$ Pa or 0.541 bar.

The stagnation pressure at exit $p_{0E1}^* = p_{E1}^* \left[1 + (\gamma - 1) M_{E1}^2/2\right]^{\gamma/(\gamma-1)} = [(\gamma + 1)/2]^{\gamma/(\gamma-1)}$ $p_{E1}^* = 1.024 \times 10^5$ Pa or 1.024 bar.

(b) Since the inlet velocity, stagnation temperature, and stagnation pressure for the reduced-length pipe are the same as for case (a), so are the inlet static temperature, static pressure, soundspeed, and mass flowrate. To find the exit conditions for the reduced-length pipe we consider a choked pipe with exit conditions identical to those of the pipe in part (a) but of length $L_2^* = 0.7 L_1^* = 15.47$ m such that $4\bar{f}_F L_2^*/D = 6.188$. The exit conditions for the reduced-length pipe are the same as the inlet conditions for the choked pipe of length $L_2^*$. The Fanno-flow Calculator gives $M_2 = 0.283$, $T_2/T_{E2}^* = 1.181$, and $p_2/p_{E2}^* = 3.841$. We thus find $T_2 = 590.5$ K, $c_2 = 487.1$ m/s, and $p_2 = 2.078$ bar. The corresponding density $\rho_2 = 1.226$ kg/m$^3$. The velocity at inlet to the short choked pipe is then $V_2 = 137.8$ m/s. Finally, from the Fanno-flow Calculator, $p_{02}/p_{0E2}^* = 2.145$ so that the stagnation pressure at inlet to the choked pipe $p_{02} = 2.196$ bar. The mass flowrate $\dot{m}_2 = \rho_2 AV_2 = 0.332$kg/s, which is the same as in part (a), as it should be.

(c) $p_{03} = 2.5 \times 10^5$ Pa, $T_0 = 600$ K, $D = 0.05$ m, $\bar{f}_F = 0.005$, $\gamma = 1.4$, $R = 287$ m$^2$/s$^2$K, $C_P = 1004.5$ m$^2$/s$^2$K, and $L_3^* = 2 L_1^*$.

The new pipe length is $L_3 = L_3^* = 2 \times 22.1 = 44.2$ m so that $4\bar{f}_F L_3^*/D = 17.68$. Using tables to find the corresponding inlet Mach number $M_3$ requires interpolation while to find $M_3$ from equation (13.14) necessitates iteration. The simplest way of determining $M_3$ is again from the Fanno-flow Calculator, which gives $M_3 = 0.184$.

From the Isentropic-flow section of the Calculator, with $M_3 = 0.184$, we find $p_3/p_{03} = 0.977$ so that $p_3 = 0.977 \times 2.5 \times 10^5 = 2.443 \times 10^5$ Pa.

Also from the Isentropic-flow Calculator, $T_3/T_0 = 0.993$ so that $T_3 = 0.993 \times 600 = 595.8$ K. The corresponding soundspeed $c_3 = \sqrt{\gamma RT_3} = 489.3$ m/s.

The flow velocity is then $V_3 = M_3 c_3 = 90.0$ m/s.

The gas density at inlet is $\rho_3 = p_3/RT_3 = 1.429$ kg/s.

The new mass flowrate $\dot{m}_3 = \rho_3 AV_3 = 1.429 \times 1.96 \times 10^{-3} \times 90.0 = 0.252$ kg/s.

## Comment:

Doubling the pipe length reduced the inlet velocity, the inlet Mach number, and the mass flowrate by about 25%.

---

### Initially supersonic Fanno flow

- If the actual pipe length $L_3 < L_1^*$, the procedure is the same as for subsonic flow.
- If the actual pipe length $L_3 > L_1^*$, use the iterative procedure of Illustrative Example 13.2 to determine the normal-shock position and all flow quantities.

Figure 13.4 illustrates the situation of adiabatic pipe flow of air with an initial Mach number $M_1 = 4$: (a) supersonic flow throughout for a pipe with $4\bar{f}_F L_1^*/D = 0.633$, where $L_1^*$ is

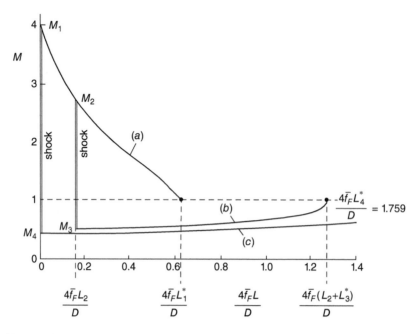

**Figure 13.4** Mach number variation along a pipe for initially supersonic Fanno flow with (a) $L = L_1^*$, (b) $L = 2\ L_1^*$ with a normal shock at $L_2$ where $M_2 = 2.70$, and (c) $L_4^* = 2.78L_1^*$ with a normal shock at the pipe inlet. In all cases the initial Mach number = 4.0.

the choking length corresponding to $M_1$; (b) a pipe of length $2\ L_1^*$ with a normal shock at $4\bar{f}_F L_2/D = 0.161$ (this location was found using the trial-and-error procedure outlined in Illustrative Example 13.2) with $M_2 = 2.70$ ahead of the shock, and $M_3 = 0.496$ behind the shock, such that $L_2 + L_3^* = 2\ L_1^*$; and (c) a pipe of length 2.78 $L_1^*$, which causes a normal shock at the pipe inlet with $M_1 = 4$, and $M_4 = 0.4$ just behind the shock, for which $4\bar{f}_F L_4^*/D = 1.759$, i.e. $L_4^* = 2.78\ L_1^*$. For a pipe with a length greater than $L_4^*$, the initial Mach number would have to reduce below 4.

---

### ILLUSTRATIVE EXAMPLE 13.2

Air at a stagnation pressure of 2.5 bar and a stagnation temperature of 600 K flows into a well-insulated 50 mm diameter pipe at a velocity of 700 m/s. The average Fanning friction factor is 0.005. Calculate (a) the maximum possible pipe length corresponding to these inlet conditions, the mass flowrate, the inlet and exit flow, and the fluid properties; (b) the mass flowrate and the inlet and exit conditions, if the pipe length is reduced by 30%; and (c) the mass flowrate and the inlet and exit conditions, if the pipe length is increased by 20%.

### Solution

As in Illustrative Example 13.1, since the pipe is well insulated, we shall assume the flow is adiabatic and can be treated as a Fanno flow. Key locations in the flow are shown in Figure E13.2, assuming supersonic flow at the inlet. It is suggested that the solution procedure is more easily followed with reference to this figure. The subscript 1 will be used to indicate conditions at the inlet, 2 for the outlet of the reduced-length pipe, and 3 for the outlet of the increased-length pipe. The subscripts 4 and 5 indicate the location of the shockwave, referring to the upstream supersonic and downstream subsonic sides of the shock, respectively.

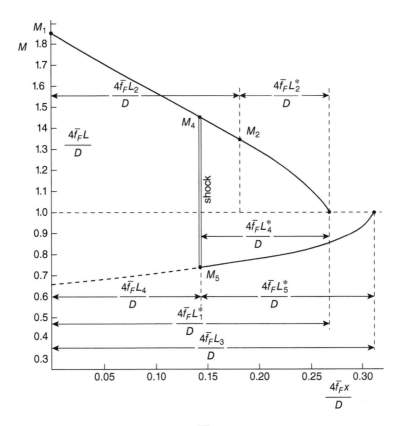

**Figure E13.2** Variation of Mach number with $4\bar{f}_F x/D$ for initially supersonic Fanno flow through (a) full-length, (b) increased-length, and (c) reduced-length pipe

(a) > $p_{01}$ = $2.5 \times 10^5$ Pa, $T_0$ = 600 K, $D$ = 0.05 m, $V_1$ = 700 m/s, $\bar{f}_F$ = 0.005, $\gamma = 1.4$, $R = 287$ m$^2$/s$^2$K, and $C_P = 1004.5$ m$^2$/s$^2$K.

For specified inlet conditions, the maximum possible pipe length in shock-free supersonic Fanno flow is the choking length $L_1^*$ where $4\bar{f}_F L_1^*/D$ can be evaluated from equation (13.16) once the inlet Mach number $M_1$ is known.

To find $M_1$ we start with $C_P T_0 = C_P T_1 + V_1^2/2$, and $V_1 = 700$ m/s, from which the inlet static temperature $T_1$ is given by

$$T_1 = T_0 - V_1^2/2C_P = 356.1 \text{ K}.$$

The corresponding soundspeed $c_1$ is obtained from

$$c_1 = \sqrt{\gamma R T_1} = 378.3 \text{ m/s}$$

so the inlet Mach number $M_1 = V_1/c_1 = 1.851$, and the flow at inlet is confirmed to be supersonic.

The value of $4\bar{f}_F L_1^*/D$ can be found by calculation from equation (13.16), by interpolation in tables, or from the Fanno-flow Calculator. As before, the simplest method is to use the Calculator, which gives $4\bar{f}_F L_1^*/D = 0.259$ so that $L_1^* = 0.647$ m.

From equation (11.22), or the Isentropic-flow Calculator, $p_1/p_{01} = 0.161$ so that $p_1 = 0.403$ bar.

The inlet gas density $\rho_1$ is found from the perfect-gas equation, $\rho_1 = p_1/RT_1 = 0.394$ kg/m$^3$.

The mass flowrate $\dot{m}_1 = \rho_1 AV_1 = 0.542$ kg/s, wherein we used $A = \pi D^2/4 = 1.963 \times 10^{-3}$ m$^2$.

The exit (choked) conditions are best found using the Fanno-flow Calculator, which gives $T_1/T_1^* = 0.712$ so that $T_1^* = 500.1$ K, $p_1/p_1^* = 0.456$ so that $p_1^* = 0.884$ bar, and $p_{01}/p_{01}^* = 1.496$ so that $p_{01}^* = 1.671$ bar.

The exit velocity is given by $V_1^* = c_1^* = \sqrt{\gamma R T_1^*} = 448.3$ m/s.

From the perfect-gas equation, the exit density is $\rho_1^* = p_1^*/RT_1^* = 0.616$ kg/m$^3$, and the mass flowrate $\dot{m} = \rho_1^* A V_1^* = 0.542$ kg/s, i.e. precisely the same as at inlet, as it should be, and so provides a check on the accuracy of the other calculations.

(b) Since the inlet velocity, stagnation temperature and stagnation pressure for the reduced-length pipe are the same as for case (a), so are the inlet static temperature, static pressure, soundspeed, and mass flowrate.

To find the exit conditions for the reduced-length pipe we consider a choked pipe with exit conditions identical to those of the pipe in part (a) (i.e. $V_2^* = V_1^*, T_2^* = T_1^*$, etc.) and of length $L_2^* = 0.3\, L_1^* = 0.194$ m such that $4\bar{f}_F L_2^*/D = 0.0776$.

For this value of $\bar{f}_F L_2^*/D$, the Fanno-flow Calculator gives $M_2 = 1.338$, $T_2/T_2^* = 0.884$, and $p_2/p_2^* = 0.703$. We thus find $T_2 = 442.1$ K, $c_2 = 421.5$ m/s, and $p_2 = 0.621$ bar.

From the perfect-gas equation, the corresponding density $\rho_2 = 0.490$ kg/m$^3$.

The velocity at inlet to the short choked pipe is $V_2 = M_2 c_2 = 564.0$ m/s.

Finally, from the Fanno-flow Calculator, $p_{02}/p_{02}^* = 1.083$ so that the stagnation pressure at inlet $p_{02} = 1.810$ bar.

The exit conditions for the reduced-length pipe must be the same as those for the short choked pipe while the mass flowrate $\dot{m} = \rho_2 AV_2$ is unchanged from 0.542 kg/s, as in part (a).

(c) $p_{02} = 2.5 \times 10^5$ Pa, $T_0 = 600$ K, $D = 0.05$ m, $V_1 = 700$ m/s, $\bar{f}_F = 0.005$, $\gamma = 1.4$, $R = 287$ m$^2$/s$^2$K, $C_P = 1004.5$ m$^2$/s$^2$K.

The pipe length is now 20% greater than $L_1^*$, i.e. $L_3 = 1.2\, L_1^* = 0.776$ m or $4\bar{f}_F L_3/D = 0.311$. Due to this increase in pipe length, there will be a shock a distance $L_4$, to be determined, from the inlet such that $4\bar{f}_F L_4/D < 4\bar{f}_F L_1^*/D \,(= 0.259)$.

We shall adopt the following trial-and-error procedure to find the shock location relative to the exit, $L_3 - L_4 = L_5^*$

- guess $4\bar{f}_F L_4/D$
- from the Fanno-flow Calculator, calculate the subsonic Mach number $M_5$ just behind the shock corresponding to $L_5^*$
- from the Normal-shock Calculator, calculate the supersonic Mach number $M_4$ just ahead of the shock corresponding to $M_5$

- from the Fanno-flow Calculator, calculate the choking length $L_4^*$ corresponding to $M_4$
- calculate $L_5^* + L_4$, the pipe length for an inlet Mach number $M_1$ with a shock a distance $L_4$ from the inlet and choked at the exit
- if $\bar{f}_F(L_5^* + L_4)/D > 1.2\bar{f}_F L_1^*/D$, we need to reduce $L_4$ and repeat the calculation until we find $\bar{f}_F(L_5^* + L_4)/D = 1.2\bar{f}_F L_1^*/D = 0.311$ within, say, $\pm 1\%$. If $\bar{f}_F(L_5^* + L_4)/D < 1.2\bar{f}_F L_1^*/D$ we need to increase $L_4$ and repeat the calculation. The logic of this sequence should be apparent from Figure E13.2. The results are set out in Table E13.2.

**Table E13.2** Trial-and-error procedure for finding the shock location relative to the exit, $L_3 - L_4 = L_5^*$

| $4\bar{f}_F L_4/D$ | $4\bar{f}_F(L_3 - L_4)/D = 4\bar{f}_F L_5^*/D$ | $M_5$ | $M_4$ | $4\bar{f}_F L_4^*/D$ | $4\bar{f}_F(L_4^* + L_5)/D$ |
|---|---|---|---|---|---|
| 1st guess 0.04 | 0.271 | 0.671 | 1.592 | 0.169 | 0.209, i.e. < 0.259 |
| 2nd guess 0.10 | 0.211 | 0.699 | 1.507 | 0.139 | 0.239, i.e. < 0.259 |
| 3rd guess 0.20 | 0.111 | 0.763 | 1.348 | 0.081 | 0.281, i.e. > 0.259 |
| 4th guess 0.15 | 0.161 | 0.727 | 1.432 | 0.111 | 0.261, i.e. > 0.259 |
| 5th guess 0.14 | 0.171 | 0.721 | 1.447 | 0.117 | 0.257, i.e. < 0.259 |
| 6th guess 0.145 | 0.166 | 0.724 | 1.439 | 0.114 | 0.259, i.e. = 0.259 |

The shock location is thus obtained from $4\bar{f}_F L_4/D = 0.145$ so that $L_4 = 0.363$ m or 0.413 m from the pipe exit. The Mach number just ahead of the shock $M_4 = 1.439$, and the Mach number just behind the shock $M_4 = 0.724$. We can now calculate all flow quantities on either side of the shock and at the pipe exit, as follows.

Just before the shock, we have supersonic Fanno flow, just as in part (b). From the supersonic Fanno-flow Calculator, for $M_4 = 1.439$ we find $T_4/T_1^* = 0.849$, $p_4/p_1^* = 0.640$, $p_{04}/p_{01}^* = 1.137$, $V_4/V_1^* = 1.326$, so that $T_4 = 0.849 \times 500.1 = 424.4$ K, $p_4 = 0.640 \times 0.884 = 0.566$ bar, $p_{04} = 1.137 \times 1.671 = 1.900$ bar, and $V_4 = 1.326 \times 448.3 = 594.3$ m/s.
From the perfect-gas equation, $\rho_4 = p_4/RT_4 = 0.465$ kg/m$^3$.
To find conditions just behind the shock, we use the Normal-shock Calculator with $M_4 = 1.439$, which gives $M_5 = 0.724$, $T_5/T_4 = 1.280$, $p_5/p_4 = 2.249$, $p_{05}/p_{04} = 0.948$, and $\rho_5/\rho_4 = 1.757$, so that $T_5 = 1.280 \times 424.4 = 543.2$ K, $p_5 = 2.249 \times 0.566 = 1.273$ bar, $p_{05} = 0.948 \times 1.900 = 1.801$ bar, and $\rho_5 = 1.757 \times 0.465 = 0.816$ kg/m$^3$.
Since $G = \rho_4 V_4 = \rho_5 V_5$ we find $V_5 = \rho_4 V_4/\rho_5 = 338.2$ m/s.
Finally, we use the subsonic Fanno-flow Calculator to find the exit conditions for the increased-length pipe. For $M_5 = 0.724$ we have $T_5/T_5^* = 1.086$, $p_5/p_5^* = 1.439$, $p_{05}/p_{05}^* = 1.078$, and $V_5/V_5^* = 0.755$, so that $T_5^* = 543.2/1.086 = 500.1$ K, $p_5^* = 1.273/1.439 = 0.884$ bar, $p_{05}^* = 1.801/1.078 = 1.671$ bar, and $V_5^* = 338.2/0.755 = 448.2$ m/s.
From the perfect-gas equation, $\rho_5^* = p_5^*/RT_5^* = 0.616$ kg/m$^3$.
As a final check, we calculate the mass flowrate $\dot{m} = \rho_5^* A V_5^* = 0.616 \times 1.963 \times 10^{-3} \times 448.2 = 0.542$ kg/s, as before.

**Comments:**

The exit temperatures for both the full-length and increased-length pipes is the same, 500.1 K, because the stagnation temperature remained constant throughout and at the exit the Mach number was unity in both cases.

It is strongly recommended that a clear schematic diagram be drawn prior to carrying out any calculation. The value of such a diagram in this case should be apparent.

___

### 13.2.2 Fanno-flow trends

The trends in all the flow variables for both subsonic and supersonic flow are most clearly revealed by differential versions[100] of the equations for $V, \rho, p, c, T, p_0$, and $s$, arranged to include $\bar{f}_F$

$$\frac{1}{V}\frac{dV}{dx} = \frac{-1}{\rho}\frac{d\rho}{dx} = \frac{\gamma M^2}{2\left(1 - M^2\right)}\frac{4\bar{f}_F}{D} \tag{13.18}$$

$$\frac{1}{p}\frac{dp}{dx} = -\frac{\gamma M^2 \left[1 + (\gamma - 1) M^2\right]}{2\left(1 - M^2\right)}\frac{4\bar{f}_F}{D} \tag{13.19}$$

$$\frac{1}{c^2}\frac{dc^2}{dx} = \frac{1}{T}\frac{dT}{dx} = -\frac{\gamma\,(\gamma - 1)\,M^4}{2\left(1 - M^2\right)}\frac{4\bar{f}_F}{D} \tag{13.20}$$

$$\frac{1}{p_0}\frac{dp_0}{dx} = -\frac{\gamma M^2}{2}\frac{4\bar{f}_F}{D} \tag{13.21}$$

$$\frac{1}{R}\frac{ds}{dx} = -\frac{1}{p_0}\frac{dp_0}{dx} = \frac{\gamma M^2}{2}\frac{4\bar{f}_F}{D}. \tag{13.22}$$

Based upon these differential equations for the flow variables, the changes that occur in adiabatic, frictional pipe flow of a perfect gas are summarised in Table 13.1.

### 13.2.3 Fanno line

We return at this point to the energy equation, equation (13.3), which, for Fanno flow, becomes

$$C_P\frac{dT_0}{dx} = C_P\frac{dT}{dx} + V\frac{dV}{dx} = 0 \tag{13.23}$$

or

$$\frac{dh_0}{dx} = \frac{dh}{dx} + V\frac{dV}{dx} = 0 \tag{13.24}$$

___

[100] The derivation of the differential equations for Fanno flow is given in Appendix 4(a).

**Table 13.1** Property changes for adiabatic, frictional pipe flow of a perfect gas under subsonic and supersonic conditions

|  | Subsonic | Supersonic |
| --- | --- | --- |
| static pressure $p$ | decreases | increases |
| static temperature, $T$ and soundspeed $c$ | decreases | increases |
| density, $\rho$ | decreases | increases |
| velocity, $V$ | increases | decreases |
| Mach number, $M$ | increases | decreases |
| stagnation pressure, $p_0$ | decreases | decreases |
| specific entropy, $s$ | increases | increases |

where $h$ is the specific enthalpy. Equation (13.24) can be integrated to give

$$h + \frac{1}{2}V^2 = h_0 = \text{constant},$$ (13.25)

where the specific stagnation enthalpy $h_0$ is constant since we are dealing with an adiabatic flow (the heat-transfer rate per unit length, $\dot{q}_S'$, in equation (13.3) was set equal to zero). From the continuity equation

$$\dot{m} = \rho A V = AG = \text{constant},$$ (6.1)

so that if we substitute $V = Gv$ in equation (13.25) we have

$$h + \frac{(Gv)^2}{2} = h_0 = \text{constant}$$ (13.26)

where $v = 1/\rho$ is the specific volume. Since $G$ and $h_0$ are constants for a given flow, this equation defines a relation between the local specific enthalpy $h$ and the local specific volume, termed the **Fanno line**. By dividing through the equation by $h_0$ we obtain the non-dimensional form

$$\frac{h}{h_0} + \frac{(Gv)^2}{2h_0} = 1$$ (13.27)

which is shown graphically in Figure 13.5(a). The local speed of sound is given by $c^2 = \gamma RT$, and also $R = c_P - c_V = (\gamma - 1)c_P/\gamma$ so that $c^2 = (\gamma - 1)c_P T = (\gamma - 1)h$. From equation (13.25) have $V^2 = 2(h_0 - h)$, so that at the sonic point $2(h_0 - h^*) = (\gamma - 1)h^*$, which leads to $h^*/h_0 = 2/(\gamma + 1)$. The dashed horizontal line in Figure 13.5(a) corresponds to this value of $h/h_0$ for $\gamma = 1.4$.

Fanno lines are more usually presented on a **Mollier diagram** of specific enthalpy $h$ versus specific entropy $s$. Since changes in specific entropy are given by

$$s - s_1 = C_P \ln\left(\frac{T}{T_1}\right) - R \ln\left(\frac{p}{p_1}\right)$$ (11.2)

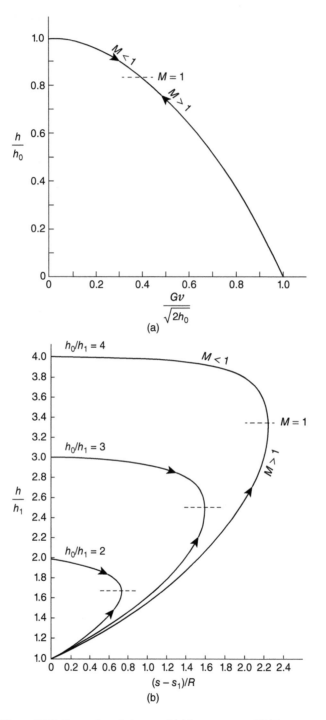

**Figure 13.5** Fanno lines in terms of (a) $h$ versus $v$ and (b) $h$ versus $s$

where $s_1$, $T_1$, and $p_1$ are reference values (e.g. at an upstream starting location), making use of the perfect-gas equation, $p = \rho RT = RT/v$, we have

$$s - s_1 = (C_P - R)\ln\left(\frac{T}{T_1}\right) + R\ln\left(\frac{v}{v_1}\right). \tag{13.28}$$

From equation (13.27)

$$\left(\frac{v}{v_1}\right)^2 = \frac{h_0 - h}{h_0 - h_1} = \frac{T_0 - T}{T_0 - T_1} \tag{13.29}$$

wherein we have used the enthalpy-temperature relation, $h = C_P T$ +constant. If the specific enthalpy is referenced to a zero value at zero absolute temperature, the constant is zero and we have $T/T_1 = h/h_1$. Substitution for $v/v_1$ and $T/T_1$ in equation (13.28), together with the relations $R = C_P - C_V$, and $\gamma = C_P/C_V$, then leads to

$$\frac{s - s_1}{R} = \left(\frac{1}{\gamma - 1}\right)\ln\left(\frac{h}{h_1}\right) + \frac{1}{2}\ln\left(\frac{h_0/h_1 - h/h_1}{h_0/h_1 - 1}\right) \tag{13.30}$$

which, for a given gas, can be plotted on the Mollier diagram in the non-dimensional form $h/h_1$ versus $(s - s_1)/R$ for specific values of the initial value $h_1/h_0$, as in Figure 13.5(b).

Figure 13.5(b) shows that each $h - s$ curve has a maximum, indicated by the short dashed lines. To find what condition the maximum corresponds to, we differentiate $s$ with respect to $h$ and set the result equal to zero

$$\frac{1}{R}\frac{ds}{dh} = \left(\frac{1}{\gamma - 1}\right)\frac{1}{h} - \frac{1}{2(h_0 - h)} = 0 \tag{13.31}$$

which leads to

$$\frac{h}{h_0} = \frac{2}{\gamma + 1}. \tag{13.32}$$

Since $h + V^2/2 = h_0$ = constant, we find that the peak corresponds with $V^2 = (\gamma - 1)h$. Since $h = C_P T$, we find $V^2 = \gamma RT = c^2$, i.e. $M = 1$. In other words, the choking or sonic condition corresponds to maximum entropy.

## 13.3 Isothermal pipe flow with wall friction

Gas flow through a long, buried, uninsulated pipeline is probably better represented as isothermal flow with wall friction[101] than the adiabatic flow just discussed. The starting point for the analysis of such a flow is the same as for the adiabatic-flow situation except that now we have the important simplification

$$T = \text{constant} = T^{*T} \tag{13.33}$$

The superscript $*T$ is used here to identify properties which correspond to the choking condition for isothermal pipe flow of a perfect gas: as we shall see (equation (13.39)), in contrast to

---

[101] Sometimes referred to as **isothermal Fanno flow**.

the adiabatic flow considered in Section 13.2, the **isothermal choking condition** corresponds to $M = M^{*T} = V^{*T}/c = 1/\sqrt{\gamma}$, i.e. $M^{*T} < 1$.

Since the temperature, and so the soundspeed $c$, is constant, the mass-conservation equation (6.1) reduces to

$$\frac{\rho^{*T}}{\rho} = \frac{V}{V^{*T}} = \frac{V}{c}\frac{c}{V^{*T}} = \sqrt{\gamma}M. \tag{13.34}$$

From equation (13.34), with the perfect-gas equation (2.9),

$$\frac{p}{p^{*T}} = \frac{\rho}{\rho^{*T}} = \frac{1}{\sqrt{\gamma}M}. \tag{13.35}$$

The stagnation pressure $p_0$ is given by

$$p_0 = p\left[1 + \left(\frac{\gamma-1}{2}\right)M^2\right]^{\frac{\gamma}{\gamma-1}} \tag{11.22}$$

so that, from equation (13.35),

$$\frac{p_0}{p_0^{*T}} = \frac{1}{\sqrt{\gamma}M}\left\{\left(\frac{2\gamma}{3\gamma-1}\right)\left[1 + \left(\frac{\gamma-1}{2}\right)M^2\right]\right\}^{\frac{\gamma}{\gamma-1}}. \tag{13.36}$$

Since $T = $ constant,

$$\frac{s - s^{*T}}{C_P} = -\left(\frac{\gamma-1}{\gamma}\right)\ln\left(\frac{p}{p^{*T}}\right) = \left(\frac{\gamma-1}{\gamma}\right)\ln\left(\sqrt{\gamma}M\right) \tag{13.37}$$

and

$$\frac{T_0}{T_0^{*T}} = \frac{2\gamma}{(3\gamma-1)}\left[1 + \left(\frac{\gamma-1}{2}\right)M^2\right]. \tag{13.38}$$

In deriving equations (13.34) to (13.38) we made no use of the momentum-conservation equation

$$-\frac{dp}{dx} - \frac{2\rho V^2 f_F}{D} = \rho V\frac{dV}{dx}. \tag{13.2}$$

Substitution for $p$, $V$, and $\rho$ in equation (13.2), using equations (13.34) and (13.35), leads to

$$\frac{1}{M^2}\frac{dM^2}{dx} = \frac{\gamma M^2}{(1-\gamma M^2)}\frac{4f_F}{D}. \tag{13.39}$$

Again, as for the adiabatic-flow situation, if we take an average value for the friction factor, $\bar{f}_F$, then equation (13.39) can be integrated to give

$$\frac{4\bar{f}_F L^{*T}}{D} = \frac{1 - \gamma M^2}{\gamma M^2} + \ln\left(\gamma M^2\right). \tag{13.40}$$

In Section 13.2, which was concerned with adiabatic (Fanno) flow, we concluded that, for both subsonic and supersonic flow, the influence of friction causes the Mach number to approach unity. Equation (13.39) reveals that for an isothermal flow, if $M < 1/\sqrt{\gamma}$, $M$ increases whereas, if $M > 1/\sqrt{\gamma}$, $M$ decreases, so that due to friction the Mach number in isothermal pipe flow of a perfect gas always tends towards the value $1/\sqrt{\gamma}$. As for Fanno flow, if the pipe is sufficiently long, after a certain length $L_{MAX}$, depending upon the value of $\bar{f}_F$, the pipe diameter, and the initial Mach number $M$ at $s = 0$, the flow becomes choked with $M \to 1/\sqrt{\gamma}$.

However, for isothermal subsonic flow the energy-conservation equation (13.3) reveals that a practical limitation is imposed by the heat-transfer rate required to achieve choking of an isothermal flow

$$\dot{q}' = \dot{m}C_P\frac{dT_0}{dx} = \dot{m}V\frac{dV}{dx} = GAV\frac{dV}{dx}. \tag{13.41}$$

If we use equation (13.34) to substitute for $V$, we find

$$\dot{q}' = \frac{GAV^2}{2M^2}\frac{dM^2}{dx} = \frac{GAc^2}{2}\frac{dM^2}{dx}. \tag{13.42}$$

We can now substitute for $dM^2/dx$ from equation (13.39), with $f_F$ replaced by $\bar{f}_F$, to find

$$\dot{q}' = \frac{\gamma GAc^2M^4}{2\left(1 - \gamma M^2\right)}\frac{4\bar{f}_F}{D} \tag{13.43}$$

so that, as $M \to 1/\sqrt{\gamma}$, $\dot{q}' \to \infty$, i.e. the heat-transfer rate required to maintain isothermal conditions becomes infinite. Equation (13.43) also shows that, for subsonic flow with $M < 1/\sqrt{\gamma}$, the heat-transfer rate to maintain isothermal flow is positive, i.e. heating is necessary, whereas, for $M > 1/\sqrt{\gamma}$, the flow must be cooled.

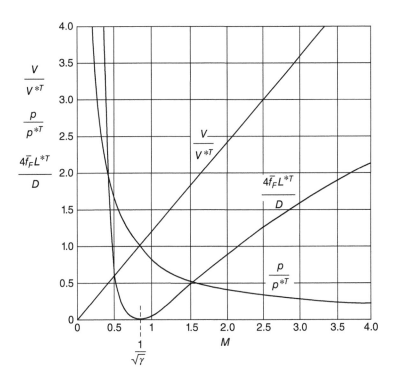

**Figure 13.6** Variation of flow properties with Mach number for isothermal flow of a perfect gas with $\gamma = 1.4$

Values for $4\bar{f}_F L^{*T}/D$ for any initial Mach number $M$, together with the ratios $V/V^{*T}$, $\rho/\rho^{*T}$, $p/p^{*T}$, $T_0/T_0^{*T}$, and $(s - s^{*T})/R$, can be calculated from the equations above, but are yet to be incorporated in the Virginia Tech Compressible Aerodynamics Calculator. Tabulated values are given in the textbook by Chapman and Walker, (1971). Figure 13.6 shows the variation with Mach number of flow properties according to the equations derived above.

---

**ILLUSTRATIVE EXAMPLE 13.3**

Air is to be pumped through a pipe of 0.5 m I.D. and 100 m in length. The inlet pressure is 7 bar and there is sufficient heat transfer through the wall to maintain the gas at 20 °C. The value of the Fanning friction factor for this flow is 0.005. Calculate the maximum possible mass flowrate and the outlet pressure.

**Solution**

Since the flowrate is a maximum, the flow must be isothermally choked at the entry to the second compressor, i.e. the pipe length $L$ must correspond with the isothermal choking length for the inlet Mach number. With $L = 100 \, \text{m} = L_1^{*T}$ we have $4\bar{f}_F L_1^{*T}/D = 4$, from which we can calculate the inlet Mach number $M_1$ using equation (13.40). This requires a trial-and-error process. Omitting the logarithmic term provides a first guess for $\gamma M_1^2$, i.e. $4\bar{f}_F L_1^{*T}/D \cong (1 - \gamma M_1^2)/\gamma M_1^2$, from which $\gamma M_1^2 \cong 0.2$.
The trial-and-error steps are set out in Table E13.3.

**Table E13.3** Trial-and-error steps for calculating the inlet Mach number $M_1$

|  | $\gamma M_1^2$ | $(1 - \gamma M_1^2)/\gamma M_1^2$ | $\ln(\gamma M^2)$ | $4\bar{f}_F L_1^{*T}/D$ |
|---|---|---|---|---|
| 1st guess | 0.20 | 4.00 | −1.60 | 2.40 |
| 2nd guess | 0.15 | 5.67 | −1.90 | 3.77 |
| 3rd guess | 0.14 | 6.14 | −1.97 | 4.17 |
| 4th guess | 0.145 | 5.90 | −1.93 | 3.97 |
| 5th guess | 0.144 | 5.94 | −1.94 | 4.00 |

With $\gamma M_1^2 \cong 0.144$ we have $M_1 = 0.321$.
The soundspeed is $c = \sqrt{\gamma RT} = 343.1$ m/s so that the inlet gas velocity $V_1 = M_1 c = 110.1$ m/s.
From the perfect-gas equation the gas density at inlet $\rho_1 = p_1/RT = 8.324 \text{kg/m}^3$ so that the mass flowrate $\dot{m} = \rho_1 A V_1 = 180$ kg/s.
The outlet density is given by $\rho_2 = \dot{m}/A V_2$. Since the flow is isothermally choked at the outlet, $M_2 = 1/\sqrt{\gamma} = 0.845$, from which $V_2 = M_2 c = 290$ m/s. The outlet density is thus 3.161 kg/m³. The outlet pressure is given by the perfect-gas equation $p_2 = \rho_2 RT = 2.658 \times 10^5$ Pa or 2.658 bar.

---

### 13.3.1 Isothermal pipe-flow trends

As for adiabatic flow, first-order ordinary differential equations can be derived[102] for $V, \rho, p, c_0, T_0, p_0, s$, and $f_F$ (note that since the flow is isothermal, $c_0$ appears instead of $c$ and $T_0$ instead of $T$) in terms of $M$ to reveal more clearly the trends for $M < 1/\sqrt{\gamma}$ and for $M > 1/\sqrt{\gamma}$:

$$\frac{1}{V}\frac{dV}{dx} = -\frac{1}{\rho}\frac{d\rho}{dx} = -\frac{1}{p}\frac{dp}{dx} = \frac{1}{2M^2}\frac{dM^2}{dx} = \frac{\gamma M^2}{2\left(1 - \gamma M^2\right)}\frac{4\bar{f}_F}{D} \tag{13.44}$$

$$\frac{1}{c_0^2}\frac{dc_0^2}{dx} = \frac{1}{T_0}\frac{dT_0}{dx} = \frac{\left(\frac{\gamma - 1}{2}\right)M^2}{\left[1 + \left(\frac{\gamma - 1}{2}\right)M^2\right]}\frac{1}{M^2}\frac{dM^2}{dx} = \frac{\gamma\left(\gamma - 1\right)M^4}{2\left(1 - \gamma M^2\right)\left[1 + \left(\frac{\gamma - 1}{2}\right)M^2\right]}\frac{4\bar{f}_F}{D} \tag{13.45}$$

$$\frac{1}{p_0}\frac{dp_0}{dx} = \frac{\gamma M^2\left[1 - \left(\frac{\gamma + 1}{2}\right)M^2\right]}{2\left(\gamma M^2 - 1\right)\left[1 + \left(\frac{\gamma - 1}{2}\right)M^2\right]}\frac{4\bar{f}_F}{D} \tag{13.46}$$

and

$$\frac{1}{R}\frac{ds}{dx} = \frac{\gamma M^2}{2}\frac{4\bar{f}_F}{D}. \tag{13.47}$$

together with

$$\frac{4\bar{f}_F}{D} = \frac{\left(1 - \gamma M^2\right)}{\gamma M^2}\frac{1}{M^2}\frac{dM^2}{dx}. \tag{13.48}$$

Based upon the equations developed here, the various changes that occur in isothermal, frictional pipe flow are summarised in Table 13.2.

**Table 13.2** Property changes for isothermal, frictional pipe flow for $M < 1/\sqrt{\gamma}$ and for $M > 1/\sqrt{\gamma}$

| | $M < 1/\sqrt{\gamma}$ (subsonic) | $M > 1/\sqrt{\gamma}$ (subsonic/supersonic) |
|---|---|---|
| static pressure, $p$ | decreases | increases |
| density, $\rho$ | decreases | increases |
| velocity, $V$ | increases | decreases |
| Mach number, $M$ | increases | decreases |
| stagnation temperature, $T_0$ | increases | decreases |
| stagnation pressure, $p_0$ | | $\begin{cases} \text{increases for } M < \sqrt{2(\gamma + 1)} \\ \text{decreases for } M > \sqrt{2(\gamma + 1)} \end{cases}$ |
| specific entropy, $s$ | increases | increases |

---

[102] The derivation of the differential equations for isothermal pipe flow is given in Appendix 4(a).

### 13.3.2  Static-pressure change in isothermal pipe flow

A useful result which can be obtained for isothermal pipe flow involves the static-pressure change, from $p_1$ to $p_2$, over a length $L$. We start with the perfect-gas equation $p = \rho RT = GRT/V$ combined with $G = \rho V$ so that

$$\frac{1}{p}\frac{dp}{dx} = -\frac{1}{V}\frac{dV}{dx}.$$  (13.49)

We can write the momentum equation (13.2) as follows

$$-\frac{1}{GV}\frac{dp}{dx} - \frac{1}{V}\frac{dV}{dx} = \frac{2\bar{f}_F}{D}.$$  (13.50)

If we now substitute in this equation for $(1/V)\,dV/dx$ from equation (13.49) we have

$$-\frac{1}{GV}\frac{dp}{dx} + \frac{1}{p}\frac{dp}{dx} = \frac{2\bar{f}_F}{D}$$  (13.51)

which leads to

$$-\frac{p}{G^2RT}\frac{dp}{dx} + \frac{1}{p}\frac{dp}{dx} = \frac{2\bar{f}_F}{D}$$  (13.52)

wherein we have made use of the relationships $V = G/\rho$, and $p = \rho RT$. Since $G, R, T$, and $f_F$ are all constant, equation (13.52) can be integrated to give

$$p_1^2 - p_2^2 - 2G^2RT\ln\left(\frac{p_1}{p_2}\right) = G^2RT\frac{4\bar{f}_F L}{D}.$$  (13.53)

Assuming $R, T, D$, and $\bar{f}_F$ are all known, equation (13.53) can be used in various ways, for example, to find

- the initial pressure $p_1$ if the exit pressure $p_2$ and pipe length $L$ are known
- the exit pressure $p_2$ if the initial pressure $p_1$ and pipe length $L$ are known
- the pipe length $L$ if the initial and exit pressures $p_1$ and $p_2$ are both known
- the mass flowrate $\dot{m} = GA$ if the pipe length $L$ and the initial and exit pressures $p_1$ and $p_2$ are all known

---

**ILLUSTRATIVE EXAMPLE 13.4**

Methane gas ($\gamma = 1.31$, gas constant 518.4 J/kg · K) is to be pumped through a pipe of 1 m I.D. connecting two compressor stations 90 km apart. At the upstream station the pressure is not to exceed 7 bar and at the downstream station it is to be at least 1.7 bar. Calculate the maximum allowable volume flowrate (at 20 °C and 1 bar), assuming there is sufficient heat transfer through the wall to maintain the gas at 20 °C. The value of the Fanning friction factor for this flow is 0.005.

## Solution

The simplest way to solve this problem is to use equation (13.53)

$$p_1^2 - p_2^2 - 2G^2RT\ln\left(\frac{p_1}{p_2}\right) = G^2RT\frac{4\bar{f}_F L}{D}$$

which can be rearranged to give

$$G^2 = \frac{p_1^2 - p_2^2}{RT\left[2\ln\left(\frac{p_1}{p_2}\right) + \frac{4\bar{f}_F L}{D}\right]}.$$

We have $R = 518.4$ J/kg·K, $p_1 = 7 \times 10^5$ Pa, $p_2 = 1.7 \times 10^5$ Pa, $T = 293$ K, $L = 9 \times 10^4$ m, $D = 1$ m, and $\bar{f}_F = 5 \times 10^{-3}$. Substitution of these values into the equation for $G^2$ gives $G = 41.03$ kg/ms so that $\dot{m} = GA = 32.2$ kg/s, wherein the cross-sectional area $A = \pi D^2/4 = 0.785$ m². To find the volumetric flowrate $\dot{Q} = \dot{m}/A$ we need the inlet gas density $\rho_1$, which can be calculated from the perfect gas equation $p_1 = \rho_1 RT$ so that $\rho_1 = 7 \times 10^5/518.4 \times 293 = 4.608$ kg/m³. The volumetric flowrate is then $32.2/4.608 = 6.99$ m³/s, or $6.04 \times 10^5$ m³/day.

The alternative approach to solving this problem is as follows. Since the flowrate is a maximum, the flow must be isothermally choked at the entry to the second compressor, i.e. the pipe length $L$ must correspond with the isothermal choking length for the inlet Mach number. With $L = 9 \times 10^4$ m $= L_1^{*T}$ we have $4\bar{f}_F L_1^{*T}/D = 1800$, from which we can calculate the inlet Mach number $M_1$ from equation (13.40). This requires a trial-and-error process. Omitting the logarithmic term provides a first guess for $M_1$, i.e. $4\bar{f}_F L_1^{*T}/D \cong \left(1 - \gamma M_1^2\right)/\gamma M_1^2$, from which $M_1 \cong 0.021$. The final value is 0.0199.

The soundspeed is $c = \sqrt{\gamma RT} = 446.1$ m/s so that the inlet gas velocity $V_1 = M_1 c = 8.863$ m/s. From the perfect-gas equation the gas density at inlet $\rho_1 = p_1/RT = 4.608$ kg/m³ so that the mass flowrate $\dot{m} = \rho_1 A V_1 = 32.2$ kg/s, in agreement with the value calculated earlier.

## Comments:

The neglect of the logarithmic term in equation (13.40) to obtain an initial guess for $M_1$ is justified by the low value of $M_1$.

The calculation reveals that friction causes the gas velocity to increase from less than 10 m/s at inlet to 390 m/s ($= M_2 c = c/\sqrt{\gamma}$) at exit.

---

## 13.4 Frictionless pipe flow with heat addition or extraction: Rayleigh flow

The third, and final, compressible pipe-flow situation to be considered is frictionless gas flow with addition of heat to, or extraction of heat from, the flowing fluid, usually referred to as Rayleigh flow. We shall not be concerned here with the mechanism by which thermal energy is added to or subtracted from the gas: combustion, evaporation, and condensation are all possibilities. Convective heat transfer at the interior surface of the pipe has been omitted from the list as it would be inconsistent with the assumption of a frictionless (i.e. inviscid) flow. The analysis is simplified by assuming the fluid is a calorically perfect gas. In reality, combustion would involve significant property variations while evaporation or condensation by definition involve phase change.

As for the adiabatic-flow situation, the mass-conservation equation reduces to

$$\frac{\rho}{\rho^*} = \frac{V^*}{V} = \frac{c^*}{c}\frac{c}{V} = \frac{\sqrt{T^*}}{\sqrt{T}}\frac{1}{M} \tag{13.54}$$

where, as was the case for adiabatic flow and as we shall show later, the flow chokes when $M = 1$. The asterisk superscript will again be used to identify flow properties corresponding with the sonic condition.

From the perfect-gas equation,

$$\frac{p}{p^*} = \frac{\rho}{\rho^*}\frac{T}{T^*} = \sqrt{\frac{T}{T^*}}\frac{1}{M} = \frac{c}{c^*}\frac{1}{M}. \tag{13.55}$$

Since Rayleigh flow is assumed to be frictionless, the momentum-conservation equation (13.2) reduces to

$$-\frac{dp}{dx} = \rho V\frac{dV}{dx} \tag{13.56}$$

which integrates to give

$$p + \rho V^2 = \text{constant} = p^* + \rho^* c^{*2} \tag{13.57}$$

from which we obtain

$$\frac{p}{p^*} + \gamma M\frac{c}{c^*} = 1 + \gamma. \tag{13.58}$$

If we combine equations (13.55) and (13.58), we find

$$\frac{c}{c^*} = \frac{(\gamma + 1)M}{1 + \gamma M^2} = \sqrt{\frac{T}{T^*}} \tag{13.59}$$

so that

$$\frac{\rho}{\rho^*} = \frac{V^*}{V} = \frac{1 + \gamma M^2}{(\gamma + 1)M^2} \tag{13.60}$$

and

$$\frac{p}{p^*} = \frac{\gamma + 1}{1 + \gamma M^2}. \tag{13.61}$$

From $T_0 = T\left[1 + (\gamma - 1)\,M^2/2\right]$ we have

$$\frac{T_0}{T_0^*} = \frac{2\,(\gamma + 1)\,M^2\left[1 + \left(\frac{\gamma - 1}{2}\right)M^2\right]}{\left(1 + \gamma M^2\right)^2} \tag{13.62}$$

and, from $p_0 = p\left[1 + (\gamma - 1)\,M^2/2\right]^{\frac{\gamma}{\gamma - 1}}$,

$$\frac{p_0}{p_0^*} = \frac{(\gamma + 1)}{\left(1 + \gamma M^2\right)}\left[\frac{2\left\{1 + \left(\frac{\gamma - 1}{2}\right)M^2\right\}}{\gamma + 1}\right]^{\frac{\gamma}{\gamma - 1}}. \tag{13.63}$$

From equation (11.2), the change in specific entropy up to the choking location is given by

$$s - s^* = C_P \ln\left(\frac{T}{T^*}\right) - R \ln\left(\frac{p}{p^*}\right). \tag{13.64}$$

After substitution in equation (13.64) for $T/T^*$, and $p/p^*$ from equations (13.59) and (13.61), respectively, and making use of the relation $R/C_P = (\gamma - 1)/\gamma$, we have, finally,

$$\frac{s - s^*}{C_P} = \ln\left[M^2\left(\frac{\gamma + 1}{1 + \gamma M^2}\right)^{\frac{\gamma + 1}{\gamma}}\right]. \tag{13.65}$$

As was the case for isothermal flow, changes in the stagnation temperature of the flow require either heat addition or extraction, i.e. either heating or cooling. According to the energy-conservation equation,

$$\dot{q}' = \dot{m} C_P \frac{dT_0}{dx} \tag{13.3}$$

where $\dot{q}'$ is the rate of heat addition to the fluid per unit length of pipe. After integration between the initial location and the choking location,

$$T_0^* - T_0 = \frac{\int_0^{L^*} \dot{q}'\,dx}{\dot{m} C_P} = \frac{\dot{q}}{\dot{m} C_P} \tag{13.66}$$

where $\dot{q}$ is the total rate of thermal-energy addition over the pipe length $L^*$. A negative value of $\dot{q}$ corresponds with heat extraction.

If $\dot{q}_{1-2}$ is the total rate of heat transfer over a length of pipe $L$, separating locations 1 and 2, we have

$$T_{02} - T_{01} = \frac{\int_{L_1}^{L_2} \dot{q}'\,dx}{\dot{m} C_P} = \frac{\dot{q}_{1-2}}{\dot{m} C_P}. \tag{13.67}$$

Values for $4\overline{f}_F L^*/D$ for any initial Mach number $M$, together with the ratios $T/T^*, p/p^*, V/V^*, \rho/\rho^*, p_0/p_0^*$, and $(s - s^*)/C_P$ can be calculated from the equations above, or obtained from the Rayleigh-flow Calculator. Figure 13.7 shows the variation with Mach number of flow properties according to the equations derived above.

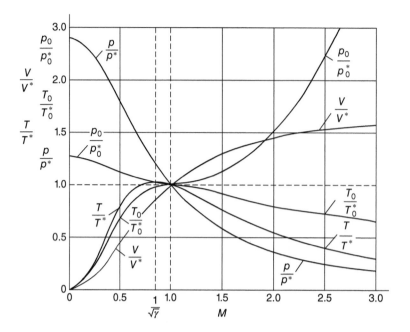

**Figure 13.7** Variation of flow properties with Mach number for Rayleigh flow of a perfect gas with $\gamma = 1.4$

---

### ILLUSTRATIVE EXAMPLE 13.5

Air enters a 75 mm-diameter pipe of length 10 m with a flow velocity of 800 m/s. The initial static temperature is 20 °C, and the static pressure is 5 bar. The air is heated at a rate of 200 kW/m. A shockwave occurs 5 m from the pipe inlet. Calculate the outlet conditions.

### Solution

Figure E13.5 shows the flow configuration, with subscript 1 denoting the inlet state, 2 and 3 the conditions upstream and downstream of the shock, respectively, and 4 the outlet state.

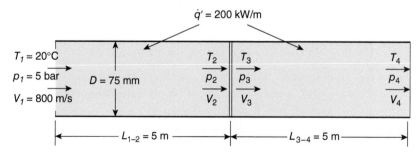

**Figure E13.5** (not to scale)

$D = 0.075$ m, $L_{1-2} = 5$ m, $L_{3-4} = 5$ m, $V_1 = 800$ m/s, $p_1 = 5 \times 10^5$ Pa, $T_1 = 293$ K, and $\dot{q}' = 2 \times 10^5$ W/m.

The cross-sectional area of the pipe $A = \pi D^2/4 = 4.418 \times 10^{-3}$ m$^2$.

From the perfect-gas equation, the inlet density $\rho_1 = p_1/RT_1 = 5.946$ kg/m$^3$.

The mass flowrate $\dot{m} = \rho_1 A V_1 = 21.01$ kg/s.

The inlet soundspeed $c_1 = \sqrt{\gamma RT_1} = 343.1$ m/s so the inlet Mach number $M_1 = V_1/c_1 = 2.332$. From the Isentropic-flow Calculator, for $M_1 = 2.332$, $T_1/T_{01} = 0.479$ so that the inlet stagnation temperature $T_{01} = 293/0.479 = 611.7$ K.

From the energy-conservation equation applied to the flow between the inlet and the shock-wave, $\dot{Q} = \dot{q}' L_{1-2} = \dot{m} C_P (T_{02} - T_{01})$, from which $T_{02} = 611.7 + 2 \times 10^5 \times 5/ (21.01 \times 1004.5) = 659.1$ K.

From the Rayleigh-flow Calculator with $M_1 = 2.332$, $T_{01}/T_0^* = 0.735$ so that $T_0^* = 832.7$ K. We then have $T_{02}/T_0^* = 0.792$ and from the Rayleigh-flow Calculator $M_2 = 2.010$.

From the Normal-shock Calculator, with $M_2 = 2.010$ we have $M_3 = 0.576$.

From the energy-conservation equation applied to the flow between the shockwave and the outlet, noting that there is no change in the stagnation temperature across the shock, $\dot{Q} = \dot{q}' L_{3-4} = \dot{m} C_P (T_{04} - T_{03})$, from which $T_{04} = 659.1 + 2 \times 10^5 \times 5/ (21.01 \times 1004.5) = 706.5$ K. We now have $T_{04}/T_0^* = 706.5/832.7 = 0.848$ and, from the Rayleigh-flow Calculator, $M_4 = 0.629$.

From the Isentropic-flow Calculator, for $M_4 = 0.629$, $T_4/T_{04} = 0.927$ so that the outlet static temperature $T_4 = 0.927 \times 706.5 = 654.7$ K.

The soundspeed at outlet is then $c_4 = \sqrt{\gamma RT_4} = 512.9$ m/s, and the flow velocity at outlet $V_4 = M_4 c_4 = 322.5$ m/s.

From the mass flowrate, the outlet density $\rho_4 = \dot{m}/A V_4 = 14.75$ kg/m$^3$.

From the perfect-gas equation, $p_4 = \rho_4 RT_4 = 2.771 \times 10^6$ Pa or 27.71 bar.

## 13.4.1 Rayleigh-flow trends

As for adiabatic and isothermal flow, first-order ordinary differential equations can be derived[103] for $V, \rho, p, c, T_0, p_0$, and $s$, but now including the heat-transfer rate per unit length $\dot{q}'$ rather than $f_F$

$$\frac{1}{V}\frac{dV}{dx} = -\frac{1}{\rho}\frac{d\rho}{dx} = \frac{1}{(1 - M^2)}\frac{\dot{q}'}{\dot{m} C_P T_0} \tag{13.68}$$

$$\frac{1}{p}\frac{dp}{dx} = -\frac{\gamma M^2}{(1 - M^2)}\frac{\dot{q}'}{\dot{m} C_P T_0} \tag{13.69}$$

$$\frac{1}{T}\frac{dT}{dx} = \frac{2}{c}\frac{dc}{dx} = \frac{(1 - \gamma M^2)}{(1 - M^2)}\frac{\dot{q}'}{\dot{m} C_P T_0} \tag{13.70}$$

---

[103] The derivation of the differential equations for Rayleigh flow is given in Appendix 4(c).

$$\frac{1}{T_0}\frac{dT_0}{dx} = \frac{\dot{q}'}{\dot{m}C_P T_0} \tag{13.71}$$

$$\frac{1}{p_0}\frac{dp_0}{dx} = -\frac{\gamma M^2}{2}\frac{\dot{q}'}{\dot{m}C_P T_0} \tag{13.72}$$

$$\frac{1}{C_P}\frac{ds}{dx} = -\frac{\gamma \dot{q}'}{\dot{m}C_P T} \tag{13.73}$$

$$\frac{1}{M^2}\frac{dM^2}{dx} = \frac{(\gamma M^2 + 1)\left[1 + \left(\frac{\gamma-1}{2}\right)M^2\right]}{(1 - M^2)}\frac{\dot{q}'}{\dot{m}C_P T}. \tag{13.74}$$

Based upon these equations, the various changes that occur in frictionless pipe flow with heating (i.e. positive heat transfer, $\dot{q}' > 0$) are summarised in Table 13.3.

For cooling ($\dot{q}' < 0$) all the changes for heating are reversed, including that for entropy which then decreases. Based upon equation (13.59), the effects of heat transfer on static temperature are illustrated in Figure 13.8 (the same curve is included in Figure 13.7).

As expected, the maximum static temperature occurs for a Mach number $M = 1/\sqrt{\gamma}$, which for air, with $\gamma = 1.4$, corresponds to $M = 0.845$, $T/T^* = 1.029$ (point $A$ on the curve). We note that, when the Mach number is in the range $1/\sqrt{\gamma} < M < 1$, the static temperature falls even though thermal energy is being added to the flow. This apparent contradiction is a consequence of the amount of thermal energy required to increase the kinetic energy of the flow. If the flow is cooled once the sonic point ($B$ on the curve) is reached, the flow becomes supersonic. The total temperature increases with the addition of thermal energy for both subsonic and supersonic flow, with a maximum when $M = 1$.

Table 13.3 Property changes for frictionless pipe flow with heating, under subsonic versus supersonic conditions

| | Subsonic | Supersonic |
|---|---|---|
| Mach number, $M$ | increases | decreases |
| static pressure, $p$ | decreases | increases |
| density, $\rho$ | decreases | increases |
| velocity, $V$ | increases | decreases |
| static temperature, $T$ | increases for $M < 1/\sqrt{\gamma}$ <br> decreases for $M > 1/\sqrt{\gamma}$ | increases |
| stagnation temperature, $T_0$ | increases | increases |
| stagnation pressure, $p_0$ | decreases | decreases |
| specific entropy, $s$ | increases | increases |

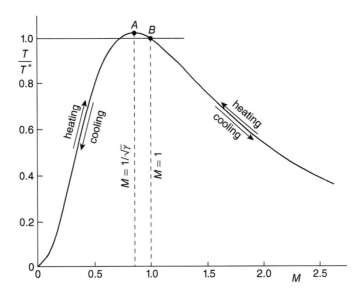

**Figure 13.8** Effect of heating and cooling on static temperature for frictionless pipe flow of a perfect gas with $\gamma = 1.4$.

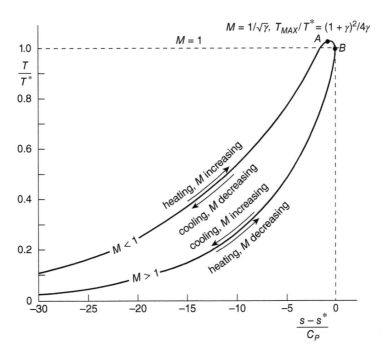

**Figure 13.9** Rayleigh line for a perfect gas with $\gamma = 1.4$

### 13.4.2 Rayleigh line

Equations (13.59) and (13.55) can be combined to give an equation for the **Rayleigh line**[104] in terms of the temperature ratio $T/T^*$, which, for a perfect gas, is equivalent to the specific-enthalpy ratio $h/h^*$

$$\frac{s - s^*}{c_P} = \ln\left(\frac{T}{T^*}\right) - \left(\frac{\gamma - 1}{\gamma}\right) \ln\left[\frac{\gamma + 1 \pm \sqrt{(\gamma + 1)^2 - 4\gamma\frac{T}{T^*}}}{2}\right]. \tag{13.75}$$

The curve representing this equation is shown in Figure 13.9.

Points $A$ and $B$ on the Rayleigh line correspond to points $A$ and $B$ in Figure 13.8.

## 13.5 SUMMARY

In this chapter we analysed gas flow through pipes, taking account of compressibility and either friction or heat exchange with the fluid. We showed that in all cases the key parameter is the Mach number. The analyses were based on the conservation laws for mass, momentum, and energy. So that significant results could be achieved, the flowing fluid was treated as a perfect gas, and the flow as one dimensional.

The student should be able to

- define the terms Mach number and choking
- explain the differences between Fanno flow, isothermal pipe flow, and Rayleigh flow
- understand the trends for each of these flows based upon the differential equations derived
- perform calculations using the equations derived in this chapter for Fanno flow, isothermal pipe flow, or Rayleigh flow
- for Fanno flow and Rayleigh flow, perform calculations using the Virginia Tech Compressible Aerodynamics Calculator
- where appropriate, in carrying out the calculations, allow for the presence of a normal shock within the flow

## 13.6 SELF-ASSESSMENT PROBLEMS

**13.1** Air flows through a pipe of 10 mm diameter and 1.2 m length. The Fanning friction factor is 0.005. The static pressure at inlet is 220 kPa and the static temperature 300 K; the static pressure at exit is 140 kPa. Calculate the mass flowrate assuming the flow is (a) isothermal (b) adiabatic.
(Answers: (a) 0.0250 kg/s; (b) 0.0254 kg/s)

**13.2** Air flows into an insulated pipe 80 mm in diameter with a stagnation pressure of 2 bar, stagnation temperature of 500 K, and velocity 200 m/s. The Fanning friction factor is 0.005.

---

[104] Sometimes referred to as the **Rayleigh curve**.

(a) Calculate the maximum possible pipe length for the specified inlet conditions, and the corresponding mass flowrate.

(b) Calculate the static pressure, static temperature, inlet velocity, and mass flowrate if the pipe length is increased to 15 m.
(Answers: (a) 6.03 m, 1.266 kg/s; (b) 1.846 bar, 487 K, 150.4 m/s, 0.999 kg/s)

13.3    Air enters an insulated pipe 50 mm in diameter and 1 m in length at 800 m/s with static pressure 10 bar and static temperature 100 °C. The Fanning friction factor is 0.005. Calculate the location of the normal shockwave required for this flow, the static pressure, temperature, and velocity at exit, and the mass flowrate.
(Answers: 0.455 m from inlet, 25.7 bar, 303.3 °C, 481.2 m/s, 14.7 kg/s)

13.4    Methane gas flows through a buried pipeline 200 m long and having a diameter of 100 mm. If the inlet velocity is 20 m/s, the inlet static pressure is 5 bar, and the gas temperature is constant at 15 °C, calculate the drop in static pressure and the mass flowrate. Assume $\gamma = 1.32$, and $R = 520$ kJ/kg $\cdot$ K, and that the Fanning friction factor is 0.004.
(Answers: 0.29 bar, 0.524 kg/s)

13.5    Air enters a pipe with cross-sectional area 0.01 m$^2$ at a static pressure of 1.5 bar, static temperature of 300 K, and a velocity of 75 m/s. If heat is added at a rate of 1.2 MW, find the static temperature, the static pressure, and the flow velocity at the pipe exit assuming frictionless flow.
(Answers: $M = 0.583$, 865 °C, 1.083 bar, 394.2 m/s)

13.6    Air flows through a pipe 200 mm in diameter. The inlet static temperature is 650 K, the inlet static pressure is 6 bar, and the inlet velocity is 100 m/s. Liquid fuel with a heating value of 60,000 kJ/kg is sprayed into the airflow such that the air:fuel ratio is 40:1. Assuming complete combustion, and that the flow can be regarded as frictionless flow of a perfect gas with $\gamma = 1.4$, and $C_P = 1004.5$ J/kg $\cdot$ K, calculate the static pressure, static temperature, and flow velocity after the combustion zone.
(Answers: 5.11 bar, 2078 K, 375.5 m/s)

# 14 Flow through axial-flow-turbomachinery blading

After some introductory remarks about **turbomachinery** in general, we show how the non-dimensional parameters used in the representation of turbomachinery data are arrived at using the principles set out in Chapter 3. As a prelude to the more complicated situation of flow through rotating **blading**, we discuss first the analysis of incompressible flow through a linear array of fixed blades (or vanes), termed a **cascade**, and then isentropic flow of a perfect gas through a cascade. Both the latter and the subsequent analysis of the flow through **axial-flow compressors** and **turbines** represent an important application of the compressible-flow theory developed in Chapter 11. Prior to the detailed analysis of flow through rotating blading, we derive **Euler's turbomachinery equation**, which is a key element in the analysis of any turbomachine.

## 14.1 Turbomachinery (general)

Machines designed to convert the energy of a fluid stream into mechanical energy, and vice versa, fall into two categories: rotary fluid machines, termed **turbomachines** or **rotodynamic**[105] **machines**, and **positive-displacement machines**, which are regarded as static. The Latin prefix *turbo*, meaning spin, whirl, or circular motion, indicates that a key component, called a **rotor**, of a turbomachine rotates. **Centrifugal pumps**, **centrifugal compressors**, **turbochargers**, **axial-flow pumps**, **axial-flow compressors**, **axial-flow turbines**, and **ducted fans** are all rotodynamic machines. An **impeller** is the rotating element of a centrifugal machine while for a turbine the rotor is often called a **runner**. Reciprocating **piston pumps**, **diaphragm pumps**, **vane pumps**, **screw pumps**, **peristaltic pumps**, **gear pumps**, and **lobe pumps**, designed primarily to operate with a liquid, are all examples of static machines although a rotating component is a central element of many such pumps. The **Roots blower** or **supercharger** is a lobe pump designed to operate with air.

This chapter is concerned primarily with axial-flow compressors and axial-flow turbines, which form the basis of the majority of engines for large commercial and military aircraft. The essential active component for both types of machine is the **blade** or **vane**, a relatively thin, tapered, and twisted 'plate' of metal or composite material with the cross section of an aerofoil. Flow passages may be incorporated within a blade to carry a cooling fluid. Both compressors and turbines generally consist of many stages (**multistage machines**), a stage comprising a **rotor** and a **stator**. A stator is a multi-bladed non-rotating disc where the blades are attached to the machine's casing. A rotor is a rotating flat disc of many blades: according to

---

[105] Sometimes referred to simply as **dynamic machines**.

*Introduction to Engineering Fluid Mechanics.* Marcel Escudier.
© Marcel Escudier 2017. Published 2017 by Oxford University Press.

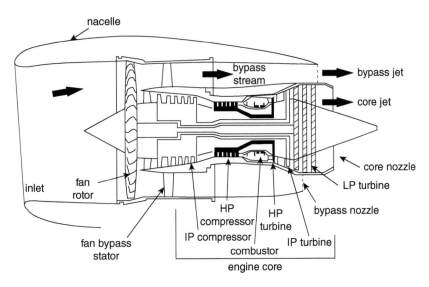

**Figure 14.1** Cross section of a turbofan engine (Diagram courtesy Rolls-Royce plc)

Cumpsty (1997), in the case of a compressor, typically between 30 and 100 blades. Rotor blades are attached at one end (the **root**) to a rotating central shaft which, in the case of an engine, runs through both the compressor and the turbine. Since the circumferential (or tangential) blade speed increases in proportion to the radial distance from the rotor axis, **blade twist** is required to maintain a constant angle of attack with respect to the **relative gas flow**. For short blades on a relatively large diameter rotor, as in the final stages of an axial-flow compressor, the degree of twist may be negligibly small. The opposite is true for the front fan of a turbofan engine or for the low-pressure stages of a gas turbine.

For an aircraft engine, such as the **bypass turbofan engine** shown schematically in Figure 14.1, air enters the compressors, and its pressure and temperature are progressively increased as it flows through each stage. This compressed air flows into **combustors**[106] where fuel is injected and burned to produce gas, at very high temperature and pressure, which flows into the turbine. As the gas flows through the turbine stages, its temperature and pressure progressively decrease. The power produced by the turbine is used to drive both the compressors and a large diameter **fan** at the front of the engine). **Thrust** is generated both by the bypass airflow through the fan and the exhaust-gas momentum (the core jet).

## 14.2 Dimensional analysis and basic non-dimensional parameters

The aim in this section is to use the principles of dimensional analysis set out in Chapter 3 to determine the non-dimensional parameters of relevance to the flow of a perfect gas through

---

[106] Combustor is a shortened form of the term **combustion chamber**.

a turbomachine. We characterise the gas by the specific gas constant $R$ and the specific-heat ratio $\gamma$. For any given design, the machine is characterised by a length scale $D$, a logical choice for which, depending upon the type of machine in question, is the diameter of the impeller or of the largest-diameter rotor blade. We assume that the stagnation pressure[107] at exit from the machine, $p_{02}$, depends upon the mass flowrate $\dot{m}$, the angular rotation speed[108] $\Omega$, the stagnation pressure at inlet $p_{01}$, and the inlet stagnation temperature $T_{01}$. We could also include the kinematic viscosity of the fluid, $\nu$, which (from Chapter 3) we know would simply add a Reynolds number, $\Omega D^2/\nu$, to the list of any non-dimensional parameters we derive. Generally speaking, the Reynolds number plays a minor role in determining the behaviour of large fluid machines and will be excluded from further consideration.

Based upon the foregoing, we may write

$$p_{02} = f\left(p_{01}, T_{01}, D, \dot{m}, \Omega, R, \gamma\right). \tag{14.1}$$

The dimensions of each physical quantity are

$$[p_{02}] = [p_{01}] = \frac{M}{LT^2},\ [T_{01}] = \theta,\ [D] = L,\ [\dot{m}] = \frac{M}{T},\ [\Omega] = \frac{1}{T},\ [R] = \frac{L^2}{T^2\theta},\ [\gamma] = 1$$

where the dimension symbols, as in Chapter 3, are $L$ for length, $T$ for time, $M$ for mass, and $\theta$ for temperature. Since we have eight physical quantities and four dimensions, from Buckingham's $\Pi$ theorem there will be four non-dimensional groups. Conventional choices are

$$\Pi_1 = \frac{p_{02}}{p_{01}},\ \Pi_2 = \frac{\dot{m}\sqrt{RT_{01}}}{p_{01}D^2},\ \Pi_3 = \frac{\Omega D}{\sqrt{RT_{01}}},\ \text{and}\ \Pi_4 = \gamma \tag{14.2}$$

so that we can write

$$\frac{p_{02}}{p_{01}} = f_1\left(\frac{\dot{m}\sqrt{RT_{01}}}{p_{01}D^2}, \frac{\Omega D}{\sqrt{RT_{01}}}, \gamma\right) \tag{14.3}$$

or

$$\Pi_1 = f_1\left(\Pi_2, \Pi_3, \Pi_4\right). \tag{14.4}$$

The first of the four parameters, $p_{02}/p_{01}$, which represents the pressure ratio across the machine, could be replaced by the non-dimensional pressure difference $\Delta p_0/p_{01}$, where $\Delta p_0 = |p_{02} - p_{01}|$ (see also Subsection 14.2.1). The second parameter is termed the **mass-flowrate coefficient** or simply the **flow coefficient**, while the third is effectively a blade-tip Mach number, since the soundspeed corresponding to $T_{01}$ is $\sqrt{\gamma RT_{01}}$, and $\Omega D$ is twice the tipspeed of a rotor of diameter $D$. The outlet stagnation temperature $T_{02}$, the power input (for a compressor) or output (for a turbine), $P$, and the torque, $G$, must also depend upon $p_{01}, T_{01}, D, \dot{m}, \Omega, R$, and $\gamma$ dependencies, which can be expressed in non-dimensional form as

---

[107] Instead of the stagnation pressure and temperature, this analysis could be formulated in terms of the static pressure and temperature.

[108] Instead of $\Omega$ we could just as well have chosen the shaft rotation speed (in rpm) $N = 60\Omega/2\pi$.

$$\Pi_5 = \frac{T_{02}}{T_{01}} = f_2\left(\frac{\dot{m}\sqrt{RT_{01}}}{p_{01}D^2}, \frac{\Omega D}{\sqrt{RT_{01}}}, \gamma\right) \tag{14.5}$$

$$\Pi_6 = \frac{G}{\rho_{01}\Omega^2 D^5} = f_3\left(\frac{\dot{m}\sqrt{RT_{01}}}{p_{01}D^2}, \frac{\Omega D}{\sqrt{RT_{01}}}, \gamma\right) \tag{14.6}$$

and

$$\Pi_7 = \frac{P}{\rho_{01}\Omega^3 D^5} = f_4\left(\frac{\dot{m}\sqrt{RT_{01}}}{p_{01}D^2}, \frac{\Omega D}{\sqrt{RT_{01}}}, \gamma\right). \tag{14.7}$$

$\Pi_6$ is termed the **torque coefficient**, and $\Pi_7$ the **power coefficient**. The symbol $\rho_{01} = p_{01}/RT_{01}$ represents the stagnation density at inlet.

We note that in the absence of any heat input or output, according to the form of the steady-flow energy equation derived in Chapter 11, equation (11.12), the stagnation temperature change $T_{02} - T_{01} = P/\dot{m}C_P = (\gamma - 1)P/\dot{m}\gamma R$. If we substitute $P = \dot{m}\gamma R(T_{02} - T_{01})/(\gamma - 1)$ into $\Pi_7$, and substitute for $\dot{m}$ and $D$ in terms of $\Pi_2$ and $\Pi_3$, we find

$$\Pi_7 = \frac{\gamma}{(\gamma - 1)} \frac{(T_{02} - T_{01})}{T_{01}} \frac{\Pi_2}{\Pi_3^3} \tag{14.8}$$

from which we conclude that $\Pi_7$ can be replaced by either $(T_{02} - T_{01})/T_{01}$ or $\Pi_5$ ($= T_{02}/T_{01}$).

Finally, therefore, we find that the flow of a perfect gas through a turbomachine can be represented by the following six parameters

$$\Pi_1 = \frac{p_{02}}{p_{01}}, \quad \Pi_2 = \frac{\dot{m}\sqrt{RT_{01}}}{p_{01}D^2}, \quad \Pi_3 = \frac{\Omega D}{\sqrt{RT_{01}}}, \quad \Pi_4 = \gamma, \quad \Pi_5 = \frac{T_{02}}{T_{01}},$$

$$\text{and} \quad \Pi_6 = \frac{G}{\rho_{01}\Omega^2 D^5}. \tag{14.9}$$

Thus, the performance of a given design of compressor or turbine can be represented in graphical form by curves of $p_{02}/p_{01}$, $T_{02}/T_{01}$, and $G/\rho_{01}\Omega^2 D^5$ plotted versus $\dot{m}\sqrt{RT_{01}}/p_{01}D^2$ for fixed values of the parameter $\Omega D/\sqrt{RT_{01}}$.

## 14.2.1 Related non-dimensional parameters

A number of related non-dimensional parameters can be defined by combining some of the groups derived above. Although generally valid, they are more usually encountered in the analysis of **hydraulic** (i.e. incompressible-flow) **machines**. We can replace $p_{01}$ in $\Pi_2$ by $\rho_{01}RT_{01}$ and introduce $\Pi_3$ to give $\Pi_2 = \dot{m}/\rho_{01}\sqrt{RT_{01}}D^2 = \dot{m}\Pi_3/\rho_{01}\Omega D^3$, from which we see that an alternative to $\Pi_2$ as a non-dimensional group, based upon the mass flowrate $\dot{m}$, is

$$\phi = \frac{\dot{m}}{\rho_{01}\Omega D^3} \tag{14.10}$$

which, like $\Pi_2$, is called the **flow coefficient**[109].

---

[109] Care should be taken to avoid confusion which can obviously arise when two different quantities, such as $\Pi_2$ and $\phi$, are given the same name.

An alternative to $\Pi_1$ is

$$\psi = \frac{(p_{02} - p_{01})}{\rho_{01} (\Omega D)^2} \tag{14.11}$$

which, like $\Delta p_0/p_{01}$, is a parameter that can be used to characterise the pressure change across a radial hydraulic machine.

Another parameter encountered in the characterisation of radial hydraulic machines is the so-called **specific speed** $N_S$, which combines $\phi$ and $\psi$ in such a way as to remove the length scale, $D$, and produce a non-dimensional parameter proportional to rotational speed, $\Omega$

$$N_S = \frac{\sqrt{\phi}}{\psi^{\frac{3}{4}}} = \frac{\sqrt{\dot{m}} \rho_{01}^{\frac{1}{4}} \Omega}{(p_{02} - p_{01})^{\frac{3}{4}}}. \tag{14.12}$$

It is also apparent that $\Pi_7$ and $\psi$ can be combined to eliminate the length scale. The result is a non-dimensional quantity also proportional to $\Omega$ and called the **power specific speed**, $N_P$

$$N_P = \frac{\sqrt{\Pi_7}}{\psi^{\frac{5}{4}}} = \frac{\sqrt{P} \rho_{01}^{\frac{3}{4}} \Omega}{(p_{02} - p_{01})^{\frac{5}{4}}}. \tag{14.13}$$

## 14.2.2 Loss coefficients and efficiencies

The term **losses** in turbomachinery refers to the reduction in stagnation pressure or stagnation enthalpy as a consequence of thermodynamic irreversibilities such as surface friction, boundary-layer separation, secondary flows, non-zero incidence angles, and shockwaves. For the most part such losses are difficult to account for analytically and are usually dealt with by applying empirical loss coefficients.

For one stage of a compressible-flow machine, a loss coefficient $K$ may be defined by

$$K = \frac{\Delta p_0}{p_0 - p} \tag{14.14}$$

where $\Delta p_0$ is the loss in stagnation pressure across the stage and $p_0 - p$ is the dynamic pressure at either inlet (for a compressor stage) or outlet (for a turbine stage).

Stagnation enthalpy provides an alternative to stagnation pressure for the definition of a loss coefficient $\xi$

$$\xi = \frac{h_{02} - h_{02,S}}{V^2/2} \tag{14.15}$$

where $h_{02}$ is the stagnation enthalpy at outlet from a stage and $h_{02,S}$ is the stagnation enthalpy at outlet for an isentropic flow with the same initial and final stagnation pressures. The denominator represents the specific kinetic energy based upon the inlet velocity, the outlet velocity, or the blade-tip speed. A similar definition is based upon static enthalpy.

The overall efficiency of a machine is related directly to these loss coefficients, and losses can also be characterised by various efficiencies defined as the ratios of the actual static or stagnation enthalpy across a stage to the change in the same quantity for an isentropic process.

For detailed information about loss coefficients and efficiencies, the reader should consult one of the specialist turbomachinery texts, such as Japikse and Baines (1994) or Cumpsty (1989).

## 14.3 Linear blade cascade: Geometry and notation

As illustrated in Figure 14.2, a linear blade cascade[110] is a row of evenly spaced, identical blades, which may be typical of axial-compressor blading (Figure 14.2(a)) or axial-turbine blading (Figure 14.2(b)). The geometry of a linear cascade is a close approximation to the rotor or stator of a turbomachine if the mean radius of the machine blading is large compared with the blade length $z$. We refer to the flow direction normal to the cascade as the axial direction, or $x$-direction, and the direction parallel to the front and back planes of the cascade as the tangential direction, or $y$-direction.

The symbols in the figure represent the following quantities

- $c$  blade chord length
- $s$  blade spacing (or pitch)

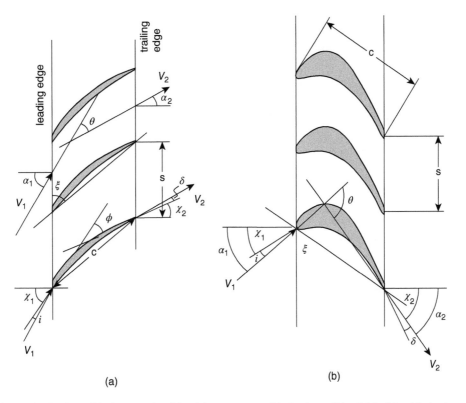

(a)     (b)

**Figure 14.2** Linear blade cascades: (a) axial-compressor blade shape (b) axial-turbine blade shape

---

[110]  The term rectilinear cascade is also used.

$V_1$   flow velocity at cascade inlet

$V_2$   flow velocity at cascade exit

$\alpha_1$   **gas-flow angle** measured between the $V_1$-direction and the axial direction

$\alpha_2$   gas-flow angle measured between the $V_2$-direction and the axial direction

$\chi_1$   **blade angle in**, measured between the axial direction and a line tangent to the camber line at the leading edge

$\chi_2$   blade angle out, measured between the axial direction and a line tangent to the camber line at the trailing edge

$i$   incidence, $i = \alpha_1 - \chi_1$

$\delta$   deviation, $\delta = \alpha_2 - \chi_2$

$\theta$   turning angle, $\theta = \alpha_1 - \alpha_2$ or $\alpha_2 - \alpha_1$

$\xi$   blade stagger angle[111]

$\phi$   camber angle, $\phi = \chi_1 - \chi_2$ or $\chi_2 - \chi_1$

The quantity $\sigma$ (Greek letter sigma) $= c/s$ is termed the **solidity**. High solidity corresponds with $\sigma \gg 1$: the blades are close together so the flow is well guided but there is a high degree of **blockage**. For low solidity, $\sigma \ll 1$: the blades are far apart, so there is little blockage, but the flow is poorly guided.

For simplicity, in the analyses developed in this book we shall assume that the gas flow follows the **camber line**[112] of a blade, i.e. a line drawn midway between the two surfaces of the blade, so that both the incidence $i$ and the deviation $\delta$ are taken as zero. For compressor blading, the camber-line shape is usually a circular arc or, sometimes, a parabola. In general, it can be seen that the flow turning angle $\theta = \phi + i - \delta$ so that if $i = \delta = 0$, $\theta = \phi$.

If we take the camber line as a circular arc of radius $R$, then the chord length $c$ can be shown to be given by $c = R\sqrt{2(1 - \cos\phi)}$, and the axial length $w$ of the cascade by $w = R(\sin\chi_1 - \sin\chi_2)$, two relations which are useful in constructing a schematic diagram of a cascade. The stagger angle $\xi$ is given by $\xi = 90° - (\chi_1 + \chi_2)/2$. If we combine the equations for $c$ and $w$ with the elimination of $R$, we have $w/c = (\sin\chi_1 - \sin\chi_2)/\sqrt{2(1 - \cos\phi)}$. Figure 14.3 shows the results of constructing diagrams with $\chi_1 = 40°$ and (a) $\chi_2 = 20°$ and (b) $\chi_2 = 60°$. In the first case, $\xi = 60°$, and $w/c = 0.866$; in the second, $\xi = 40°$, and $w/c = 0.643$.

Although the channel between adjacent blades is of constant width $s$, a consequence of the blade curvature is that the effective flow area (taken normal to the flow direction) varies. If $\chi_1 > \chi_2$, as in Figure 14.3(a), the flow area increases. Assuming the flow is subsonic, from what we learned in Section 7.6, the consequence of an increase in cross-sectional area is a decrease in flow velocity and an increase in static pressure. The opposite is true if $\chi_1 < \chi_2$, as in Figure 14.3(b). As we shall learn in Chapter 18, increasing static pressure in the flow direction is referred to as an **adverse pressure gradient** because it can lead to **separation** of the flow from the surface, **stall**, **instability**, and, in the case of a compressor, **surge**, all of which are undesirable. The pressure ratio across a compressor stage is therefore usually relatively low (1.05 to 2) as are the forces acting on an individual blade so that the blades are quite thin. This reasoning also explains why there are so many stages (20 to 30 is typical for an aeroengine compressor) in a multistage axial-flow compressor. Turbine blades, on the other hand, are

---

[111] Although the symbol $\gamma$ is commonly used for the stagger angle, to avoid confusion with the specific-heat ratio, we use the symbol $\xi$.

[112] The term **blade centre line** is sometimes used.

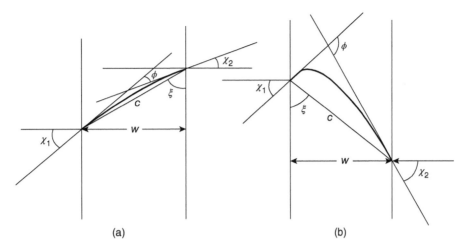

**Figure 14.3** Construction of a cascade geometry for (a) $\chi_1 > \chi_2$ and (b) $\chi_1 < \chi_2$

generally much thicker because the flow in a blade channel leads to a **favourable pressure gradient**, i.e. decreasing pressure, with little danger of large-scale separation. The pressure ratio for a turbine stage is thus quite large, and the number of stages in a multistage axial-flow turbine relatively low (in an aircraft engine, a high-pressure turbine may have as few as two stages, a low-pressure turbine less than ten).

## 14.4  Incompressible flow through a linear cascade

The broken lines in Figure 14.4(a) show the control volume, centred on a blade camber line, that will be used in the analysis of flow through a single cascade channel. The upper and lower boundaries also have the shape of the camber line, and each boundary is located midway between two adjacent blades. The upstream and downstream boundaries coincide with the front and back faces of the cascade. The analysis will assume one-dimensional, incompressible flow through the control volume. In addition to the notation introduced in Figure 14.2, Figure 14.4(a) includes the components of the flow velocity $V$ in the axial direction, $V_A$, and in the tangential direction, $V_T$, the static pressures $p$ acting on the upstream and downstream faces of the control volume, and the $x$- and $y$-components of the force acting on the fluid within the control volume, $F_X$ and $F_Y$. As shown in Figure 14.4(b), the force components acting on a blade, $F_{BX}$ and $F_{BY}$, will be equal in magnitude but opposite in sign and direction to $F_X$ and $F_Y$, i.e. $F_{BX} = -F_X$, and $F_{BY} = -F_Y$. The subscripts 1 and 2 refer to the inlet and outlet of the control volume, respectively, and the subscript $B$ to the blade. If the blade length is $z$, and the blade spacing is $s$, then the mass flowrate $\dot{m}'$ through the control volume (i.e. through a single flow channel between adjacent blades) is given by the **continuity equation** derived in Chapter 6 (equation (6.1)) as

$$\dot{m}' = \rho s z V_{A1} = \rho s z V_{A2}. \tag{14.16}$$

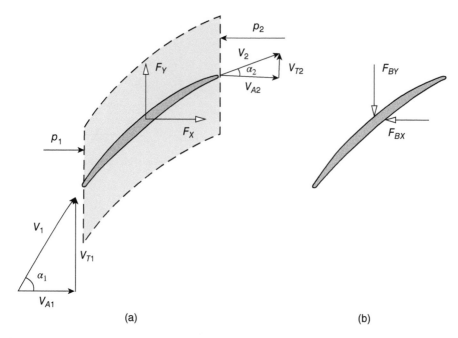

**Figure 14.4** Cascade-flow analysis: (a) control volume (b) force components acting on a blade

Since the fluid is incompressible, with density $\rho$, we conclude that the axial component of velocity is constant, i.e.

$$V_{A1} = V_{A2} = V_A. \tag{14.17}$$

From the flow geometry,

$$V_A = V_1 \cos\alpha_1 = V_2 \cos\alpha_2. \tag{14.18}$$

If the upstream static pressure is $p_1$, and the downstream static pressure is $p_2$, then, from the **linear momentum equation** applied to the control volume (see Section 9.2) in the $x$-direction, the net axial force acting on the fluid within the control volume is given by the equation

$$F_X + (p_1 - p_2)\,sz = \dot{m}'\,(V_{A2} - V_{A1}) = 0 \tag{14.19}$$

where $F_X$ is the $x$-component of the force exerted on the fluid by a single blade (the **fluid-structure interaction** force). Because the axial component of velocity is constant, there is no change in the axial component of the momentum flowrate, and $F_X$ is given by

$$F_X = (p_2 - p_1)\,sz. \tag{14.20}$$

From the linear momentum equation applied to the control volume in the $y$-direction we have

$$F_Y = \dot{m}'\,(V_{T2} - V_{T1}) = \rho sz V_A\,(V_{T2} - V_{T1}). \tag{14.21}$$

We have already $V_A = V \cos\alpha$ and from the flow geometry we also see that

$$V_T = V \sin\alpha. \tag{14.22}$$

If we combine equations (14.18) and (14.22) to eliminate $V$, we have

$$V_T = V_A \tan \alpha \tag{14.23}$$

so that, from equation (14.21)

$$F_Y = \rho s z V_A^2 (\tan \alpha_2 - \tan \alpha_1). \tag{14.24}$$

To determine the static-pressure change across the control volume, we introduce an extended form of Bernoulli's equation, which we derived in Chapter 7 as equation (7.5), neglecting altitude changes

$$p_{01} = p_{02} + \Delta p_{0F} \tag{14.25}$$

where $p_{01}$ and $p_{02}$ are the stagnation pressures upstream and downstream of the control volume, respectively, and $\Delta p_{0F}$ is the reduction in stagnation pressure due to surface friction.

Since stagnation pressure is given by

$$p_0 = p + \frac{1}{2} \rho V^2 \tag{14.26}$$

from equation (14.25) we have

$$p_1 + \frac{1}{2} \rho V_1^2 = p_2 + \frac{1}{2} \rho V_2^2 + \Delta p_{0F} \tag{14.27}$$

or

$$p_2 - p_1 = \frac{1}{2} \rho \left( V_1^2 - V_2^2 \right) - \Delta p_{0F}. \tag{14.28}$$

From equations (14.18) and (14.22), we have

$$V^2 = V_A^2 + V_T^2 \tag{14.29}$$

so that, since $V_{A1} = V_{A2}$, equation (14.28) can be rewritten in terms of the change in the square of the tangential component of velocity, $V_T$:

$$p_2 - p_1 = \frac{1}{2} \rho \left( V_{T1}^2 - V_{T2}^2 \right) - \Delta p_{0F}. \tag{14.30}$$

The two components of force, $F_X$ and $F_Y$, can thus be written as

$$F_X = \left[ \frac{1}{2} \rho \left( V_{T1}^2 - V_{T2}^2 \right) - \Delta p_{0F} \right] s z \tag{14.31}$$

and

$$F_Y = \rho s z V_A (V_{T2} - V_{T1}). \tag{14.32}$$

The force components acting on an individual blade are thus

$$F_{BX} = \left[ \frac{1}{2} \rho \left( V_{T2}^2 - V_{T1}^2 \right) + \Delta p_{0F} \right] s z \tag{14.33}$$

and

$$F_{BY} = \rho s z V_A (V_{T1} - V_{T2}). \tag{14.34}$$

## 14.5  Compressible flow through a linear cascade

We consider now isentropic flow of a perfect gas through a linear cascade of blades. As we pointed out in Section 14.4, although the blades in a cascade are uniformly spaced, the effective flow area varies through each channel of the cascade as the blade angle $\chi$ changes from its inlet value $\chi_1$ to its outlet value $\chi_2$. Changes in flow properties are a consequence of this area change, much as was the case in Section 11.7, where we analysed isentropic flow of a perfect gas through a convergent-divergent nozzle. If the width of the blade channel (i.e. the distance between adjacent blade surfaces measured normal to the flow) is $t$ at a location where the blade angle is $\chi$ then $t = s \cos \chi$. As stated in Section 14.3, we shall assume that the flow follows the blade camber line so that the gas-flow angle $\alpha$ is equal to $\chi$. The inlet area is then $A_1 = sz \cos \alpha_1$ and the outlet area $A_2 = sz \cos \alpha_2$. The area ratio is thus $A_2/A_1 = \cos \alpha_2 / \cos \alpha_1$. Although the analysis of flow through a cascade could be written in terms of the area ratio, we shall retain the flow angles.

We shall seek to express the outlet Mach number $M_2$ in terms of the inlet Mach number $M_1$. We start with the continuity equation (6.1), which may be written as

$$\dot{m}' = \rho_1 szV_{A1} = \rho_2 szV_{A2} \tag{14.35}$$

where the symbols are the same as for the incompressible case but now reflect the fact that the density at inlet to the cascade, $\rho_1$, will be different from that at outlet, $\rho_2$.
Equation (14.35) leads to

$$\rho_1 V_{A1} = \rho_2 V_{A2}. \tag{14.36}$$

The axial-velocity component $V_A = V \cos \alpha$ so that equation (14.36) becomes

$$\rho_1 V_1 \cos \alpha_1 = \rho_2 V_2 \cos \alpha_2. \tag{14.37}$$

From the definition of the Mach number, we have $V = Mc = M\sqrt{\gamma RT}$ so that equation (14.37) becomes

$$\rho_1 M_1 \sqrt{T_1} \cos \alpha_1 = \rho_2 M_2 \sqrt{T_2} \cos \alpha_2. \tag{14.38}$$

From the perfect-gas equation, equation (2.9)

$$p = \rho RT \tag{14.39}$$

which can be used to eliminate the gas densities from equation (14.38) so that

$$\frac{p_1 M_1 \cos \alpha_1}{\sqrt{T_1}} = \frac{p_2 M_2 \cos \alpha_2}{\sqrt{T_2}}. \tag{14.40}$$

The stagnation and static properties are related to the Mach number as follows (see Chapter 11)

$$\frac{p_0}{p} = \left[ 1 + \frac{(\gamma - 1)}{2} M^2 \right]^{\gamma/(\gamma-1)} \tag{11.22}$$

and

$$\frac{T_0}{T} = 1 + \frac{(\gamma - 1)}{2} M^2. \tag{11.20}$$

We can substitute for $p$ and $T$ in equation (14.40) using these two equations, recognising that, since the flow is assumed to be isentropic, both the stagnation pressure, $p_0$, and the stagnation temperature, $T_0$, remain constant as the gas passes through the cascade. Equation (14.40) then leads to

$$\frac{M_2 \cos \alpha_2}{\left[1 + \frac{(\gamma - 1)}{2} M_2^2\right]^{(\gamma+1)/2(\gamma-1)}} = \frac{M_1 \cos \alpha_1}{\left[1 + \frac{(\gamma - 1)}{2} M_1^2\right]^{(\gamma+1)/2(\gamma-1)}}. \tag{14.41}$$

For a given cascade, the flow angles $\alpha_1$ and $\alpha_2$ are known. If $\alpha_1 > \alpha_2$ the flow channel is divergent and we should expect $M_2 < M_1$, and $p_2 > p_1$, if the inlet flow is subsonic, and the reverse if the flow is supersonic. If $\alpha_1 < \alpha_2$ the flow channel is convergent and we should expect $M_2 > M_1$, and $p_2 < p_1$, if the inlet flow is subsonic, and again the reverse if the flow is supersonic. With $M_1, \alpha_1$, and $\alpha_2$ specified, $M_2$ can be calculated from equation (14.41), unfortunately necessitating an iterative procedure. Once $M_2$ is known, all other flow quantities are easily calculated. Figure 14.5 shows how the outlet Mach number $M_2$ and overall static pressure ratio $p_2/p_1$ vary with the outlet-flow angle $\alpha_2$ for a cascade with inlet-flow angle $\alpha_1 = 20°$, for a subsonic flow with inlet Mach number $M_1 = 0.5$ (Figure 14.5(a)), and for a supersonic

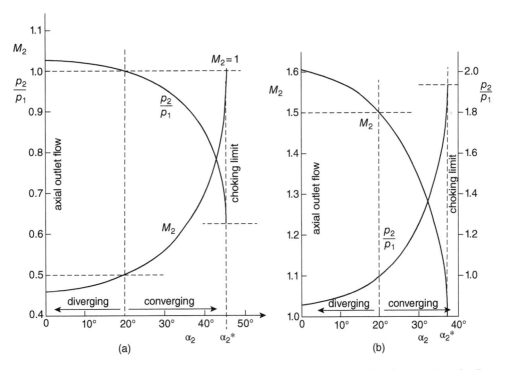

**Figure 14.5** Outlet Mach number $M_2$ and pressure ratio $p_2/p_1$ versus outlet-flow angle $\alpha_2$ for flow through a cascade with inlet-flow angle $\alpha_1 = 20°$ and inlet Mach number (a) $M_1 = 0.5$ and (b) $M_1 = 1.5$

flow with $M_1 = 1.5$ (Figure 14.5(b)). We note that no account has been taken of the influence of the blade cross section in the calculations underlying Figure 14.5, only of the inlet- and outlet-flow angles.

Just as with a convergent duct, as discussed in Section 11.6, for a given inlet Mach number $M_1$ and inlet-flow angle $\alpha_1$, as the outlet-flow angle $\alpha_2$ increases, the outlet Mach number $M_2$ increases until it reaches unity and the flow channel is choked. It is easily seen from equation (14.41) that the choking limit ($M_2 = 1$) is reached when the outlet-flow angle $\alpha_2^*$ is given by

$$\cos \alpha_2^* = \left[ \frac{\gamma + 1}{2 + (\gamma - 1) M_1^2} \right]^{(\gamma+1)/2(\gamma-1)} M_1 \cos \alpha_1. \tag{14.42}$$

Another limiting case is when the outlet-flow angle falls to zero (i.e. the outflow is in the axial direction) and the channel area is then a maximum. The corresponding outlet Mach number $M_2$ can be calculated from

$$\frac{M_2}{\left[1 + \frac{(\gamma - 1)}{2} M_2^2\right]^{(\gamma+1)/2(\gamma-1)}} = \frac{M_1 \cos \alpha_1}{\left[1 + \frac{(\gamma - 1)}{2} M_1^2\right]^{(\gamma+1)/2(\gamma-1)}}. \tag{14.43}$$

The choking limit and the axial-outflow limit are both identified in Figure 14.5.

---

**ILLUSTRATIVE EXAMPLE 14.1**

Air at 500°C and 10 bar flows through a linear cascade with blade spacing 100 mm, blade height 50 mm, and an initial flow direction of 30° measured from the axial-flow direction. Determine the Mach number, flow velocity, static pressure, and static temperature at outlet from the cascade for the following conditions: (a) mass flowrate between adjacent blades of 6 kg/s and an outflow direction of 20°, (b) mass flowrate of 6 kg/s and an outflow direction of 40°, and (c) mass flowrate of 11 kg/s and an outflow direction of 20°.

Solution

(a) $s = 0.1$ m, $z = 0.05$ m, $\dot{m}' = 6$ kg/s, $T_1 = 773$ K, $p_1 = 10^6$ Pa, $\alpha_1 = 30°$, $\alpha_2 = 20°$, $R = 287$ m²/s² K, and $\gamma = 1.4$.
From the perfect-gas equation, $\rho_1 = p_1/RT_1 = 4.508$ kg/m³.
From the continuity equation, $V_{A1} = \dot{m}'/\rho_1 s z = 266.2$ m/s.
Since $V_{A1} = V_1 \cos \alpha_1$, we have $V_1 = 307.4$ m/s.
Since $c_1 = \sqrt{\gamma R T_1}$ we have $c_1 = 557.3$ m/s, and $M_1 = V_1/c_1 = 0.552$.
From the Isentropic-flow Calculator, for $M_1 = 0.552$, we find $T_1/T_0 = 0.943$, and $p_1/p_0 = 0.813$, so that $T_0 = 820.0$ K, and $p_0 = 1.230 \times 10^6$ Pa.
With $M_1 = 0.552$, the right-hand side of equation (14.41) has the value 0.400 and we need to find the value of $M_2$ which leads to the same value for the left-hand side. The calculation is aided if we recognise that $(\gamma + 1)/2(\gamma - 1) = 3$. We guess $M_2 = 0.5$, for which the left-hand side equals 0.406. As a second guess we take $M_2 = 0.49$, for which the left-hand side equals 0.400.
From the Isentropic-flow Calculator with $M_2 = 0.490$, we find $T_2/T_0 = 0.954$, and $p_2/p_0 = 0.849$, so that $T_2 = 782.5$ K, and $p_2 = 1.044 \times 10^6$ Pa, i.e. as expected, both the static temperature

and the static pressure have been increased by compression of the flow as it passed through the cascade. The outlet-gas soundspeed is $c_2 = \sqrt{\gamma R T_2} = 560.7$ m/s so the outlet-gas velocity $V_2 = M_2 c_2 = 274.7$ m/s.

(b) $s = 0.1$ m, $z = 0.05$ m, $\dot{m}' = 6$ kg/s, $T_1 = 773$ K, $p_1 = 10^6$ Pa, $\alpha_1 = 30°$, $\alpha_2 = 40°$, $R = 287$ m$^2$/s$^2$K, and $\gamma = 1.4$.

The inlet-flow conditions are the same as for part (a), i.e. $\rho_1 = 4.508$ kg/m$^3$, $V_{A1} = 266.2$ m/s, $V_1 = 307.4$ m/s, $c_1 = 557.3$ m/s, $M_1 = 0.552$, $T_0 = 820.0$ K, $p_0 = 1.230 \times 10^6$ Pa, and the right-hand side of equation (14.41) again equals 0.400. The outlet angle $\alpha_2$ is now 40°, so as an initial guess for $M_2$ we choose 0.600 for which the left-hand side equals 0.373. As a second guess we take $M_2 = 0.65$ for which the left-hand side equals 0.390. After a few more iterations, we find $M_2 = 0.682$.

From the Isentropic-flow Calculator with $M_2 = 0.682$, we find $T_2/T_0 = 0.915$, and $p_2/p_0 = 0.732$, so that $T_2 = 750.2$ K, and $p_2 = 9.01 \times 10^5$ Pa, i.e. both the static temperature and the static pressure have now been decreased by expansion of the flow as it passed through the cascade. The outlet-gas soundspeed is $c_2 = \sqrt{\gamma R T_2} = 549.0$ m/s so the outlet-gas velocity $V_2 = M_2 c_2 = 374.4$ m/s.

(c) $s = 0.1$ m, $z = 0.05$ m, $T_1 = 773$ K, $\dot{m}' = 11$ kg/s, $p_1 = 10^6$ Pa, $\alpha_1 = 30°$, $\alpha_2 = 20°$, $R = 287$ m$^2$/s$^2$K, and $\gamma = 1.4$.

The inlet density $\rho_1 = 4.508$ kg/m$^3$, as before.

From the continuity equation, $V_{A1} = \dot{m}'/\rho_1 s z = 488.1$ m/s.

Since $V_{A1} = V_1 \cos\alpha_1$, we have $V_1 = 563.6$ m/s.

The inlet soundspeed is unchanged, $c_1 = 557.3$ m/s, so $M_1 = V_1/c_1 = 1.011$.

From the Isentropic-flow Calculator, for $M_1 = 1.011$, we find $T_1/T_0 = 0.830$, and $p_1/p_0 = 0.521$, so that $T_0 = 931.1$ K, and $p_0 = 1.918 \times 10^6$ Pa.

With $M_1 = 1.011$, the right-hand side of equation (14.41) has the value 0.501. We guess $M_2 = 1.2$, for which the left-hand side equals 0.528. As a second guess we take $M_2 = 1.3$, for which the left-hand side equals 0.510. After a few more iterations we have $M_2 = 1.342$, for which the left-hand side is 0.501.

From the Isentropic-flow Calculator with $M_2 = 1.342$, we find $T_2/T_0 = 0.735$, and $p_2/p_0 = 0.341$, so that $T_2 = 684.5$ K, and $p_2 = 6.535 \times 10^5$ Pa, i.e. in contrast to the subsonic case, both the static temperature and the static pressure have been decreased by expansion of the flow as it passed through the cascade.

---

### 14.5.1  Blade forces

As for the incompressible-flow case, discussed in Section 14.4, in order to determine the forces exerted by the gas flow on a single blade, or by a blade on the gas flow, we must now apply the linear momentum equation to the fluid flowing through the control volume (the flow situation is identical to that shown in Figure 14.4).

The momentum equation for the $x$-direction is

$$F_X + (p_1 - p_2)\, sz = \dot{m}' (V_{A2} - V_{A1}) \tag{14.44}$$

but, in contrast to the situation for incompressible flow, it is now the case that the axial component of velocity $V_A$ changes.

The linear momentum equation for the $y$-direction is identical to that for the incompressible-flow situation

$$F_Y = \dot{m}' \left( V_{T2} - V_{T1} \right). \tag{14.45}$$

The force components acting on a single blade are equal in magnitude but opposite in sign to the force components exerted by the blade on the gas, i.e.

$$F_{BX} = -F_X = \dot{m}' \left( V_{A1} - V_{A2} \right) + \left( p_1 - p_2 \right) sz \tag{14.46}$$

and

$$F_{BY} = -F_Y = \dot{m}' \left( V_{T1} - V_{T2} \right). \tag{14.47}$$

As before, the subscript $B$ indicates that the force is that acting on the blade. The resultant force acting on a blade is then

$$R_B = \sqrt{F_{BX}^2 + F_{BY}^2} \tag{14.48}$$

and the angle $\theta_B$ between $R_B$ and the axial direction is

$$\theta_B = \tan^{-1} \left( \frac{F_{BY}}{F_{BX}} \right). \tag{14.49}$$

---

## ILLUSTRATIVE EXAMPLE 14.2

Calculate the resultant force and its direction acting on a single blade in the cascade considered in Illustrative Example 14.1 for all three flow conditions.

### Solution

(a) From Illustrative Example 14.1, we have $\dot{m}' = 6$ kg/s, $s = 0.1$ m, $z = 0.05$ m, $\alpha_1 = 30°$, $\alpha_2 = 20°$, $p_1 = 10^6$ Pa, $p_2 = 1.044 \times 10^6$ Pa, $V_1 = 307.4$ m/s, and $V_2 = 274.8$ m/s.
We also have $V_{A1} = 266.2$ m/s, $V_{A2} = V_2 \cos \alpha_2 = 258.3$ m/s, $V_{T1} = V_1 \sin \alpha_1 = 153.7$ m/s, and $V_{T2} = V_2 \sin \alpha_2 = 93.97$ m/s.
From equation (14.46) $F_{BX} = -169.3$ N and from equation (14.47) $F_{BY} = 358.4$ N, so that $R_B = \sqrt{F_{BX}^2 + F_{BY}^2} = 396.4$ N. The direction is an angle given by $\theta_B = \tan^{-1} (F_{BY}/F_{BX}) = -64.7°$.

(b) From Illustrative Example 14.1, we have $\dot{m}' = 6$ kg/s, $s = 0.1$ m, $z = 0.05$ m, $\alpha_1 = 30°$, $\alpha_2 = 40°$, $p_1 = 10^6$ Pa, $p_2 = 9.01 \times 10^5$ Pa, $V_1 = 307.4$ m/s, and $V_2 = 374.4$ m/s.
We also have $V_{A1} = 266.2$ m/s, $V_{A2} = V_2 \cos \alpha_2 = 286.8$ m/s, $V_{T1} = V_1 \sin \alpha_1 = 153.7$ m/s, and $V_{T2} = V_2 \sin \alpha_2 = 240.7$ m/s.
From equation (14.46) $F_{BX} = 372.7$ N and from equation (14.47) $F_{BY} = -521.9$ N,
so that $R_B = \sqrt{F_{BX}^2 + F_{BY}^2} = 641.3$ N. The direction is given by $\theta_B = \tan^{-1} (F_{BY}/F_{BX}) - 54.5°$

(c) From Illustrative Example 14.1, we have $\dot{m}' = 11$ kg/s, $s = 0.1$ m, $z = 0.05$ m, $\alpha_1 = 30°$, $\alpha_2 = 20°$, $p_1 = 10^6$ Pa, $p_2 = 6.535 \times 10^5$ Pa, $V_1 = 563.6$ m/s, and $V_2 = 703.8$ m/s.
We also have $V_{A1} = 488.1$ m/s, $V_{A2} = V_2 \cos \alpha_2 = 661.4$ m/s, $V_{T1} = V_1 \sin \alpha_1 = 281.8$ m/s, and $V_{T2} = V_2 \sin \alpha_2 = 240.7$ m/s.
From equation (14.46) $F_{BX} = -173.8$ N and from equation (14.47) $F_{BY} = 451.8$ N, so that $R_B = \sqrt{F_{BX}^2 + F_{BY}^2} = 484.1$ N. The direction is given by $\theta_B = \tan^{-1} (F_{BY}/F_{BX}) = -69°$.
The results of parts (a) and (b) are illustrated through the vector triangles in Figure E14.2.

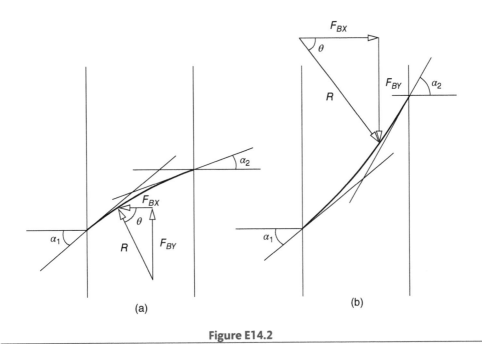

**Figure E14.2**

## 14.6 **Rotor-flow velocity triangles**

Until now in this chapter we have considered flow through a stationary array of blades. We turn now to flow through a ring of blades attached to a rotating disc, i.e. a **compressor** or **turbine rotor**. To analyse such a flow, it is convenient to consider flow relative to the rotor. If the blade length $z$ is much less than the rotor radius $R$, then it is reasonable to treat the flow as two-dimensional and very similar to cascade flow. The circumferential velocity of the rotor $U$ is determined at the mean radius of the rotor, $\bar{R}$, so that $U = \Omega\bar{R}$, $\Omega$ being the angular velocity of the rotor in rad/s. If the mean radius changes from $\bar{R}_1$ at entry to the rotor to $\bar{R}_2$ at exit, there will be a change in $U$ from $U_1$ to $U_2$. If the rotational speed of the rotor in rpm is $N$, then $\Omega = 2\pi N/60$. If the fluid absolute velocity is $V$, and the fluid velocity relative to the rotor blade is $W$, then $\vec{V} = \vec{W} + \vec{U}$, the arrows indicating that these are vector quantities and that the addition is a vector addition, as illustrated by the velocity triangles in Figure 14.6: (a) represents a turbine rotor, and (b) a compressor rotor. The velocity $W$ is also referred to as the velocity in a rotating frame of reference, and $V$ as the velocity in an absolute or stationary frame of reference. The subscripts 1 and 2 refer to the inlet flow and the outlet flow, respectively. The direction of the absolute velocity with respect to the axial direction is measured by the angle $\alpha$, and that of the relative velocity by the angle $\beta$. For consistency with our analysis of cascade flow, it is assumed that the relative flow follows the blade camber line, i.e. $\beta = \chi$.

Both $V$ and $W$ can be resolved into axial components, $V_A$ and $W_A$, and tangential components, $V_T$ and $W_T$, such that

$$V_A = V\cos\alpha, \quad V_T = V\sin\alpha, \quad W_A = W\cos\beta, \quad \text{and} \quad W_T = W\sin\beta. \tag{14.50}$$

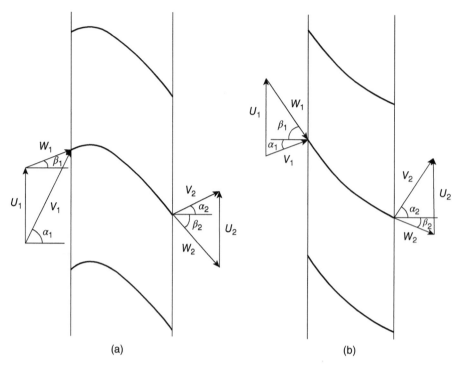

**Figure 14.6** Velocity triangles for flow through (a) a turbine rotor and (b) a compressor rotor

It can be seen from the velocity triangles that $W_A = V_A$, as must be the case since the blade speed $U$ has no axial component. Also, from the velocity triangles we have

$$W^2 = V^2 + U^2 - 2UV \sin \alpha. \tag{14.51}$$

An axial-flow turbine stage comprises a ring of fixed guidevanes or nozzles, termed a stator, followed by a rotor. An axial-flow compressor stage consists of a rotor followed by a stator. The first stage of a multistage compressor is preceded by inlet guidevanes which accelerate the flow rather than diffuse it. For a particular stage in a multistage machine, the purpose of the stator is to accept the flow from the rotor of the previous stage and turn it into the appropriate direction for the succeeding rotor.

## 14.7 Euler's turbomachinery equation for an axial-flow rotor

We consider flow through an idealised rotor such as that shown in either Figure 14.6(a) or Figure 14.6(b). For an axial-flow turbomachine, it can be assumed that the radial component of flow velocity is negligible so that the fluid velocity has two non-zero components: tangential $V_T$ and axial $V_A$. The flow enters at mean radius $\bar{R}_1$ with tangential velocity $V_{T1}$ and leaves at radius $\bar{R}_2$ with tangential velocity $V_{T2}$. The **moment of momentum** (or **angular momentum**) per unit mass about the rotor axis is thus $\bar{R}_1 V_{T1}$ at entry, and $\bar{R}_2 V_{T2}$ at outlet. If the mass flowrate is $\dot{m}$, the torque $T$ exerted on the flow by the rotor to produce this change in the moment of momentum is given by

$$T = \dot{m}\left(\bar{R}_2 V_{T2} - \bar{R}_1 V_{T1}\right) \tag{14.52}$$

which is known as **Euler's turbomachinery equation** and is valid for both axial-flow compressors ($T > 0$) and axial-flow turbines ($T < 0$), and for both incompressible and compressible flow. If the angular velocity of the rotor is $\Omega$, the rate of work input[113] to the flow $\dot{W}$, i.e. the power input $P$, is

$$P = \dot{W} = \Omega T = \Omega \dot{m}\left(\bar{R}_2 V_{T2} - \bar{R}_1 V_{T1}\right). \tag{14.53}$$

Since the **blade speed** $U$ at radius $\bar{R}$ is given by $U = \Omega\bar{R}$, equation (14.53) can be written as

$$P = \dot{W} = \dot{m}\left(U_2 V_{T2} - U_1 V_{T1}\right) \tag{14.54}$$

another form of Euler's turbomachinery equation. If the flow is assumed to be adiabatic, the power input into the flow must equal the change in the stagnation enthalpy flowrate, i.e.

$$P = \dot{W} = \dot{m}\left(U_2 V_{T2} - U_1 V_{T1}\right) = \dot{m}\left(h_{02} - h_{01}\right) \tag{14.55}$$

where the stagnation enthalpy $h_0 = h + V^2/2$, $h$ being the specific enthalpy of the fluid, and $V$ the absolute velocity. From equation (14.42) we see that

$$h_2 + \frac{1}{2}V_2^2 - U_2 V_{T2} = h_1 + \frac{1}{2}V_1^2 - U_1 V_{T1}. \tag{14.56}$$

From the velocity triangle we see that $V^2 = V_A^2 + V_T^2$, and $W^2 = W_A^2 + W_T^2$. Since the axial components of velocity, $V_A$ and $W_A$, are unaltered by the tangential blade velocity $U$, we have $V^2 - W^2 = V_T^2 - W_T^2$. We also have $V_T + W_T = U$ so that $V^2 - W^2 = V_T^2 - W_T^2 = (V_T + W_T)(V_T - W_T) = U(2V_T - U)$, or $V^2 - 2UV_T = W^2 - U^2$. If we use this result to substitute for $V_1^2$ and $V_2^2$ in equation (14.56), we have

$$h_2 + \frac{1}{2}W_2^2 - \frac{1}{2}U_2^2 = h_1 + \frac{1}{2}W_1^2 - \frac{1}{2}U_1^2. \tag{14.57}$$

The combination $h + W^2/2$ is referred to as the **relative stagnation enthalpy**, $h_{0,REL}$ while the combination of terms $h + W^2/2 - U^2/2$ has been termed the **rothalpy**, $I$. Equation (14.57) shows that for an adiabatic flow, as assumed in the derivation of this equation, the rothalpy remains constant through a turbomachine. For a perfect gas $h = C_P T$, so that equation (14.57) can be written in terms of the static temperature $T$ as

$$C_P T_2 + \frac{1}{2}W_2^2 - \frac{1}{2}U_2^2 = C_P T_1 + \frac{1}{2}W_1^2 - \frac{1}{2}U_1^2 \tag{14.58}$$

and in terms of the **relative stagnation temperature** $T_{0,REL}$ as

$$C_P T_{02,REL} - \frac{1}{2}U_2^2 = C_P T_{01,REL} - \frac{1}{2}U_1^2 \tag{14.59}$$

---

[113] For a turbine, the purpose of which is to convert flow energy into mechanical energy, the power output will be $-P$.

the relative stagnation temperature being defined as

$$T_{0,REL} = T + \frac{W^2}{2C_P} = T\left[1 + \left(\frac{\gamma - 1}{2}\right)M_{REL}^2\right] \tag{14.60}$$

where the **relative Mach number** $M_{REL} = W/c$.

We can also define a **relative stagnation pressure** $p_{0,REL}$ as

$$p_{0,REL} = \left[1 + \left(\frac{\gamma - 1}{2}\right)M_{REL}^2\right]^{\gamma/(\gamma-1)}. \tag{14.61}$$

We observe that, for an axial compressor or turbine, the mean blade radius for a single stage is approximately constant, so that $U_1 \approx U_2$, and both the relative stagnation enthalpy and the relative stagnation temperature then remain constant. The relative stagnation pressure also remains constant across the rotor if the flow is assumed to be isentropic. In essence the rotor problem has been reduced to the cascade problem dealt with in Section 14.5, provided the stagnation temperature, $T_0$, is replaced by the relative stagnation temperature, $T_{0,REL}$, and the absolute velocity, $V$, by the relative velocity, $W$.

As already stated, Euler's turbomachinery equation (14.52) is valid for both compressible and incompressible flow, and therefore so is equation (14.55). For an incompressible flow, it is more usual to see the latter equation written as

$$P = \dot{W} = \dot{m}\left(U_2 V_{T2} - U_1 V_{T1}\right) = \dot{Q}\left(p_{02} - p_{01}\right). \tag{14.62}$$

If we introduce the first $Tds$, or **Gibbs, equation**, the justification for replacing $\dot{m}\left(h_{02} - h_{01}\right)$ in equation (14.53) by $\dot{Q}\left(p_{02} - p_{01}\right)$, and the associated limitations, become clear:

$$Tds = du + pd\left(\frac{1}{\rho}\right) \tag{14.63}$$

wherein $s$ is the specific entropy and $u$ is the specific internal energy. We observe that, for an incompressible flow, equation (14.63) shows that $Tds = du$, i.e. an increase in the specific internal energy is consistent with an increase in specific entropy. From the outset we assumed that the flow is adiabatic. If, in addition, we now assume that it is reversible (i.e. no frictional effects) then $ds = 0$, i.e. the flow is isentropic, and also for an incompressible flow there is no change in the internal energy.

By definition, the specific enthalpy $h = u + p/\rho$ so that equation (14.63) leads to the **second** $Tds$ **equation**, which was introduced in Chapter 11

$$Tds = dh - \frac{1}{\rho}dp. \tag{11.1}$$

For an incompressible, isentropic flow, equation (11.1) may be integrated and we have

$$h_2 - h_1 = \frac{1}{\rho}\left(p_2 - p_1\right). \tag{14.64}$$

If we add $V_2^2/2 - V_1^2/2$ to both sides of equation (14.52), we have

$$h_2 + \frac{1}{2}V_2^2 - \left(h_1 + \frac{1}{2}V_1^2\right) = \frac{p_2}{\rho} + \frac{1}{2}V_2^2 - \left(\frac{p_1}{\rho} + \frac{1}{2}V_1^2\right) \tag{14.65}$$

which may be rewritten (since the flow is incompressible) as

$$h_{02} - h_{01} = \frac{1}{\rho} \left( p_{02} - p_{01} \right). \tag{14.66}$$

Finally, if we use equation (14.66) to substitute for $h_{02} - h_{01}$ in equation (14.55), we have

$$P = \dot{W} = \dot{m} \left( U_2 V_{T2} - U_1 V_{T1} \right) = \frac{\dot{m}}{\rho} \left( p_{02} - p_{01} \right) = \dot{Q} \left( p_{02} - p_{01} \right) \tag{14.67}$$

which is the equation for the power input or output of an axial-flow turbomachine where the fluid is incompressible. In more general applications for turbomachinery handling an incompressible fluid, we have

$$P = \dot{W} = \frac{\dot{m}}{\rho} \left( p_{02} - p_{01} \right) = \dot{Q} \left( p_{02} - p_{01} \right). \tag{14.68}$$

## 14.8 Compressible flow through an axial turbomachine stage

We complete this chapter by applying many of the principles of compressible flow of a perfect gas, established in Chapter 11 and previous sections of this chapter, to gas flow through a stator-rotor stage, such as the axial-flow turbine stage shown in Figure 14.7.

At any axial location within either an axial-flow compressor or a turbine, the gas flows through the annular space between concentric surfaces of outer radius $R_O$ and inner radius $R_I$. The corresponding blade length[114] $z = R_O - R_I$, and the flow area $A$ is given by $A = \pi \left( R_O^2 - R_I^2 \right) = \pi \left( R_O - R_I \right) \left( R_O + R_I \right)$. If we define a mean diameter $\overline{D}$ by $\overline{D} = (D_O + D_I)/2 = R_O + R_I$, then

$$A = \pi z \overline{D}. \tag{14.69}$$

We shall assume that the mean diameter of the stator at inlet is $\overline{D}_1$, and $\overline{D}_2$ at outlet, while the blade length increases from $z_1$ to $z_2$. The diameter of the rotor at inlet is also taken to be $\overline{D}_2$, increasing to $\overline{D}_3$ at outlet, while the corresponding blade length increases from $z_2$ to $z_3$. As shown in Figure 14.7, the flow angles and flow velocities are $\alpha_1$ and $V_1$, respectively, at inlet to the stator, and $\alpha_2$ and $V_2$, respectively, at outlet. For the rotor the absolute flow angle and gas velocity at inlet are $\alpha_2$ and $V_2$, respectively, while the relative flow angle and velocity are $\beta_2$ and $W_2$, respectively. At outlet from the rotor the angles are $\alpha_3$ and $\beta_3$, respectively, and velocities are $V_3$ and $W_3$, respectively. The angular velocity of the rotor is $\Omega$ so that the circumferential speed is $U_2 = \Omega D_2/2$ at inlet, and $U_3 = \Omega D_3/2$ at outlet. The mass flowrate of the gas is $\dot{m}$.

(a) **Stator inlet**

If the static pressure and static temperature of the gas at inlet to the stator are $p_1$ and $T_1$, respectively, then, from the perfect-gas equation, the inlet gas density $\rho_1$ is

$$\rho_1 = p_1/RT_1.$$

---

[114] In practice there has to be a small gap or clearance between the blade tips and the casing (for the rotor) and the shaft (for the stator).

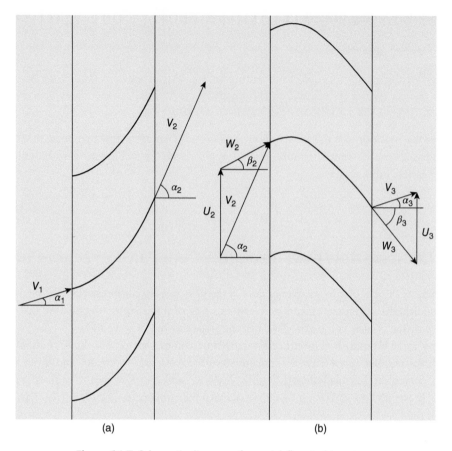

**Figure 14.7** Schematic diagram of an axial-flow turbine stage

From equation (14.70) the inlet area $A_1$ is given by

$$A_1 = \pi \overline{D}_1 z_1$$

and from the continuity equation, the axial-velocity component at inlet to the stator is

$$V_{A1} = \dot{m}/\rho_1 A_1 = \dot{m}/\rho_1 \pi \overline{D}_1 z_1.$$

From the velocity triangle at inlet to the stator, shown in Figure 14.7(a), the inlet velocity $V_1$ is

$$V_1 = V_{A1}/\cos \alpha_1$$

and the tangential component of velocity $V_{T1}$ is

$$V_{T1} = V_1 \sin \alpha_1.$$

It is useful to introduce the flow area $A_{1N}$ normal to $V_1$ from $\dot{m} = \rho_1 A_{1N} V_1$

$$A_{1N} = A_1 \cos \alpha_1.$$

The soundspeed at inlet $c_1$ is given by

$$c_1 = \sqrt{\gamma R T_1}$$

so the inlet Mach number $M_1$ is

$$M_1 = V_1/c_1.$$

The stagnation temperature $T_{01}$ and stagnation pressure $p_{01}$ are then

$$T_{01} = T_1 + \frac{V_1^2}{2C_p} = T_1\left[1 + \left(\frac{\gamma - 1}{2}\right)M_1^2\right]$$

and

$$p_{01} = p_1\left[1 + \left(\frac{\gamma - 1}{2}\right)M_1^2\right]^{\gamma/(\gamma-1)}.$$

These expressions for $T_{01}$ and $p_{01}$ were derived in Section 11.3, and numerical values for the ratios $T_1/T_{01}$ and $p_1/p_{01}$, given $M_1$, can be determined using the Isentropic-flow Calculator. The ratio $A_{1N}/A_1^*$, $A_1^*$ being the choking area corresponding with $M_1$, can also be found using the Calculator or calculated from equation (11.50):

$$A_1^* = \frac{\dot{m}\sqrt{RT_{01}}}{p_{01}}\sqrt{\frac{1}{\gamma}\left(\frac{\gamma + 1}{2}\right)^{(\gamma+1)/(\gamma-1)}}.$$

**(b) Stator outlet**

We shall assume that the flow through the stator is isentropic so that the stagnation temperature and pressure remain unchanged, i.e. at outlet $T_{02} = T_{01}$, and $p_{02} = p_{01}$.

The Mach number $M_2$ at outlet from the stator can be determined through the Calculator from the ratio of the outlet area $A_{2N}$, normal to the outflow velocity $V_2$, to the choking area $A_2^*$, which is unchanged from $A_1^*$ since the flow is isentropic. The Isentropic-flow Calculator also gives values for $p_2/p_{02}$ and $T_2/T_{02}$ so that the static temperature $T_2$ and static pressure $p_2$ can be calculated.

From the perfect-gas equation $\rho_2 = p_2/RT_2$, the soundspeed $c_2 = \sqrt{\gamma RT_2}$, the outlet velocity $V_2 = M_2 c_2$, and the tangential component of velocity $V_{T2} = V_2 \sin\alpha_2$.

**(c) Rotor inlet**

Since we are now concerned with flow through a rotor, it is appropriate to consider the flow relative to the rotor. The velocity triangle for the inlet flow is shown in Figure 14.6(a), with appropriate changes to the subscripts. The absolute velocity at inlet is $V_2$ at angle $\alpha_2$ to the axial direction, while the relative velocity is $W_2$ at angle $\beta_2$ which we assume is the same as the blade angle $\chi_2$ at the rotor inlet.

From the velocity triangle for the flow at the rotor inlet, the axial components of the absolute and relative velocities are equal, i.e. $W_{A2} = V_{A2} = V_2\cos\alpha_2 = W_2\cos\chi_2$, from which the relative velocity $W_2$ at inlet to the rotor can be obtained.

The relative Mach number at inlet to the rotor $M_{2,REL} = W_2/c_2$, and from the Isentropic-flow Calculator we can find $p_2/p_{02,REL}$ and $T_2/T_{02,REL}$ and hence the relative stagnation pressure $p_{02,REL}$ and the relative stagnation temperature $T_{02,REL}$.

Since both the relative and absolute velocities at the rotor inlet, $W_2$ and $V_2$, are now both known together with the angle between them, $\alpha_2 - \beta_2$ (see the velocity triangle in Figure 14.7 at the rotor inlet), we can calculate the blade speed $U_2$ from

$$U_2^2 = V_2^2 + W_2^2 - 2V_2 W_2 \cos(\alpha_2 - \beta_2).$$

The rotational speed of the rotor is then $\Omega = 2U_2/\overline{D}_2$.

The relative stagnation temperature $T_{02,REL}$ is given by

$$T_{02,REL} = T_2 + W_2^2/2C_P$$

and the relative stagnation pressure $p_{02,REL}$ by

$$p_{02,REL} = p_2 \left[ 1 + \left( \frac{\gamma - 1}{2} \right) M_{2REL}^2 \right]^{\gamma/(\gamma-1)}.$$

Both $T_{02,REL}$ and $p_{02,REL}$ can be calculated from these equations but are more easily determined using the Isentropic-flow Calculator from $T_2/T_{02,REL}$ and $p_2/p_{02,REL}$.

The relative stagnation enthalpy $h_{02,REL} = C_P T_{02,REL}$ and the rothalpy $I_2$ at entry to the rotor is given by

$$I_2 = h_{02,REL} - \frac{1}{2} U_2^2 = C_P T_2 + \frac{1}{2} W_2^2 - \frac{(\Omega \overline{D}_2)^2}{8}.$$

(d) **Rotor outlet**

Since the flow is adiabatic, the rothalpy remains constant so that at the rotor outlet $I_3 = I_2$, i.e.

$$C_P T_3 + \frac{1}{2} W_3^2 - \frac{1}{2} U_3^2 = C_P T_2 + \frac{1}{2} W_2^2 - \frac{1}{8} \left( \Omega \overline{D}_2 \right)^2.$$

and the blade speed at outlet from the rotor is given by $U_3 = \Omega \overline{D}_3/2$ so that

$$C_P T_3 + \frac{1}{2} W_3^2 = C_P T_2 + \frac{1}{2} W_2^2 + \frac{\Omega^2}{8} \left( \overline{D}_3^2 - \overline{D}_2^2 \right). \tag{14.70}$$

The relative Mach number $M_{3,REL} = W_3/c_3$, where the soundspeed $c_3 = \sqrt{\gamma R T_3}$. Equation (14.70) can thus be written as

$$c_3 \sqrt{\left[ 1 + \left( \frac{\gamma - 1}{2} \right) M_{3,REL}^2 \right]} = \sqrt{(\gamma - 1) \left[ C_P T_2 + \frac{1}{2} W_2^2 + \frac{\Omega^2}{8} \left( \overline{D}_3^2 - \overline{D}_2^2 \right) \right]}. \tag{14.71}$$

The mass flowrate $\dot{m}$ is again unchanged, so that

$$\dot{m} = \rho_3 \pi z_3 \overline{D}_3 W_{A3} = \rho_3 \pi z_3 \overline{D}_3 W_3 \cos \beta_3 \tag{14.72}$$

from which

$$\rho_3 W_3 = \frac{\dot{m}}{\pi z_3 \overline{D}_3 \cos \chi_3} \tag{14.73}$$

wherein we have assumed that the relative flow direction at outlet from the rotor coincides with the blade angle, i.e. $\beta_3 = \chi_3$.

From the perfect-gas equation, $\rho_3 = p_3/R T_3$ so that equation (14.73) can be written as

$$\frac{p_3 M_{3,REL}}{c_3} = \frac{\dot{m}}{\gamma \pi z_3 \overline{D}_3 \cos \chi_3}. \tag{14.74}$$

Equations (14.71) and (14.74) can be combined to eliminate $c_3$. After some rearrangement we find

$$p_3 M_{3,REL} \sqrt{\left[ 1 + \left( \frac{\gamma - 1}{2} \right) M_{3,REL}^2 \right]} = \frac{\dot{m} \sqrt{(\gamma - 1) \left[ C_P T_2 + \frac{1}{2} W_2^2 + \frac{\Omega^2}{8} \left( \overline{D}_3^2 - \overline{D}_2^2 \right) \right]}}{\gamma \pi z_3 \overline{D}_3 \cos \chi_3}. \tag{14.75}$$

To progress further requires information about the static pressure $p_3$ at the rotor outlet. For simplicity, we shall assume that the relative stagnation pressure remains unchanged through the rotor, i.e. $p_{03REL} = p_{02REL}$. We note that $p_{03REL} = p_3 \left[ 1 + (\gamma - 1) M_{3,REL}^2/2 \right]^{\gamma/\gamma-1}$ so that equation (14.75) can be written as

$$\frac{M_{3,REL}}{\left[1 + \left(\frac{\gamma - 1}{2}\right) M_{3,REL}^2\right]^{(\gamma+1)/2(\gamma-1)}} = \frac{\dot{m}\sqrt{(\gamma - 1)\left[C_P T_2 + \frac{1}{2} W_2^2 + \frac{\Omega^2}{8}\left(\overline{D}_3^2 - \overline{D}_2^2\right)\right]}}{\gamma \pi z_3 \overline{D}_3 \cos \chi_3 p_{02,REL}}.$$

(14.76)

Equation (14.76) can be solved (iteratively) for $M_{3,REL}$, since all terms on the right-hand side of the equation are known. Once the value of $M_{3,REL}$ has been determined, it is straightforward to calculate $p_3, \rho_3, T_3, c_3$, and $W_3$.

The final flow velocity $V_3$ and direction $\alpha_3$ can be found using the velocity triangle at the rotor outlet

$$V_3^2 = W_3^2 + U_3^2 - 2 W_3 U_3 \sin \beta_3$$

(14.77)

and

$$\cos \alpha_3 = \frac{W_3 \cos \beta_3}{V_3}.$$

(14.78)

The stagnation temperature at outlet is found from $T_{03} = T_3 + V_3^2/2C_P$, and the stagnation enthalpy from $h_{03} = C_P T_{03}$.

Finally, the power output from the turbine $P$ can be calculated from either the steady-flow-energy equation

$$P = -\dot{m}(h_{03} - h_{02})$$

(14.79)

or Euler's turbomachinery equation

$$P = -\dot{m}(U_3 V_{T3} - U_2 V_{T2}).$$

(14.54)

## 14.9 Degree of reaction $\Lambda$

The power generated in a turbine is invariably accompanied by expansion of the working fluid. For any stage of a multistage turbine the expansion is shared by the rotor and the stator. An important design parameter which quantifies this sharing is the **degree of reaction** (or **reaction**) $\Lambda$ defined by

$$\Lambda = \frac{\Delta h_{ROTOR}}{\Delta h_{STAGE}} = \frac{\Delta T_{ROTOR}}{\Delta T_{STAGE}}$$

(14.80)

i.e. the ratio of the static enthalpy (or temperature) change across the rotor to the change in static enthalpy[115] (or temperature) across the stage.

---

[115] The denominator in the definition of the reaction is sometimes taken as the change in stagnation enthalpy across the stage.

In principle the value of $\Lambda$ can be anywhere between zero (i.e. all the enthalpy change occurs across the stator) and unity (i.e. all the enthalpy change occurs across the rotor and the stator simply turns the flow without expanding it). A zero-reaction turbine is commonly referred to as an impulse turbine. According to Euler's turbomachinery equation, power is produced as a result of turning the fluid through an angle, the turning angle $\theta = \alpha_2 - \alpha_1$, in the stator-blade passages thereby changing the tangential component of velocity. Values for $\Lambda$ close to 0.5 are more usual than the two extremes.

---

**ILLUSTRATIVE EXAMPLE 14.3**

Air, at a static temperature of 700°C and static pressure of 3.75 bar, enters an axial-flow gas-turbine stage at a mass flowrate of 25 kg/s. The stator-blade angles are 20° at entry and 50° at exit. For the rotor, the blade angle at entry is 25° and –45° at exit. The mean blade diameter is 0.5 m for both stator and rotor, and the blade length is 50 mm for both. Calculate the flow conditions (static pressure, static temperature, flow velocity and direction, Mach number) at entry to and exit from the rotor, the reaction, and the power developed by the turbine. Sketch the velocity triangles at entry to, and exit from the stator/entry to the rotor, and at exit from the rotor, including both the absolute and relative velocities and their components. Assume the flow is adiabatic throughout, is isentropic through the stator, and the relative stagnation pressure is constant through the rotor. Assume too that the flow direction always matches the blade angles. The schematic diagram of an axial gas-turbine stage in Figure 14.7 shows the geometric arrangement for this example.

**Solution**

$p_1 = 3.75 \times 10^5$ Pa, $T_1 = 973$K, $\alpha_1 = 20°$, $\alpha_2 = 50°$, $\beta_2 = 25°$, $\beta_3 = -45°$, $\overline{D} = 0.5$ m, and $z = 0.05$ m.

(a) **Stator entry**

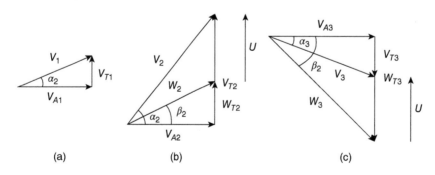

**Figure E14.3**

Annulus area $A = \pi \overline{D} z = 0.0785$m$^2$.
From the perfect-gas equation, entry density $\rho_1 = p_1/RT_1 = 1.343$ kg/m$^3$.
From the continuity equation, the axial-velocity component $V_{A1} = \dot{m}/\rho_1 A =$
   237.0 m/s $= V_1 \cos\alpha_1$.

From the velocity triangle at entry to the stator, Figure E14.3(a), absolute velocity
$V_1 = V_{A1}/\cos\alpha_1 = 252.2$ m/s.
Soundspeed $c_1 = \sqrt{\gamma R T_1} = 625.3$ m/s.
Entry Mach number $M_1 = V_1/c_1 = 0.4034$.
From the Isentropic-flow Calculator with this value of $M_1$, $T_1/T_{01} = 0.9685$, and
$p_1/p_{01} = 0.8939$, so that stagnation temperature $T_{01} = 1004.7$ K, and stagnation pressure
$p_{01} = 4.195$ bar.
It is convenient to introduce the choking area $A_1^*$ corresponding to $M_1$ : from the
Isentropic-flow Calculator, the area ratio $A_{1N}/A_1^* = 1.579$, where $A_{1N}$ is the flow area
normal to the flow direction at entry, i.e. $A_{1N} = A\cos\alpha_1$.

(b) **Stator exit/rotor entry**
The flow area normal to the flow direction at exit from the stator, $A_{2N} = A\cos\alpha_2$,
and the choking area remains unchanged since the flow is isentropic, so the area ratio
$A_{2N}/A_2^* = (A_{1N}/A_1^*)(A_{2N}/A_{1N}) = (A_{1N}/A_1^*)(\cos\alpha_2/\cos\alpha_1) = 1.080$.
From the Isentropic-flow Calculator with this area ratio, $M_2 = 0.7206$, $p_2/p_{02} = 0.7077$,
and $T_2/T_{02} = 0.9059$.
Since the flow through the stator is isentropic, $T_{02} = T_{01} = 1004.7$ K, and $p_{02} = p_{01} = 4.195$ bar, so that static temperature $T_2 = 910.2$ K, and static pressure $p_2 = 2.969$ bar.
From the perfect-gas equation, the stator exit density $\rho_2 = p_2/RT_2 = 1.136$ kg/m$^3$.
From the continuity equation, the axial-velocity component $V_{A2} = \dot{m}/\rho_2 A = 280.1$ m/s
$= V_2\cos\alpha_2$.
From the velocity triangle at the stator exit, Figure E14.3(b), the absolute velocity $V_2 = V_{A2}/\cos\alpha_2 = 435.7$ m/s, and the tangential component of velocity $V_{T2} = V_2\sin\alpha_2 = 333.8$ m/s.
The soundspeed $c_2 = \sqrt{\gamma RT_2} = 604.7$ m/s, and the Mach number $M_2 = V_2/c_2 = 0.7206$.
From the velocity triangle for the rotor entry, the axial component of relative velocity
$W_{A2} = V_{A2} = 280.1$ m/s $= W_2\cos\beta_2$, so that the relative velocity $W_2 = 309.0$ m/s.
The relative-velocity Mach number at entry to the rotor $M_{2,REL} = W_2/c_2 = 0.5110$.
From the Isentropic-flow Calculator, for this value of the relative Mach number,
$T_2/T_{02,REL} = 0.9504$, and $p_2/p_{02,REL} = 0.8368$, so that the relative stagnation temperature
$T_{02,REL} = 957.7$ K, and the relative stagnation pressure $p_{02,REL} = 3.548$ bar.
From the velocity triangle at rotor entry, $U^2 = V_2^2 + W_2^2 - 2V_2 W_2\cos(\alpha_2 - \beta_2)$, from
which the blade speed $U = 203.2$ m/s.
The rothalpy at entry to the rotor $I_2 = C_P T_2 + W_2^2/2 - U^2/2 = 9.414 \times 10^5$ m$^2$/s$^2$.

(c) **Rotor exit**
Since the relative stagnation pressure remains constant through the rotor, the relative
stagnation pressure at exit $p_{03,REL} = p_{02,REL} = 3.548$ bar.
The rothalpy also remains unchanged through the rotor, so that at the rotor exit $I_2 = I_3 = C_P T_3 + W_3^2/2 - U^2/2 = 9.414$ m$^2$/s$^2$.
The relative stagnation enthalpy is then $h_{03,REL} = I_3 + U^2/2 = C_P T_{03,REL} = C_P T_3 + W_3^2/2 = 9.620 \times 10^5$ m$^2$/s$^2$, and the relative stagnation temperature $T_{03,REL} = 957.7$ K.
As shown in Section 14.8, we can combine all the information necessary to calculate the
relative outlet Mach number $M_{3,REL}$ into the equation, based upon equation (14.76):

$$M_{3,REL} = \frac{\dot{m}\sqrt{(\gamma-1)\left(I + \frac{1}{2}U^2\right)}\left[1 + \left(\frac{\gamma-1}{2}\right)M_{3,REL}^2\right]^{(\gamma+1)/2(\gamma-1)}}{\gamma p_{0,REL}\pi \overline{D} z \cos\beta_3}.$$

Note that since $D_2 = D_3$, here, the term $D_3^2 - D_2^2$ in equation (14.76) vanishes.

If we substitute $I = I_3$, $p_{0,REL} = p_{02,REL}$, and the values for $\dot{m}, \gamma, \pi, \overline{D}, z$, and $\beta_3$, we find

$$M_{3,REL} = 0.5622 \left[ 1 + \left( \frac{\gamma - 1}{2} \right) M_{3,REL}^2 \right]^{(\gamma+1)/2(\gamma-1)}$$

which, after several iterations starting with $M_{3,REL} = 0.5622$, leads to $M_{3,REL} = 0.8233$.
From the Isentropic-flow Calculator, $p_3/p_{03,REL} = 0.6409$, and $T_3/T_{03,REL} = 0.8806$, from which $p_3 = 2.274$ bar, and $T_3 = 843.4$ K.
The exit density $\rho_3 = p_3/RT_3 = 0.9393$ kg/m³.
The exit soundspeed $c_3 = \sqrt{\gamma RT_3} = 582.1$ m/s.
The relative exit velocity $W_3 = M_{3,REL}c_3 = 479.2$ m/s.
From the velocity triangle at the rotor exit, Figure E14.3(c), the axial component of $W_3$ is $W_{A3} = W_3 \cos\beta_3 = 338.9$ m/s $= V_{A3}$, and the tangential component of $W_3$ is $W_{T3} = -W_3 \sin\beta_3 = -338.9$ m/s.
Also from the velocity triangle at the rotor exit, $V_3^2 = W_3^2 + U^2 - 2W_3 U \sin\beta_3$, from which the flow velocity at exit from the rotor $V_3 = 365.0$ m/s.
The tangential component of velocity $V_{T3} = U - W_{T3} = -135.7$ m/s.
The flow direction for $V_3$ is then $\alpha_3 = \sin^{-1}(V_{T3}/V_3) = -21.82°$.
The Mach number $M_3 = V_3/c_3 = 0.6271$.
From the Isentropic-flow Calculator, $p_3/p_{03} = 0.7672$, and $T_3/T_{03} = 0.9271$, from which $p_{03} = 2.963$ bar, and $T_{03} = 909.7$ K.
The reaction $\Lambda = (T_2 - T_3)/(T_1 - T_3) = 0.5152$.
The power output of the turbine is then $P = \dot{m}C_P(T_{02} - T_{03}) = 2.385 \times 10^6$ W, with the same value given by $P = \dot{m}U(V_{T2} - V_{T3})$.

**Comments:**

Although this was a lengthy calculation, the reader should be aware that it consisted primarily of numerous small steps using basic relations, many of them repeated several times. The important point to note is that the procedure involved was systematic with each step leading to the next.

 14.10 SUMMARY

This chapter has been concerned primarily with the flow of a compressible fluid through stationary and moving blading, for the most part using the analysis introduced in Chapter 11. The principles of dimensional analysis were applied to determine the appropriate non-dimensional parameters to characterise the performance of a turbomachine. The analysis of incompressible flow through a linear cascade was followed by the analysis of compressible flow. Velocity triangles and Euler's turbomachinery equation were introduced to analyse flow through a rotor. The concepts introduced were applied to the analysis of an axial turbomachine stage.
    The student should

- be able to apply the principles of dimensional analysis to derive the non-dimensional parameters used to characterise the performance of turbomachinery

- in the context of turbomachinery, understand what is meant by a cascade, a stator, a rotor, and a stage
- understand how changes in the flow direction (turning) for flow through a cascade, stator, or rotor result in a change in flow area
- be able to analyse both incompressible and compressible flow through a linear cascade of blades or an axial-flow stator
- understand the principle of relative flow through moving blading and be able to construct velocity triangles at entry to and exit from a rotor
- understand the basis for Euler's turbomachinery equation and be able to apply it to a turbomachine rotor
- understand the concepts and definitions of relative stagnation temperature, relative stagnation pressure, relative Mach number, and the property rothalpy
- be able to analyse both incompressible and compressible flow through an axial-flow rotor to determine both flow properties at entry and exit as well as the power input (for a compressor) or output (for a turbine)
- understand the concept and definition of the reaction

## ? 14.11 SELF-ASSESSMENT PROBLEMS

**14.1** An axial-flow compressor is designed to compress hydrogen gas (specific gas constant $R = 4124 \text{ m}^2/\text{s}^2 \cdot \text{K}$) from a temperature $T_1$ of 25°C and a pressure $p_1$ of 1 bar when running at a rotation speed $N$ of 5000 rpm. The design mass flowrate $\dot{m}$ is 75 kg/s. Calculate the rotation speed and mass flowrate for a half-scale model compressor tested with air at an initial temperature of 20°C and initial pressure 1 bar, assuming the flow coefficient $(\dot{m}\sqrt{RT_1}/p_1 D^2)$ and the non-dimensional rotation speed $(\Omega D/\sqrt{RT_1})$ are the same for the actual compressor and the model. The ratio of specific heats $\gamma$ for both gases may be taken as 1.4.
(Answers: 2616 rpm, 71.7 kg/s)

**14.2** The mean blade diameter at inlet to the inlet of one stage of an axial-flow turbine is 0.75 m, and the blade length 150 mm. At exit from the stator and entry to the rotor the mean blade diameter is 0.77 m, and the blade length is 170 mm. At exit from the rotor the mean blade diameter is 0.79 m, and the blade length is 190 mm. The mass flowrate of gas through the stage is 35 kg/s. The static temperature at entry to the stage is 930°C, and the static pressure is 2.5 bar. The flow enters the stage with a flow angle of 20°. The flow exit angle from the stator is 65° while the blade angles are 30° at entry to the rotor, and 60° at exit. All angles are with respect to the axial-flow direction. It may also be assumed that the flow processes within the stator are isentropic, while within the rotor the rothalpy and stagnation pressure remain constant. The gas can be treated as a perfect gas with the same properties as air.

Calculate the flow conditions (static pressure, static temperature, flow velocity, and flow direction) at (a) the stator exit and (b) the rotor exit. Also calculate the power output from the turbine and the reaction.
(Answers: (a) 2.262 bar, 1169.1 K, 298.7 m/s, 65°; (b) 2.161 bar, 1154.9 K, 114.0 m/s, 2.90°, 1.834 MW, 0.296)

**14.3**  An axial-flow compressor with mean blade diameter 0.75 m is designed so that the axial component of the gas velocity is constant throughout the machine and the flow enters the rotor of each stage at 20° to the axial direction (i.e. this is the angle at which the flow leaves each stator). The entry-flow angle for each stator is 40°. The mass flowrate is 100 kg/s, and the rotational speed is 7500 rpm. For a stage where the inlet static temperature is 20°C, the static pressure is 2 bar, and the blade length is 120 mm, sketch the velocity triangles at entry to and exit from the rotor, and calculate

(a) the axial component of velocity

(b) the static pressure, static temperature, absolute velocity, and Mach number at exit from the rotor

(c) the static pressure, static temperature, absolute velocity, and Mach number at exit from the stator

(d) the blade length at exit from the rotor and from the stator

(e) the reaction and the power input into the stage

Assume that the gas has the same properties as air, that the relative stagnation pressure and the rothalpy are both constant within the rotor, that the flow is adiabatic, and that the flow through the stator is isentropic.

(Answers: (a) 148.7 m/s; (b) 2.366 bar, 307.4 K, 194.1 m/s, 0.5523; (c) 2.5402 bar, 313.7 K, 0.4457; (d) 106 mm, 101 mm; (e) 0.696, 2.081 MW)

# 15

# Basic equations of viscous-fluid flow

In this chapter we derive the partial differential equations, based upon the principles of mass and momentum conservation, which describe unsteady (i.e. time-dependent), three-dimensional viscous-fluid flows. The principle of mass conservation leads to the **continuity equation**, while **Cauchy's equations of motion** result from Newton's second law of motion applied to a fluid. Cauchy's equations lead to the **Navier-Stokes equations** when we introduce **Stokes' constitutive equations**, which, for a Newtonian fluid, connect the **normal** and **shear** stresses[116] with the **strain rates** within the fluid. If flows with density and viscosity variations are to be considered, it is also necessary to introduce the energy-conservation equation and an equation of state, but this is not the case here, where we shall ultimately limit attention to flows with physical properties independent of pressure and temperature. For completeness and reference, we present the continuity equation, Cauchy's equations, and the Navier-Stokes equations not only in rectangular-Cartesian-coordinate form, but also in cylindrical-coordinate form and in vector form. We conclude the chapter by extending consideration to the flow of a **generalised Newtonian fluid**.

## 15.1 Equations of motion in Cartesian-coordinate form

In this section we derive the basic equations for the flow of a viscous fluid in a rectangular Cartesian-coordinate system with velocity components $u$ in the $x$-direction, $v$ in the $y$-direction, and $w$ in the $z$-direction. At any instant of time $t$ the fluid is assumed to have density $\rho(t, x, y, z)$ and dynamic viscosity $\mu(t, x, y, z)$.

### 15.1.1 Continuity equation

We consider flow through an infinitesimal cubic volume element fixed in space and shown in Figure 15.1. For convenience we identify the cube dimensions as $\delta x$, $\delta y$, and $\delta z$, although, being a cube, it is obvious that the side lengths are all equal in magnitude. In words, the **principle of mass conservation** is

$$[rate\ of\ mass\ accumulation] = [mass\ flowrate\ in] - [mass\ flowrate\ out]. \qquad (15.1)$$

The mass of the volume element at time $t$ is $\rho\,\delta x\,\delta y\,\delta z$ so that the rate of mass accumulation is $\delta x\,\delta y\,\delta z\,(\partial\rho/\partial t)$.

---

[116] The term **shearing stress** is sometimes used.

*Introduction to Engineering Fluid Mechanics*. Marcel Escudier.
© Marcel Escudier 2017. Published 2017 by Oxford University Press.

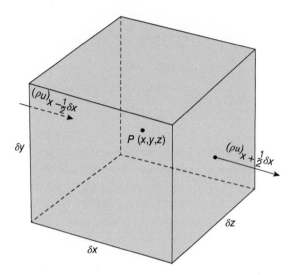

**Figure 15.1** Infinitesimal volume element fixed in space. Only the $x$-direction mass flux is shown.

We take the location of the cube centre as the point $P(x, y, z)$ so that the faces perpendicular to the $x$-direction are located at $x - \delta x/2$, and $x + \delta x/2$. The mass flowrate into the cube through the face at $x - \delta x/2$ is

$$(\rho u)_{x-\delta x/2}\, \delta y\, \delta z,$$

the product $\rho u$ being the **mass flux**, and the subscript $x - \delta x/2$ denoting the $x$-location of the cube face. The mass flowrate out of the cube through the opposite face at $x + \delta x/2$ is

$$(\rho u)_{x+\delta x/2}\, \delta y\, \delta z.$$

Similar expressions may be written for the other two pairs of faces.

According to equation (15.1), we thus have

$$\delta x\, \delta y\, \delta z \frac{\partial \rho}{\partial t} = \left[ (\rho u)_{x-\delta x/2} - (\rho u)_{x+\delta x/2} \right] \delta y\, \delta z + \left[ (\rho v)_{y-\delta y/2} - (\rho v)_{y+\delta y/2} \right] \delta x\, \delta z$$

$$+ \left[ (\rho w)_{z-\delta z/2} - (\rho w)_{z+\delta z/2} \right] \delta y\, \delta x. \tag{15.2}$$

If we divide through by $\delta x\, \delta y\, \delta z$ and take the limit as $\delta x$, $\delta y$, and $\delta z$ approach zero, we find

$$\frac{\partial \rho}{\partial t} + \frac{\partial}{\partial x}(\rho u) + \frac{\partial}{\partial y}(\rho v) + \frac{\partial}{\partial z}(\rho w) = 0. \tag{15.3}$$

This general **continuity equation**, a reduced form of which was derived in Section 6.8 for steady, one-dimensional flow, can also be written as

$$\frac{\partial \rho}{\partial t} + u\frac{\partial \rho}{\partial x} + v\frac{\partial \rho}{\partial y} + w\frac{\partial \rho}{\partial z} + \rho \left( \frac{\partial u}{\partial x} + \frac{\partial v}{\partial y} + \frac{\partial w}{\partial z} \right) = 0, \tag{15.4}$$

or

$$\frac{D\rho}{Dt} + \rho \left( \frac{\partial u}{\partial x} + \frac{\partial v}{\partial y} + \frac{\partial w}{\partial z} \right) = 0 \tag{15.5}$$

where we have introduced the **derivative following the fluid**[117]

$$\frac{D}{Dt} = \frac{\partial}{\partial t} + u\frac{\partial}{\partial x} + v\frac{\partial}{\partial y} + w\frac{\partial}{\partial z}.$$ (15.6)

This derivative is also variously referred as the **particle, substantial** (or **substantive**), **total, Lagrangian, Eulerian,** or **material derivative**. In this text we shall use the term material derivative. The first term on the right-hand side of equation (15.6) represents change due to any **unsteadiness** (i.e. time dependence) at the point $P(x, y, z)$. The final three terms are associated with changes due to fluid particles changing their position and are referred to as the **advective** or **convective**[118] terms.

It is easily seen from equation (15.5) that, for the flow (steady or unsteady) of a constant-density fluid, the continuity equation in Cartesian-coordinate form reduces to

$$\frac{\partial u}{\partial x} + \frac{\partial v}{\partial y} + \frac{\partial w}{\partial z} = 0.$$ (15.7)

Compact forms of the continuity equation (15.3), independent of coordinate system, are

$$\frac{\partial \rho}{\partial t} + div\,(\rho V) = \frac{\partial \rho}{\partial t} + \nabla \cdot (\rho V) = 0$$ (15.8)

and

$$\frac{D\rho}{Dt} + \rho\,div\,V = \frac{D\rho}{Dt} + \rho\nabla \cdot V = 0,$$ (15.9)

where $V$ is the vector velocity

$$V = iu + jv + kw,$$ (15.10)

$i$, $j$, and $k$ being the unit vectors in the $x$-, $y$-, and $z$-directions, respectively, while $div$ represents the vector operator[119] of **divergence**, and the symbol $\nabla$ represents the del (or gradient) operator

$$\nabla = i\frac{\partial}{\partial x} + j\frac{\partial}{\partial y} + k\frac{\partial}{\partial z}.$$ (15.11)

For a Cartesian-coordinate system, the divergence of a vector $A\,(= iA_x + jA_y + kA_z)$ is the dot product of the vectors $\nabla$ and $A$

$$div\,A = \nabla \cdot A = \nabla \cdot (iA_x + jA_y + kA_z) = \frac{\partial A_x}{\partial x} + \frac{\partial A_y}{\partial y} + \frac{\partial A_z}{\partial z}.$$ (15.12)

It should be clear that in equation (15.9) we have replaced the partial derivatives of the velocity components in equations (15.4) and (15.5) by the divergence ($div$) of the vector velocity $V$, i.e. $\nabla \cdot V$.

---

[117] The symbol $d/dt$ is sometimes used for this quantity instead of $D/Dt$ although this can obviously lead to confusion as $d/dt$ normally represents a simple time derivative.

[118] The term 'convective' is more commonly encountered in the study of heat transfer.

[119] The vector form of the continuity equation, together with a number of other equations later on in this chapter, are included here for completeness and reference although it is appreciated that many engineering students in the early years of study will not be familiar with vector algebra.

For a constant-density flow we have

$$\nabla \cdot V = 0. \tag{15.13}$$

### 15.1.2 Cauchy's equations of motion

In Subsection 15.1.1, we showed how the principle of mass conservation applied to flow through an elemental cube fixed in space led to the continuity equation. We now apply the **principle of linear momentum conservation**, which is derived directly from **Newton's second law of motion**, to the same volume element. In words, the principle of momentum conservation, for any given direction, may be written for the volume element as

$$\left[net\ force\ acting\ on\ fluid\ volume\right] = \left[rate\ of\ momentum\ accumulation\right]$$
$$+ \left[momentum\ flowrate\ out\right] - \left[momentum\ flowrate\ in\right]. \tag{15.14}$$

The forces arise from the shear stresses, $\tau_{xy}, \tau_{yx}, \tau_{yz}, \tau_{zy}, \tau_{zx}$, and $\tau_{xz}$, and the normal stresses, $\sigma_{xx}, \sigma_{yy}$, and $\sigma_{zz}$, acting at the centre of the cube $P(x, y, z)$, as shown in Figure 15.2(b). These stresses act on the three orthogonal plane surfaces (within the fluid) which intersect at $P(x, y, z)$, as shown in Figure 15.2(a). The symbol $X$ denotes the surface with sides $\delta y$ and $\delta z$, and $Y$ the surface with sides $\delta z$ and $\delta x$, while $Z$ denotes the surface with sides $\delta x$ and $\delta y$. The point $P$ also corresponds with the centroids of the surfaces $X$, $Y$, and $Z$. The first subscript attached to a stress indicates the direction of the normal of the fluid surface on which the stress acts, while the second subscript indicates the direction in which the stress itself acts.

Figure 15.3 shows the shear and normal stresses acting in the $x$-direction on all six faces of the element, which has side lengths $\delta x, \delta y$, and $\delta z$. The normal forces in the $x$-direction are due to the normal stress $\sigma_{xx}$, which includes the fluid pressure $p$, as discussed in Subsection 15.1.4, together with $f_x$, the $x$-component of the body force per unit mass (not included in Figure 15.3), which acts through the centre $P$. Typical body forces which arise are those due to gravitational or centripetal acceleration, or electromagnetism (in **magnetohydrodynamics**). On the $x - \delta x/2$ face of the volume the normal force is

$$- \left( \sigma_{xx} - \frac{1}{2} \frac{\partial \sigma_{xx}}{\partial x} \delta x \right) \delta y\, \delta z,$$

and on the opposite face it is

$$+ \left( \sigma_{xx} + \frac{1}{2} \frac{\partial \sigma_{xx}}{\partial x} \delta x \right) \delta y\, \delta z.$$

The normal-force difference in the $x$-direction is thus

$$\frac{\partial \sigma_{xx}}{\partial x} \delta x\, \delta y\, \delta z$$

while the body force is $\rho f_x\, \delta x\, \delta y\, \delta z$.

The $x$-direction forces exerted on the remaining faces of the volume are due to the shear stresses $\tau_{xy}, \tau_{yx}, \tau_{xz}$, and $\tau_{zx}$.

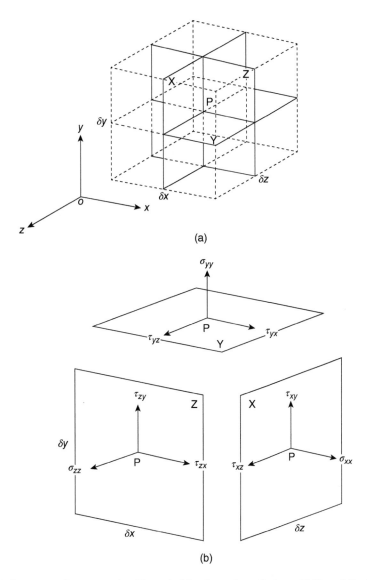

**Figure 15.2** Stresses acting at a point $P(x, y, z)$: (a) orientation of planes $X$, $Y$, and $Z$ within volume element (b) normal stresses $\sigma_{xx}$, $\sigma_{yy}$, and $\sigma_{zz}$; shear stresses $\tau_{xy}$, $\tau_{yx}$, $\tau_{yz}$, $\tau_{zy}$, $\tau_{zx}$, and $\tau_{xz}$

It is evident that the shear forces

$$- \left( \tau_{yx} - \frac{1}{2} \frac{\partial \tau_{yx}}{\partial y} \, \delta y \right) \delta x \, \delta z \quad \text{exerted on the } y - \frac{\delta y}{2} \text{ face}$$

and

$$+ \left( \tau_{yx} + \frac{1}{2} \frac{\partial \tau_{yx}}{\partial y} \, \delta y \right) \delta x \, \delta z \quad \text{exerted on the } y + \frac{\delta y}{2} \text{ face}$$

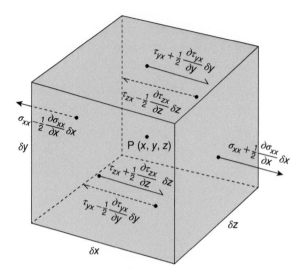

**Figure 15.3** $x$-Direction normal and shear stresses acting on the faces of an elemental cube of fluid

generate moments on the volume element acting clockwise about an axis parallel to the $z$-axis and passing through $P$, while the forces

$$-\left(\tau_{xy} - \frac{1}{2}\frac{\partial \tau_{xy}}{\partial x}\delta x\right)\delta y\,\delta z \quad \text{exerted on the } x - \frac{\delta x}{2} \text{ face}$$

and

$$+\left(\tau_{xy} + \frac{1}{2}\frac{\partial \tau_{xy}}{\partial x}\delta x\right)\delta y\,\delta z \quad \text{exerted on the } x + \frac{\delta x}{2} \text{ face}$$

generate moments acting anticlockwise. The net clockwise moment is

$$\frac{1}{2}\left(\tau_{yx} - \tau_{xy}\right)\delta x\,\delta y\,\delta z = \rho\delta x\,\delta y\,\delta z\, k^2\,\ddot{\theta}$$

where $k$ is the radius of gyration of the volume element about the $z$-axis and $\ddot{\theta}$ is the angular acceleration of the element. For the elemental cube, $k^2 = \delta x^2/6$ so that, as the size of the volume approaches zero, it must be that $\tau_{yx} - \tau_{xy}$ also approaches zero and so $\tau_{yx} = \tau_{xy}$. Similar arguments lead to the conclusion that $\tau_{zx} = \tau_{xz}$, and $\tau_{yz} = \tau_{zy}$.

The shear forces exerted on the volume element in the $x$-direction are

$$-\left(\tau_{yx} - \frac{1}{2}\frac{\partial \tau_{yx}}{\partial y}\delta y\right)\delta x\,\delta z \quad \text{acting on the } y - \frac{\delta y}{2} \text{ face}$$

and

$$+\left(\tau_{yx} + \frac{1}{2}\frac{\partial \tau_{yx}}{\partial y}\delta y\right)\delta x\,\delta z \quad \text{acting on the } y + \frac{\delta y}{2} \text{ face,}$$

as before. In addition, there are shear forces exerted on the volume element in the $x$-direction

$$-\left(\tau_{zx} - \frac{1}{2}\frac{\partial \tau_{zx}}{\partial z}\delta z\right)\delta x\,\delta y \quad \text{acting on the } z - \frac{\delta z}{2} \text{ face}$$

and

$$+ \left( \tau_{zx} + \frac{1}{2} \frac{\partial \tau_{zx}}{\partial z} \, \delta z \right) \delta x \, \delta z \quad \text{acting on the } z + \frac{\delta z}{2} \text{ face.}$$

The net shear force in the $x$-direction acting on the volume is thus

$$+ \left( \frac{\partial \tau_{yx}}{\partial y} + \frac{\partial \tau_{zx}}{\partial z} \right) \delta x \, \delta y \, \delta z.$$

The $x$-momentum of the volume element at time $t$ is $\rho u \, \delta x \, \delta y \, \delta z$ so that the rate of momentum accumulation, the first term on the right-hand side of equation (15.14), is

$$\delta x \, \delta y \, \delta z \frac{\partial}{\partial t} (\rho u).$$

Momentum is advected into, and out of, the volume by each of the velocity components. The momentum fluxes are $\rho u^2$, $\rho v u$, and $\rho w u$, so that the $x$-momentum flowrate out of the volume is

$$\left( \rho u^2 \right)_{x + \delta x/2} \delta y \, \delta z + (\rho v u)_{y + \delta y/2} \, \delta x \, \delta z + (\rho w u)_{z + \delta z/2} \, \delta x \, \delta y$$

and the $x$-momentum flowrate into the volume is

$$\left( \rho u^2 \right)_{x - \delta x/2} \delta y \, \delta z + (\rho v u)_{y - \delta y/2} \, \delta x \, \delta z + (\rho w u)_{z - \delta z/2} \, \delta x \, \delta y.$$

The difference between the outflow and the inflow of momentum is thus

$$\left[ \left( \rho u^2 \right)_{x + \delta x/2} - \left( \rho u^2 \right)_{x - \delta x/2} \right] \delta y \, \delta z + \left[ (\rho v u)_{y + \delta y/2} - (\rho v u)_{y - \delta y/2} \right] \delta x \, \delta z$$
$$+ \left[ (\rho w u)_{z + \delta z/2} - (\rho w u)_{z - \delta z/2} \right] \delta x \, \delta y.$$

In the limit, as $\delta x$, $\delta y$, and $\delta z$ approach zero, the difference becomes

$$\left[ \frac{\partial}{\partial x} \left( \rho u^2 \right) + \frac{\partial}{\partial y} (\rho u v) + \frac{\partial}{\partial z} (\rho u w) \right] \delta x \, \delta y \, \delta z.$$

If we substitute for all the terms in equation (15.14) and divide through by $\delta x \, \delta y \, \delta z$ we have

$$\frac{\partial}{\partial t} (\rho u) + \left[ \frac{\partial}{\partial x} \left( \rho u^2 \right) + \frac{\partial}{\partial y} (\rho u v) + \frac{\partial}{\partial z} (\rho u w) \right] = \left( \frac{\partial \sigma_{xx}}{\partial x} + \frac{\partial \tau_{yx}}{\partial y} + \frac{\partial \tau_{zx}}{\partial z} \right) + f_x$$

which leads to

$$\rho \frac{Du}{Dt} = \rho \left( \frac{\partial u}{\partial t} + u \frac{\partial u}{\partial x} + v \frac{\partial u}{\partial y} + w \frac{\partial u}{\partial z} \right) = \left( \frac{\partial \sigma_{xx}}{\partial x} + \frac{\partial \tau_{yx}}{\partial y} + \frac{\partial \tau_{zx}}{\partial z} \right) + \rho f_x. \tag{15.15}$$

Equation (15.15) is one of the three equations of motion for a fluid, the other two being similar equations for $v$, in the $y$-direction, and for $w$, in the $z$-direction

$$\rho \frac{Dv}{Dt} = \rho \left( \frac{\partial v}{\partial t} + u \frac{\partial v}{\partial x} + v \frac{\partial v}{\partial y} + w \frac{\partial v}{\partial z} \right) = \left( \frac{\partial \sigma_{yy}}{\partial y} + \frac{\partial \tau_{zy}}{\partial z} + \frac{\partial \tau_{xy}}{\partial x} \right) + \rho f_y \tag{15.16}$$

$$\rho \frac{Dw}{Dt} = \rho \left( \frac{\partial w}{\partial t} + u \frac{\partial w}{\partial x} + v \frac{\partial w}{\partial y} + w \frac{\partial w}{\partial z} \right) = \left( \frac{\partial \sigma_{zz}}{\partial z} + \frac{\partial \tau_{xz}}{\partial x} + \frac{\partial \tau_{yz}}{\partial y} \right) + \rho f_z. \tag{15.17}$$

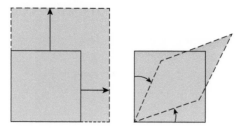

**Figure 15.4** Distortion of a fluid element: (a) volumetric (b) angular

Equations (15.15) to (15.17), known as **Cauchy's equations**, are quite general with no limitations on the fluid considered other than it satisfies the **continuum hypothesis** (discussed in Section 2.5).

### 15.1.3 Strain rates and shear rates

The motion of an infinitesimal volume of flowing fluid can be described in terms of translation, rotation, **volumetric distortion**, and **angular distortion**. Only distortion of a fluid element, but not translation or rotation, is associated with stress: Figure 15.4 illustrates the two forms of distortion.

Volumetric distortion is a consequence of the **longitudinal** or **extensional strain** (i.e. stretching) in the $x$-, $y$-, and $z$-directions, the three components of **extensional** (or **normal**) **strain rate** being

$$\dot{e}_{xx} = \frac{\partial u}{\partial x}, \qquad \dot{e}_{yy} = \frac{\partial v}{\partial y}, \qquad \text{and} \quad \dot{e}_{zz} = \frac{\partial w}{\partial z}. \tag{15.18}$$

The total extensional-strain rate, i.e. the rate of change in volume per unit volume, $\Theta$ (upper case Greek letter theta), termed the **dilation**[120], is given by

$$\Theta = \dot{e}_{xx} + \dot{e}_{yy} + \dot{e}_{zz} = \frac{\partial u}{\partial x} + \frac{\partial v}{\partial y} + \frac{\partial w}{\partial z} = div\ V = \nabla \cdot V. \tag{15.19}$$

The continuity equation (15.9) can be written in terms of $\Theta$ as

$$\frac{D\rho}{Dt} + \rho\Theta = 0 \tag{15.20}$$

from which it is evident that the dilation is identically zero for a constant-density flow.

The three components of the rate of angular distortion, often referred to as **shear rates** or **shear-strain rates**, are

$$\dot{\gamma}_{xy} = \dot{\gamma}_{yx} = \frac{1}{2}\left(\frac{\partial u}{\partial y} + \frac{\partial v}{\partial x}\right),\ \dot{\gamma}_{xz} = \dot{\gamma}_{zx} = \frac{1}{2}\left(\frac{\partial u}{\partial z} + \frac{\partial w}{\partial x}\right),\ \text{and}\ \dot{\gamma}_{yz} = \dot{\gamma}_{yz} = \frac{1}{2}\left(\frac{\partial w}{\partial y} + \frac{\partial v}{\partial z}\right).$$

$$\tag{15.21}$$

---

[120] The spelling **dilatation** is also used.

It should be noted that the **dynamics** of fluid flow is not involved in these descriptions of fluid motion, which are purely **kinematic**. Also, the temporal rather than spatial aspect of strain rates is apparent from their dimension, 1/T, and emphasised by the dot in the symbols indicating per unit time.

### 15.1.4 Stokes' constitutive equations

The normal- and shear-stress components, introduced in Subsection 15.1.2, are related to the strain rates of Subsection 15.1.3 by so-called **constitutive equations**. For a **Newtonian fluid** these are based upon three assumptions

  (a) the stress components are proportional to the strain rates, a consequence of which is that in the absence of angular distortion the shear stresses vanish,
  (b) within the fluid there are no preferred directions, i.e. the fluid is **isotropic**, and
  (c) in the absence of dilation, the normal stresses must reduce to the pressure.

The equations, introduced in a paper by Stokes in 1845, corresponding to these assumptions may be written, for the normal-stress components, as

$$\sigma_{xx} = -p - \frac{2}{3}\mu\,\Theta + 2\mu\,\dot{e}_{xx} = -p - \frac{2}{3}\mu\Theta + 2\mu\frac{\partial u}{\partial x} \tag{15.22}$$

$$\sigma_{yy} = -p - \frac{2}{3}\mu\,\Theta + 2\mu\,\dot{e}_{yy} = -p - \frac{2}{3}\mu\Theta + 2\mu\frac{\partial v}{\partial y} \tag{15.23}$$

$$\sigma_{xx} = -p - \frac{2}{3}\mu\,\Theta + 2\mu\,\dot{e}_{zz} = -p - \frac{2}{3}\mu\Theta + 2\mu\frac{\partial w}{\partial z} \tag{15.24}$$

where $\Theta$ is the dilation given by equation (15.19), $\mu$ is the dynamic viscosity or, more precisely, the **first coefficient of viscosity**, and $p$ is the **average (mechanical) pressure**. The latter is the arithmetic average of the three normal stresses, i.e.

$$p = -\frac{1}{3}\left(\sigma_{xx} + \sigma_{yy} + \sigma_{zz}\right). \tag{15.25}$$

For an incompressible fluid and for a perfect monatomic gas, the average mechanical pressure at a point is identically equal to the **thermodynamic pressure** at that point. More generally, the difference between the average pressure and the thermodynamic pressure is equal to $-\lambda\Theta$, where $\lambda$ is the **bulk viscosity** (or **second coefficient of viscosity**). This difference is negligible except for extreme cases of compressible flow, such as within a shockwave or in a high-frequency sound field. **Stokes' assumption** is that the thermodynamic and average pressures are equal, i.e. $\lambda\Theta = 0$.

For the shear-stress components

$$\tau_{xy} = \tau_{yx} = 2\mu\dot{\gamma}_{xy} = \mu\left(\frac{\partial v}{\partial x} + \frac{\partial u}{\partial y}\right) \tag{15.26}$$

$$\tau_{yz} = \tau_{zy} = 2\mu\dot{\gamma}_{yz} = \mu\left(\frac{\partial w}{\partial y} + \frac{\partial v}{\partial z}\right) \tag{15.27}$$

$$\tau_{zx} = \tau_{xz} = 2\mu \dot{\gamma}_{xz} = \mu \left( \frac{\partial u}{\partial z} + \frac{\partial w}{\partial x} \right). \tag{15.28}$$

Equation (15.26) is a more general version, for a flow with velocity components $u$ and $v$ (and $w$) of $\tau = \mu \, \partial u/\partial y$ introduced in Section 2.8 for a simple unidirectional, developed flow where $u$ is independent of $x$ and $z$ so that $u = u(y)$.

### 15.1.5 Navier-Stokes equations

In rectangular Cartesian coordinates, after substitution for $\sigma_{xx}, \tau_{xy}$, and $\tau_{xz}$ in Cauchy's equations, equations (15.15) to (15.17), we have:

$x$-component

$$\rho \frac{Du}{Dt} = -\frac{\partial p}{\partial x} - \frac{\partial}{\partial x} \left( \frac{2}{3} \mu \Theta \right) + \frac{\partial}{\partial x} \left( 2\mu \frac{\partial u}{\partial x} \right) + \frac{\partial}{\partial y} \left[ \mu \left( \frac{\partial u}{\partial y} + \frac{\partial v}{\partial x} \right) \right]$$

$$+ \frac{\partial}{\partial z} \left[ \mu \left( \frac{\partial u}{\partial z} + \frac{\partial w}{\partial x} \right) \right] + \rho f_x \tag{15.29}$$

$y$-component

$$\rho \frac{Dv}{Dt} = -\frac{\partial p}{\partial y} - \frac{\partial}{\partial y} \left( \frac{2}{3} \mu \Theta \right) + \frac{\partial}{\partial y} \left( 2\mu \frac{\partial u}{\partial y} \right) + \frac{\partial}{\partial x} \left[ \mu \left( \frac{\partial u}{\partial y} + \frac{\partial v}{\partial x} \right) \right]$$

$$+ \frac{\partial}{\partial z} \left[ \mu \left( \frac{\partial v}{\partial z} + \frac{\partial w}{\partial y} \right) \right] + \rho f_y \tag{15.30}$$

$z$-component

$$\rho \frac{Dw}{Dt} = -\frac{\partial p}{\partial z} - \frac{\partial}{\partial z} \left( \frac{2}{3} \mu \Theta \right) + \frac{\partial}{\partial z} \left( 2\mu \frac{\partial w}{\partial z} \right) + \frac{\partial}{\partial y} \left[ \mu \left( \frac{\partial v}{\partial z} + \frac{\partial w}{\partial y} \right) \right]$$

$$+ \frac{\partial}{\partial x} \left[ \mu \left( \frac{\partial u}{\partial z} + \frac{\partial w}{\partial x} \right) \right] + \rho f_z. \tag{15.31}$$

This set of three partial differential equations for a Newtonian fluid, established independently by Louis Navier, in 1827, and by George Stokes, in 1845, are known as the **Navier-Stokes equations**. They are quite general and apply to both steady and unsteady flow, to one-, two-, and three-dimensional flows, and to flows where the viscosity and density are temperature and/or pressure dependent.

### 15.1.6 Constant- and uniform-property flow

It might be thought that, since we now have a complete set of equations governing the flow of a Newtonian fluid, it would now possible to solve any such flow problem. In fact, the Navier-Stokes equations are a set of coupled, unsteady, non-linear, second-order, partial differential equations to which there are as yet no general analytical or numerical solutions: less than a hundred exact analytical solutions have been found to highly simplified versions of the equations. Some of these solutions for internal laminar flow will be discussed in Chapter 16.

The main source of difficulty stems not from the inclusion of viscosity in the equations but from the non-linearity of the advective terms, which arises from terms like $u(\partial u/\partial x) = (\partial u^2/\partial x)/2$. Because the viscous terms are linear, and the advective terms non-linear, the term **quasi-linear** is sometimes used to describe the Navier-Stokes equations.

If we restrict attention to flow of a fluid with constant and uniform[121] viscosity, the Navier-Stokes equations reduce to:

$x$-component

$$\rho\frac{Du}{Dt} = -\frac{\partial p}{\partial x} + \mu\left[\left(\frac{\partial^2 u}{\partial x^2} + \frac{\partial^2 u}{\partial y^2} + \frac{\partial^2 u}{\partial z^2}\right) + \frac{1}{3}\frac{\partial\Theta}{\partial x}\right] + \rho f_x \qquad (15.32)$$

$y$-component

$$\rho\frac{Dv}{Dt} = -\frac{\partial p}{\partial y} + \mu\left[\left(\frac{\partial^2 v}{\partial x^2} + \frac{\partial^2 v}{\partial y^2} + \frac{\partial^2 v}{\partial z^2}\right) + \frac{1}{3}\frac{\partial\Theta}{\partial y}\right] + \rho f_y \qquad (15.33)$$

$z$-component

$$\rho\frac{Dw}{Dt} = -\frac{\partial p}{\partial z} + \mu\left[\left(\frac{\partial^2 w}{\partial x^2} + \frac{\partial^2 w}{\partial y^2} + \frac{\partial^2 w}{\partial z^2}\right) + \frac{1}{3}\frac{\partial\Theta}{\partial z}\right] + \rho f_z. \qquad (15.34)$$

In vector form these equations may be written as

$$\rho\frac{DV}{Dt} = -\nabla p + \mu\left[\nabla^2 V + \frac{1}{3}\nabla\Theta\right] + \rho f \qquad (15.35)$$

where

$$\nabla^2 = \frac{\partial^2}{\partial x^2} + \frac{\partial^2}{\partial y^2} + \frac{\partial^2}{\partial z^2} \qquad (15.36)$$

is the **Laplacian operator**.

If, in addition to restricting attention to constant-viscosity fluids, we also limit ourselves to fluid with constant and uniform density, then the dilation $\Theta = 0$.

## 15.2 **Equations of motion in cylindrical-coordinate form**

For certain problems, particularly internal flows through ducts of circular or annular cross section, it is convenient to work with the equations of motion in cylindrical-coordinate[122] form in which $x$ is the distance along the axis, $r$ is the distance from the axis, and $\theta$ is the azimuthal angle about the axis. The velocity components are now $u$ in the $x$-direction, referred to as the axial-direction, $v$ in the $r$-, or radial, direction, and $w$ in the $\theta$-, or tangential, direction. The equations are arrived at through a transformation of the equations in Section 15.1.

---

[121] In principle, a fluid could have a viscosity that varied with time and depended upon position in a flow, i.e. $\mu(t, x, y, z)$. The restriction to uniform viscosity means that the viscosity is the same at all points throughout the flowfield, and constant means unchanging with time.

[122] The term **cylindrical-polar coordinate** is also used.

### 15.2.1 Continuity equation

The continuity equation can be written as

$$\frac{\partial \rho}{\partial t} + \frac{\partial}{\partial x}(\rho u) + \frac{1}{r}\frac{\partial}{\partial r}(\rho r v) + \frac{1}{r}\frac{\partial}{\partial \theta}(\rho w) = 0 \tag{15.37}$$

or

$$\frac{\partial \rho}{\partial t} + u\frac{\partial \rho}{\partial x} + v\frac{\partial \rho}{\partial r} + \frac{w}{r}\frac{\partial \rho}{\partial \theta} + \rho\left[\frac{\partial u}{\partial x} + \frac{1}{r}\frac{\partial}{\partial r}(rv) + \frac{1}{r}\frac{\partial w}{\partial \theta}\right] = 0 \tag{15.38}$$

or

$$\frac{D\rho}{Dt} + \rho\left[\frac{\partial u}{\partial x} + \frac{1}{r}\frac{\partial}{\partial r}(rv) + \frac{1}{r}\frac{\partial w}{\partial \theta}\right] = 0. \tag{15.39}$$

Note that the material derivative in cylindrical coordinates is

$$\frac{D}{Dt} = \frac{\partial}{\partial t} + u\frac{\partial}{\partial x} + v\frac{\partial}{\partial r} + \frac{w}{r}\frac{\partial}{\partial \theta}. \tag{15.40}$$

For a constant-density fluid, equation (15.39) reduces to

$$\frac{\partial u}{\partial x} + \frac{1}{r}\frac{\partial}{\partial r}(rv) + \frac{1}{r}\frac{\partial w}{\partial \theta} = 0. \tag{15.41}$$

### 15.2.2 Cauchy's equations of motion

The normal-stress components are now $\sigma_{xx}$ in the $x$-direction, $\sigma_{rr}$ in the $r$-direction, and $\sigma_{\theta\theta}$ in the $\theta$-direction; the shear-stress components are $\tau_{xr}, \tau_{r\theta}$, and $\tau_{\theta x}$; and the components of the body force per unit mass are $f_x, f_r$, and $f_\theta$. The subscripts $x, r$, and $\theta$ are to be interpreted as explained in Subsection 15.1.2. Cauchy's equations in cylindrical coordinates are:

$x$-component

$$\rho\frac{Du}{Dt} = \rho\left(\frac{\partial u}{\partial t} + u\frac{\partial u}{\partial x} + v\frac{\partial u}{\partial r} + \frac{w}{r}\frac{\partial u}{\partial \theta}\right) = \frac{\partial \sigma_{xx}}{\partial x} + \frac{1}{r}\frac{\partial}{\partial r}(r\tau_{rx}) + \frac{1}{r}\frac{\partial \tau_{\theta x}}{\partial \theta} + \rho f_x \tag{15.42}$$

$r$-component

$$\rho\left(\frac{Dv}{Dt} - \frac{w^2}{r}\right) = \rho\left(\frac{\partial v}{\partial t} + u\frac{\partial v}{\partial x} + v\frac{\partial v}{\partial r} + \frac{w}{r}\frac{\partial v}{\partial \theta} - \frac{w^2}{r}\right)$$

$$= \frac{\partial \tau_{xr}}{\partial x} + \frac{1}{r}\frac{\partial}{\partial r}(r\sigma_{rr}) + \frac{1}{r}\frac{\partial \tau_{\theta r}}{\partial \theta} - \frac{\sigma_{\theta\theta}}{r} + \rho f_r \tag{15.43}$$

$\theta$-component

$$\rho\left(\frac{Dw}{Dt} + \frac{vw}{r}\right) = \rho\left(\frac{\partial w}{\partial t} + u\frac{\partial w}{\partial x} + v\frac{\partial w}{\partial r} + \frac{w}{r}\frac{\partial w}{\partial \theta} + \frac{vw}{r}\right)$$

$$= \frac{\partial \tau_{x\theta}}{\partial x} + \frac{1}{r^2}\frac{\partial}{\partial r}(r^2\tau_{r\theta}) + \frac{1}{r}\frac{\partial \sigma_{\theta\theta}}{\partial \theta} + \rho f_\theta.. \tag{15.44}$$

The second term on the left-hand side of equation (15.43), $w^2/r$, arises in the transformation from rectangular Cartesian to cylindrical coordinates[123]. It represents the **centripetal force** per unit mass and is an effective force in the radial direction as a consequence of fluid motion in the tangential direction.

The second term on the left-hand side of equation (15.44), $vw/r$, also arises in the transformation. It represents the **Coriolis force** per unit mass and is an effective force in the tangential direction as a consequence of fluid motion in both the radial and tangential directions.

### 15.2.3 Strain rates

The longitudinal strain rates are

$$\dot{e}_{xx} = \frac{\partial u}{\partial x}, \quad \dot{e}_{rr} = \frac{\partial v}{\partial r}, \quad \text{and} \quad \dot{e}_{\theta\theta} = \frac{1}{r}\frac{\partial w}{\partial \theta} + \frac{v}{r} \tag{15.45}$$

and the angular strain rates are

$$\dot{\gamma}_{r\theta} = \dot{\gamma}_{\theta r} = \frac{1}{2}\left[r\frac{\partial}{\partial r}\left(\frac{w}{r}\right) + \frac{1}{r}\frac{\partial v}{\partial \theta}\right], \quad \dot{\gamma}_{\theta x} = \dot{\gamma}_{x\theta} = \frac{1}{2}\left(\frac{1}{r}\frac{\partial u}{\partial \theta} + \frac{\partial w}{\partial x}\right),$$

$$\dot{\gamma}_{xr} = \dot{\gamma}_{rx} = \frac{1}{2}\left(\frac{\partial v}{\partial x} + \frac{\partial u}{\partial r}\right). \tag{15.46}$$

The dilation is

$$\Theta = \dot{e}_{xx} + \dot{e}_{rr} + \dot{e}_{\theta\theta} = \frac{\partial u}{\partial x} + \frac{1}{r}\frac{\partial}{\partial r}(rv) + \frac{1}{r}\frac{\partial w}{\partial \theta}. \tag{15.47}$$

### 15.2.4 Stokes' constitutive equations

The **constitutive equations** relating the normal- and shear-stress components to the strain rates for a Newtonian fluid in cylindrical coordinates are as follows:

For the normal-stress components

$$\sigma_{rr} = -p - \frac{2}{3}\mu\Theta + 2\mu\,\dot{e}_{rr} = -p - \frac{2}{3}\mu\,\Theta + 2\mu\frac{\partial v}{\partial r} \tag{15.48}$$

$$\sigma_{\theta\theta} = -p - \frac{2}{3}\mu\Theta + 2\mu\,\dot{e}_{\theta\theta} = -p - \frac{2}{3}\mu\,\Theta + 2\mu\left(\frac{1}{r}\frac{\partial w}{\partial \theta} + \frac{v}{r}\right) \tag{15.49}$$

$$\sigma_{xx} = -p - \frac{2}{3}\mu\Theta + 2\mu\,\dot{e}_{xx} = -p - \frac{2}{3}\mu\,\Theta + 2\mu\frac{\partial u}{\partial x} \tag{15.50}$$

and, for the shear-stress components

$$\tau_{xr} = \tau_{rx} = 2\mu\,\dot{\gamma}_{xr} = \mu\left(\frac{\partial v}{\partial x} + \frac{\partial u}{\partial r}\right) \tag{15.51}$$

---

[123] This term, and the corresponding term in equation (15.44), is sometimes, arguably incorrectly, included in the material derivative itself.

$$\tau_{r\theta} = \tau_{\theta r} = 2\mu \, \dot{\gamma}_{r\theta} = \mu \left[ r \frac{\partial}{\partial r} \left( \frac{w}{r} \right) + \frac{1}{r} \frac{\partial v}{\partial \theta} \right] \tag{15.52}$$

$$\tau_{\theta x} = \tau_{x\theta} = 2\mu \, \dot{\gamma}_{rx} = \mu \left( \frac{1}{r} \frac{\partial u}{\partial \theta} + \frac{\partial w}{\partial x} \right). \tag{15.53}$$

### 15.2.5 Navier-Stokes equations

In cylindrical coordinates, after substitution for the normal and shear-stress components in Cauchy's equations (15.42) to (15.44), we have:

$x$-component

$$\rho \frac{Du}{Dt} = -\frac{\partial p}{\partial x} - \frac{\partial}{\partial x} \left( \frac{2}{3} \mu \Theta \right) + \frac{\partial}{\partial x} \left( 2\mu \frac{\partial u}{\partial x} \right) + \frac{1}{r} \frac{\partial}{\partial r} \left[ \mu r \left( \frac{\partial u}{\partial r} + \frac{\partial v}{\partial x} \right) \right]$$

$$+ \frac{1}{r} \frac{\partial}{\partial \theta} \left[ \mu \left( \frac{1}{r} \frac{\partial u}{\partial \theta} + \frac{\partial w}{\partial x} \right) \right] + \rho f_x \tag{15.54}$$

$r$-component

$$\rho \left( \frac{Dv}{Dt} - \frac{w^2}{r} \right) = -\frac{\partial p}{\partial r} - \frac{\partial}{\partial y} \left( \frac{2}{3} \mu \Theta \right) + \frac{\partial}{\partial r} \left( 2\mu \frac{\partial v}{\partial r} \right) + \frac{1}{r} \frac{\partial}{\partial \theta} \left[ \mu \left( \frac{1}{r} \frac{\partial v}{\partial \theta} + \frac{\partial w}{\partial r} - \frac{w}{r} \right) \right]$$

$$+ \frac{\partial}{\partial x} \left[ \mu \left( \frac{\partial v}{\partial x} + \frac{\partial w}{\partial r} \right) \right] + \frac{2\mu}{r} \left( \frac{\partial v}{\partial r} - \frac{v}{r} - \frac{1}{r} \frac{\partial w}{\partial \theta} \right) + \rho f_r \tag{15.55}$$

$\theta$-component

$$\rho \left( \frac{Dw}{Dt} + \frac{vw}{r} \right) = -\frac{1}{r} \frac{\partial p}{\partial \theta} - \frac{1}{r} \frac{\partial}{\partial \theta} \left( \frac{2}{3} \mu \Theta \right) + \frac{1}{r} \frac{\partial}{\partial \theta} \left( \frac{2\mu}{r} \frac{\partial w}{\partial \theta} + \frac{2\mu v}{r} \right)$$

$$+ \frac{\partial}{\partial r} \left[ \mu \left( \frac{1}{r} \frac{\partial v}{\partial \theta} + \frac{\partial w}{\partial r} - \frac{w}{r} \right) \right] + \frac{2\mu}{r} \left( \frac{1}{r} \frac{\partial v}{\partial \theta} + \frac{\partial w}{\partial r} - \frac{w}{r} \right)$$

$$+ \frac{\partial}{\partial x} \left[ \mu \left( \frac{1}{r} \frac{\partial u}{\partial \theta} + \frac{\partial w}{\partial x} \right) \right] + \rho f_\theta. \tag{15.56}$$

### 15.2.6 Constant- and uniform-property flow

In cylindrical coordinates, the Navier-Stokes equations for a fluid with constant and uniform properties are:

$x$-component

$$\rho \frac{Du}{Dt} = -\frac{\partial p}{\partial x} + \mu \left[ \frac{\partial^2 u}{\partial x^2} + \frac{1}{r} \frac{\partial}{\partial r} \left( r \frac{\partial u}{\partial r} \right) + \frac{1}{r^2} \frac{\partial^2 u}{\partial \theta^2} \right] \tag{15.57}$$

$r$-component

$$\rho \left( \frac{Dv}{Dt} - \frac{w^2}{r} \right) = -\frac{\partial p}{\partial r} + \mu \left[ \frac{\partial^2 v}{\partial x^2} + \frac{\partial}{\partial r} \left\{ \frac{1}{r} \frac{\partial}{\partial r} (rv) \right\} + \frac{1}{r^2} \frac{\partial^2 v}{\partial \theta^2} - \frac{2}{r^2} \frac{\partial w}{\partial \theta} \right] \tag{15.58}$$

$\theta$-component

$$\rho \left( \frac{Dw}{Dt} + \frac{vw}{r} \right) = -\frac{1}{r} \frac{\partial p}{\partial \theta} + \mu \left[ \frac{\partial^2 w}{\partial x^2} + \frac{\partial}{\partial r} \left\{ \frac{1}{r} \frac{\partial}{\partial r} (rw) \right\} + \frac{1}{r^2} \frac{\partial^2 w}{\partial \theta^2} + \frac{2}{r^2} \frac{\partial v}{\partial \theta} \right]. \tag{15.59}$$

## 15.3 Boundary conditions

For flow over a solid surface, the velocity component normal to the surface $v_S$, say, must be zero but there are also practical problems where the surface is porous and there is flow through it. For example, the turbine blades of a gas turbine are commonly cooled by **blowing** cool gas through tiny holes in the surface into the flow over the surface, $v_S > 0$, a process called **transpiration cooling**. In other situations, **flow control** involves **suction** through the surface, $v_S < 0$. For almost all the problems we shall consider, our first **boundary condition** is $v_S = 0$.

Although it is obvious that fluid cannot pass through a solid surface, it is primarily a matter of experimental observation that the components of fluid velocity tangential to the surface, $u$ and $w$, are also zero and this is our second boundary condition. According to this **no-slip condition**, introduced in Section 6.4, in the immediate vicinity of a solid surface a consequence of viscosity is that the fluid is brought to rest (or, more generally, if the surface is itself moving, to the same velocity as the surface so that the relative velocity is zero). In essence, the fluid adheres to the surface. In an external flow, the change from zero to the external (**free-stream**) velocity takes place across a relatively thin layer of fluid called the viscous **boundary layer**[124]. In an internal flow, as the flow develops from the flow inlet, velocity changes initially occur across a boundary layer but ultimately the flow becomes **fully developed** (i.e. unchanging with streamwise location) and the entire cross section is influenced by viscosity. Chapter 16 is concerned primarily with the analysis of fully-developed internal flows, while laminar boundary layers are the subject of Chapter 17.

## 15.4 Non-dimensional form of the Navier-Stokes and continuity equations

If we can identify a length scale, $L$, and a velocity scale, $U$, for any given flow, then we can transform the Navier-Stokes and continuity equations to non-dimensional form. For simplicity, we restrict consideration to constant- and uniform-property flows with zero body force so that, in Cartesian coordinates, the relevant equations are those of Subsection 15.1.6. We introduce the non-dimensional variables $t^* = tU/L$, $x^* = x/L$, $y^* = y/L$, $z^* = z/L$, $u^* = u/U$, $v^* = v/U$, $w^* = w/U$, and $p^* = p/\rho U^2$. The non-dimensional form of the Navier-Stokes equations is then

$$\frac{\partial u^*}{\partial t^*} + u^* \frac{\partial u^*}{\partial x^*} + v^* \frac{\partial u^*}{\partial y^*} + w^* \frac{\partial u^*}{\partial z^*} = -\frac{\partial p^*}{\partial x^*} + \frac{\mu}{\rho UL} \left( \frac{\partial^2 u^*}{\partial x^{*2}} + \frac{\partial^2 u^*}{\partial y^{*2}} + \frac{\partial^2 u^*}{\partial z^{*2}} \right) \tag{15.60}$$

---

[124] In situations where there is heat transfer between a flowing fluid and a surface, termed **forced convection**, the temperature change from the surface value to the free-stream value occurs across a **thermal boundary layer**.

$$\frac{\partial v^*}{\partial t^*} + u^*\frac{\partial v^*}{\partial x^*} + v^*\frac{\partial v^*}{\partial y^*} + w^*\frac{\partial v^*}{\partial z^*} = -\frac{\partial p^*}{\partial y^*} + \frac{\mu}{\rho UL}\left(\frac{\partial^2 v^*}{\partial x^{*2}} + \frac{\partial^2 v^*}{\partial y^{*2}} + \frac{\partial^2 v^*}{\partial z^{*2}}\right) \tag{15.61}$$

and

$$\frac{\partial w^*}{\partial t^*} + u^*\frac{\partial w^*}{\partial x^*} + v^*\frac{\partial w^*}{\partial y^*} + w^*\frac{\partial w^*}{\partial z^*} = -\frac{\partial p^*}{\partial z^*} + \frac{\mu}{\rho UL}\left(\frac{\partial^2 w^*}{\partial x^{*2}} + \frac{\partial^2 w^*}{\partial y^{*2}} + \frac{\partial^2 w^*}{\partial z^{*2}}\right), \tag{15.62}$$

while the continuity equation in non-dimensional form is

$$\frac{\partial u^*}{\partial x^*} + \frac{\partial v^*}{\partial y^*} + \frac{\partial w^*}{\partial z^*} = 0. \tag{15.63}$$

Transformation of the usual boundary conditions, $u_S = 0$, $w_S = 0$, and $v_S = 0$, is obviously trivial.

The important result of this section is that the Reynolds number $Re = \rho UL/\mu$ emerges as the single parameter in the set of equations and it can be concluded that, for two geometrically similar situations with the same Reynolds number (dynamic similarity), the solutions to equations (15.60) to (15.63) will be the same, and the corresponding flow patterns identical (i.e. kinematically similar) (see Section 3.14). It is crucially important to understand that a Reynolds number defined for, say, flow through a pipe, based upon pipe diameter and average flow velocity, has no direct relevance to flow at the same flow velocity over a cylinder of the same diameter or over an aerofoil with thickness equal to the pipe diameter.

We have already established that any flow of a fluid satisfying Stokes' constitutive equations is governed by the Navier-Stokes equations. Such viscous-fluid flows are often categorised as being either **laminar** or **turbulent**, terms we shall define more precisely in Chapters 16 and 18 than we did in Sections 3.12 and 3.16, with an appropriately defined Reynolds number differentiating between the two: laminar flow below a **critical Reynolds number** and turbulent flow for higher values. This crude characterisation is adequate for some basic engineering situations, such as steady flow through long pipes, but in general more careful consideration of the flow condition is necessary.

## 15.5 Flow of a generalised Newtonian fluid

Up to this point we have been concerned with the governing equations for the flow of a Newtonian fluid. In Section 2.10 we discussed briefly the characteristics of so-called **non-Newtonian** fluids, for which the **apparent** or **effective** dynamic viscosity $\mu_{EFF}$ is no longer a fluid parameter independent of any distortion of the fluid and which may exhibit time-dependent elastic effects or normal-stress effects. The effective viscosity is defined in the same way as the Newtonian viscosity $\mu$, through the equation

$$\tau_{yx} = \mu_{EFF}\dot{\gamma}. \tag{15.64}$$

For many real fluids, known as **generalised Newtonian fluids**, the viscosity is shear dependent but other effects are negligible, at least in steady laminar flow. In Cartesian coordinates the equations describing the flow are equations (15.29) to (15.31) and, in polar coordinates, equations (15.54) to (15.56), with $\mu_{EFF} = \mu_{EFF}(\dot{\gamma})$, $\dot{\gamma}$ being the magnitude of the total strain rate, which can be calculated from the shear-strain-rate[125] components as

$$\dot{\gamma} = \sqrt{\dot{\gamma}_{xy}^2 + \dot{\gamma}_{yz}^2 + \dot{\gamma}_{zx}^2} \tag{15.65}$$

in Cartesian coordinates, with a similar expression (replacing $\dot{\gamma}_{xy}$ with $\dot{\gamma}_{xr}$, etc.) in cylindrical coordinates. A practical example in which two orthogonal strain-rate components play a role is that of the flow of a non-Newtonian fluid, such as drilling mud, through a concentric annulus (a borehole) with centrebody (the drill stem) rotation. The magnitude of the total strain rate is then

$$\dot{\gamma} = \sqrt{\dot{\gamma}_{xr}^2 + \dot{\gamma}_{r\theta}^2}. \tag{15.66}$$

For many real fluids of industrial importance it is found that the apparent viscosity is constant at low and very-high shear rates with decreasing (or increasing) viscosity at intermediate shear rates, as illustrated in Figure 15.5. Such a plot of viscosity versus shear rate is known as a **flow curve**. The **Ostwald-de Waele power-law model** is widely used as an empirical representation of the intermediate region

$$\mu_{EFF} = K\dot{\gamma}^{n-1}, \tag{15.67}$$

the constant $K$ being called the consistency index, with units $\text{Pa} \cdot \text{s}^n$, and $n$ being the power-law exponent. Clearly, if $n < 1$ equation (15.67) corresponds to an effective viscosity which decreases with increasing shear rate, termed **shear-thinning** (or **pseudoplastic**) behaviour, but

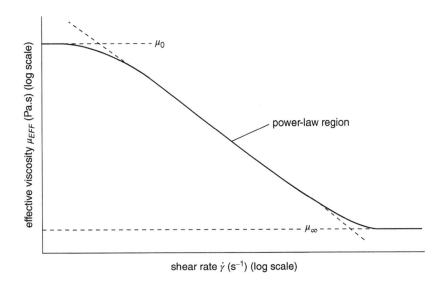

**Figure 15.5** Effective viscosity versus shear rate for a shear-thinning liquid

[125] In more general situations the total strain rate would have to include the extensional strain-rate components, so that $\dot{\gamma} = \sqrt{\dot{\gamma}_{xy}^2 + \dot{\gamma}_{yz}^2 + \dot{\gamma}_{zx}^2 + \frac{1}{2}\left(\dot{e}_{xx}^2 + \dot{e}_{yy}^2 + \dot{e}_{zz}^2\right)}.$

if $n > 1$ the viscosity increases with shear rate, and the fluid is said to be **shear thickening** (or **dilatant**). The power-law model has been widely used for the polymeric liquids commonly found in the food and cosmetic industries.

The simple power-law equation has the obvious drawback that (for $n < 1$), as $\dot{\gamma} \to 0$, $\mu_{EFF} \to \infty$. Nevertheless, analytical solutions to the flow equations are possible for a number of practical flow geometries and provide some insight into the qualitative effects of a shear-dependent viscosity. In Chapter 16 we present the solution for fully-developed flow of a power-law fluid between parallel plates.

The **Sisko model** provides a better description of most emulsions and suspensions over a wide range of shear rates (0.1 to $10^3$ s$^{-1}$). This model overcomes the high-shear-rate deficiency ($\mu_{EFF} \to 0$) of the power-law equation by adding an appropriate constant viscosity $\mu_\infty$ so that

$$\mu_{EFF} = K\dot{\gamma}^{n-1} + \mu_\infty. \tag{15.68}$$

Many liquids exhibit a yield stress, i.e. a shear stress below which there is no shear flow. The list is extensive and includes fine aqueous nuclear fuel slurries, rocket-propellant pastes, drilling muds, toothpaste, ketchup, certain inks and paints, mayonnaise, and yoghurt. A simple model which for many purposes adequately describes such **yield-stress** (or **viscoplastic**) liquids is the **Bingham plastic**

$$\tau_{yx} = \pm\tau_Y + \mu_P\dot{\gamma} \tag{15.69}$$

where $\tau_Y$ is the **yield stress** of the liquid and $\mu_P$ is the constant **plastic viscosity**[126].

The **Herschel-Bulkley model** is a more general equation used to characterise a viscoplastic liquid

$$\tau_{yx} = \pm\tau_Y + K\dot{\gamma}^n \tag{15.70}$$

where $K$ is the consistency index and $n$ the power-law index. Equation (15.70) has the merit of reducing to the power-law or Bingham models with appropriate choices for $\tau_Y, K$, and $n$.

The **Casson equation** also represents an alternative model for a viscoplastic liquid

$$\sqrt{-\tau_{yx}} = \sqrt{\tau_Y} + \sqrt{\mu_P\dot{\gamma}} \tag{15.71}$$

and has been found to represent the viscosity of molten chocolate and blood. In Chapter 16 we derive an equation for pipe flow of a Bingham plastic and state the equivalent equation for a Casson liquid.

There are several $\mu_{EFF}(\dot{\gamma})$ models for shear-thinning liquids involving more parameters than the power-law model, such as the five-parameter **Carreau-Yasuda model**, which are useful in curve fitting experimental data

$$\frac{\mu_{EFF} - \mu_\infty}{\mu_0 - \mu_\infty} = \left[1 + (\lambda\dot{\gamma})^a\right]^{(n-1)/a}. \tag{15.72}$$

The Carreau-Yasuda model parameters are the zero-shear-rate viscosity, $\mu_0$; the infinite-shear-rate viscosity, $\mu_\infty$; a time constant, $\lambda$; a power-law exponent, $n$; and a parameter, $a$, which

---

[126] The positive sign in equation (15.68) is used when $\tau_{yx}$ is positive, and the negative sign when it is negative.

quantifies the transition from the zero-shear-rate plateau and the power-law region. Unfortunately, the Carreau-Yasuda model is unsuitable for analytical work but can be used in numerical simulations.

The reader is reminded that most synthetic, and some natural, liquids exhibit elastic behaviour (**viscoelasticity**) and time-dependent behaviour (termed **thixotropy** if there is a viscosity decrease with time when a stress is suddenly applied, and **rheopecty** when the opposite occurs). Analysis of the flow of such fluids is outside the scope of this text.

 ## 15.6 SUMMARY

In this chapter we showed how the momentum-conservation equation (Newton's second law of motion) applied to an infinitesimal cube of fluid leads to a set of partial differential equations, Cauchy's equations of motion, which govern the flow of any fluid satisfying the continuum hypothesis. Any fluid flow must also satisfy the continuity equation, another partial differential equation, which is derived from the mass-conservation equation. It was shown that distortion of a flowing fluid can be split into elongational distortion and angular distortion or shear strain. For a Newtonian fluid, the normal and shear stresses in Cauchy's equations are related to the elongational and shear-strain rates through Stokes' constitutive equations. Substitution of these constitutive equations into Cauchy's equations leads to the Navier-Stokes equations, which govern steady or unsteady flow of a fluid which may have density and viscosity which vary with pressure and temperature. A minor modification of the constitutive equations for a Newtonian fluid allows consideration of fluids for which the viscosity depends upon the shear-strain rates, so-called generalised Newtonian fluids.

Self-assessment problems relevant to this chapter are included in Chapters 16 and 17.

At the end of this chapter, the student should be able to

- interpret the individual terms in Cauchy's equations of motion
- distinguish between extensional (or longitudinal) strain and angular strain
- explain the difference between a Newtonian and a non-Newtonian fluid
- understand the concept of a yield stress

# 16 Internal laminar flow

In this chapter we present analytical solutions of the Navier-Stokes equations for the flow of pressure- and shear-driven flows of a viscous fluid through long ducts of constant cross section. Pressure-driven flow of this type is referred to as Poiseuille flow, while the shear-driven flow is known as Couette flow. For the fully-developed flows under consideration, analytical solutions are possible because the non-linear terms in the Navier-Stokes equations are identically zero. Although for the most part we limit our attention to constant-property Newtonian fluids, we also include limited consideration of some generalised Newtonian liquids. The chapter concludes with a presentation of the principles underlying the design of instruments used to measure viscosity.

## 16.1 General remarks

A **laminar flow**[127] is one in which all transport processes occur at a molecular level, the rate of transport (or movement) being proportional to the gradient of a flow property. **Viscosity** is the transport property that relates momentum flux to velocity gradient, already discussed in Section 2.8; **thermal conductivity** relates heat flux to temperature gradient; and the **diffusion coefficient** relates mass transport to concentration gradient. We shall be concerned almost exclusively with the effects of viscosity.

For a steady (i.e. time independent) laminar flow, the fluid can be visualised as flowing smoothly in infinitesimally thin layers[128] such that the velocity varies smoothly from one layer to the next. A laminar flow in which disturbances introduced from outside (such as from surface imperfections) are attenuated is said to be **stable**. For a given flow geometry, a flow remains stable up to a certain **critical Reynolds number**[129] $Re_C$ or, for rotating flow, the **Taylor number** (see Section 16.5). As $Re$ is progressively increased above $Re_C$, new stable flow patterns may appear until a point is reached when $Re = Re_U$, say, at which any disturbances introduced into the flow are amplified because the flow has become **unstable**. Demonstrating that solutions to the Navier-Stokes equations are unstable for $Re \geq Re_U$ is the subject of **stability analysis**, which is beyond the scope of this book. We shall assume that the flows for which we present solutions in this chapter are in a Reynolds-number regime where they are stable.

---

[127] The term **streamline flow** is sometimes used.

[128] The word laminar stems from *lamina*, which is New Latin meaning 'thin plate' or 'layer'.

[129] It is crucially important for the reader to understand that the numerical value of the critical Reynolds number for one geometry is normally quite different from that from another, not least because of different choices for the length and velocity scales in each case.

*Introduction to Engineering Fluid Mechanics*. Marcel Escudier.
© Marcel Escudier 2017. Published 2017 by Oxford University Press.

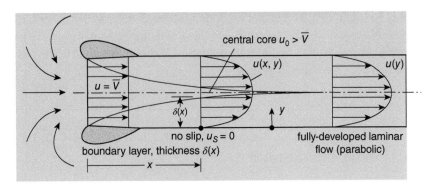

**Figure 16.1** Developing flow between parallel plates

It was pointed out in Chapter 15 that, when a viscous fluid flows steadily through a pipe or other duct, streamwise-velocity changes occur normal to the duct wall as the flow develops from the duct inlet, initially across a relatively thin **boundary layer**, as illustrated in Figure 16.1 for flow between two parallel plates. Ultimately, the entire cross section is influenced by viscosity, and the flow is said to be **fully developed** (i.e. unchanging with streamwise location). The streamwise distance over which this process occurs is called the **development length**. This chapter is concerned with the analysis of steady, fully-developed, internal, laminar flows. If the streamwise coordinate is $x$, and the corresponding velocity component is $u$, then for a fully-developed flow we require $\partial u/\partial x = 0$ so that, in rectangular Cartesian coordinates, $u = u\,(y,z)$ or, in polar-cylindrical coordinates, $u = u\,(r,\theta)$. If there is no relative movement between the duct walls, we also have $v = 0$ in the $y$-direction, and $w = 0$ in the $z$-direction (or $v = 0$ in the $r$-direction, and $w = 0$ in the $\theta$-direction). As we shall show, the assumption that a laminar flow is steady and fully developed reduces the Navier-Stokes equations to a level where straightforward analytical solutions are possible. Not only are such solutions of practical relevance, for example, to flow through long pipelines and capillary tubes, and to viscometry, but they also provide insight into laminar flows more generally.

Two principal types of steady, fully-developed, laminar duct flow may be identified. The first is **Poiseuille flow**, where the flow through a long **cylindrical duct**[130] is driven by a pressure difference applied between the ends of the duct. The second is **Couette flow**, where flow of the fluid contained between two parallel or concentric surfaces occurs when there is relative tangential movement between the two surfaces. A third possibility involves a combination of Couette and Poiseuille flows for a given geometry.

Whether the most appropriate starting point for a detailed analysis of any flow is the continuity and Navier-Stokes equations in Cartesian- or polar-cylindrical-coordinate form depends upon the duct geometry. The Cartesian-coordinate form of the equations is chosen in Section 16.2, which concerns Poiseuille flow in general, although either form could be chosen. The detailed analysis of Poiseuille flow through rectangular, triangular, and other non-circular

[130] Although the term cylindrical is commonly used to refer to an object of circular cross section, more generally it applies to any object for which the surface is generated by lines parallel to a straight axis.

duct shapes is generally complicated and beyond the scope of this text. However, if the equations are posed in polar-cylindrical-coordinate form, flow through or within axisymmetric ducts is more amenable to analysis and is the subject of Sections 16.3 and 16.5. In all cases we shall assume that body forces are negligible.

## 16.2 Poiseuille flow of a Newtonian fluid, hydraulic diameter, and Poiseuille number

We discuss first the general case of Poiseuille flow of a Newtonian fluid through a cylindrical duct of arbitrary cross section. Our starting point is the continuity and Navier-Stokes equations in rectangular-Cartesian-coordinate form for uniform-property flow, which were derived in Subsection 15.1.6:

continuity

$$\frac{\partial u}{\partial x} + \frac{\partial v}{\partial y} + \frac{\partial w}{\partial z} = 0 \tag{16.1}$$

$x$-component

$$\rho \left( \frac{\partial u}{\partial t} + u\frac{\partial u}{\partial x} + v\frac{\partial u}{\partial y} + w\frac{\partial u}{\partial z} \right) = -\frac{\partial p}{\partial x} + \mu \left( \frac{\partial^2 u}{\partial x^2} + \frac{\partial^2 u}{\partial y^2} + \frac{\partial^2 u}{\partial z^2} \right) \tag{16.2}$$

$y$-component

$$\rho \left( \frac{\partial v}{\partial t} + u\frac{\partial v}{\partial x} + v\frac{\partial v}{\partial y} + w\frac{\partial v}{\partial z} \right) = -\frac{\partial p}{\partial y} + \mu \left( \frac{\partial^2 v}{\partial x^2} + \frac{\partial^2 v}{\partial y^2} + \frac{\partial^2 v}{\partial z^2} \right) \tag{16.3}$$

$z$-component

$$\rho \left( \frac{\partial w}{\partial t} + u\frac{\partial w}{\partial x} + v\frac{\partial w}{\partial y} + w\frac{\partial w}{\partial z} \right) = -\frac{\partial p}{\partial z} + \mu \left( \frac{\partial^2 w}{\partial x^2} + \frac{\partial^2 w}{\partial y^2} + \frac{\partial^2 w}{\partial z^2} \right). \tag{16.4}$$

For steady Poiseuille flow, $\partial/\partial t = 0$, $\partial u/\partial x = 0$, $v = w = 0$, so that equation (16.2) reduces to

$$0 = -\frac{\partial p}{\partial x} + \mu \left( \frac{\partial^2 u}{\partial y^2} + \frac{\partial^2 u}{\partial z^2} \right) \tag{16.5}$$

while all that remains of equations (16.3) and (16.4) is

$$\frac{\partial p}{\partial y} = \frac{\partial p}{\partial z} = 0 \tag{16.6}$$

from which we conclude that $p = p(x)$. Since $u$ is independent of $x$, the parenthetical term on the right-hand side of equation (16.5) must be independent of $x$ while $\partial p/\partial x$ can only be a function of $x$ or a constant. For both conditions to be satisfied we conclude that the pressure gradient $dp/dx$ must be a constant independent of $x$.

We thus have

$$\frac{dp}{dx} = \mu \left( \frac{\partial^2 u}{\partial y^2} + \frac{\partial^2 u}{\partial z^2} \right) = \text{constant}. \tag{16.7}$$

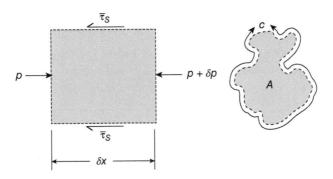

**Figure 16.2** Pressure difference $\delta p$ and average shear stress $\bar{\tau}_S$ acting on a control volume of infinitesimal length $\delta x$, circumference $c$, and cross-sectional area $A$

From Subsection 15.2.6 we find that the corresponding equation in polar-cylindrical coordinates is

$$\frac{dp}{dx} = \mu \left[ \frac{1}{r} \frac{\partial}{\partial r} \left( r \frac{\partial u}{\partial r} \right) + \frac{1}{r^2} \frac{\partial^2 u}{\partial \theta^2} \right]. \tag{16.8}$$

If the cross-sectional area of the duct is $A$, and the circumference of the **wetted perimeter**[131] is $c$, then it is easily seen, from the force balance for the elemental control volume of length $\delta x$ shown in Figure 16.2[132], that, for fully-developed duct flow,

$$-\delta p\, A - \bar{\tau}_S\, c\, \delta x = 0 \tag{16.9}$$

or, as $\delta x \to 0$,

$$\bar{\tau}_S = -\frac{A}{c} \frac{dp}{dx} = \text{constant.} \tag{16.10}$$

The quantity $\bar{\tau}_S$ represents the average shear stress exerted on the fluid by the **wetted surface** of the duct while $\delta p$ is the change in static pressure between the ends of the control volume. The right-hand side of equation (16.9) is zero because it represents the change in momentum flowrate through the control volume, which must be zero for fully-developed flow.

We can now use equation (16.10) to substitute for $dp/dx$ in equation (16.7) in terms of $\bar{\tau}_S$ to give

$$-\frac{c\bar{\tau}_S}{\mu A} = \frac{\partial^2 u}{\partial y^2} + \frac{\partial^2 u}{\partial z^2}. \tag{16.11}$$

To convert equation (16.11) to non-dimensional form, we need to identify suitable length and velocity scales. For the latter we choose the average flow velocity $\bar{V}$, which can be determined from the volumetric flowrate $\dot{Q}$

$$\bar{V} = \frac{\dot{Q}}{A}. \tag{16.12}$$

---

[131] The term wetted refers to any surface in contact with a fluid, whether a liquid or a gas.

[132] Note that the cross section of the control volume is that of the duct with cross-sectional area $A$.

The choice of a length scale is less straightforward since there is no 'natural' <u>axial</u> length scale for a fully-developed flow. Instead, it is conventional to define a length scale in terms of the geometric properties of the cross section, $A$ and $c$. The ratio $A/c$, which appears in equation (16.10), has the dimensions of length and is easily seen to equal $D/4$ for a circular pipe of diameter $D$, i.e. $4A/c = D$. More generally, the combination $4A/c$ is used to define a length scale

$$D_H = \frac{4A}{c} \qquad\qquad (16.13)$$

which is referred to as the **hydraulic diameter**[133]. The hydraulic diameter is easily calculated for any cross section: e.g. for a square duct of side length $a$, we find $D_H = a$ and, for a duct of annular cross section, with inner radius $R_I$ and outer radius $R_O$, we have $D_H = 2\delta$, where $\delta = R_O - R_I$ is the average width of the gap between the inner and outer surfaces. The latter example reveals a deficiency of the hydraulic-diameter concept: in the case of an annulus, it takes no account of **eccentricity**, i.e. the situation where the two cylinders are not concentric.

If we define the non-dimensional variables $u^* = u/\bar{V}, y^* = y/D_H$, and $z^* = z/D_H$, equation (16.11) can be written in non-dimensional form as

$$\frac{4\bar{\tau}_S D_H}{\mu \bar{V}} = -\left(\frac{\partial^2 u^*}{\partial y^{*2}} + \frac{\partial^2 u^*}{\partial z^{*2}}\right) = \text{constant}. \qquad\qquad (16.14)$$

Although the term on the left-hand side of equation (16.14), is a natural non-dimensional parameter to characterise a Poiseuille flow, it is conventional for any duct flow of a viscous fluid to introduce a **Reynolds number**, defined here as

$$Re_H = \frac{\rho \bar{V} D_H}{\mu} \qquad\qquad (16.15)$$

and the **Fanning friction factor**[134]

$$f_F = \frac{2\bar{\tau}_S}{\rho \bar{V}^2} \qquad\qquad (16.16)$$

where $\rho$ is the fluid density.

In terms of $Re_H$ and $f_F$, equation (16.14) may be written as

$$-\left(\frac{\partial^2 u^*}{\partial y^{*2}} + \frac{\partial^2 u^*}{\partial z^{*2}}\right) = \text{constant} = 2f_F \, Re_H = 2Po \qquad\qquad (16.17)$$

where $Po$ is the **Poiseuille number** $Po$, defined as

$$Po = f_F \, Re_H = \frac{2\tau_S D_H}{\mu \bar{V}}. \qquad\qquad (16.18)$$

Although the appearance of a Reynolds number is to be expected in any analysis of viscous fluid flow, it can be argued that it has been introduced here artificially (and unnecessarily). The

---

[133] It should be noted that the hydraulic radius, $R_H$, is defined as $A/c$ and so is equal to $D_H/4$, not $D_H/2$ as would seem logical.

[134] The Darcy friction factor $f_D = 8\tau_S/\rho \bar{V}^2$, also referred to as the Darcy-Weisbach friction factor, is also in common use, particularly in hydraulics.

artificiality is associated with the introduction of the fluid density, $\rho$, which does not appear in either equation (16.7) or equation (16.10) as it plays no role in a fully-developed flow, where no fluid particle undergoes either spatial or temporal acceleration since $Du/Dt = 0$. However, in the majority of viscous-fluid flow problems the fluid density does play a significant role, so the Reynolds number then arises naturally as the non-dimensional flow parameter, and it is conventional to introduce it in the way we have done here, even for fully-developed flows where the Poiseuille number is arguably more appropriate.

Solutions of equation (16.17), or its polar-cylindrical form, have been obtained for a wide variety of duct cross sections leading to velocity distributions, $u(y, z)$ or $u(r, \theta)$, and corresponding values of the Poiseuille number. A selection of the latter have been plotted in Figure 16.3 against an aspect ratio which characterises the relevant cross section. We note that the result $Po = 16$ for a pipe of circular cross section, discussed in detail in Subsection 16.3.1,

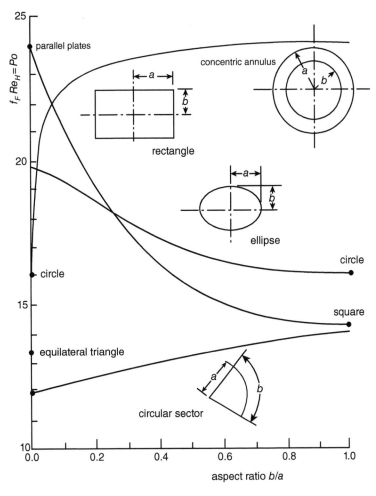

**Figure 16.3** Poiseuille number $Po$ versus aspect ratio $b/a$ for Poiseuille flow through ducts with the cross sections shown

appears both as a limiting case for flow through a duct of elliptical cross section as $b/a \to 1$ and for flow through a concentric annulus, discussed in Subsection 16.3.2, as $b/a \to 0$. The result $Po = 24$ for flow between parallel plates, discussed in Section 16.4, also appears as a limiting case, both for flow through a rectangular duct as $b/a \to 0$ and for flow through a concentric annulus as $b/a \to 1$.

Figure 16.3 is based upon data reported in the book by Shah and London (1978), which also gives data for a wide variety of other cross sections.

## 16.3 Poiseuille flow through an axisymmetric cylindrical duct

In this section we derive solutions for Poiseuille flow through both a circular pipe and a concentric-annular duct from simplified forms of the flow equations in polar-cylindrical-coordinate form. From Section 15.2 we have, for the flow of a uniform-property Newtonian fluid:

continuity equation

$$\frac{\partial u}{\partial x} + \frac{1}{r}\frac{\partial}{\partial r}(rv) + \frac{1}{r}\frac{\partial w}{\partial \theta} = 0 \tag{16.19}$$

$x$-component

$$\rho\left(\frac{\partial u}{\partial t} + u\frac{\partial u}{\partial x} + v\frac{\partial u}{\partial r} + \frac{w}{r}\frac{\partial u}{\partial \theta}\right) = -\frac{\partial p}{\partial x} + \mu\left[\frac{\partial^2 u}{\partial x^2} + \frac{1}{r}\frac{\partial}{\partial r}\left(r\frac{\partial u}{\partial r}\right) + \frac{1}{r^2}\frac{\partial^2 u}{\partial \theta^2}\right] \tag{16.20}$$

$r$-component

$$\rho\left(\frac{\partial v}{\partial t} + u\frac{\partial v}{\partial x} - v\frac{\partial v}{\partial r} + \frac{w}{r}\frac{\partial v}{\partial \theta} - \frac{w^2}{r}\right)$$

$$= -\frac{\partial p}{\partial r} + \mu\left[\frac{\partial^2 v}{\partial x^2} + \frac{\partial}{\partial r}\left\{\frac{1}{r}\frac{\partial}{\partial r}(rv)\right\} + \frac{1}{r^2}\frac{\partial^2 v}{\partial \theta^2} - \frac{2}{r^2}\frac{\partial w}{\partial \theta}\right] \tag{16.21}$$

$\theta$-component

$$\rho\left(\frac{\partial w}{\partial t} + u\frac{\partial w}{\partial x} + v\frac{\partial w}{\partial r} + \frac{w}{r}\frac{\partial w}{\partial \theta} + \frac{vw}{r}\right)$$

$$= -\frac{1}{r}\frac{\partial p}{\partial \theta} + \mu\left[\frac{\partial^2 w}{\partial x^2} + \frac{\partial}{\partial r}\left\{\frac{1}{r}\frac{\partial}{\partial r}(rw)\right\} + \frac{1}{r^2}\frac{\partial^2 w}{\partial \theta^2} + \frac{2}{r^2}\frac{\partial v}{\partial \theta}\right]. \tag{16.22}$$

The boundary conditions are $v_S = 0$ at an impervious surface and, from the no-slip condition, $u_S = w_S = 0$ at any stationary, solid surface[135]. The subscript $S$ is used to denote conditions at a wetted surface.

Since the flow is fully developed, $\partial u/\partial x = 0$ and, in the absence of rotation of either surface, the azimuthal velocity $w = 0$, so that the continuity equation simplifies to

$$\frac{d}{dr}(rv) = 0 \tag{16.23}$$

---

[135] More generally, if the surface itself is moving, $u_S$ and $w_S$ equal the surface velocity components in the $x$- and/or $\theta$ directions, respectively.

which may be integrated to give $rv = $ constant. Since $v_S = 0$, we conclude that $v = 0$ throughout the flowfield.

With $v = w = 0$, equation (16.21) reduces to

$$\frac{\partial p}{\partial r} = 0 \tag{16.24}$$

and equation (16.22) to

$$\frac{\partial p}{\partial \theta} = 0. \tag{16.25}$$

The only possibility for the static pressure $p$ which satisfies both equations (16.24) and (16.25) is $p = p(x)$. Equation (16.2) then reduces to

$$\frac{1}{r}\frac{\partial}{\partial r}\left(r\frac{\partial u}{\partial r}\right) + \frac{1}{r^2}\frac{\partial^2 u}{\partial \theta^2} = \frac{1}{\mu}\frac{dp}{dx}. \tag{16.26}$$

For fully-developed flow, the left-hand side of this equation must be independent of $x$, while the right-hand side can only be a function of $x$ or is a constant. For both to be satisfied, we conclude that the pressure gradient $dp/dx$ must be a constant (i.e. independent of $x$).

Since we are restricting attention to axisymmetric geometries, such as a circular pipe or a concentric annulus, $u = u(r)$ and equation (16.26) can be integrated once to give

$$r\frac{du}{dr} = \frac{r^2}{2\mu}\frac{dp}{dx} + A \tag{16.27}$$

where $A$ is a constant of integration, while a second integration leads to

$$u = \frac{r^2}{4\mu}\frac{dp}{dx} + A\ln r + B. \tag{16.28}$$

To proceed further and determine the constants of integration, $A$ and $B$, we have to consider specific geometries.

### 16.3.1  Poiseuille flow through a circular pipe: Hagen-Poiseuille flow

Poiseuille flow through a cylindrical pipe of circular cross section is referred to as **Hagen-Poiseuille flow**. If the pipe has radius $R$ (diameter $D$), the two constants in equation (16.28) have the values

$$A = 0 \quad \text{and} \quad B = -\frac{R^2}{4\mu}\frac{dp}{dx}.$$

The value for $A$ is a consequence of the fact that equation (16.27) must be valid for all values of the radial position $r$, including $r = 0$ (the centreline). The value for $B$ is a consequence of the no-slip condition at the pipe surface, $r = R$. Thus, the **velocity distribution** (or **profile**) for a circular pipe is given by

$$u = \frac{(R^2 - r^2)}{4\mu}\left(-\frac{dp}{dx}\right), \tag{16.29}$$

with the minus sign in the second parenthesis emphasising that <u>for the flow to be in the positive x-direction the **pressure gradient** $dp/dx$ must be negative</u>, i.e. the static pressure decreases with streamwise distance $x$.

The centreline velocity is thus

$$u_0 = \frac{R^2}{4\mu}\left(-\frac{dp}{dx}\right)$$

(16.30)

so that the velocity distribution can be written as

$$\frac{u}{u_0} = 1 - \left(\frac{r}{R}\right)^2$$

(16.31)

i.e. the velocity variation is parabolic with $r$.

The shear stress at any radial position within the fluid is given by

$$\tau_{rx} = \mu\frac{du}{dr} = \frac{r}{2}\frac{dp}{dx} = \frac{-2\mu r u_0}{R^2}$$

(16.32)

so that the shear stress exerted by the fluid on the pipe surface $\tau_S$ is given by

$$\tau_S = -\left.\tau_{rx}\right|_{r=R} = \frac{2\mu u_0}{R}.$$

(16.33)

Equation (16.32) shows that the shear stress variation within the fluid is proportional to distance $r$ from the centreline. The radial distributions of shear stress and velocity are shown in Figure 16.4.

As we have emphasised in previous chapters, it is advantageous to present formulae in non-dimensional form. In the case of equation (16.33) this is achieved by rearranging the terms, leading to

$$\frac{\tau_S R}{\mu u_0} = 2.$$

(16.34)

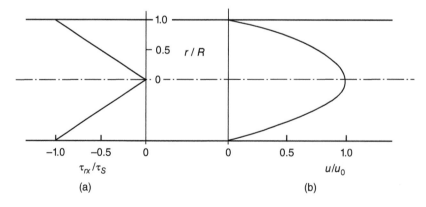

**Figure 16.4** Shear stress and velocity distributions in fully-developed laminar pipe flow

From a practical point of view it is more convenient to express a formula for the wall shear stress $\tau_S$ in terms of an average flow velocity $\bar{V}$, based upon the **volumetric flowrate** $\dot{Q}$, rather than the centreline velocity $u_0$. We define $\bar{V}$ as follows

$$\bar{V} = \frac{\dot{Q}}{A} = \frac{4\dot{Q}}{\pi D^2} \tag{16.35}$$

where $A = \pi D^2/4$ is the cross-sectional area of the pipe, $D = 2R$ being its diameter. Since the velocity distribution is now known, we can calculate $\dot{Q}$ from:

$$\dot{Q} = \int_0^R 2\pi r u \, dr \tag{16.36}$$

so that, after substitution for $u$ from equation (16.31), we have

$$\bar{V} = \frac{\int_0^R 2\pi r u \, dr}{\pi R^2} = \frac{1}{2} u_0 \tag{16.37}$$

and equation (16.34) can be rewritten as

$$\frac{\tau_S D}{\mu \bar{V}} = 8. \tag{16.38}$$

The conclusion that the average flow velocity is 50% of the peak velocity, shown by equation (16.37), is peculiar to fully-developed laminar flow in a circular pipe and is not a general result applicable to any cross section.

In terms of the Fanning friction factor $f_F$ and Reynolds number $Re$, defined in Section 16.2 but noting that in this case $D_H = D$ and so the subscript $H$ is unnecessary, equation (16.38) leads to

$$\frac{\tau_S D}{\mu \bar{V}} = \frac{1}{2} \frac{2\tau_S}{\rho \bar{V}^2} \frac{\rho D \bar{V}}{\mu} = \frac{1}{2} f_F \, Re = 8 \tag{16.39}$$

or

$$f_F \, Re = Po = 16 = Po \tag{16.40}$$

where $Po$ is the Poiseuille number, also defined in Section 16.2.

Above a **critical Reynolds number**, usually taken as 2100, Hagen-Poiseuille flow becomes unstable and eventually turbulent so that the analysis presented here is no longer valid. Poiseuille and Couette flow through other geometries also remain laminar up to values of critical Reynolds (or Taylor) numbers, which will be different in each case.

## 16.3.2 Poiseuille flow through a concentric annulus

The geometry of a concentric annulus is illustrated in Figure 16.5. The fluid flows through the annular space between an inner circular pipe of radius $R_I$ and an outer circular pipe, co-axial with the inner pipe, of radius $R_O$. Equation (16.28) is still valid, and the constants of integration, $A$ and $B$, are again determined from the boundary conditions, which are now $u = 0$ for both $r = R_I$, and $r = R_O$, so that

$$0 = \frac{R_I^2}{4\mu} \frac{dp}{dx} + A \ln R_I + B$$

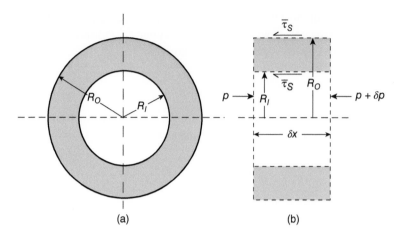

**Figure 16.5** (a) Geometry of a concentric annulus (b) Elemental control volume

and

$$0 = \frac{R_O^2}{4\mu}\frac{dp}{dx} + A \ln R_O + B.$$

If we subtract one equation from the other, after some rearrangement we find

$$A = \frac{\left(R_O^2 - R_I^2\right)}{4\mu \ln(R_O/R_I)}\left(-\frac{dp}{dx}\right).$$

Substitution of this expression for $A$ into either of the equations including $B$ leads to

$$B = \left[ R_O^2 - \frac{\left(R_O^2 - R_I^2\right)\ln R_O}{\ln(R_O/R_I)} \right] \frac{1}{4\mu}\left(-\frac{dp}{dx}\right)$$

or

$$B = \left[ R_I^2 - \frac{\left(R_O^2 - R_I^2\right)\ln R_I}{\ln(R_O/R_I)} \right] \frac{1}{4\mu}\left(-\frac{dp}{dx}\right).$$

It is easily shown that the two equations for $B$ are equivalent.

If the expressions for $A$ and $B$ are substituted into equation (16.28) we have the following equation for the velocity distribution in the annular region $R_O \geq r \geq R_I$

$$u = \left[ \frac{R_O^2 - r^2}{R_O^2 - R_I^2} - \frac{\ln(r/R_O)}{\ln(R_I/R_O)} \right] \frac{\left(R_O^2 - R_I^2\right)}{4\mu}\left(-\frac{dp}{dx}\right). \tag{16.41}$$

The volumetric flowrate $\dot{Q}$ through the annulus can now be obtained from

$$\dot{Q} = \int_{R_I}^{R_O} 2\pi r u \, dr = \left[ R_O^2 + R_I^2 - \frac{\left(R_O^2 - R_I^2\right)}{\ln(R_O/R_I)} \right] \frac{\pi\left(R_O^2 - R_I^2\right)}{8\mu}\left(-\frac{dp}{dx}\right). \tag{16.42}$$

The surface shear stresses acting on the cylindrical surfaces can be determined individually from equation (16.41) but, from a practical point of view, a value $\bar{\tau}_S$ averaged over the two annular surfaces is adequate. From a force balance for the elemental control volume shown in Figure 16.5(b)

$$\bar{\tau}_S = \frac{1}{2} (R_O - R_I) \left( -\frac{dp}{dx} \right) \tag{16.43}$$

so that

$$\dot{Q} = \left[ R_O^2 + R_I^2 - \frac{(R_O^2 - R_I^2)}{\ln (R_O/R_I)} \right] \frac{\pi (R_O + R_I) \bar{\tau}_S}{4\mu} \tag{16.44}$$

and the average flow velocity $\bar{V}$ is

$$\bar{V} = \frac{\dot{Q}}{A} = \frac{\dot{Q}}{\pi (R_O^2 - R_I^2)} = \left[ R_O^2 + R_I^2 - \frac{(R_O^2 - R_I^2)}{\ln (R_O/R_I)} \right] \frac{\bar{\tau}_S}{4\mu (R_O - R_I)}. \tag{16.45}$$

We can now convert this result to the Poiseuille-number form, as we did for the circular pipe, with the result

$$Po = f_F Re_H = \frac{16 (R_O - R_I)^2}{\left[ R_O^2 + R_I^2 - \frac{(R_O^2 - R_I^2)}{\ln (R_O/R_I)} \right]} \tag{16.46}$$

where the Reynolds number has been defined as $Re_H = \rho \bar{V} D_H / \mu$, $D_H$ being the hydraulic diameter $2 (R_O - R_I)$.

We note that, if the radius of the inner cylinder is negligibly small, i.e. $R_I \ll R_O$, equation (16.46) reduces to $Po = 16$, the result for a circular pipe. This conclusion is not quite as straightforward as it seems at first since, no matter how small the radius of the inner cylinder, the noslip condition holds at its surface, i.e. for $r = R_I$, $u = 0$ even when $R_I \to 0$ but, in the absence of an inner cylinder, $u$ is a maximum for $r = 0$. This apparent contradiction is resolved when it is realised that, as $R_I \to 0$, the location of the velocity maximum in the annulus also tends to $r = 0$, and the radial extent of the region affected by the inner cylinder becomes negligibly small. If the annular gap $\delta = R_O - R_I$ is small compared with $R_O$, then $Po \to 24$, the value for Poiseuille flow between infinite parallel plates, discussed in Section 16.4. In fact, if the radius ratio exceeds 0.5, a good approximation (within 1%) for annular Poiseuille flow is $Po = 24$.

## 16.4 Combined plane Couette and Poiseuille flow between infinite parallel plates: Couette-Poiseuille flow

Plane (or linear) Couette flow arises in the fluid between two parallel plates when one of them moves at constant tangential velocity relative to the other. A pressure difference imposed on the fluid between the ends of two fixed parallel plates produces a Poiseuille flow if the plates are sufficiently long for the flow to become fully developed. Analysis of a combination of these two flows provides some insight into the influence of a streamwise pressure gradient on a

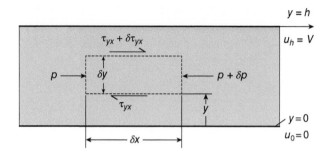

**Figure 16.6** Control volume for analysis of plane Couette-Poiseuille flow

boundary layer, which is the subject of Section 17.4 and Subsection 17.6.2. The situation is illustrated in Figure 16.6. The streamwise coordinate is taken as $x$, and the plates are assumed to be of infinite extent in the $z$-direction. The plate spacing is $h$, the velocity of the moving plate is $V$, and the fluid viscosity is $\mu$. The fluid velocity a distance $y$ from the stationary plate is $u$, which, since the flow is fully developed, is a function only of $y$, i.e. $u = u(y)$.

We could analyse this flow starting with the continuity and Navier-Stokes equations, but for such a simple flow we shall develop the solution from first principles. We refer to the elemental control volume shown in Figure 16.6, which has length $\delta x$ and height[136] $\delta y$. The static pressure $p$ changes by an amount $\delta p$ across the control volume in the $x$-direction, and the shear stress $\tau_{yx}$ changes by an amount $\delta \tau_{yx}$ in the $y$-direction. A force balance for the control volume gives

$$\delta \tau_{yx}\, \delta x - \delta p\, \delta y = 0$$

the right-hand side being zero because there is no change in fluid momentum for a fully-developed flow. This equation can be rearranged as

$$\frac{\delta p}{\delta x} - \frac{\delta \tau_{yx}}{\delta y} = 0$$

so that, in the limit $\delta x \to 0, \delta y \to 0$, we have

$$\frac{\partial \tau_{yx}}{\partial y} = \frac{\partial p}{\partial x}. \tag{16.47}$$

From Subsection 15.1.4, Stokes' constitutive equations here reduce to

$$\tau_{yx} = \mu \frac{du}{dy} \tag{16.48}$$

so that, from equation (16.47)

$$\mu \frac{d^2 u}{dy^2} = \frac{dp}{dx} = \text{constant.} \tag{16.49}$$

Partial derivatives have been replaced by ordinary derivatives, and the pressure gradient must be a constant because the flow is fully developed, i.e. $u = u(y)$.

---

[136] The reader is reminded that elemental control volumes such as that depicted in Figure 16.6 have to be imagined to be infinitesimally small and are shown disproportionately large here and elsewhere.

Equation (16.49) can be integrated once to give

$$\mu \frac{du}{dy} = y \frac{dp}{dx} + A$$

and a second integration leads to

$$\mu u = \frac{y^2}{2} \frac{dp}{dx} + Ay + B$$

where $A$ and $B$ are constants of integration, determined from the no-slip boundary conditions $y = 0, u = 0$, and $y = h, u = V$. We find

$$B = 0 \qquad \text{and} \qquad A = \frac{\mu V}{h} - \frac{h}{2} \frac{dp}{dx}$$

so that, after some rearrangement, the velocity distribution is given by

$$u = \frac{Vy}{h} - \frac{y}{2\mu} (h - y) \frac{dp}{dx}. \tag{16.50}$$

The first (linear) term on the right-hand side of equation (16.50) represents the Couette-flow contribution to $u$, while the second (parabolic) term is the Poiseuille-flow contribution.

A non-dimensional form of equation (16.50) is then

$$\frac{u}{V} = \left[ 1 - \frac{1}{2} (1 - \xi) \lambda_P \right] \xi \tag{16.51}$$

where $\xi = y/h$, and

$$\lambda_P = -\frac{h^2}{\mu V} \frac{dp}{dx} \tag{16.52}$$

is a non-dimensional **pressure-gradient parameter** for Poiseuille flow.

From equation (16.50), the shear stress, $\tau_{yx}$, at any distance from the lower plate, $y$, is given by

$$\tau_{yx} = \mu \frac{du}{dy} = \frac{\mu V}{h} + \left( y - \frac{h}{2} \right) \frac{dp}{dx} \tag{16.53}$$

so that the shear stress exerted on the lower plate $\tau_0$ is given by

$$\tau_0 = \frac{\mu V}{h} - \frac{h}{2} \frac{dp}{dx}. \tag{16.54}$$

The volumetric flowrate per unit width $\dot{Q}'$ is given by

$$\dot{Q}' = \int_0^h u \, dy = \frac{Vh}{2} - \frac{h^3}{12\mu} \frac{dp}{dx}. \tag{16.55}$$

An average velocity $\bar{V}$ can be defined from $\dot{Q}' = \bar{V}h$ so that, from equation (16.55), we have

$$\bar{V} = \frac{V}{2} - \frac{h^2}{12\mu}\frac{dp}{dx}. \tag{16.56}$$

As before, the Fanning friction factor $f_F$ is defined by

$$f_F = \frac{2\tau_0}{\rho\bar{V}^2} \tag{16.57}$$

and the Reynolds number $Re_H$ by

$$Re_H = \frac{2\rho\bar{V}h}{\mu}, \tag{16.58}$$

since the hydraulic diameter here is $2h$ The Poiseuille number $Po$ is then given by

$$Po = f_F\, Re_H = \frac{4\tau_0 h}{\mu\bar{V}} = 24\left(\frac{2 - \lambda_P}{6 - \lambda_P}\right) \tag{16.59}$$

from which for plane Couette flow with $\lambda_P = 0$ we find $Po = 8$. To find the value of $Po$ for plane Poiseuille flow we observe that the definition of $\lambda_P$ with $V \to 0$ leads to $\lambda_P \to \infty$ and, from equation (16.59), $Po \to 24$.

Velocity distributions according to equation (16.51) are shown in Figure 16.7 for values of the pressure-gradient parameter in the range $-10 < \lambda_P < 10$.

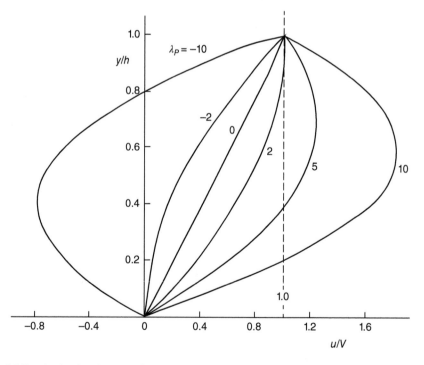

**Figure 16.7** Velocity distributions for Couette-Poiseuille flow for different values of the pressure-gradient parameter $\lambda_P$.

We note the following

- $\lambda_P = 0$ corresponds to plane Couette flow, with $u$ proportional to $y$, and constant shear stress $\tau_{yx} = \mu V/h$
- for $\lambda_P = 2$ the shear stress at $y = h$ is zero, and the corresponding velocity distribution $u/V = y/h(2 - y/h)$ can be used as an approximate model for a zero pressure-gradient boundary layer (see Section 17.6)
- for $\lambda_P > 2$ the shear stress exerted on the upper plate is positive, and the velocity distribution exhibits a maximum at $y/h = 1/\lambda_P + 1/2$, at which location the shear stress is zero
- for $\lambda_P = -2$ the shear stress at $y = 0$ is zero and corresponds to the condition in boundary-layer analysis referred to as **flow separation**
- for $\lambda_P < -2$ both the shear stress exerted on the lower plate and the near-surface flow velocity are negative: there is so-called **backflow** near the lower surface

---

**ILLUSTRATIVE EXAMPLE 16.1**

A fluid with viscosity $\mu$ is contained within the annular space between two long concentric cylinders, the outer of which has radius $R_O$, while the inner has radius $R_I$. The outer cylinder moves axially at velocity $u_O$ while the inner cylinder is stationary. There is zero pressure difference between the ends of the cylinders.

Derive a reduced form of the Navier-Stokes equations for the flow, stating all the assumptions made. Solve the equations to find the radial variation of the axial velocity $u(r)$ within the annular gap and the corresponding distribution of the shear stress $\tau_{rx}$ acting on the fluid. Derive an equation for volumetric flowrate $\dot{Q}$ through the annulus.

### Solution

Since the cylinders are long, we can assume the flow is fully developed, and $\partial u/\partial x = 0$. In the absence of rotation of either cylinder, it must also be that $w = 0$. From the continuity equation with $\partial u/\partial x = 0$, and $w = 0$, it must be that $v = 0$.

The boundary conditions are $r = R_I$, $u = 0$, and $r = R_O$, $u = u_O$. We also have $\partial p/\partial x = 0$.
The $r$-component of the Navier-Stokes equations then reduces to

$$\frac{\partial p}{\partial r} = 0,$$

the $\theta$-component reduces to

$$\frac{\partial p}{\partial \theta} = 0,$$

and we have also $\partial p/\partial x = 0$, so it must be that the static pressure $p$ is the same at all points within the annulus.

The $x$-component of the Navier-Stokes equations is thus

$$\frac{\mu}{r}\frac{d}{dr}\left(r\frac{du}{dr}\right) = 0$$

or

$$\frac{d}{dr}\left(r\frac{du}{dr}\right) = 0$$

which can be integrated to give

$$r\frac{du}{dr} = A$$

or

$$\frac{du}{dr} = \frac{A}{r}.$$

A second integration gives

$$u = A\ln r + B$$

where $A$ and $B$ are constants of integration to be determined from the boundary conditions

$$0 = A\ln R_I + B$$

and

$$u_O = A\ln R_O + B.$$

We thus find

$$A = \frac{u_O}{\ln(R_O/R_I)} \text{ and } B = -\frac{u_O\ln R_I}{\ln(R_O/R_I)}.$$

Thus, for the velocity distribution we have

$$u = u_O\frac{\ln(r/R_I)}{\ln(R_O/R_I)}.$$

The shear stress $\tau_{rx}$ is given by

$$\tau_{rx} = \mu\frac{du}{dr} = \frac{\mu A}{r} = \frac{\mu u_O}{r}\frac{\ln(r/R_I)}{\ln(R_O/R_I)}.$$

The volumetric flowrate $\dot{Q}$ is found from

$$\dot{Q} = \int_{R_I}^{R_O} 2\pi r u\, dr = \frac{2\pi u_O R_I^2}{\ln\xi_O}\int_1^{\xi_O} \xi\ln\xi\, d\xi$$

where, for convenience, we have introduced the variable $\xi = r/R_I$.
From tables of integrals,

$$\int^{\xi} \ln\xi\, d\xi = \frac{\xi^2}{2}\ln\xi - \frac{\xi^2}{4}$$

so that

$$\dot{Q} = \frac{\pi u_O R_I^2}{\ln \xi_O} \left( \xi_O^2 \ln \xi_O - \frac{\xi_O^2}{2} + \frac{1}{4} \right)$$

or

$$\dot{Q} = \pi u_O R_O^2 \left[ 1 - \frac{1}{2 \ln (R_O/R_I)} \left( \frac{R_O^2 - R_I^2}{R_O^2} \right) \right].$$

## 16.5 Taylor-Couette flow

As we stated in Section 16.1, if fluid is contained between two parallel or concentric surfaces and there is relative tangential movement between the two surfaces, the resulting flow is termed a **Couette flow**. The two Couette-flow possibilities for the concentric-cylinder geometry are flow resulting from axial movement of one cylinder relative to the other, as in Illustrative Example 16.1, or from differential rotation of the two cylinders. The latter, known as **Taylor-Couette** flow, is straightforward to analyse. If the axes of the two cylinders are parallel but not coincident, the geometry is said to be **eccentric**, the **offset** between the two axes defining the **eccentricity**. Streamwise flow through an eccentric annulus with inner cylinder rotation is of great practical interest, particularly if the fluid is a liquid with **shear-thinning** (non-Newtonian) properties, which is a good model for the flow of drilling fluid during drilling of an oil or gas well. For a Newtonian fluid a series solution has been found for the eccentric geometry but for a non-Newtonian fluid this problem can only be solved numerically.

The analysis in Subsection 16.3.2 of fully-developed axial flow through a concentric annulus assumed no rotation of either cylinder. If we now allow rotation of the inner cylinder, at angular velocity $\Omega$, but retain the restriction to steady, fully-developed flow, the flow equations in polar-cylindrical coordinates reduce to:

$x$-component

$$0 = -\frac{\partial p}{\partial x} + \frac{\mu}{r} \frac{d}{dr} \left( r \frac{du}{dr} \right) \tag{16.60}$$

$r$-component

$$\frac{\rho w^2}{r} = -\frac{\partial p}{\partial r} \tag{16.61}$$

$\theta$-component

$$0 = \frac{d}{dr} \left[ \frac{1}{r} \frac{d}{dr} (rw) \right] \tag{16.62}$$

with boundary conditions $r = R_I, u = 0, w = \Omega R_I; r = R_O, u = 0, w = 0$.

Since we are restricting consideration to fully-developed flow, the second term on the right-hand side of equation (16.60) can depend only upon $r$. If we differentiate with respect to $x$, it must be the case that $\partial^2 p/\partial x^2 = 0$ and so $\partial p/\partial x$ is a constant (which could be zero) with respect to $x$ (as will become apparent shortly the static pressure $p$ varies with $r$ if there is centrebody rotation).

Integration of equation (16.62) leads to

$$\frac{1}{r}\frac{d}{dr}(rw) = A$$

where $A$ is a constant of integration. Integration again leads to

$$rw = \frac{Ar^2}{2} + B$$

where $B$ is again a constant of integration. The boundary conditions for the tangential velocity component $w$ lead to

$$A = \frac{-2\Omega R_I^2}{R_O^2 - R_I^2} \qquad \text{and} \qquad B = \frac{\Omega R_O^2 R_I^2}{R_O^2 - R_I^2}$$

so that

$$w = \left(\frac{\Omega R_O R_I^2}{R_O^2 - R_I^2}\right)\left(\frac{R_O}{r} - \frac{r}{R_O}\right). \tag{16.63}$$

Equation (16.63) can also be written as

$$w = \left(\frac{\Omega R_I^2}{R_O^2 - R_I^2}\right)\left(\frac{R_O^2 - r^2}{r}\right) = \left(\frac{\Omega R_I^2}{R_O^2 - R_I^2}\right)\frac{(R_O - r)(R_O + r)}{r} \tag{16.64}$$

from which it is straightforward to show that, for geometries where the annular gap between the inner and outer cylinders $\delta = R_O - R_I$ is small (i.e. $\delta/R_I \ll 1$)

$$w = \frac{Vy}{\delta} \tag{16.65}$$

where $y = R_O - r$ is the distance measured from the inner surface of the outer cylinder and $V = \Omega R_I$ is the peripheral velocity of the outer surface of the inner cylinder. Evidently, there is no influence of curvature, and the proportional variation of flow velocity is identical to that for **plane** (or **linear**) **Couette** flow, which we analysed in Section 16.4.

Now that the dependence of the tangential velocity component on $r$, $w(r)$, is known, the static-pressure variation $p(r)$ can be obtained by substituting equation (16.63) for $w$ into equation (16.61)

$$\frac{\partial p}{\partial r} = \rho \left(\frac{\Omega R_O R_I^2}{R_O^2 - R_I^2}\right)^2 \left(\frac{R_O^2}{r^3} - \frac{2}{r} + \frac{r}{R_O^2}\right)$$

which can be integrated to give

$$p - p_I = \rho \left(\frac{\Omega R_O R_I^2}{R_O^2 - R_I^2}\right)^2 \left[\frac{1}{2}\left(\frac{R_O^2}{R_I^2} - \frac{R_O^2}{r^2}\right) - 2\ln\left(\frac{r}{R_I}\right) + \frac{1}{2}\left(\frac{r^2}{R_O^2} - \frac{R_I^2}{R_O^2}\right)\right] \tag{16.66}$$

where $p_I$ is the static pressure at $r = R_I$. As already demonstrated, both $p$ and $p_I$ are dependent upon the axial location $x$ so that $\partial p_I/\partial x = \partial p/\partial x = $ constant.

Also of interest is the shear stress, $\tau_{r\theta}$, at any radius $r$, which (from equation (15.53), derived in Chapter 15) is given by

$$\tau_{r\theta} = \mu r \frac{d}{dr}\left(\frac{w}{r}\right) = \frac{-2\mu}{r^2}\left(\frac{\Omega R_O^2 R_I^2}{R_O^2 - R_I^2}\right). \tag{16.67}$$

The shear stress exerted by the fluid on the inner (rotating) cylinder is thus

$$\tau_{S_I} = 2\mu\left(\frac{\Omega R_O^2}{R_O^2 - R_I^2}\right). \tag{16.68}$$

Finally, if the length of the inner cylinder is $L$, the torque $T_I$ exerted on its surface is

$$T_I = \frac{4\pi\mu\Omega R_I^2 R_O^2 L}{R_O^2 - R_I^2}. \tag{16.69}$$

The torque exerted on the inner surface of the outer cylinder is identical in magnitude but opposite in sign.

As we discuss in Section 16.7, equation (16.69) is the basis for the design of **concentric-cylinder rheometers** and **viscometers**[137], which are instruments used for the measurement of viscosity.

The solution to the problem of flow through a concentric annulus with inner-cylinder rotation, so far, reveals three significant features for this flow

- the variation of the tangential- (or swirl-) velocity component with $r$ is not directly dependent on viscosity, although the no-slip condition was involved in the derivation
- the radial variations of the axial $u(r)$ and tangential $w(r)$ velocity components are independent
- the $w(r)$ variation has a linear component and a component where $w$ is inversely dependent upon $r$

As stated in Section 16.1, laminar flows become unstable if an appropriately defined Reynolds or Taylor number exceeds a critical value. One of the earliest flows for which such a stability criterion was established is that of Taylor-Couette flow where only the inner cylinder rotates. The relevant non-dimensional parameter is the **Taylor number**, defined by

$$Ta = \left(\frac{R_O}{R_I} - 1\right)^3\left(\frac{\rho\Omega R_I^2}{\mu}\right)^2 \tag{16.70}$$

which can be viewed as a modified Reynolds number. If the annular gap is small, it is found that the flow becomes unstable if the Taylor number exceeds about 1700.

---

**ILLUSTRATIVE EXAMPLE 16.2**

A viscous fluid fills the annular space between two long concentric cylinders, the outer of which has radius $R_O$, and the inner has radius $R_I$. The outer cylinder rotates clockwise at an angular velocity $\omega$ while the inner cylinder rotates anticlockwise at angular velocity $-\omega$. There is zero pressure difference between the ends of the annulus.

[137] The term **viscosimeter** is also used.

State the boundary conditions for the flow. Derive a reduced form of the Navier-Stokes equations for the flow, stating all the assumptions made. Solve the equations to find the radial variation of the circumferential velocity $w(r)$ and shear stress $\tau_{r\theta}$ within the annular gap and find the location of zero velocity. Derive an equation for the torque per unit length $T'$ exerted on the fluid.

### Solution

The boundary conditions are $w = \omega R_O$ at $r = R_O$, and $w = -\omega R_I$ at $r = R_I$.

We have $\partial p/\partial x = 0$ and, from symmetry, $\partial p/\partial\theta = 0$ so $p = p(r)$. It must also be that $u = v = 0$, and $w = w(r)$.

The Navier-Stokes equations reduce to

$$\frac{dp}{dr} = \frac{\rho w^2}{r}$$

and

$$\frac{d}{dr}\left[\frac{1}{r}\frac{d}{dr}(rw)\right] = 0.$$

The second equation can be integrated to give

$$\frac{d}{dr}(rw) = Ar$$

and a second integration gives

$$w = \frac{Ar}{2} + \frac{B}{r}$$

where $A$ and $B$ are constants of integration.

From the boundary conditions we have

$$\omega R_O = \frac{AR_O}{2} + \frac{B}{R_O}$$

and

$$-\omega R_I = \frac{AR_I}{2} + \frac{B}{R_I}$$

from which

$$A = 2\omega\left(\frac{R_O^2 + R_I^2}{R_O^2 - R_I^2}\right)$$

and

$$B = \frac{-2\omega(R_O R_I)^2}{(R_O^2 - R_I^2)}$$

so that

$$w = \frac{\omega}{(R_O^2 - R_I^2)}\left[r\left(R_O^2 + R_I^2\right) - \frac{2(R_O R_I)^2}{r}\right].$$

We can check the result by substituting $r = R_I$, which leads to $w = -\omega R_I$, and by substituting $r = R_O$, which leads to $w = \omega R_O$, both of which are correct.

The location of zero velocity, i.e. $w = 0$, is then

$$r_0 = \frac{\sqrt{2} R_O R_I}{\sqrt{\left(R_O^2 + R_I^2\right)}}.$$

The shear stress is given by

$$\tau_{r\theta} = \mu r \frac{d}{dr}\left(\frac{w}{r}\right) = \frac{4\mu\omega\left(R_O R_I\right)^2}{\left(R_O^2 - R_I^2\right)r^2}$$

and the torque per unit length is then

$$T'\left(r\right) = \tau_{r\theta} 2\pi r^2 = \frac{8\pi\mu\omega\left(R_O R_I\right)^2}{\left(R_O^2 - R_I^2\right)}.$$

**Comments:**

The shear stress decreases monotonically with $r$ from $4\mu\omega R_O^2/\left(R_O^2 - R_I^2\right)$ at the surface of the inner cylinder to $4\mu\omega R_I^2/\left(R_O^2 - R_I^2\right)$ at the inner surface of the outer cylinder.
The torque is independent of $r$.

---

## 16.6 Poiseuille flow of generalised Newtonian fluids between infinite parallel plates

In this section we analyse Poiseuille flow for four model fluids: the **Ostwald-de Waele power-law fluid**, the **Bingham plastic fluid**, the **Casson fluid**, and the **Herschel-Bulkley fluid**. Each is an example of a **generalised Newtonian fluid** for which the effective viscosity $\mu_{EFF}$ depends upon the shear rate $\dot\gamma_{xy}$ (see Section 15.5).

### 16.6.1 Power-law fluid

We consider pressure-driven flow between parallel plates separated by a distance $2hh$ as shown in Figure 16.8 (only half the channel is shown). The plates are assumed to be of infinite length and width so that the flow can be considered as fully developed and two dimensional. From the symmetry of the geometry, the flow will be symmetrical about the centreplane, from which we measure the normal distance $y$ at which location the velocity is $u$ and the shear stress is $\tau_{yx}$. Also, from symmetry, the shear stress must be zero on the centreplane.

Since we are considering fully-developed flow with $u = u\left(y\right)$, the shear rate is given by $\dot\gamma_{xy} = du/dy$, $\dot\gamma_{xz} = \dot\gamma_{yz} = 0$, so that, for a power-law fluid

$$\mu_{EFF} = K\left(\frac{du}{dy}\right)^{n-1}. \tag{16.71}$$

For the shear stress we then have

$$\tau_{yx} = \mu_{EFF}\frac{du}{dy} = K\left(\frac{du}{dy}\right)^n. \tag{16.72}$$

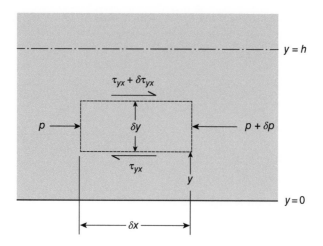

**Figure 16.8** Elemental control volume for Poiseuille flow between parallel plates

The elemental control volume in Figure 16.8 is of infinitesimal height $\delta y$ and infinitesimal length $\delta x$ (not shown to scale in the diagram). The static pressure changes from $p$ on the upstream face to $p + \delta p$ on the downstream face, while the shear stress changes from $\tau_{yx}$ on the lower face to $\tau_{yx} + \delta\tau_{yx}$ on the upper face. For fully-developed flow, the forces acting on the control volume must be in balance so that

$$\delta\tau_{yx}\,\delta x - \delta p\,\delta y = 0$$

from which

$$\frac{\delta\tau_{yx}}{\delta y} = \frac{\delta p}{\delta x}.$$

In the limit $\delta x \to 0$, $\delta y \to 0$:

$$\frac{d\tau_{yx}}{dy} = \frac{dp}{dx} = \text{constant} \tag{16.73}$$

the full, rather than partial, derivatives recognising the fact that we are considering a fully-developed flow.

Equation (16.73) can be integrated to give

$$\tau_{yx} = y\frac{dp}{dx} + A$$

where the constant of integration $A$ is given by

$$A = -h\frac{dp}{dx}$$

since, from symmetry, $\tau_{yx}$ must be zero on the centreplane $(y = h)$. We thus have

$$\tau_{yx} = \left(h - y\right)\left(-\frac{dp}{dx}\right) = K\left(\frac{du}{dy}\right)^{n}. \tag{16.74}$$

Equation (16.74) can be rearranged as

$$\frac{du}{dy} = (h - y)^{1/n} \left( -\frac{1}{K}\frac{dp}{dx} \right)^{1/n}$$

which can be integrated to give

$$-u = \left( \frac{n}{n+1} \right) (h - y)^{(n+1)/n} \left( -\frac{1}{K}\frac{dp}{dx} \right)^{1/n} + B.$$

The constant of integration $B$ is obtained from the no-slip condition, $u(0) = 0$

$$B = - \left( \frac{n}{n+1} \right) h^{(n+1)/n} \left( -\frac{1}{K}\frac{dp}{dx} \right)^{1/n}$$

so that for the velocity distribution we have

$$u = \left( \frac{n}{n+1} \right) \left[ h^{(n+1)/n} - (h - y)^{(n+1)/n} \right] \left( -\frac{1}{K}\frac{dp}{dx} \right)^{1/n}. \tag{16.75}$$

With $y = h$ the centreline velocity $u_0$ is thus

$$u_0 = \left( \frac{n}{n+1} \right) \left( -\frac{1}{K}\frac{dp}{dx} \right)^{1/n} h^{(n+1)/n} \tag{16.76}$$

and equation (16.75) can be written as

$$\frac{u}{u_0} = 1 - \xi^{(n+1)/n} \tag{16.77}$$

where the non-dimensional quantity $\xi = (h - y)/h$.

The volumetric flowrate per unit width $\dot{Q}'$ is given by

$$\dot{Q}' = 2 \int_0^h u\,dy = 2u_0 h \int_0^1 \left[ 1 - \xi^{(n+1)/n} \right] d\xi = 2 \left( \frac{n+1}{2n+1} \right) u_0 h \tag{16.78}$$

and the average flow velocity $\bar{V}$ by

$$\bar{V} = \frac{\dot{Q}'}{2h} = \left( \frac{n+1}{2n+1} \right) u_0 = \left( \frac{n}{2n+1} \right) \left( -\frac{1}{K}\frac{dp}{dx} \right)^{1/n} h^{(n+1)/n} \tag{16.79}$$

wherein we have used equation (16.76) to substitute for $u_0$.

Table 16.1 shows how the ratio of the average flow velocity $\bar{V}$ to the peak velocity $u_0$ changes with the exponent $n$.

The conclusion to be drawn from Table 16.1 is that, as the fluid becomes more and more shear thinning, i.e. as $n$ decreases, the velocity profile flattens out until for $n = 0$ the velocity is completely uniform. For $n = 1$, i.e. for a Newtonian fluid, equation (16.77) shows that the velocity variation is parabolic, as for the flow of a Newtonian fluid through a circular pipe, but now $\bar{V}/u_0 = 2/3$ compared with $1/2$ in the latter case (a consequence of the difference in cross section). As Table 16.1 shows, $\bar{V}/u_0 = 1/2$ now corresponds to the situation where the fluid is infinitely shear thickening.

As always it is convenient to define a Reynolds number but we now have the complication of a varying effective viscosity, so the question to be answered is, at what location in the duct

**Table 16.1** Ratio of average flow velocity $\bar{V}$ to peak velocity $u_0$ as a function of exponent $n$ for Poiseuille flow of a power-law fluid between infinite parallel plates

| $n$ | $\bar{V}/u_0$ | |
|---|---|---|
| $\infty$ | 0.500 | shear thickening |
| 10 | 0.524 | " |
| 5 | 0.545 | " |
| 2 | 0.600 | " |
| 1 | 0.667 | Newtonian |
| 0.8 | 0.692 | shear thinning |
| 0.5 | 0.750 | " |
| 0.2 | 0.850 | " |
| 0.1 | 0.917 | " |
| 0 | 1.000 | " |

should $\mu_{EFF}$ be evaluated? From the velocity distribution, equation (16.77), we have

$$\frac{du}{dy} = \left(\frac{n+1}{n}\right)\frac{u_0}{h}\left(1-\frac{y}{h}\right)^{1/n} \tag{16.80}$$

so that, from equation (16.71),

$$\mu_{EFF} = K\left(-\frac{du}{dy}\right)^{n-1} = K\left[\left(\frac{n+1}{n}\right)\frac{u_0}{h}\right]^{n-1}\left(1-\frac{y}{h}\right)^{(n-1)/n}. \tag{16.81}$$

As for the flow of a Newtonian fluid through a cylindrical duct, the average velocity $\bar{V}$ is a more logical choice for the velocity scale than the centreline velocity $u_0$. We can substitute for $u_0$ in terms of $\bar{V}$ from equation (16.79) to find

$$\mu_{EFF} = K\left[\left(\frac{2n+1}{n}\right)\frac{\bar{V}}{h}\right]^{n-1}\left(1-\frac{y}{h}\right)^{(n-1)/n}. \tag{16.82}$$

It would make no sense to evaluate $\mu_{EFF}$ from this equation on the centreline ($y = h$), as this would yield either $\mu_{EFF} = 0$ for $n > 1$, or $\mu_{EFF} = \infty$ for $n < 1$, so we choose the channel wall, $y = 0$, at which location

$$\mu_{EFF,S} = K\left[\left(\frac{2n+1}{n}\right)\frac{\bar{V}}{h}\right]^{n-1}. \tag{16.83}$$

We can again introduce the hydraulic diameter, which in this case, is given by $D_H = 4h$. An appropriate definition for the Reynolds number is thus

$$Re_H = \frac{\rho \bar{V} D_H}{\mu_{EFF,S}} = \frac{4\rho \bar{V}^{2-n} h^n}{K} \left( \frac{2n+1}{n} \right)^{n-1}. \tag{16.84}$$

From equation (16.74) the shear stress $\tau_S$ exerted on the lower plate is given by

$$\tau_S = -h \frac{dp}{dx}$$

and if we substitute for $dp/dx$ using equation (16.79) we find

$$\tau_S = K \left[ \left( \frac{2n+1}{n} \right) \frac{\bar{V}}{h} \right]^n. \tag{16.85}$$

Division by $\rho \bar{V}^2/2$ leads to

$$f_F = \frac{2\tau_S}{\rho \bar{V}^2} = \frac{2K}{\rho \bar{V}^{2-n}} \left[ \left( \frac{2n+1}{n} \right) \frac{1}{h} \right]^n. \tag{16.86}$$

Finally, the Poiseuille number is given by

$$Po = f_F \, Re_H = 8 \left( \frac{2n+1}{n} \right)^{2n-1}. \tag{16.87}$$

With $n = 1$, equation (16.87) gives $f_F \, Re = 24$, in agreement with what we found in Section 16.4 for Poiseuille flow of a Newtonian fluid between parallel plates (equation (16.59) with $\lambda_P \to \infty$).

A similar analysis for fully-developed flow of a power-law fluid through a circular pipe leads to

$$Po = 4 \left( \frac{3n+1}{n} \right). \tag{16.88}$$

With $n = 1$, equation (16.88) gives $Po = 16$, in agreement with what we found in Subsection 16.3.1.

## 16.6.2 Bingham plastic

We consider Poiseuille flow of a Bingham plastic through a circular tube of radius $R$. The shear stress-shear rate equation is now

$$\tau_{rx} = \tau_Y + \mu_P \frac{du}{dr} \tag{16.89}$$

where $\tau_Y$ is the **yield stress** and $\mu_P$ is the **plastic viscosity**.

The starting point for the analysis is Cauchy's equation of motion for steady axial flow, with $v = w = 0$, and the normal stress $\sigma_{xx} = -p$:

$$-\frac{dp}{dx} + \frac{1}{r} \frac{d}{dr} (r\tau_{rx}) = 0 \tag{16.90}$$

where $dp/dx$ is the imposed pressure gradient. As in previous analyses of fully-developed duct flow, it must be that the two terms of this equation are constant. After integration we have

$$-\frac{r^2}{2}\frac{dp}{dx} + r\tau_{rx} = A.$$

Since this equation must be valid for all values of $r$, including $r = 0$, we conclude that the constant of integration $A = 0$, and

$$\tau_{rx} = \tau_Y + \mu_P\frac{du}{dr} = \frac{r}{2}\frac{dp}{dx} \tag{16.91}$$

where we have substituted for $\tau_{rx}$ from equation (16.89).

If we define a radius $r_P$ through the equation

$$\tau_Y = \frac{-r_P}{2}\frac{dp}{dx} \tag{16.92}$$

we see that, if $r < r_{Pt}$, then $-\tau_{rx} < \tau_Y$, i.e. the shear stress is below the level for shear flow. This situation is interpreted to mean that there is a central 'plastic plug' of radius $r_P$ within which the velocity gradient is zero.

For $r \geq r_P$ we can integrate equation (16.92) to find

$$\tau_Y r + \mu_P u = \frac{r^2}{4}\frac{dp}{dx} + B$$

or, after rearrangement

$$u = \frac{r^2}{4\mu_P}\frac{dp}{dx} + \frac{\tau_Y r}{\mu_P} + \frac{B}{\mu_P}.$$

At $r = R$ we have the no-slip condition, $u = 0$, so that

$$\frac{B}{\mu_P} = \frac{-R^2}{4\mu_P}\frac{dp}{dx} - \frac{\tau_Y R}{\mu_P} \tag{16.93}$$

and, after some rearrangement

$$u = \left[1 - \left(\frac{r}{R}\right)^2\right]\left(-\frac{R^2}{4\mu_P}\frac{dp}{dx}\right) - \frac{\tau_Y R}{\mu_P}\left[1 - \frac{r}{R}\right] \quad \text{for} \quad r \geq r_P. \tag{16.94}$$

Since this equation is valid at the edge of the plug, $r = r_P$, and the velocity gradient is zero within it, we have therefore

$$u_P = \left[1 - \sigma_P^2\right]\left(-\frac{R^2}{4\mu_P}\frac{dp}{dx}\right) - \frac{\tau_Y R}{\mu_P}[1 - \sigma_P] \quad \text{for} \quad r \leq r_P \tag{16.95}$$

for the fluid velocity $u_P$ within the plug. The quantity $\sigma_P$ is the non-dimensional plug radius defined by

$$\sigma_P = \frac{r_P}{R}. \tag{16.96}$$

From the definition of $r_P$, equation (16.92), it is seen that the pressure-gradient term in equations (16.94) and (16.95)

$$\left(-\frac{R^2}{4\mu_P}\frac{dp}{dx}\right) = \frac{1}{2\sigma_P}\frac{\tau_Y R}{\mu_P} \tag{16.97}$$

so that the equations for the velocity distribution can be written as

$$u = \left(1 - \frac{r}{R}\right)\left[\frac{1}{2\sigma_P}\left(1 + \frac{r}{R}\right) - 1\right]\frac{\tau_Y R}{\mu_P} \text{ for } r \geq r_P \tag{16.98}$$

and

$$u_P = \frac{[1 - \sigma_P]^2}{2\sigma_P}\frac{\tau_Y R}{\mu_P} \qquad\qquad \text{for } r \leq r_P. \tag{16.99}$$

From equation (16.10), with $c = 2\pi R$ and $A = \pi R^2$:

$$\tau_S = -\frac{R}{2}\frac{dp}{dx} \tag{16.100}$$

so that, from equation (16.92)

$$\frac{\tau_Y}{\tau_S} = \frac{r_P}{R} = \sigma_P \tag{16.101}$$

and another form for the velocity distribution is

$$u = \left(1 - \frac{r}{R}\right)\left[1 + \frac{r}{R} - \sigma_P\right]\frac{\tau_S R}{\mu_P} \qquad\qquad \text{for } r \geq r_P \tag{16.102}$$

and

$$u_P = [1 - \sigma_P]^2 \frac{\tau_S R}{\mu_P} \qquad\qquad \text{for } r \leq r_P. \tag{16.103}$$

The volumetric flowrate $\dot{Q}$ and the average flow velocity $\bar{V}$ are found from

$$\dot{Q} = \int_0^R 2\pi r u \, dr = \pi R^2 \bar{V}.$$

After some algebra we arrive at

$$\frac{4\mu_P \dot{Q}}{\pi R^3 \tau_Y} = \frac{1}{\xi_P}\left[1 - \frac{4}{3}\sigma_P + \frac{1}{3}\sigma_P^4\right] \tag{16.104}$$

which is known as the **Buckingham-Reiner equation**. Equation (16.104) can be rewritten as

$$Po = f_F \, Rep = 16\left[1 + \frac{1}{6}Bi - \frac{1}{3}\frac{Bi^4}{(f_F \, Rep)^3}\right] \tag{16.105}$$

where the non-dimensional combination

$$Bi = \frac{\tau_Y D}{\mu_P \bar{V}} \tag{16.106}$$

is the **Bingham number**[138], and

---

[138] Equation (16.105) can also be written in terms of the **Hedstrom number**, $He$, defined by $He = Bi \, Rep$.

$$Rep = \frac{\rho \bar{V} D}{\mu_P} \tag{16.107}$$

is the Reynolds number based upon the plastic viscosity. It can be seen from equation (16.105) that the Poiseuille number here is a function of the Bingham number and, for small values of $Bi$, e.g. where the yield stress is small and the plastic viscosity is large, a useful approximation is

$$Po \approx 16 \left( 1 + \frac{1}{6} Bi \right). \tag{16.108}$$

### 16.6.3 Casson fluid

A fluid which obeys the shear-stress equation

$$\sqrt{\tau_{rx}} = \sqrt{\tau_Y} + \sqrt{\mu_P \frac{du}{dr}}, \tag{16.109}$$

where the symbols have the same meanings as for the flow of a Bingham plastic, is known as a **Casson fluid**. An analysis similar to that for a Bingham plastic for flow through a circular pipe leads to

$$\frac{4\mu_P \dot{Q}}{\pi R^3 \tau_Y} = \frac{1}{\xi_P} \left[ 1 - \frac{16}{7} \sqrt{\sigma_P} + \frac{4}{3} \sigma_P - \frac{1}{21} \sigma_P^4 \right]. \tag{16.110}$$

### 16.6.4 Herschel-Bulkley fluid

The **Herschel-Bulkley** shear-stress **equation** is

$$\tau_{rx} = \tau_Y + K\dot{\gamma}^n$$

and the resulting equation for the volumetric flowrate in pipe flow is

$$\frac{\dot{Q}}{n\pi R^3} \left( \frac{K}{\tau_Y} \right)^{1/n} = \frac{(1 - \sigma_P)^{(n+1)/n}}{\sigma_P^{1/n}} \left[ \frac{(1 - \sigma_P)^2}{1 + 3n} + \frac{2\sigma_P (1 - \sigma_P)}{1 + 2n} + \frac{\sigma_P^2}{1 + n} \right]. \tag{16.111}$$

In the equations for the volumetric flowrate of the three viscoplastic liquids, the right-hand side is a function of $\sigma_P \, (= \tau_Y/\tau_S)$ and $n$, while the left-hand side is independent of $\tau_S$. It is therefore straightforward to calculate the flowrate if the surface shear stress is known, whereas the inverse situation requires a numerical approach.

## 16.7 Viscometer equations

An instrument used to measure the viscosity of a fluid is called a **viscometer**, while a **rheo-meter** is a viscometer that can also be used to measure the viscoelastic moduli, normal-stress coefficients, etc., of a non-Newtonian liquid. The assumption of constant viscosity greatly sim-plifies the measurement problem and for Newtonian fluids, whether liquid or gas, numerous viscometer designs are available, including the **capillary viscometer**, in which the dynamic viscosity $\mu$ is determined by measuring the volumetric flowrate $\dot{Q}$ produced by imposing a pressure difference $\Delta p$ across the ends of a capillary tube of length $L$ and internal diameter $D$. Assuming Hagen-Poiseuille flow through the tube, it can be shown that

$$\mu = \frac{\pi D^4 \Delta p}{128 \dot{Q} L}. \tag{16.112}$$

The dependence of $\mu$ on $D^4$ shows that a capillary tube for this application must be produced to high precision with a diameter $D$ which is constant along its length and accurately known. To minimise end effects, particularly entrance effects to the location where the flow is fully developed, it is essential that $L/D \gg 1$.

For a shear-thinning or shear-thickening liquid, the effective viscosity is shear rate dependent so that ideally the liquid should be subjected continuously to a uniform shear rate and a uniform shear stress. Figure 16.9 shows two rotational configurations in which this can be achieved to a high degree of accuracy. In the concentric-cylinder geometry (Figure 16.9(a)), the liquid is contained within the annular space between the two cylinders. The effective viscosity is determined by measuring the torque exerted on the rotating inner cylinder (or **bob**, the outer cylinder being referred to as the **cup**) over a range of rotation rates, or the rotation rate is measured over a range of torques, depending upon whether torque or rotation speed is controlled[139]. The second geometry shown, Figure 16.9(b), is the cone-and-plate arrangement, in which an inverted cone of large included angle ($> 172°$ is typical) rotates about an axis normal to a plate. The liquid occupies the region between the cone and the plate, and the effective viscosity is determined by measuring the torque required to rotate the cone over a range of rotation rates.

In Section 16.5 we showed that, for the concentric-cylinder geometry, the variation of tangential shear stress $\tau_{r\theta}$ with radius $r$ for a Newtonian fluid is given by

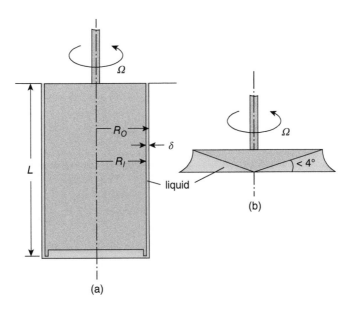

**Figure 16.9** Viscometer geometries: (a) concentric cylinder (b) cone and plate

[139] In some instruments, the inner cylinder is stationary, and the outer cylinder rotates.

$$\tau_{r\theta} = \frac{-2\mu}{r^2}\left(\frac{\Omega R_O^2 R_I^2}{R_O^2 - R_I^2}\right). \tag{16.67}$$

where $\Omega$ is the constant angular velocity of the inner cylinder, $R_I$ is the outer diameter of the inner cylinder, $R_O$ is the inner diameter of the outer cylinder, and $w$ is the tangential velocity at radius $r$.

From equation (16.67) we see that, if the annular gap width is $\delta = R_O - R_I$, and the mean radius is $\bar{R} = (R_O + R_I)/2$, the ratio of the value of $\tau_{r\theta}$ at $r = R_O$ to the value at $r = R_I$ is given by

$$\left(\frac{1 + \varepsilon/2}{1 - \varepsilon/2}\right)^2$$

where $\varepsilon = \delta/\bar{R}$. This ratio shows that the shear stress is constant across the annular gap to within 1% if $\varepsilon < 0.005$ so that, for an inner cylinder of diameter 30 mm, the gap should be less than 0.075 mm (or 75 $\mu$m). A value for $\varepsilon$ of 0.02 is more typical of a commercially available laboratory instrument, and then the shear-stress ratio is about 4%. Although this analysis assumed the fluid was Newtonian, it is reasonable to assume that the conclusions are valid for liquid with shear-dependent properties.

For the concentric-cylinder geometry, the shear rate is given by

$$\dot{\gamma}_{r\theta} = \frac{1}{2}r\frac{d}{dr}\left(\frac{w}{r}\right)$$

and the radial variation of the tangential velocity $w$ was shown to be

$$w = \left(\frac{\Omega R_I^2}{R_O^2 - R_I^2}\right)\left(\frac{R_O^2 - r^2}{r}\right). \tag{16.64}$$

The shear rate $\dot{\gamma}_{r\theta}$ is thus given by

$$\dot{\gamma}_{r\theta} = -\left(\frac{\Omega R_I^2 R_O^2}{R_O^2 - R_I^2}\right)\frac{1}{r^2} \tag{16.113}$$

which can be rewritten in terms of $\varepsilon$ as

$$\dot{\gamma}_{r\theta} = -\frac{V}{2\delta}\left(\frac{1 - \frac{1}{4}\varepsilon^2}{1 + \frac{1}{2}\varepsilon - \frac{y}{R}}\right)^2 \approx -\frac{V}{2\delta} \tag{16.114}$$

where $y = R_O - r$ is the distance measured from the inner surface of the outer cylinder and $V = \Omega R_I$ is the peripheral velocity of the outer surface of the inner cylinder. Evidently, the shear rate within the fluid is practically constant, again provided $\varepsilon \ll 1$. Equation (16.114) shows that with $\varepsilon = 0.02$ the shear rate varies by about $\pm 2\%$ from its value at the midpoint of the annulus, and the latter is within 0.02% of $V/2\delta$.

Finally, for the velocity within the gap we can show that

$$w = \frac{Vy}{\delta}\left(1 - \frac{1}{2}\varepsilon\right)\frac{\left(1 + \frac{1}{2}\varepsilon - \frac{1}{2}\frac{y}{R}\right)}{\left(1 + \frac{1}{2}\varepsilon - \frac{y}{R}\right)} \approx \frac{Vy}{\delta}. \tag{16.115}$$

Evidently, there is no influence of curvature in the final approximation, and the proportional variation of flow velocity with $y$ is identical to that for **plane** (or **linear**) **Couette** flow, which we analysed in Section 16.4.

As shown in Section 16.5, if the cylinder length is $L$, for a Newtonian fluid the magnitude of the torque $T$ exerted on either cylinder is

$$T = \frac{4\pi \mu \Omega R_I^2 R_O^2 L}{R_O^2 - R_I^2} \qquad (16.69)$$

from which the viscosity $\mu$ is easily calculated from a torque measurement on either cylinder. For a Newtonian fluid, the viscosity is independent of the shear rate and so the rotation speed. The shear-rate dependence of the **effective viscosity** of a non-Newtonian liquid can be found in a concentric-cylinder rheometer by varying the rotation speed $\Omega$ and, if $\varepsilon \ll 1$, the value of $\mu_{EFF}$ will correspond very closely to the shear rate at the midpoint of the annulus.

Once again we can introduce the ratio $\varepsilon = \delta/\bar{R}$ and rewrite the torque equation as

$$T = \frac{2\pi \mu \Omega \bar{R}^3 L}{\delta} \left(1 - \frac{1}{4}\varepsilon^2\right)^2 \qquad (16.116)$$

which reveals a number of important features of the concentric-cylinder flow:

- the dependence of the torque $T$ on the mean radius $\bar{R}$ is very strong (essentially cubic if $\varepsilon$ is less than about 0.3)
- the dependence of the torque $T$ on the annular gap $\delta$ is also strong (essentially inverse, again if $\varepsilon$ is less than about 0.3)

An important conclusion to be drawn from these two observations is that both the cup and the bob of a concentric-cylinder viscometer must be manufactured to very high tolerances. In fact, it is usual for manufacturers of precision instruments to supply these two components as a matched pair.

- If, as is usual in a narrow-gap, concentric-cylinder viscometer, $\varepsilon \ll 1$, then an excellent approximation to equation (16.116) is

$$T \approx \frac{2\pi \mu \Omega \bar{R}^3 L}{\delta}. \qquad (16.117)$$

With $\varepsilon = 0.02$, we see that the term in equation (16.116) involving $\varepsilon$ has the value 0.9998 so that the error in the approximation is less than 0.02%, which for most practical purposes is acceptable. This value of $\varepsilon$ is satisfied with $\bar{R} = 25$ mm, and $\delta = 500$ $\mu$m, values typical of a laboratory viscometer.

A practical consideration which we have not mentioned is that, for a simple flat-bottomed cylinder, the measured torque would include a contribution from the viscous torque exerted on the lower end of the bob. In principle this can be accounted for, but introduces additional uncertainty into the measurement. To minimise this influence, the lower end of the bob is usually either conical in shape or recessed, as shown in Figure 16.9(a), the recess being gas filled.

A viscometer intended for use as a rheometer typically uses the **cone-and-plate geometry** shown in Figure 16.9(b). If the cone rotates at constant angular velocity $\Omega$ the shear rate exerted on the liquid sample within the gap is practically independent of radius for cone semi-angles close to 90° (gap angle $\leq$ 4°). For such small gap angles the tangential velocity $w$ can be assumed to vary linearly with normal distance $y$ from the plate according to $w = Vy/\alpha r$, where the velocity $V$ at radius $r$ on the cone surface is $V = \Omega r$ so that $w = \Omega y/\alpha$, $\alpha$ being the gap angle in radians. The corresponding shear rate is $\dot{\gamma}_{y\theta} = dw/dy = \Omega/\alpha$, and the shear stress $\tau_{y\theta} = \mu_{EFF}\Omega/\alpha$. Evidently, within the small-angle approximation, the shear rate and shear stress are uniform throughout the liquid sample. The torque $T$ exerted on the fluid by a cone of radius $R$ is given by

$$ T = \int_0^R 2\pi r^2 \tau_{y\theta}\, dr = \frac{2\pi R^3 \mu_{EFF}\Omega}{3\alpha}. \tag{16.118}$$

The shear-dependence of the effective viscosity for a non-Newtonian liquid is obtained by measuring the torque as a function of angular velocity.

Among the practical considerations that have to be taken into account for both a concentric-cylinder and a cone-and-plate rheometer are limiting the rotation speed to avoid secondary flows in the gap, avoidance of surface evaporation and drying of the sample liquid, and, in the case of suspensions, avoiding sedimentation of the solid phase.

 ## 16.8 SUMMARY

In this chapter we have shown how solutions to the Navier-Stokes equations can be derived for steady, fully-developed flow of a constant-viscosity Newtonian fluid through a cylindrical duct. The pressure-driven flow of generalised Newtonian fluids was also discussed. Solutions were also derived for shear-driven flow within the annular space between two concentric cylinders or in the space between two parallel plates when there is relative tangential movement between the wetted surfaces.

The student should be able to

- understand what is meant by the terms cylindrical duct, fully-developed flow, Poiseuille flow, and Couette flow
- simplify the Navier-Stokes equations in either Cartesian-coordinate or polar-cylindrical-coordinate form and so derive the equations for fully-developed flow
- solve the simplified Navier-Stokes equations for fully-developed flow: through a circular pipe; through a concentric annulus, with or without rotation of one or both wetted surfaces; and between parallel plates, with or without relative tangential movement between the wetted surfaces
- solve the equations of fluid motion for fully-developed flow of a power-law or Bingham plastic liquid between parallel plates or through a circular pipe, to find equations for the distribution of velocity within the flow channel
- derive an expression for the volumetric flowrate in terms of the surface shear stress for any of the flows for which velocity distributions were found
- for liquids obeying either the Casson or Herschel-Bulkley equations, be able to calculate the flowrate, knowing the model parameters and the surface shear stress
- understand the principles underlying the design of a capillary tube, a concentric cylinder, or a cone-and-plate viscometer

## ? 16.9 SELF-ASSESSMENT PROBLEMS

**16.1** A laminar-flow heat exchanger is made up of 3000 channels each having the cross section of an equilateral triangle of side length 10 mm. Each channel is of length 500 mm. Paraffin, with a viscosity of $1.92 \times 10^{-3}$ Pa·s and a density of 804 kg/m$^3$, flows through the heat exchanger with an average velocity of 0.41 m/s. Calculate the pressure drop across the heat exchanger, the mass flowrate of paraffin, and the power required to pump the paraffin.
The Poiseuille number for this channel shape is 13.33333.
(Answers: 78.72 Pa; 85.64 kg/s; 8.385 W)

**16.2** A liquid with constant viscosity 0.9 Pa·s and density 900 kg/m$^3$ is pumped through a pipeline of radius 1 m and length 5 km at a mass flowrate of 2900 kg/s. If the pressure at the downstream end of the pipeline is 1 bar, calculate the pressure at the upstream end. Assume that the flow is laminar and fully developed so that the relationship between the surface shear stress $\tau_S$, the pipe diameter $D$, the fluid viscosity $\mu$, and the mean velocity $\bar{V}$ is

$$\frac{\tau_S D}{\mu \bar{V}} = 8.$$

Calculate the flow Reynolds number and state whether the assumption of laminar flow is valid.
(Answers: 1.369 bar, 2051, just ($Re < 2100$))

**16.3** The velocity $u$ at radius $r$ for fully-developed Poiseuille flow through a concentric annulus of a fluid with constant viscosity $\mu$ is given by

$$u = \left[ \frac{R_O^2 - r^2}{R_O^2 - R_I^2} - \frac{\ln(r/R_O)}{\ln(R_I/R_O)} \right] \frac{(R_O^2 - R_I^2)}{4\mu} \left( -\frac{dp}{dx} \right),$$

where $R_I$ is the radius of the inner cylinder, $R_O$ is the radius of the outer cylinder, and $dp/dx$ is the axial pressure gradient.
If the inner cylinder is rotating at angular velocity $\Omega$, the tangential velocity at radius $r$ is given by

$$w = \left( \frac{\Omega R_I^2}{R_O^2 - R_I^2} \right) \left( \frac{R_O^2 - r^2}{r} \right).$$

The shear stresses in the axial and tangential directions are given by

$$\tau_{rx} = \mu \frac{du}{dr} \quad \text{and} \quad \tau_{r\theta} = \mu r \frac{d}{dr} \left( \frac{w}{r} \right),$$

respectively.
Derive expressions for
(a) the radial distributions of $\tau_{rx}$ and $\tau_{r\theta}$,
(b) the location of the peak axial velocity, and
(c) the torque exerted on the fluid at the location of peak axial velocity.

**16.4** A fluid with constant viscosity $\mu$ is contained between parallel plates separated by a distance $h$. The upper plate moves in the positive $x$-direction with constant velocity $V$, while the lower plate is stationary. There is a negative streamwise pressure gradient $dp/dx$ imposed upon the fluid. As shown in Section 16.4, the velocity distribution between the plates is given by the equation

$$\frac{u}{V} = \frac{y}{h} + \frac{\lambda_P}{2}\frac{y}{h}\left(1 - \frac{y}{h}\right)$$

where $y$ is the normal distance from the lower plate and $\lambda$ is the non-dimensional pressure-gradient parameter defined as

$$\lambda_P = -\frac{h^2}{\mu V}\frac{dp}{dx}.$$

If $\lambda_P < -2$, there is forward flow in the vicinity of the upper (moving) plate, and backflow near the lower stationary plate.

(a) Show that, in addition to $y = 0$, the location of zero velocity $y_0$ for $\lambda_P < -2$ is given by

$$\frac{y_0}{h} = 1 + \frac{2}{\lambda_P}.$$

(b) Show that at the location of zero velocity the shear stress $\tau_0$ is given by

$$\tau_0 = -\left(1 + \frac{\lambda_P}{2}\right)\frac{\mu V}{h}.$$

(c) Show that the volumetric flowrate in the forward direction $\dot{Q}_F'$ is given by

$$\dot{Q}_F' = -\frac{1}{\lambda_P}\left(1 + \frac{2}{3\lambda_P}\right)Vh$$

and the volumetric backflow $\dot{Q}_B'$ is given by

$$\dot{Q}_B' = \frac{2}{3\lambda_P^2}\left(1 + \frac{\lambda_P}{2}\right)^3.$$

**16.5** (a) The annular gap in a concentric-cylinder apparatus is filled with a liquid of constant viscosity $\mu$. If the outer radius of the inner cylinder is $R_I$, the inner radius of the outer cylinder is $R_O$, and the outer cylinder rotates with angular velocity $\Omega$ while the inner cylinder is stationary, show that the tangential velocity $w$ of the liquid at radius $r$ is given by the equation

$$w = \frac{\Omega R_O^2}{(R_O - R_I)}\left[1 - \left(\frac{R_O}{r}\right)\right].$$

(b) If the liquid of part (a) is an oil with a viscosity of 0.5 Pa · s, calculate the torque per unit length exerted on the outer cylinder if the radius of the inner cylinder is 50 mm, the gap width is 2 mm, and the angular velocity of the outer cylinder is 20 rad/s.
(Answer: 8.495 N · m)

**16.6** If the liquid in problem 16.5 is replaced by a Bingham plastic with yield stress $\tau_Y$ and plastic viscosity $\mu_P$, show that the tangential velocity variation is given by[140]

$$w = \frac{T'r}{4\pi R_I^2 \mu_P}\left[1 - \left(\frac{R_I}{r}\right)^2\right] + \frac{\tau_Y r}{\mu_P}\ln\left(\frac{R_I}{r}\right)$$

where $T'$ is the torque per unit length imposed on the liquid. Assume that the shear stress at all radii exceeds the yield stress.

---

[140] The equation for $w$ for this flow is often written in terms of $R_O$, rather than $R_I$, as $w = \Omega r + T'r\left[1 - (R_O/r)^2\right]/\left(4\pi R_O^2\mu_P\right) - \tau_Y r\ln(r/R_O)/\mu_P$, which is known as the **Reiner-Rivlin equation**.

# 17 Laminar boundary layers

This chapter is concerned with laminar boundary layers, by which we mean the near-surface region for developing laminar flow of a viscous fluid over a solid surface[141]. We introduce the simplifying assumptions for thin boundary layers, which reduce the Navier-Stokes equations to the so-called **boundary-layer equations**. In the case of a **zero-pressure-gradient (flat-plate) boundary layer**, the velocity profiles exhibit self-similarity, which allows the partial differential boundary-layer equations to be reduced to a single ordinary differential equation known as **Blasius' equation**. It is shown that so-called **wedge-flow** boundary layers, where $U_\infty \propto x^m$, also exhibit self-similarity and lead to the **Falkner-Skan equation**. Because exact solution of any of these equations is possible only numerically, a number of approximate procedures, involving an integrated form of the boundary-layer equations known as **von Kármán's momentum-integral equation**, have been developed which allow useful information to be obtained with relatively little effort. These procedures include methods in which a simple form, such as a polynomial function, is assumed for the velocity profile. When substituted into the momentum-integral equation, **Pohlhausen's** quartic **velocity profile** leads to an ordinary differential equation, incorporating a pressure-gradient parameter, which can be used to calculate the development of any laminar boundary layer. Great simplification without major loss of accuracy results when the full differential equation is replaced by a linear correlation based upon the similarity solutions and other exact calculations.

## 17.1 Introductory remarks

In Chapter 16 we analysed a variety of internal laminar flows which were strongly influenced by fluid viscosity and the no-slip boundary condition, but where there were no changes in velocity in either the streamwise or the azimuthal direction. In the case of flow through a long duct this meant that beyond a certain location there were no velocity changes in the axial direction, or $x$, -direction, and such flows were said to be **fully developed**. This chapter is also concerned with flows strongly affected by viscosity and the no-slip boundary condition but which develop in the streamwise direction, or $x$, -direction, starting from a uniform approach velocity, $U_\infty$. The situation for a uniform flow approaching a stationary thin flat plate aligned with the approach flow is shown schematically in Figure 17.1 (only the flow above the surface is shown). For convenience, a Cartesian-coordinate system has been adopted and it is assumed

---

[141] A boundary layer can also develop over a porous surface or a liquid surface, but such flows will not be considered in this book.

*Introduction to Engineering Fluid Mechanics*. Marcel Escudier.
© Marcel Escudier 2017. Published 2017 by Oxford University Press.

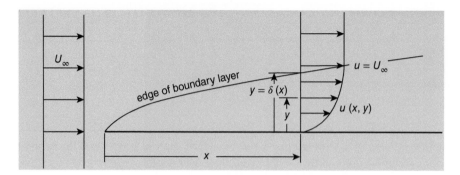

**Figure 17.1** Viscous flow over an aligned flat plate

that there is no flow in the $z$, -direction, a restriction that will apply throughout this and the following chapter. Such flows are said to be **two dimensional**. We shall also restrict ourselves to the study of steady, constant-property, flows in the absence of body forces.

A number of features are included in Figure 17.1 which will becoming increasingly familiar as this chapter develops

- at the surface ($y = 0$), as a consequence of the **no-slip condition**, the streamwise velocity component $u = u_S = 0$
- changes in the streamwise velocity component $u$ occur both in the streamwise direction, or $x$, -direction, and in the normal direction, or $y$, -direction, i.e. $u = u(x, y)$
- with increasing normal distance $y$ from the surface, the streamwise velocity component $u$ tends asymptotically to the approach velocity $U_\infty$
- there is a relatively thin viscosity-affected region $y < \delta$, where $u < U_\infty$, referred to as the boundary layer, which 'grows' in thickness with streamwise distance $x$, i.e. $\delta = \delta(x)$
- for values of the normal distance $y$ greater than $\delta$ it is assumed that $u = U_\infty$
- although it cannot be defined precisely, the length $\delta$ is referred to as the **boundary-layer thickness**
- the region $y > \delta$ where $u = U_\infty$ is referred to as the **free stream**[142] in which it is assumed that all velocity gradients with respect to $y$ are zero, i.e. $\partial^n u/\partial y^n = 0, n = 1, 2, 3 \ldots$
- within the boundary layer (i.e. $y < \delta$) the normal component of velocity $v \ll u$ (for flow over a solid surface, $v = v_S = 0$ at $y = 0$)
- consistent with the previous statement is that streamlines within the boundary layer are almost parallel to the solid surface
- there is negligible change in static pressure $p$ across the boundary layer, i.e. $\partial p/\partial y \approx 0$

Some of the characteristics of flat-plate-boundary-layer flow we have just listed are identical with, or closely related to, those of fully-developed flow. With minor modification, these characteristics are typical of all boundary layers. The concept of a boundary layer, the crucial idea that explained many features of real (i.e. viscous) flows, was introduced by Ludwig Prandtl in 1904.

---

[142] The term **mainstream** is also used.

## 17.2 Two-dimensional laminar boundary-layer equations

Our starting point is the dimensional form of the continuity and Navier-Stokes equations in rectangular-Cartesian coordinates, derived in Section 15.1, assuming steady, constant-property, two-dimensional flow with zero body forces:

continuity

$$\frac{\partial u}{\partial x} + \frac{\partial v}{\partial y} = 0. \tag{17.1}$$

$x$-component

$$u\frac{\partial u}{\partial x} + v\frac{\partial u}{\partial y} = -\frac{1}{\rho}\frac{\partial p}{\partial x} + \nu\left(\frac{\partial^2 u}{\partial x^2} + \frac{\partial^2 u}{\partial y^2}\right) \tag{17.2}$$

$y$-component

$$u\frac{\partial v}{\partial x} + v\frac{\partial v}{\partial y} = -\frac{1}{\rho}\frac{\partial p}{\partial y} + \nu\left(\frac{\partial^2 v}{\partial x^2} + \frac{\partial^2 v}{\partial y^2}\right). \tag{17.3}$$

In spite of considerable simplification compared with the full Navier-Stokes equations, these equations still represent a major challenge, even to numerical solution. We now introduce the key **boundary-layer approximations** which further simplify the equations to be solved

- there is negligible change in the static pressure $p$ across the boundary layer, i.e. $\partial p/\partial y = 0$
- streamwise gradients of $u$ are much less than gradients with respect to the normal distance $y$, in particular $\partial^2 u/\partial x^2 \ll \partial^2 u/\partial y^2$

The continuity equation is unchanged, but the static pressure $p$ is now dependent only upon $x$, and equation (17.2) reduces to

$$u\frac{\partial u}{\partial x} + v\frac{\partial u}{\partial y} = -\frac{1}{\rho}\frac{dp}{dx} + \nu\frac{\partial^2 u}{\partial y^2}. \tag{17.4}$$

The essential difficulty of the non-linearity of the terms on the left-hand side of the partial differential equation (17.4) remains. These terms are identically zero for the fully-developed flows of Chapter 16. However, as we shall see in Sections 17.3 and 17.4, under some circumstances equations (17.1) and (17.4) can be combined and reduced to a single ordinary differential equation which can be solved numerically with relatively little effort[143]. As we shall also see, in Sections 17.5 and 17.6, a great deal of insight can be obtained from quite simple approximate solution methods.

For $y > \delta$, $\partial u/\partial y \to 0$, $\partial^2 u/\partial y^2 \to 0$, and $u \to U_\infty(x)$, the free-stream velocity, so that equation (17.4) reduces to

$$U_\infty\frac{dU_\infty}{dx} = -\frac{1}{\rho}\frac{dp}{dx}, \tag{17.5}$$

---

[143] Shortly after their formulation in the early 20th century such numerical calculations were carried out by hand, requiring considerable effort.

i.e. the differential form of **Bernoulli's equation**. It is usually the case in practice that either $p(x)$ or $U_\infty(x)$ is specified. If $dp/dx$ is zero, the flow is usually described as a **flat-plate boundary layer** even though the plate is not necessarily flat[144]. The term **zero-pressure-gradient boundary layer** is also used. If $dp/dx < 0$, the outer flow accelerates, and as we shall see the surface shear stress $\tau_S$ increases: the **pressure gradient** is said to be **favourable**. The opposite is true for $dp/dx > 0$, termed an **adverse pressure gradient**, and **flow separation**, where the surface shear stress $\tau_S \to 0$, is possible.

The question should be asked 'under what conditions are the boundary-layer approximations valid?' To answer this we introduce the following non-dimensional variables: $x^* = x/L$, $y^* = y\sqrt{Re}/L$, $u^* = u/U_0$, $v^* = v\sqrt{Re}/U_0$, and $p^* = p/\rho U_0^2$, with $L$ being a length scale characteristic of the streamwise flow direction, such as the length of the surface over which the boundary layer develops, and $U_0$ being a velocity scale, such as the velocity far upstream of the region influenced by viscosity. The reason for incorporating a **Reynolds number**, $Re = U_0 L/\nu$, in the definitions of $v^*$ and $y^*$ will become apparent shortly. In non-dimensional form, equations (17.1) to (17.3) may be written as:

continuity

$$\frac{\partial u^*}{\partial x^*} + \frac{\partial v^*}{\partial y^*} = 0 \qquad (17.6)$$

$x$-component

$$u^* \frac{\partial u^*}{\partial x^*} + v^* \frac{\partial u^*}{\partial y^*} = -\frac{\partial p^*}{\partial x^*} + \frac{\partial^2 u^*}{\partial y^{*2}} + \frac{1}{Re}\frac{\partial^2 u^*}{\partial x^{*2}} \qquad (17.7)$$

$y$-component

$$\frac{1}{Re}\left(u^* \frac{\partial v^*}{\partial x^*} + v^* \frac{\partial v^*}{\partial y^*}\right) = -\frac{\partial p^*}{\partial y^*} + \frac{1}{Re^2}\frac{\partial^2 v^*}{\partial y^{*2}} + \frac{1}{Re}\frac{\partial^2 v^*}{\partial x^{*2}} \qquad (17.8)$$

with boundary conditions: $y^* = 0$, $u^* = 0$, $v^* = 0$, and $y^* \to \infty$, $u^* \to U_\infty^*(x)$. We see that, if $Re \to \infty$, equation (17.8) reduces to

$$0 = -\frac{\partial p^*}{\partial y^*}, \qquad (17.9)$$

and equation (17.7) to

$$u^* \frac{\partial u^*}{\partial x^*} + v^* \frac{\partial u^*}{\partial y^*} = -\frac{\partial p^*}{\partial x^*} + \frac{\partial^2 u^*}{\partial y^{*2}}, \qquad (17.10)$$

while equation (17.6) is unchanged.

If we now revert to dimensional variables we again arrive at equations (17.1) and (17.4), with equation (17.9) justifying the change from $\partial p/\partial x$ to $dp/dx$. The answer to our question, evidently, is that we are concerned with flows for which the Reynolds number is high. Typically, this means $Re > 10^3$ but it turns out that as $Re$ approaches $10^6$ laminar boundary layers become unstable and eventually turbulent, just as for pipe and channel flows[145].

---

[144] The boundary-layer equations can be extended to strongly curved surfaces, but that is beyond the scope of this book.

[145] Note that, whereas for duct flows it was appropriate to define the Reynolds number in terms of the hydraulic diameter, here we are using a streamwise length scale.

It is also informative to make estimates of the orders of magnitude of each of the velocity-gradient terms in equations (17.1) to (17.3). We shall again select $U_0$ and $L$ as the velocity and length scales, respectively, for the streamwise direction, but $V$ and $\delta$ for the transverse direction, with the assumption that $\delta \ll L$. So far as the continuity equation is concerned, we have

$$\frac{\partial u}{\partial x} + \frac{\partial v}{\partial y} = 0 \tag{17.1}$$

with orders of magnitude[146]

$$\frac{\partial u}{\partial x} = O\left(\frac{U_0}{L}\right), \quad \frac{\partial v}{\partial y} = O\left(\frac{V}{\delta}\right).$$

Since there are only two terms in the continuity equation, they must be not only of the same order of, but equal in, magnitude[147] so that

$$V = \frac{U_0 \delta}{L},$$

a result we shall use to replace $V$ from now on.

For the $x$-direction we have

$$u\frac{\partial u}{\partial x} + v\frac{\partial u}{\partial y} = -\frac{1}{\rho}\frac{\partial p}{\partial x} + v\left(\frac{\partial^2 u}{\partial x^2} + \frac{\partial^2 u}{\partial y^2}\right) \tag{17.2}$$

with orders of magnitude

$$u\frac{\partial u}{\partial x} = O\left(\frac{U_0^2}{L}\right), \quad v\frac{\partial u}{\partial y} = O\left(\frac{VU_0}{\delta}\right) = O\left(\frac{U_0^2}{L}\right),$$

$$v\frac{\partial^2 u}{\partial x^2} = O\left(\frac{vU_0}{L^2}\right), \quad v\frac{\partial^2 u}{\partial y^2} = O\left(\frac{vU_0}{\delta^2}\right),$$

from which we can conclude that, since $\delta \ll L$, it must be that $\partial^2 u/\partial x^2 \ll \partial^2 u/\partial y^2$, just as we postulated earlier in this section, and we can neglect $\partial^2 u/\partial x^2$. Assuming all remaining velocity terms are of equal magnitude, we conclude that

$$\frac{vU_0}{\delta^2} = \frac{U_0^2}{L} \quad \text{or} \quad \frac{\delta}{L} = \sqrt{\frac{v}{U_0 L}} = \frac{1}{\sqrt{Re}},$$

i.e. the assumption that $\delta \ll L$ is consistent with saying $Re \gg 1$. Note too that our conclusion that $\delta/L = 1/\sqrt{Re}$ reveals why it was appropriate to define $y^*$ and $v^*$ in terms of $\sqrt{Re}$.
For the $y$-direction we have

$$u\frac{\partial v}{\partial x} + v\frac{\partial v}{\partial y} = -\frac{1}{\rho}\frac{\partial p}{\partial y} + v\left(\frac{\partial^2 v}{\partial x^2} + \frac{\partial^2 v}{\partial y^2}\right) \tag{17.3}$$

---

[146]  $O(X)$ is used to mean of order of magnitude $X$.
[147]  Note that signs are not relevant to orders of magnitude estimates.

with orders of magnitude

$$u\frac{\partial v}{\partial x} = O\left(\frac{U_0 V}{L}\right) = O\left(\frac{U_0^2 \delta}{L^2}\right), \quad v\frac{\partial v}{\partial y} = O\left(\frac{V^2}{\delta}\right) = O\left(\frac{U_0^2 \delta}{L^2}\right),$$

$$\nu\frac{\partial^2 v}{\partial x^2} = O\left(\frac{\nu V}{L^2}\right) = O\left(\frac{\nu U_0 \delta}{L^3}\right), \quad \nu\frac{\partial^2 v}{\partial y^2} = O\left(\frac{\nu V}{\delta^2}\right) = O\left(\frac{\nu U_0}{\delta L}\right).$$

Clearly, the first three velocity terms are negligible compared with the fourth term and can be neglected from now on.

So far we have avoided any statement about the pressure gradients in the $x$- and $y$-directions. From equation (17.2) for the $x$-direction it must be that

$$\frac{1}{\rho}\frac{\partial p}{\partial x} = O\left(\frac{U_0^2}{L}\right)$$

while from equation (17.3), for the $y$-direction,

$$\frac{1}{\rho}\frac{\partial p}{\partial y} = O\left(\frac{\nu U_0}{\delta L}\right) = O\left(\frac{1}{\sqrt{Re}}\frac{U_0^2}{L}\right)$$

and we conclude that $\partial p/\partial y \ll \partial p/\partial x$, again consistent with our original postulate.

Equations (17.1) and (17.4) constitute what are referred to as the constant-property, two-dimensional boundary-layer equations with surface boundary conditions $y = 0$, $u = u_S = 0$, for a stationary surface, and $y = 0$, $v = v_S = 0$, for an impermeable surface. If the surface has velocity $u_S$ in the $x$-direction, and **mass transfer** through a porous surface with velocity $v_S$ in the $y$-direction, then the surface boundary conditions become $y = 0$, $u = u_S$, $v = v_S$. **Transpiration cooling** is the term used when cooler fluid is **blown** through a porous surface into hotter boundary-layer fluid, i.e. $v_S > 0$. **Boundary-layer control**, for example to prevent or delay **boundary-layer separation**, can be achieved by **suction**, when $v_S < 0$. For $y \to \infty$, the boundary conditions are $u \to U_\infty$, and $\partial^n u/\partial y^n \to 0$.

Since the boundary-layer approximations lead to $\partial u/\partial y = O(U_0/\delta)$, and $\partial u/\partial x = O(U_0/L)$, it should be apparent that a consequence is that the shear stress $\tau$ at any location $x$ within a boundary layer is given by

$$\tau = \mu\frac{\partial u}{\partial y}. \tag{17.11}$$

A number of commercial software packages are available which can be used for the numerical integration of the laminar boundary-layer equations, and such calculations are now regarded as routine. The purpose of this chapter is to give the reader some insight into the properties and behaviour of laminar boundary layers, based upon exact solutions of the equations for a flat-plate boundary layer and for boundary layers where the free-stream velocity is proportional to $x^m$, or from more general approximate solutions. The accuracy of such solutions will be adequate for many engineering applications and often provide useful confirmation that a numerical solution has not been compromised by the input of faulty data or an error in the computer program.

## 17.3 Flat-plate laminar boundary layer: Blasius' solution

### 17.3.1 Derivation and solution of Blasius' equation

For the flat-plate boundary layer, where the **pressure gradient**[148] $dp/dx = 0$, the boundary-layer equations are

$$\frac{\partial u}{\partial x} + \frac{\partial v}{\partial y} = 0 \tag{17.1}$$

and

$$u\frac{\partial u}{\partial x} + v\frac{\partial u}{\partial y} = \nu\frac{\partial^2 u}{\partial y^2} \tag{17.12}$$

with boundary conditions $u(x, 0) = 0$, $v(x, 0) = 0$, $u(x, \infty) = U_\infty = $ constant, and

$$y \to \infty, \partial^n u / \partial y^n \to 0.$$

Figure 17.2 shows a schematic diagram of a flat-plate boundary layer in which are sketched **velocity profiles**[149] at three arbitrary $x$-locations. Although sketched only roughly, the three profiles show several important features: in each case $u$ changes from zero to the same free-stream value $U_\infty$ in a distance $\delta$ which increases with $x$; the velocity gradient $\partial u / \partial y$ decreases

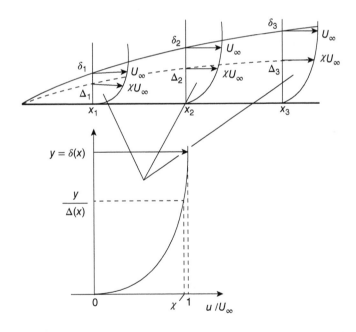

**Figure 17.2** Similar velocity profiles for a flat-plate laminar boundary layer

[148] Since for most boundary layers $\partial p / \partial y = 0$, it is usual to refer to $dp/dx$ as the pressure gradient.

[149] A velocity profile or **distribution** is a curve showing the variation of the axial velocity $u$ with normal distance from the surface $y$ at any streamwise location $x$.

to zero as the free stream is approached; and with increasing distance $x$ the boundary layer thickens and the near-surface velocity gradient $\partial u/\partial y|_0$ decreases. We can summarise these qualitative observations by saying the profiles are **similar** in shape[150], which suggests that, if we plot all three profiles in the form $u/U_\infty$ versus $y/\Delta(x)$, they might fall on a single curve, the length $\Delta(x)$ being the $y$-value at which $u$ is the same fraction $\chi$ of $U_\infty$ at every $x$-location. Mathematically, we can state this speculation as

$$\frac{u}{U_\infty} = f\left[\frac{y}{\Delta(x)}\right] \tag{17.13}$$

with

$$\chi = f(1) \tag{17.14}$$

and the question which must be asked is 'can we find a **scaling length** $\Delta$ which varies with $x$ in such a way that $u/U_\infty$ is a function of $y/\Delta$ only?'

To answer this question, we start by defining the non-dimensional variable $\eta$

$$\eta = \frac{y}{\Delta(x)} \tag{17.15}$$

so that according to equation (17.13) we anticipate that

$$u = U_\infty f(\eta). \tag{17.16}$$

Because we are trying to demonstrate **similarity** between one velocity profile and the next, $\eta$ is termed a **similarity variable**.

From the continuity equation, we have

$$\frac{\partial v}{\partial y} = -\frac{\partial u}{\partial x}$$

so that

$$v = -\int_0^y \frac{\partial u}{\partial x} dy. \tag{17.17}$$

From equation (17.16), we have

$$\left.\frac{\partial u}{\partial x}\right|_y = \left.\frac{\partial\left[U_\infty f(\eta)\right]}{\partial x}\right|_y = U_\infty \left.\frac{\partial f(\eta)}{\partial x}\right|_y = U_\infty \left.\frac{df}{d\eta}\frac{\partial\eta}{\partial x}\right|_y$$

and, from the definition of $\eta$, equation (17.15),

$$\left.\frac{\partial\eta}{\partial x}\right|_y = -\frac{y}{\Delta^2}\frac{d\Delta}{dx} = -\frac{\eta}{\Delta}\frac{d\Delta}{dx},$$

so that

$$\left.\frac{\partial u}{\partial x}\right|_y = -\frac{U_\infty}{\Delta}\frac{d\Delta}{dx}\eta f' \tag{17.18}$$

---

[150] The term **congruent** is also used.

and, from equation (17.17),

$$v = U_\infty \frac{d\Delta}{dx} \int_0^\eta \eta f' \, d\eta = U_\infty \frac{d\Delta}{dx} \int_0^f \eta \, df = U_\infty \frac{d\Delta}{dx} \left( \eta f - \int_0^\eta f \, d\eta \right). \tag{17.19}$$

In equation (17.18) we have introduced **Lagrange's dash-notation**[151] for differentiation, i.e. $f' = df/d\eta$, a second derivative is denoted by two dashes, e.g. $f'' = d^2 f/d\eta^2$, etc. An important feature of the present analysis, because it has consequences so far as establishing $\Delta(x)$ is concerned, is the separation into products of $x$-dependent and $\eta$-dependent quantities as evidenced, for example, by equations (17.18) and (17.19).

For those readers unfamiliar with manipulating partial differentials, a few comments may be helpful. First, we are dealing with partial differentials because the function $f$ depends upon two variables, $\Delta$ and $y$. The fact that $\Delta$ is itself a function of a third variable, $x$, has also to be borne in mind. Second, the subscript $y$ alongside a vertical line as in

$$\left. \frac{\partial u}{\partial x} \right|_y$$

is a reminder that the partial differentiation of $u$ with respect to $x$ is carried out treating $y$ as though it were a constant. Third,

$$\left. \frac{\partial f(\eta)}{\partial x} \right|_y$$

is evaluated recognising that we are dealing with a function, in this case $f$, which depends upon the variable $\eta$, which in turn is defined in terms of $y$ and $\Delta$, although it is only $\Delta$ that varies directly with $x$. Although this process may at first seem confusing, and possibly intimidating, if the differentiation process is dealt with systematically, it should ultimately prove straightforward. We should also point out that the final step in arriving at equation (17.19) involved integration by parts, taking into account the boundary condition $y = 0, u = 0$.

Since it is convenient to avoid functions involving integrals, we introduce the variable $F(\eta)$ defined by

$$F = \int_0^\eta f \, d\eta \tag{17.20}$$

so that, from equation (17.19), we have

$$v = U_\infty \frac{d\Delta}{dx} \left( \eta F' - F \right). \tag{17.21}$$

Apart from convenience, the quantity represented by $F$ has an important physical significance, which becomes apparent if we revert to dimensional quantities

$$U_\infty \Delta \int_0^\eta f \, d\eta = U_\infty \Delta F = \psi$$

where the quantity $\psi$, known as the **stream function**, is a measure of the flowrate within the boundary layer between the surface and $y$, i.e.

---

[151] Also referred to as **prime notation**.

$$\psi = \int_0^y u\, dy \tag{17.22}$$

and $F$ is the **non-dimensional stream function**.

From equation (17.20) we have

$$f = \frac{dF}{d\eta} = F' \tag{17.23}$$

so that

$$u = U_\infty f = U_\infty F', \tag{17.24}$$

$$\left.\frac{\partial u}{\partial x}\right|_y = U_\infty \left.\frac{\partial F'}{\partial x}\right|_y = U_\infty F'' \left.\frac{\partial \eta}{\partial x}\right|_y = -\eta F'' \frac{U_\infty}{\Delta}\frac{d\Delta}{dx}, \tag{17.25}$$

$$\left.\frac{\partial u}{\partial y}\right|_x = U_\infty \left.\frac{\partial F'}{\partial y}\right|_x = U_\infty F'' \left.\frac{\partial \eta}{\partial y}\right|_x = F'' \frac{U_\infty}{\Delta}, \tag{17.26}$$

and

$$\left.\frac{\partial^2 u}{\partial y^2}\right|_x = \frac{U_\infty}{\Delta} \left.\frac{\partial F''}{\partial y}\right|_x = F''' \frac{U_\infty}{\Delta^2}. \tag{17.27}$$

Finally, we can substitute for the various terms in the flat-plate boundary-layer equation (17.12)

$$U_\infty F' \left( -\eta F'' \frac{U_\infty}{\Delta}\frac{d\Delta}{dx} \right) + \left( U_\infty \frac{d\Delta}{dx} \right) (\eta F' - F) F'' \frac{U_\infty}{\Delta} = \nu F''' \frac{U_\infty}{\Delta^2}$$

which simplifies to

$$F''' + \frac{U_\infty}{2\nu}\frac{d\Delta^2}{dx}FF'' = 0. \tag{17.28}$$

Since $f$, and so $F$, are functions of $\eta$ only, and $\Delta$ is a function of $x$ only, we conclude that

$$\frac{U_\infty}{\nu}\frac{d\Delta^2}{dx} = \text{constant} = \alpha \tag{17.29}$$

where $\alpha$ is an arbitrary constant. Integration of equation (17.29) leads to

$$\Delta = \sqrt{\frac{\alpha\nu x}{U_\infty}}, \tag{17.30}$$

wherein we have set $\Delta = 0$ at $x = 0$, i.e. the boundary layer has zero thickness at its origin, and we then have

$$\eta = \frac{y}{\Delta} = y\sqrt{\frac{U_\infty}{\alpha\nu x}} \tag{17.31}$$

together with

$$\frac{u}{U_\infty} = f\left(\frac{y}{\Delta}\right) = f\left(y\sqrt{\frac{U_\infty}{\alpha\nu x}}\right). \tag{17.32}$$

Equation (17.30) provides the answer to our original question: 'can we find a scaling length $\Delta$ which varies with $x$ in such a way that $u/U_\infty$ is a function of $y/\Delta$?' It should be clear that the answer is unchanged whatever the value chosen for $\alpha$: Heinrich Blasius, who first investigated the problem of the flat-plate boundary layer, effectively chose $\alpha = 2$, which is what we adopt here, while others have preferred $\alpha = 1$, or more complicated forms when the pressure gradient is non-zero (see Section 17.4).

According to equations (17.28) and (17.30) with $\alpha = 2$, the ordinary differential equation to be solved to find the form of the velocity distribution within a flat-plate boundary layer is

$$F''' + FF'' = 0 \tag{17.33}$$

subject to the boundary conditions

$F(0) = 0$
$F'(0) = 0, \quad$ from the no – slip condition
$F'(\infty) = 1, \quad i.e.\ y \to \infty, \quad u \to U_\infty$
$F''(\infty) = 0$
$F'''(\infty) = 0.$

There is no analytical solution to equation (17.33), which is referred to as **Blasius' equation**[152]. Early solutions were found using series approximations, while today the equation is easily solved numerically. Values for the velocity ratio $u/U_\infty = f$, the non-dimensional stream function $F = \int_0^\eta f d\eta$, and the non-dimensional velocity gradient $f'$, are listed in Table 17.1 for values of non-dimensional distance from the **wall**[153], $\eta = y\sqrt{U_\infty/(2vx)}$. For the most part values of calculated quantities are given to 4 d.p. The exceptions are values given to 5 d.p. for $\eta > 4.6$ as $F'$ and $F''$ approach their asymptotes. The velocity distribution calculated from Blasius' equation is included in Figure 17.6 in Subsection 17.4.1.

Values for dimensional quantities are calculated from $y = \eta\sqrt{(2vx)/U_\infty}$, $\Psi = F\sqrt{(2vxU_\infty)}$, $u = fU_\infty$, and $\tau = f'\rho\sqrt{(vU_\infty^3)/2x}$.

### 17.3.2 Comments on Blasius' solution

(a) From equations (17.25) and (17.26)

$$\left.\frac{\partial u}{\partial x}\right|_y = -\eta F''\frac{U_\infty}{\Delta}\frac{d\Delta}{dx} \quad \text{and} \quad \left.\frac{\partial u}{\partial y}\right|_x = F''\frac{U_\infty}{\Delta},$$

taken together with equation (17.30) for $\Delta$, we see that

$$\left.\frac{\partial u}{\partial x}\right|_y = -\eta\frac{d\Delta}{dx}\left.\frac{\partial u}{\partial y}\right|_x = -\frac{y}{x}\left.\frac{\partial u}{\partial y}\right|_x. \tag{17.34}$$

[152] Although in work subsequent to that of Blasius the value $\alpha = 1$ was frequently used, the corresponding differential equation was still referred to as Blasius' equation.
[153] A surface over which there is fluid flow is commonly referred to as the wall.

**Table 17.1** Blasius' solution with $\alpha = 2$ (based upon Table V.1 in Rosenhead (1963))

| $\eta$ | $F = \int_0^\eta f\, d\eta$ | $F' = f$ | $F'' = \dfrac{df}{d\eta}$ | $\eta$ | $F = \int_0^\eta f\, d\eta$ | $F' = f$ | $F'' = \dfrac{df}{d\eta}$ |
|---|---|---|---|---|---|---|---|
| 0 | 0 | 0 | 0.4696 | 2.0 | 0.8870 | 0.8167 | 0.2557 |
| 0.1 | 0.0023 | 0.0470 | 0.4696 | 2.2 | 1.0549 | 0.8633 | 0.2106 |
| 0.2 | 0.0094 | 0.0939 | 0.4693 | 2.4 | 1.2315 | 0.9011 | 0.1677 |
| 0.3 | 0.0211 | 0.1408 | 0.4686 | 2.6 | 0.4148 | 0.9306 | 0.1286 |
| 0.4 | 0.0375 | 0.1876 | 0.4673 | 2.8 | 0.6033 | 0.9529 | 0.0951 |
| 0.5 | 0.0586 | 0.2342 | 0.4650 | 3.0 | 0.7056 | 0.9691 | 0.0677 |
| 0.6 | 0.0844 | 0.2806 | 0.4617 | 3.2 | 1.9906 | 0.9804 | 0.0464 |
| 0.7 | 0.1147 | 0.3265 | 0.4572 | 3.4 | 2.1875 | 0.9880 | 0.0305 |
| 0.8 | 0.1497 | 0.3720 | 0.4512 | 3.6 | 2.3856 | 0.9929 | 0.0193 |
| 0.9 | 0.1891 | 0.4167 | 0.4436 | 3.8 | 2.5845 | 0.9959 | 0.0118 |
| 1.0 | 0.2330 | 0.4606 | 0.4344 | 4.0 | 2.7839 | 0.9978 | 0.0069 |
| 1.1 | 0.2812 | 0.5035 | 0.4234 | 4.2 | 2.9836 | 0.9988 | 0.0039 |
| 1.2 | 0.3337 | 0.5452 | 0.4106 | 4.4 | 3.1834 | 0.9994 | 0.0021 |
| 1.3 | 0.3902 | 0.5856 | 0.3960 | 4.6 | 3.3833 | 0.9997 | 0.0011 |
| 1.4 | 0.4507 | 0.6244 | 0.3797 | 4.8 | 3.5833 | 0.99986 | 0.00054 |
| 1.5 | 0.5150 | 0.6615 | 0.3618 | 5.0 | 3.7932 | 0.99994 | 0.00026 |
| 1.6 | 0.5830 | 0.6967 | 0.3425 | 5.2 | 3.9832 | 0.99997 | 0.00012 |
| 1.7 | 0.6543 | 0.7299 | 0.3220 | 5.4 | 4.1832 | 0.99999 | 0.00005 |
| 1.8 | 0.7289 | 0.7611 | 0.3004 | 5.6 | 4.3832 | 1.00000 | 0.00002 |
| 1.9 | 0.8064 | 0.7900 | 0.2783 | 6.0 | 4.7832 | 1.00000 | 0.00000 |

Since $y/x \ll 1$, we confirm that the streamwise velocity gradient $\partial u/\partial x$ is much smaller than the transverse (or cross-stream) velocity gradient $\partial u/\partial y$.

(b) From equation (17.21) combined with the definitions of $f$, $F$, and $F'$, and the equation for $\Delta$, we have

$$\frac{v}{u} = \frac{1}{2}\sqrt{\frac{2v}{U_\infty x}}\left(\eta - \frac{F}{F'}\right) = \frac{1}{2}\sqrt{\frac{2v}{U_\infty x}}\left(\eta - \frac{1}{f}\int_0^\eta f\, d\eta\right). \tag{17.35}$$

Close to the surface,

$$f \approx \eta f'(0) \text{ so that } \int_0^\eta f\, d\eta \approx \frac{1}{2}\eta^2 f'(0)$$

and, from equation (17.35)

$$\frac{v}{u} \approx \frac{\eta}{4}\sqrt{\frac{2v}{U_\infty x}} = \frac{\eta}{4}\sqrt{\frac{2}{Re_x}}.$$

Far from the surface $f \to 1$ and from the Blasius table we observe that for $\eta \to \infty$

$$\eta - \int_0^\eta f d\eta \to \text{constant} = 1.2168$$

so that, from equation (17.35)

$$\frac{v}{u} = \frac{V_\infty}{U_\infty} \to \frac{0.8604}{\sqrt{Re_x}}. \tag{17.36}$$

We conclude therefore that, for all values of $\eta$, $v > 0$ and $v \ll u$.

(c) The shear stress $\tau$ at any location within the boundary layer is given by

$$\tau = \mu \left.\frac{\partial u}{\partial y}\right|_x = f' \mu \frac{U_\infty}{\Delta} = f' \rho U_\infty^{3/2}\sqrt{\frac{v}{2x}},$$

at the surface ($\eta = 0$) therefore

$$\tau_S = f'(0)\rho U_\infty^{3/2}\sqrt{\frac{v}{2x}} = 0.3321\rho U_\infty^{3/2}\sqrt{\frac{v}{2x}} = 0.2348\rho U_\infty^{3/2}\sqrt{\frac{v}{x}}. \tag{17.37}$$

Equation (17.37) confirms that $\tau_S$ decreases with increasing $x$. From the usual definition for the **skin-friction coefficient** (or **friction factor**[154]) $c_f$ for a boundary layer we have

$$\frac{c_f}{2} = \frac{\tau_S}{\rho U_\infty^2} = \frac{f'(0)}{\sqrt{2Re_x}} = \frac{0.3321}{\sqrt{Re_x}}. \tag{17.38}$$

From equation (17.37) the total **drag force** $D'$ per unit width exerted by the fluid on one side of a plate of length $L$ is given by

$$D' = \int_0^L \tau_S \, dx = 0.6642\rho U_\infty^{3/2}\sqrt{vL}$$

from which we can define an average drag coefficient $C_F$

$$C_F = \frac{D'}{\rho U_\infty^2 L} = \frac{0.6642}{\sqrt{Re_L}} = 2c_f(L). \tag{17.39}$$

(d) At the outset of our analysis of the flat-plate boundary layer, we introduced the 'scaling length' $\Delta$, which we subsequently showed satisfies equation (17.30)

$$\Delta = \sqrt{\frac{\alpha v x}{U_\infty}}.$$

From the tabulated solution, with $\alpha = 2$, we now see that, when $\eta = 1$, (i.e. $y = \Delta$), the velocity ratio $u/U_\infty = 0.4606$, which is thus the value of $\chi$ in Figure 17.2. Had we chosen $\alpha = 1$ rather than $\alpha = 2$ at the conclusion of the derivation of Blasius' equation, we would have found $u/U_\infty = 0.3298$ at $y = \Delta$, i.e. a different point on the same velocity profile.

---

[154] Note that $c_f$ is defined in a similar way as the Fanno friction factor for duct flow $f_F = 2\tau_S/\rho\overline{V}^2$ but it is conventional to write equations including the skin-friction coefficient in terms of $c_f/2$ (spoken as 'see eff over two'). The symbol $f$ is sometimes used instead of $c_f/2$.

(e)  From the tabulated solution to Blasius' equation, we see that, for $\eta \geq 5$, $f = 1$ to within 0.01% so it would be reasonable to define the boundary-layer thickness as $\delta = 5\Delta$, i.e.

$$\frac{\delta}{x} = \frac{5\Delta}{x} = \frac{5}{x}\sqrt{\frac{2vx}{U_\infty}} = \frac{7.07}{\sqrt{Re_x}}. \qquad (17.40)$$

Since the velocity $u$ approaches the free-stream velocity $U_\infty$ asymptotically, it is impossible to put a precise value on the boundary-layer thickness. A common choice is the location where $u/U_\infty = 0.99$ (corresponding with $\eta \approx 3.5$), possibly because until relatively recently this corresponded roughly with the precision to which a flow velocity could be measured with a Pitot tube or hot-wire anemometer. Whatever choice is made, we see that $\delta/x \sim 1/\sqrt{Re_x}$.

(f)  If, as in part (e) we take the edge of the boundary layer to correspond with $\eta = 5$, the non-dimensional flowrate within the boundary layer is given by $F(5) = 3.7832$. From the definition of $F$, equation (17.20), we then see that the volumetric flowrate per unit width is the stream function $\psi$ corresponding to $y = \delta$

$$\psi(\delta) = U_\infty \Delta F(5) = 5.3503\sqrt{vU_\infty x}. \qquad (17.41)$$

Essential features of a streamline are that there is no flow across it (Section 6.3) and that the stream function is constant along it. Since $\psi(\delta) \sim \sqrt{x}$ we can draw four important conclusions

• the flowrate within the boundary layer increases with streamwise distance $x$
• the edge of the boundary layer corresponding to $y = \delta(x)$ is not a streamline
• there is flow from the free stream into the boundary layer, a phenomenon called **entrainment**[155]
• streamlines originating in the uniform flow upstream of the plate are deflected away from the plate before passing into it, consistent with the earlier conclusion that $v_\infty > 0$ (equation (17.36))

(g)  As illustrated in Figure 17.3, the deflection of any streamline can be quantified as follows. Consider a streamline upstream of the plate a distance $Y$ above the plane of the plate. The value of the stream function $\psi_0$ for this streamline is

$$\psi_0 = U_\infty Y. \qquad (17.42)$$

At some streamwise location a distance $x$ from the leading edge of the plate, for the same streamline, the value of the stream function $\psi(x, \delta)$ at the edge of the boundary layer is given by

$$\psi(x, \delta) = \int_0^\delta u \, dy \qquad (17.43)$$

which must equal $\psi_0$. Since $u < U_\infty$ throughout the boundary layer and since $\psi_0$ and $\psi(x, \delta)$ correspond to the same streamline, it must be that $\delta > Y$, i.e. the streamline has been deflected (or displaced) away from the plate by an amount

$$\delta - Y = \delta - \frac{\psi_0}{U_\infty} = \delta - \frac{1}{U_\infty}\int_0^\delta u \, dy = \int_0^\delta \left(1 - \frac{u}{U_\infty}\right) dy = \delta^*. \qquad (17.44)$$

Note that in deriving equation (17.44) we have made use of the relation $\int_0^\delta dy = \delta$.

---

[155] Entrainment is a feature of all shear layers, i.e. boundary layers, jets, wakes, wall jets, etc.

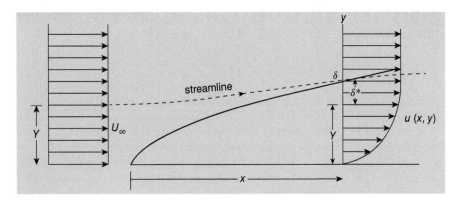

**Figure 17.3** Deflection of a streamline by the boundary layer: The displacement thickness

The quantity $\delta^*(x)$, defined by equation (17.44), is termed the **displacement thickness**[156]. The upper limit of the integral in equation (17.44) can be replaced by $\infty$ since, for $y > \delta$, $u = U_\infty$, the integrand falls to zero, and there is no further contribution to the integral. The displacement thickness is one of a number of **integral thicknesses** which arise in boundary-layer theory. Although the boundary-layer thickness $\delta$ cannot be defined precisely, a virtue of these integral thicknesses is that they are.

As illustrated in Figure 17.4, the area representing the **velocity deficit** between the velocity distribution $u(x, y)$ and the free-stream velocity $U_\infty$ corresponds with the displacement thickness, a quantity which increases with streamwise distance $x$ from the leading edge of the plate.

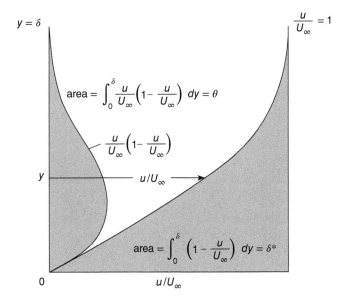

**Figure 17.4** Geometric interpretation of the displacement and momentum thicknesses

[156] The symbol $\delta_1$ is often used instead of $\delta^*$.

If we again take $\eta = 5$ to define the thickness of the boundary layer then, from Table 17.1, the corresponding value $F = 3.7832$ and, from equation (17.44),

$$\frac{\delta^*}{\delta} = \int_0^5 \left(1 - f\right) d\eta = 5 - F(5) = 1.2168. \tag{17.45}$$

A consequence of the **streamline-displacement effect** is an increase in the free-stream velocity for developing flow in a duct of constant cross-sectional area. That this must be the case is easily seen from the results of Chapter 16, where the peak velocity in Poiseuille flow through a circular pipe is greater than the average velocity by a factor of 2.

As a final comment on the displacement thickness, it should be noted that the $\delta^*$ definition of equation (17.44) is not limited to flat-plate flows but applies generally for constant-property boundary layers.

(h) Although the **momentum thickness** (or **momentum-deficit thickness**)[157] $\theta(x)$ does not arise directly in the analysis of the flat-plate boundary layer, it is also illustrated graphically in Figure 17.4 and introduced here as it plays an important role in boundary-layer theory generally (see Section 17.5). The definition of $\theta$, which is again quite general for constant-property boundary-layer flows, is

$$\theta = \int_0^\delta \frac{u}{U_\infty} \left(1 - \frac{u}{U_\infty}\right) dy, \tag{17.46}$$

and, as is the case for $\delta^*, \theta$ can be evaluated precisely. For the flat-plate, laminar boundary layer, we have

$$\frac{\theta}{\delta} = \int_0^5 f \left(1 - f\right) d\eta = F(5) - \int_0^5 f^2 d\eta.$$

Numerical integration of the last term leads to

$$\frac{\theta}{\delta} = 0.09393. \tag{17.47}$$

(i) The reader will have noticed that $\sqrt{Re_x}$ appears in several results which arise from the solution of Blasius' equation. This partially explains the incorporation of $\sqrt{Re}$ into the non-dimensionalisation of the boundary-layer equations discussed in Section 17.2.

(j) The validity of Blasius' equation requires that the flow is steady and laminar. As already remarked in Section 17.2, for duct flow, once the Reynolds number exceeds a critical value, the flow becomes unstable, and eventually transition to turbulent flow occurs. Since a boundary layer is always developing with distance $x$ along the surface, the Reynolds number $Re_x = U_\infty x / \nu$ is often chosen as the appropriate parameter, and a value of about $3.5 \times 10^5$ can be taken as the upper limit for stability. For a direct analogy with channel flow, $U_\infty \delta / \nu$ suggests itself as the appropriate Reynolds number, with the boundary-layer thickness $\delta$ replacing the hydraulic diameter $D_H$ as the length scale. However, given the uncertainty in determining $\delta$, the momentum thickness $\theta$ is a better choice. With a **momentum-thickness Reynolds number** $Re_\theta = U_\infty \theta / \nu$ it is found that a flat-plate laminar boundary layer is stable, provided $Re_\theta$ is less than about 300.

---

[157] If the symbol $\delta_1$ is used for the displacement thickness rather than $\delta^*$, it is usual for the momentum thickness to be represented by $\delta_2$ rather than $\theta$.

A cautionary note has to be sounded as flow stability and transition to turbulence are influenced by such factors as surface roughness, unsteadiness of the free stream (**free-stream turbulence**), and the streamwise pressure gradient, particularly an adverse gradient.

## 17.4 Wedge-flow laminar boundary layers: Falkner and Skan's equation

In this section we are concerned with boundary layers which develop on wedge surfaces such as that shown in Figure 17.5(a), which illustrates two-dimensional flow over a symmetric wedge of included angle $\beta$ (measured in radians). It can be shown that in the absence of viscosity, the fluid velocity at the wedge surface $U_\infty(x)$ is given by

$$U_\infty = Kx^m \tag{17.48}$$

where $x$ is the distance along the wedge surface measured from the apex and $K$ is a dimensional constant. The exponent $m$ is related to the wedge angle[158] by

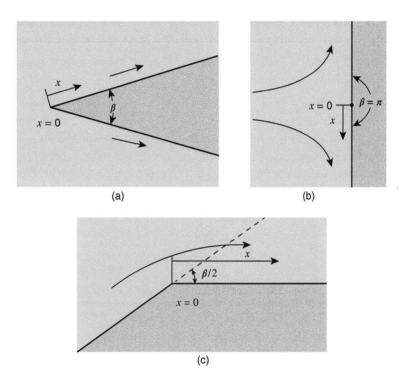

(a)        (b)

(c)

**Figure 17.5** (a) Flow over a symmetrical wedge (b) Flow approaching a stagnation point (c) Flow on the leeward side of a corner

---

[158] In some treatments of wedge flows the wedge angle is taken as $\pi\beta$ rather than $\beta$ as here. It is also the case that the symbol $\beta$ sometimes represents the semi-included (or half angle) of the wedge rather than the included angle as here.

$$m = \frac{\beta}{2\pi - \beta} \quad \text{or} \quad \frac{\beta}{2\pi} = \frac{m}{m+1}. \tag{17.49}$$

Two special cases incorporated in this formulation are $m = 0$, which corresponds with the zero pressure-gradient or flat-plate flow we have just dealt with, and $m = 1$, $\beta = \pi$, which is a good approximation to the flow near a forward stagnation point[159], as illustrated in Figure 17.5(b). Wedge flows with $\beta < 0$ are referred to as **diffusion flows** and realised physically by the flow on the leeward side of a sharp corner, as shown in Figure 17.5(c).

### 17.4.1 Derivation and solution of Falkner and Skan's equation

The pressure gradient is now non-zero, and the boundary-layer equations are

$$\frac{\partial u}{\partial x} + \frac{\partial v}{\partial y} = 0 \tag{17.1}$$

and

$$u\frac{\partial u}{\partial x} + v\frac{\partial u}{\partial y} = U_\infty \frac{dU_\infty}{dx} + v\frac{\partial^2 u}{\partial y^2}, \tag{17.50}$$

the latter corresponding with equation (17.4) with $-dp/dx$ replaced by $\rho U_\infty dU_\infty/dx$.

The similarity variable which reduced the partial differential equations to a single ordinary differential equation for a flat-plate boundary layer was

$$\eta = y\sqrt{\frac{U_\infty}{\alpha v x}} \tag{17.32}$$

where $\alpha$ is an arbitrary constant. As already indicated, Blasius chose $\alpha = 2$, many workers since have adopted $\alpha = 1$, while another common choice when $m \neq 0$ is $\alpha = 2/m + 1$ (so that $\alpha = 2$ for $m = 0$). Whichever value is used leads to an ordinary differential equation for wedge-flow boundary layers using essentially the same analytical procedure as in Section 17.3. We shall continue to use $\alpha = 2$.

Since the free-stream velocity obeys $U_\infty = Kx^m$, the pressure-gradient term in equation (17.50) may be written as

$$U_\infty \frac{dU_\infty}{dx} = m\frac{U_\infty^2}{x}. \tag{17.51}$$

We can transform all terms in equation (17.50) into the variables $F$ (or $f$) and $\eta$ as follows

$$u = U_\infty f = U_\infty F' \tag{17.52}$$

$$\left.\frac{\partial u}{\partial x}\right|_y = \frac{U_\infty}{x}\left[\left(\frac{m-1}{2}\right)\eta F'' + mF'\right] \tag{17.53}$$

$$v = -\sqrt{\frac{\alpha v U_\infty}{x}}F''\left[\left(\frac{m-1}{2}\right)\eta F' + \left(\frac{m+1}{2}\right)F\right] \tag{17.54}$$

$$\left.\frac{\partial u}{\partial y}\right|_x = \frac{U_\infty^{3/2}}{\sqrt{\alpha v x}}F'' \tag{17.55}$$

---

[159] The flow with $m = 1$ is known as **Hiemenz flow**.

and

$$\nu \left. \frac{\partial^2 u}{\partial y^2} \right|_x = \frac{U_\infty^2}{\alpha x} F'''.$$

(17.56)

Substitution of equations (17.52) to (17.56) into equation (17.50) leads to

$$F''' + \alpha \left( \frac{m+1}{2} \right) FF'' + \alpha m \left( 1 - F'^2 \right) = 0$$

(17.57)

and we see that with $\alpha = 1$ we have

$$F''' + \left( \frac{m+1}{2} \right) FF'' + m \left( 1 - F'^2 \right) = 0$$

(17.58)

which is commonly referred to as the **Falkner-Skan equation.**
With $\alpha = 2$, equation (17.57) becomes

$$F''' + (m+1) FF'' + 2m \left( 1 - F'^2 \right) = 0$$

(17.59)

and with $\alpha = 2/(m+1)$ we have

$$F''' + FF'' + \frac{2m}{m+1} \left( 1 - F'^2 \right) = 0$$

(17.60)

or

$$F''' + FF'' + \frac{\beta}{\pi} \left( 1 - F'^2 \right) = 0$$

(17.61)

where $\beta$ is the wedge included angle in radians.
With $m = 0$ and $\alpha = 2$ in equation (17.57) we recover Blasius' equation

$$F''' + \frac{1}{2}\alpha FF'' = 0.$$

(17.33)

The boundary conditions for the wedge-flow boundary-layer problem are

$$F(0) = 0$$
$$F'(0) = 0$$

and

$$F'(\infty) = 1.$$

As was the case for Blasius' equation, the Falkner-Skan equation (17.58) can be solved only numerically. The most important results of such numerical calculations are the values of the integral boundary-layer thicknesses, $\delta^*$ and $\theta$, and the skin-friction coefficient $c_f/2$.
From the definition of $\delta^*$, equation (17.44), we have

$$\delta^* = \int_0^\infty \left( 1 - \frac{u}{U_\infty} \right) dy = \sqrt{\frac{\alpha \nu x}{U_\infty}} \int_0^{\eta_\delta} \left( 1 - f \right) d\eta$$

(17.62)

so that

$$\frac{\delta^*}{x} \sqrt{Re_x} = \sqrt{\alpha}[\eta_\delta - F(\eta_\delta)].$$

(17.63)

Note that in equation (17.62) the upper limit of the integral was changed from $\infty$ to $\eta_\delta$, the value of $\eta$ at which $1 - f$, or $\eta - F(\eta)$, is no longer measurably different from zero. For the Blasius problem ($m = 0$) we took $\eta_\delta = 5$ but, for other values of $m$, somewhat different values are appropriate. From equation (17.46) for $\theta$ we have

$$\theta = \int_0^\infty \frac{u}{U_\infty}\left(1 - \frac{u}{U_\infty}\right) dy = \sqrt{\frac{\alpha \nu x}{U_\infty}} \int_0^\infty f\left(1 - f\right) d\eta \tag{17.64}$$

so that

$$\frac{\theta}{x}\sqrt{Re_x} = \sqrt{\alpha}\left[F\left(\eta_\delta\right) - \int_0^{\eta_\delta} f^2 d\eta\right]. \tag{17.65}$$

The surface shear stress $\tau_S$ is given by

$$\tau_S = \mu \left.\frac{\partial u}{\partial y}\right|_0 = \rho\sqrt{\frac{\nu}{\alpha x}}U_\infty^{3/2} f'(0) \tag{17.66}$$

from which

$$\frac{c_f}{2}\sqrt{Re_x} = \frac{F''(0)}{\sqrt{\alpha}}. \tag{17.67}$$

Note that both the skin-friction coefficient $c_f/2 = \tau_S/\rho U_\infty^2$ and the Reynolds number $Re_x = U_\infty x/\nu$ include the free-stream velocity $U_\infty(x)$, which is no longer constant (as it was for the Blasius problem) but varies with the streamwise distance $x$ for wedge flows according to $U_\infty = Kx^m$.

Table 17.2 lists values of $(\delta^*/x)\sqrt{Re_x}$, $(\theta/x)\sqrt{Re_x}$, $(c_f/2)\sqrt{Re_x}$, and $(c_f/2)Re_\theta$ for various values of the wedge angle $\beta$ and the exponent $m$ (the parameter $\alpha = 2$).

It is sometimes convenient to express both analytical results and experimental data in terms of a Reynolds number $Re_\theta$ based upon the momentum thickness $\theta$ rather than the streamwise distance $x$

$$Re_\theta = \frac{U_\infty \theta}{\nu}. \tag{17.68}$$

For wedge flows we can use equation (17.65) to relate $Re_\theta$ and $Re_x$

$$Re_\theta = \frac{U_\infty \theta}{\nu} = \frac{U_\infty x}{\nu}\frac{\theta}{x} = \left(\frac{\theta}{x}\sqrt{Re_x}\right)\sqrt{Re_x} = \sqrt{\alpha}\left[F\left(\eta_\delta\right) - \int_0^{\eta_\delta} f^2 d\eta\right]\sqrt{Re_x}. \tag{17.69}$$

If we combine equation (17.67) with equation (17.69) we have

$$\frac{c_f}{2} = \frac{F''(0)}{\sqrt{\alpha Re_x}} = \frac{F''(0)\left[F\left(\eta_\delta\right) - \int_0^{\eta_\delta} f^2 d\eta\right]}{Re_\theta} \tag{17.70}$$

or

$$\frac{c_f}{2}Re_\theta = F''(0)\left[F\left(\eta_\delta\right) - \int_0^{\eta_\delta} f^2 d\eta\right] = C\left(m\right), \tag{17.71}$$

where the constant $C$ is a function of the exponent $m$, a result which is reminiscent of equation (16.18) for Poiseuille flow.

**Table 17.2** Results from the solution of Falkner and Skan's equation with $\alpha = 2$

| $\beta$ | $m$ | $(\delta^*/x)\sqrt{Re_x}$ | $(\theta/x)\sqrt{Re_x}$ | $\left(c_f/2\right)\sqrt{Re_x}$ | $\left(c_f/2\right)Re_\theta$ | |
|---|---|---|---|---|---|---|
| (rad) | | | | | | |
| $\pi$ | 1 | 0.6479 | 0.2923 | 1.2326 | 0.3603 | stagnation point |
| $\pi/2$ | 1/3 | 0.9854 | 0.4290 | 0.7575 | 0.3250 | |
| $\pi/5$ | 1/9 | 1.3204 | 0.5477 | 0.5118 | 0.2803 | |
| 0.5711 | 0.1 | 1.3478 | 0.5566 | 0.4966 | 0.2764 | |
| 0 | 0 | 1.7208 | 0.6641 | 0.3321 | 0.2205 | flat plate |
| −0.0635 | −0.01 | 1.7800 | 0.6789 | 0.3115 | 0.2115 | |
| $-\pi/10$ | −0.0476 | 2.0907 | 0.7464 | 0.2203 | 0.1644 | |
| −0.3307 | −0.05 | 2.1174 | 0.7515 | 0.2135 | 0.1600 | |
| $-3\pi/20$ | −0.06977 | 2.4149 | 0.7994 | 0.1475 | 0.1180 | |
| −0.6247 | −0.0904 | 3.4978 | 0.8681 | 0 | 0 | separtion |

Velocity profiles corresponding to Falkner and Skan's wedge-flow solutions, again with $\alpha = 2$, including that corresponding to Blasius' solution for $m = 0$, are shown in Figure 17.6 for $-0.0904 < m < 1$.

### 17.4.2 Comments on Falkner-Skan solutions

(a) For a wedge flow, where $U_\infty = Kx^m = Kx^{\beta/(2\pi-\beta)}$, we see that $dU_\infty/dx = mU_\infty/x = \beta U_\infty/[(2\pi - \beta)x]$ so that if $m > 0$ (and $\beta > 0$) the free-stream is accelerating and the static pressure is decreasing, i.e. the pressure gradient is negative. If $m < 0$ (and $\beta < 0$) the free-stream is decelerating, the static pressure is increasing, and the pressure gradient is positive.

(b) For a positive pressure gradient (i.e. $m < 0, \beta < 0$), we observe from Table 17.2 that $\left(c_f/2\right)\sqrt{Re_x}$ is lower than its value for a flat-plate boundary layer ($m = 0$). In fact, for $m = -0.09042854$ ($\beta = -0.62466699$), the skin-friction coefficient $c_f/2$, and so the wall-shear stress $\tau_S$, fall to zero, and the boundary layer is said to **separate** from the wall. A positive pressure gradient is thus said to be an **adverse pressure gradient**.

It has to be pointed out that, for boundary layers subjected to strong adverse pressure gradients, the boundary layer thickens rapidly, there is significant divergence of the near-wall streamlines, the normal velocity $v$ may become comparable with $u$, and the normal pressure gradient is no longer negligible, i.e. the boundary-layer approximation is no longer valid. A sketch illustrating near-wall streamlines in the

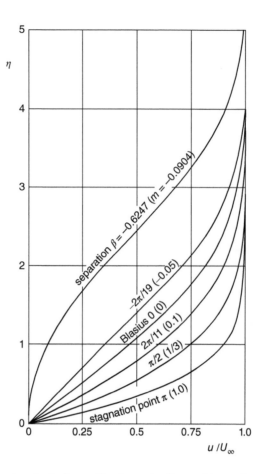

**Figure 17.6** Non-dimensional velocity profiles for wedge flows with $\pi \geq \beta \geq -0.6247$ (parameter on curves is $\beta$ followed by $m$ in parentheses) and $\alpha = 2$

vicinity of a separation point is shown in Figure 17.7. Included in the sketch are streamlines indicating reverse flow downstream of this point.

For a negative pressure gradient (i.e. $m > 0$, $\beta > 0$), Table 17.2 shows that $\left(c_f/2\right)\sqrt{Re_x}$ is always higher than its value for a flat-plate boundary layer. A negative pressure gradient is thus said to be a **favourable pressure gradient**.

In Chapter 14 we pointed out that the design of compressor blades is dictated by the adverse pressure gradients generated by the progressive pressure increase through a compressor, while turbine blades generally experience favourable pressure gradients.

(c) In Subsection 17.6.3 we show that an appropriate parameter[160] to quantify the pressure gradient is

$$\lambda_\theta = \frac{\theta^2}{\nu}\frac{dU_\infty}{dx} = -\frac{\theta^2}{\mu U_\infty}\frac{dp}{dx}. \tag{17.72}$$

[160] When spoken the parameter is 'lamda theta'.

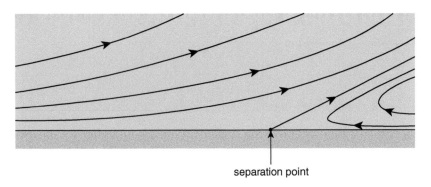

separation point

**Figure 17.7** Near-wall streamlines in the vicinity of a separation point

For the wedge flows we thus find

$$\lambda_\theta = m \left(\frac{\theta}{x}\right)^2 Re_x \qquad (17.73)$$

so that $\lambda_\theta > 0$ corresponds with a favourable pressure gradient, and $\lambda_\theta < 0$ with an adverse pressure gradient. Since $(\theta/x)\sqrt{Re_x}$ is tabulated in Table 17.2 as a function of $m$, $\lambda_\theta$ is readily obtained for the given values of $m$.

(d) The special case $\beta = \pi(m = 1)$ corresponds to flow at a stagnation point (Hiemenz flow). Since $(\theta/x)\sqrt{Re_x} = \theta\sqrt{Cx^{m-1}/\nu}$, we find the curious result that for flow near a stagnation point the momentum thickness $\theta$ is independent of $x$ and equal to the value at $x = 0$. The same is true for $\delta^*$, $\Delta$, and $\delta$, i.e. all the boundary-layer thicknesses are non-zero at the stagnation point and independent of distance from it.

(e) Another interesting case is that for $m = 1/3$, for which we see that

$$\frac{c_f}{2}\sqrt{Re_x} = \frac{\tau_S}{\rho U_\infty^2}\sqrt{\frac{U_\infty x}{\nu}} = \frac{\tau_S}{\rho}\sqrt{\frac{x}{U_\infty^3}}, \qquad (17.74)$$

i.e. the wall shear stress $\tau_S$ is independent of $x$ since $U_\infty \propto x^{1/3}$.

---

**ILLUSTRATIVE EXAMPLE 17.1**

(a) Show that, for a flow with $U_\infty$ proportional to $x^{1/3}$, the wall shear stress $\tau_S$ for a laminar boundary layer is constant.

(b) Calculate $\tau_S$ for an airflow for which

$$\frac{U_\infty}{U_0} = B\left(\frac{x}{L}\right)^{1/3}$$

where $B$ = constant, $U_0 L/\nu = 10^3$, $\rho U_0^2/2 = 240$ Pa, and, at $x/L = 10$, the free-stream velocity $U_\infty = 2U_0$.

For either (a) or (b) make use of any relevant information from the Falkner-Skan solutions.

## Solution

(a)  For any wedge flow, we found

$$\frac{c_f}{2}\sqrt{Re_x} = c \tag{17.67}$$

where $c$ is a constant. We thus have

$$\frac{\tau_S}{\rho U_\infty^2}\sqrt{\frac{U_\infty x}{\nu}} = c$$

and, with $U_\infty = Kx^m$, where $K$ is a constant,

$$\frac{\tau_S}{\rho K^2 x^{2m}}\sqrt{\frac{Kx^{m+1}}{\nu}} = c$$

which can be rearranged to give

$$\tau_S = \rho c\sqrt{K^3\nu}x^{(3m-1)/2}.$$

If $m = 1/3$, then $\tau_S = \rho c\sqrt{K^3\nu} = $ constant.

(b)  $U_0 L/\nu = 10^3$, $\rho U_0^2/2 = 240$ Pa, $x/L = 10$, $U_\infty = 2U_0$, and

$$\frac{U_\infty}{U_0} = B\left(\frac{x}{L}\right)^{1/3}$$

so that

$$B = \frac{2}{10^{1/3}} = 0.9283.$$

---

## 17.5  von Kármán's momentum-integral equation

We have seen that even for the most basic boundary layer, that which develops along a flat plate where we can reduce the partial differential equations governing the flow to a single ordinary differential equation, an analytical solution to the problem is not possible and numerical integration is necessary. By today's standards such calculations can be regarded as routine, and much the same can be said for the wedge-flow solutions. In fact commercial software is widely available which allows the solution of boundary-layer problems where variation of the free-stream velocity $U_\infty(x)$, or static pressure $p(x)$, can take any specified form. However, it is frequently the case that an approximate solution is either all that is needed or is helpful in interpreting a complete solution of high accuracy. An integrated form of the boundary-layer equations, known as **von Kármán's momentum-integral equation** (or just the **momentum-integral equation**[161]), is the starting point for many approximate solutions. The momentum-integral equation can be derived by formal integration of the boundary-layer form of the $x$-momentum equation (17.4) together with the continuity equation (17.1) or, by considering the forces acting on, and the momentum flowrates flowing through, a control volume of infinitesimal width $\delta x$ and height $\delta$. We shall adopt the second approach.

---

[161]  The momentum-integral equation is sometimes referred to as the **integral momentum equation**.

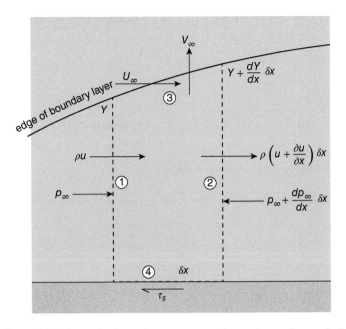

**Figure 17.8** Control volume for momentum-integral-equation analysis

The control volume is shown in Figure 17.8. In this section, to avoid confusion with small quantities such as $\delta x$, the symbol $Y$ rather than $\delta$ will be used to represent the boundary-layer thickness. The width $\delta x$, in the streamwise direction, is greatly exaggerated for clarity. We shall assume the surface width is $Z$ but that the flow is two-dimensional with zero velocity in the $z$-direction. There is flow of mass and $x$-momentum into the control volume across face ①, outflow across face ②, and zero flow across face ④ since the surface is assumed to be impermeable. Across face ③, which corresponds to the edge of the boundary layer, flow of mass and $x$-momentum associated with both the free-stream velocity $U_\infty(x)$ and the $y$-component of velocity $V_\infty(x)$ has to be accounted for. So far as forces are concerned, there are pressure forces acting on faces ①, ②, and ③ due to the static pressure $p(x)$, which is assumed to be constant across the boundary layer, i.e. $\partial p/\partial y = 0$, and a shear force acting on face ④ due to the surface shear stress $\tau_S(x)$.

### 17.5.1 Mass-conservation equation

We consider first the mass flowrate $\dot{m}_Y$ within the boundary layer, which is given by

$$\dot{m}_Y = \int_0^Y \rho\, u\, dy\, Z. \tag{17.75}$$

For steady flow, the mass flowrates into and out of the control volume must be in balance, so that we can write the **mass-conservation equation** as

$$\left[ \dot{m}_Y + \left( \frac{d\dot{m}_Y}{dx} \right) \delta x \right] - \dot{m}_Y + \rho_\infty V_\infty \delta x Z - \rho_\infty U_\infty \frac{dY}{dx} \delta x Z = 0. \tag{17.76}$$
$$\quad\;\;② \qquad\qquad ① \qquad\quad ③ \qquad\qquad ③$$

The encircled numerals below each term in equation (17.76) indicate the face of the control volume to which the term corresponds. After substitution for $\dot{m}_Y$ from equation (17.75) and some rearrangement we find

$$\frac{d}{dx}\int_0^Y \rho u\, dy = \rho_\infty U_\infty \frac{dY}{dx} - \rho_\infty V_\infty \tag{17.77}$$

which will be used later to eliminate $V_\infty$ from the momentum equation.

### 17.5.2 x-Momentum-conservation equation

The basic equation is

> *net force acting on the control volume in the x-direction*
> *= x-momentum flowrate leaving control volume*
> *−x-momentum flowrate entering control volume* (17.78)

from which we have

$$pYZ - \left[ pYZ + \frac{d}{dx}(pYZ)\,\delta x \right] + \bar{p}\frac{dY}{dx}\delta xZ - \tau_S \delta xZ \tag{17.79}$$
$$\quad ① \qquad\qquad ② \qquad\qquad\qquad ③ \qquad\quad ④$$

$$= \left[ \int_0^Y \rho u^2\, dyZ + \frac{d}{dx}\int_0^Y \rho u^2\, dyZ\delta x \right] - \int_0^Y \rho u^2\, dyZ + \rho_\infty V_\infty U_\infty \delta xZ - \rho_\infty U_\infty^2 \frac{dY}{dx}\delta xZ.$$
$$\qquad\qquad ② \qquad\qquad\qquad\qquad\qquad\qquad ① \qquad\qquad ③ \qquad\qquad ③$$

The encircled numerals below each term in equation (17.79) indicate the face of the control volume to which the term corresponds. Most of the terms in this equation should require no explanation. In evaluating the force exerted on face ③ by the static pressure, we have taken a mean static pressure $\bar{p}$ while $(dY/dx)\,\delta x$ represents the projection of face ③ onto a plane normal to the surface. The terms involving $U_\infty$ take into account the flow of fluid across face ③ with a component of velocity $U_\infty$ in the x-direction and a mass flowrate due to velocity components in both the x- and y-directions.

A number of terms in equation (17.79) cancel each other out and, after some simplification, we find

$$\tau_S + Y\frac{dp}{dx} + (p - \bar{p})\frac{dY}{dx} = -\frac{d}{dx}\int_0^Y \rho u^2\, dy - \rho U_\infty \left( V_\infty - U_\infty \frac{dY}{dx} \right). \tag{17.80}$$

For the mean static pressure $\bar{p}$ we can write

$$\bar{p} = \frac{1}{2}\left[ p(x) + p(x + \delta x) \right]$$

so that

$$p - \bar{p} = \frac{1}{2}\left[ p(x + \delta x) - p(x) \right] \approx \frac{1}{2}\frac{dp}{dx}\delta x + O(\delta x^2).$$

Since the control volume is infinitesimally thin, we can neglect terms of second order and higher in this expression for $p - \bar{p}$. If we now substitute for $p - \bar{p}$ in equation (17.80), we have

$$\tau_S + Y\frac{dp}{dx} + \frac{1}{2}\frac{dp}{dx}\frac{dY}{dx}\delta x = -\frac{d}{dx}\int_0^Y \rho u^2 dy - \rho_\infty U_\infty \left( V_\infty - U_\infty \frac{dY}{dx}\right) \tag{17.81}$$

and we see that as $\delta x \to 0$ the third term must disappear leaving

$$\tau_S + Y\frac{dp}{dx} = -\frac{d}{dx}\int_0^Y \rho u^2 dy - \rho_\infty U_\infty \left( V_\infty - U_\infty \frac{dY}{dx}\right). \tag{17.82}$$

Substitution from equation (17.77) for the parenthetical term in equation (17.82), which includes $V_\infty$, leads to

$$\tau_S + Y\frac{dp}{dx} = -\frac{d}{dx}\int_0^Y \rho u^2\, dy + U_\infty \frac{d}{dx}\int_0^Y \rho u dy. \tag{17.83}$$

We can now introduce Bernoulli's equation in the form

$$\frac{dp}{dx} + \rho_\infty U_\infty \frac{dU_\infty}{dx} = 0$$

so that, after some rearrangement, equation (17.83) may be written as

$$\tau_S = U_\infty \frac{d}{dx}\int_0^Y \rho u dy - \frac{d}{dx}\int_0^Y \rho u^2\, dy + \rho_\infty U_\infty \frac{dU_\infty}{dx}\int_0^Y dy. \tag{17.84}$$

The final term on the right-hand side of equation (17.84) is arrived at by observing that

$$Y = \int_0^Y dy.$$

### 17.5.3 von Kármán's momentum-integral equation

So far we have put no restriction on the density $\rho$ so that equation (17.84) is valid for both compressible and incompressible flow. If we now limit our attention to constant-density flow and introduce the displacement thickness $\delta^*$ and the momentum thickness $\theta$, which were defined in Subsection 17.3.2, equation (17.84) leads to

$$\frac{\tau_S}{\rho} = \frac{d\left(\theta U_\infty^2\right)}{dx} + \delta^* U_\infty \frac{dU_\infty}{dx} \tag{17.85}$$

which is one of a number of forms of **von Kármán's momentum-integral equation**. In terms of the skin-friction coefficient $c_f/2$ defined by equation (17.38), we can write equation (17.85) as

$$\frac{c_f}{2} = \frac{d\theta}{dx} + \left(\delta^* + 2\theta\right)\frac{1}{U_\infty}\frac{dU_\infty}{dx} \tag{17.86}$$

or

$$\frac{c_f}{2} = \frac{d\theta}{dx} + (H + 2)\frac{\theta}{U_\infty}\frac{dU_\infty}{dx} \tag{17.87}$$

where

$$H = \frac{\delta^*}{\theta} \tag{17.88}$$

defines a non-dimensional quantity known as the **boundary-layer shape factor**[162], or simply the **shape factor**. As the name suggests, the value of $H$ is closely related to the shape of the velocity profile. The definitions of $\delta^*$ and $\theta$ are given by equations (17.44) and (17.46), respectively

$$\delta^* = \int_0^\delta \left(1 - \frac{u}{U_\infty}\right) dy \tag{17.44}$$

$$\theta = \int_0^\delta \frac{u}{U_\infty}\left(1 - \frac{u}{U_\infty}\right) dy. \tag{17.46}$$

### 17.5.4 Comments on the momentum-integral equation

(a) The arbitrariness associated with the boundary-layer thickness $Y (= \delta)$ is removed once $\delta^*$ and $\theta$ are introduced.

(b) For a flat-plate boundary layer, equation (17.87) simplifies to

$$\frac{c_f}{2} = \frac{d\theta}{dx} \tag{17.89}$$

which, if $\theta(0) = 0$, can be integrated with respect to $x$ to give

$$\theta(x) = \int_0^L \frac{c_f}{2} dx = \frac{1}{\rho U_\infty^2} \int_0^L \tau_S \, dx$$

so that the total drag force per unit width $D'$ exerted on a length $L$ of the surface is given by

$$D' = \rho U_\infty^2 \theta(L), \tag{17.90}$$

i.e. for the flat-plate boundary layer the momentum thickness at $x = L$ is a direct measure of the drag force over the region $0 < x < L$.

(c) For a constant-density boundary layer, equation (17.77) can be rearranged as an equation for the $y$-direction velocity at the edge of the boundary layer, $V_\infty$

$$V_\infty = U_\infty \frac{dY}{dx} - \frac{d}{dx}\int_0^Y u \, dy \tag{17.91}$$

which can be written in terms of the displacement thickness $\delta^*$ as

$$V_\infty = \frac{d(U_\infty \delta^*)}{dx} - Y\frac{dU_\infty}{dx}. \tag{17.92}$$

For a flat-plate boundary layer, equation (17.92) simplifies to

$$V_\infty = U_\infty \frac{d\delta^*}{dx} = \frac{U_\infty}{2}\frac{\delta^*}{x} = \frac{0.8604}{\sqrt{Re_x}} \tag{17.93}$$

wherein we have used the result for $\delta^*$ in Table 17.2 for $m = 0$. Equation (17.93) leads to the same conclusion regarding $V_\infty$ as equation (17.36), which we commented upon in Subsection 17.3.2 following the analysis of Blasius' equation.

---

[162] When $\delta_1$ and $\delta_2$ are used instead of $\delta^*$ and $\theta$, it is usual for the shape factor to be represented by the symbol $H_{12}$ rather than $H$. Also, the term **form factor** is sometimes used instead of shape factor.

## 17.6 Profile methods of solution

We saw in Section 17.3 that to arrive at practically useful numerical results requires significant effort, including the numerical solution of a non-linear ordinary differential equation, even for flow over a flat plate. We now show how results of acceptable accuracy can be obtained with far less effort if we make reasonable assumptions concerning the shape of the velocity profile.

Any assumed velocity profile should satisfy as many of the boundary conditions stated at the end of Section 17.2 as possible

$$\text{at the surface: no} - \text{slip condition}: y = 0, \ u = 0, \ \left.\frac{\partial^2 u}{\partial y^2}\right|_S = -\frac{U_\infty}{\nu}\frac{dU_\infty}{dx} \tag{17.94}$$

$$\text{approach to the free-stream velocity}: y \to \infty, \ u \to U_\infty, \ \frac{\partial^n u}{\partial y^n} \to 0,$$

$$n = 1, 2, 3, \ldots .. \tag{17.95}$$

At the surface the no-slip condition is a fundamental requirement and it should be clear that the first derivative, $\partial u/\partial y|_S = \tau_S/\mu$, is the main quantity of engineering interest and so cannot be assumed. Equation (17.94) for the second derivative at the surface arises directly from the $x$-momentum equation with $u = v = 0$. Nothing can be said about higher derivatives at the surface whereas, ideally, all derivatives should become zero with approach to the free stream.

### 17.6.1 Flat-plate laminar boundary layer with cubic velocity profile

We assume that the velocity profile has the cubic-polynomial form

$$u = a + by + cy^2 + dy^3 \tag{17.96}$$

where $a, b, c,$ and $d$ are constants to be determined from the boundary conditions.

It will be found that the algebra is simplified if we use non-dimensional variables, defined by

$$f = \frac{u}{U_\infty} \quad \text{and} \quad \xi = \frac{y}{\delta} \tag{17.97}$$

so that from equations (17.94) and (17.95) the boundary conditions to be satisfied are

$$\xi = 0, \quad f = 0, \quad \left.\frac{d^2 f}{d\xi^2}\right|_S = -\frac{\delta^2}{\nu}\frac{dU_\infty}{dx} = -\lambda \tag{17.98}$$

and

$$\xi \to 1, \quad f \to 1, \quad \frac{d^n f}{d\xi^n} \to 0. \tag{17.99}$$

Note that we have reverted to using the symbol $\delta$ rather than $Y$ to represent the boundary-layer thickness. Equation (17.98) defines the non-dimensional quantity $\lambda$, which is seen to be a **pressure-gradient parameter** since, from Bernoulli's equation, $p + \rho U_\infty^2/2 = $ constant. For the flat-plate boundary layer, $\lambda = 0$.

In non-dimensional form, the equation for the velocity profile is

$$f = A + B\xi + C\xi^2 + D\xi^3 \tag{17.100}$$

where $A, B, C$, and $D$ have replaced $a, b, c$, and $d$ as the constants to be determined from the boundary conditions.

The no-slip condition requires that $A = 0$, and the free-stream boundary condition (at $\xi = 1$) for $f$ then requires

$$1 = B + C + D. \tag{17.101}$$

If we differentiate equation (17.100) with respect to $\xi$ we have

$$\frac{df}{d\xi} = B + 2C\xi + 3D\xi^2 \tag{17.102}$$

and the free-stream boundary condition for $df/d\xi$ requires

$$0 = B + 2C + 3D. \tag{17.103}$$

If we differentiate equation (17.100) again, we have

$$\frac{d^2f}{d\xi^2} = 2C + 6D\xi. \tag{17.104}$$

For the flat-plate boundary layer $\lambda = 0$ and from equation (17.98) $d^2f/d\xi^2|_S = 0$ so that $C = 0$. From equations (17.101) and (17.103), the remaining constants are found to be $B = 3/2$, and $D = -1/2$, and the velocity-profile equation becomes

$$f = \tfrac{1}{2}\xi\left(3 - \xi^2\right) \tag{17.105}$$

or

$$\frac{u}{U_\infty} = \frac{1}{2}\frac{y}{\delta}\left[3 - \left(\frac{y}{\delta}\right)^2\right]. \tag{17.106}$$

We can now use equation (17.105) to evaluate $\delta^*, \theta, H$, and $c_f/2$ as follows

$$\frac{\delta^*}{\delta} = \int_0^1 \left(1 - f\right)d\xi = \frac{3}{8}, \tag{17.107}$$

$$\frac{\theta}{\delta} = \int_0^1 f\left(1 - f\right)d\xi = \frac{39}{280}, \tag{17.108}$$

$$H = \frac{\delta^*}{\theta} = \frac{35}{13} = 2.69, \tag{17.109}$$

and

$$\frac{c_f}{2} = \frac{\tau_S}{\rho U_\infty^2} = \frac{\mu\frac{\partial u}{\partial y}\big|_S}{\rho U_\infty^2} = \frac{\nu}{U_\infty\delta}\frac{df}{d\xi}\bigg|_0 = \frac{3\nu}{2U_\infty\delta}. \tag{17.110}$$

Since the boundary-layer thickness $\delta$ has yet to be quantified, at this stage only the value for the shape factor $H$ can be compared with the exact value from the numerical solution of Blasius' equation: 2.69 compared with 2.59 calculated from the values listed in Table 17.2.

To find an equation for the boundary-layer thickness $\delta(x)$ we use the momentum-integral equation which, for the flat plate, is given by equation (17.89)

$$\frac{d\theta}{dx} = \frac{c_f}{2}.$$

Substitution for $c_f/2$ from equation (17.110) and for $\theta/\delta$ from equation (17.108) leads to

$$\frac{39}{280}\frac{d\delta}{dx} = \frac{3v}{2U_\infty\delta} \tag{17.111}$$

which can be rearranged as

$$\delta\frac{d\delta}{dx} = \frac{140v}{13U_\infty}. \tag{17.112}$$

If we assume the boundary layer has zero thickness at $x = 0$, i.e. $\delta(0) = 0$, after integration of equation (17.112) we have

$$\delta = \sqrt{\frac{280vx}{13U_\infty}} \tag{17.113}$$

so that

$$\frac{c_f}{2} = \frac{3v}{2U_\infty\delta} = \sqrt{\frac{117}{1120Re_x}} = \frac{0.323}{\sqrt{Re_x}} \tag{17.114}$$

which is within 3% of the value for $m = 0$ in Table 17.2

$$\frac{c_f}{2} = \frac{0.331}{\sqrt{Re_x}}. \tag{17.115}$$

The final comparison is with the result for the momentum thickness:

$$\theta = \frac{39}{280}\delta = \frac{39}{280}\sqrt{\frac{280vx}{13U_\infty}} \tag{17.116}$$

or

$$\frac{\theta}{x}\sqrt{Re_x} = 0.646 \tag{17.117}$$

while the value in the table is 0.664, again within 3%.

In Table 17.3 we compare with the results of Blasius' analysis values for $(\delta^*/x)$ $\sqrt{Re_x}$, $(\theta/x)\sqrt{Re_x}$, $H$, and $(c_f/2)\sqrt{Re_x}$ for a zero-pressure-gradient boundary layer determined from linear, quadratic[163], cubic, and quartic profiles, plus a sine-form profile. It is noticeable that the results are relatively insensitive to the profile assumed, even $f = \xi$ leading to a value for $c_f/2$ within 13% of the exact (Blasius) value.

---

[163] As shown in Section 16.4, the quadratic profile corresponds to Couette-Poiseuille flow with $\lambda_P = 2$.

**Table 17.3** Flat-plate boundary-layer values for various velocity profiles

| Profile | $(\delta^*/x)\sqrt{Re_x}$ | $(\theta/x)\sqrt{Re_x}$ | $H$ | $\left(c_f/2\right)\sqrt{Re_x}$ |
|---|---|---|---|---|
| $f = \xi$ | 1.732 | 0.577 | 3.000 | 0.289 |
| $f = 2\xi - \xi^2$ | 1.826 | 0.730 | 2.501 | 0.365 |
| $f = 3\left(\xi - \xi^3\right)/2$ | 1.740 | 0.646 | 2.700 | 0.323 |
| $f = 2\xi - 2\xi^3 + \xi^4$ | 1.752 | 0.686 | 2.550 | 0.343 |
| $f = \sin\left(\pi\xi/2\right)$ | 1.741 | 0.654 | 2.660 | 0.327 |
| Blasius | 1.721 | 0.664 | 2.590 | 0.332 |

## 17.6.2 Pohlhausen's quartic velocity profile for boundary layers with $dp/dx \neq 0$

For flows with a streamwise pressure gradient $dp/dx$, a quartic equation has been found to represent a wide range of velocity profiles for both accelerating (i.e. favourable-pressure-gradient) and decelerating (adverse-pressure-gradient) flows

$$f = A + B\xi + C\xi^2 + D\xi^3 + E\xi^4. \tag{17.118}$$

From equation (17.94), the boundary condition for $d^2f/d\xi^2$ at the surface, $\xi = 0$, is

$$\left.\frac{d^2f}{d\xi^2}\right|_S = -\frac{\delta^2}{\nu}\frac{dU_\infty}{dx} = -\lambda,$$

$\lambda$ being the pressure-gradient parameter (now non-zero) defined by equation (17.98), while the other boundary conditions remain the same

$$\xi \to 1, \quad \frac{d^2f}{d\xi^2} \to 0, f \to 1 \tag{17.119}$$

As in Subsection 17.6.1, we can determine the constants $A, B, C,$ and $D$, with the result now that the velocity-profile equation (17.118) is given by

$$f = 2\xi - 2\xi^3 + \xi^4 + \frac{\lambda}{6}\xi\left(1 - \xi\right)^3. \tag{17.120}$$

This is known as **Pohlhausen's equation**, and the pressure-gradient parameter $\lambda$ as **Pohlhausen's parameter**.

The wall shear stress $\tau_S$ is given by

$$\tau_S = \mu\left.\frac{\partial u}{\partial y}\right|_0 = \frac{\mu U_\infty}{\delta}\left.\frac{df}{d\xi}\right|_0 = \frac{\mu U_\infty}{\delta}\left(2 + \frac{\lambda}{6}\right) \tag{17.121}$$

so that the skin-friction coefficient $c_f/2$ is given by

$$\frac{c_f}{2} = \frac{\tau_S}{\rho U_\infty^2} = \left(2 + \frac{\lambda}{6}\right)\frac{\nu}{U_\infty\delta}. \tag{17.122}$$

As for the flat-plate situation, we can derive expressions for $\delta^*/\delta$ and $\theta/\delta$ by substituting the velocity-profile equation (121) for $f(\xi)$ in the definitions of $\delta^*$ and $\theta$ (equations (17.44) and (17.46)). The results are

$$\frac{\delta^*}{\delta} = \frac{1}{10}\left(3 - \frac{\lambda}{12}\right) = \tilde{\Phi}\,(\lambda) \tag{17.123}$$

and

$$\frac{\theta}{\delta} = \frac{1}{63}\left(\frac{37}{5} - \frac{\lambda}{15} - \frac{\lambda^2}{144}\right) = \tilde{\Theta}\,(\lambda). \tag{17.124}$$

Equations (17.123) and (17.124) define the two non-dimensional quantities[164], $\tilde{\Phi}$ and $\tilde{\Theta}$, both of which depend only upon $\lambda$. From the definition of the shape factor $H$ we have

$$H = \frac{\delta^*}{\theta} = \frac{\tilde{\Phi}}{\tilde{\Theta}} = \frac{63}{10}\frac{\left(3 - \frac{\lambda}{12}\right)}{\left(\frac{37}{5} - \frac{\lambda}{15} - \frac{\lambda^2}{144}\right)}. \tag{17.125}$$

Equations (17.122), (17.123), (17.124), and (17.125) can be substituted into the momentum-integral equation (17.87) to give the following ordinary differential equation for the momentum thickness $\theta$

$$\frac{U_\infty\theta}{\nu}\frac{d\theta}{dx} = \left(2 + \frac{\lambda}{6}\right)\tilde{\Theta} - (2+H)\tilde{\Theta}^2\lambda = \frac{1}{2}F_1\,(\lambda). \tag{17.126}$$

Equation (17.126) defines the function $F_1(\lambda)$, which will be needed later. The quantity $\tilde{\Theta}^2\lambda$ provides the connection between $\theta$ and $\lambda$ through the identity

$$\frac{\theta^2}{\nu}\frac{dU_\infty}{dx} = \tilde{\Theta}^2\frac{\delta^2}{\nu}\frac{dU_\infty}{dx} = \tilde{\Theta}^2\lambda \tag{17.127}$$

so that equations (17.126) and (17.127) can be combined to give

$$\frac{d}{dx}\left(\frac{U_\infty\theta^2}{\nu}\right) = \left(2 + \frac{\lambda}{6}\right)\tilde{\Theta} - (1+H)\tilde{\Theta}^2\lambda. \tag{17.128}$$

Unfortunately, once again, we have arrived at a differential equation that can be solved only numerically, except for the trivial (flat-plate) case of $\lambda = 0$. A simplifying feature of equations (17.126), (17.127), and (17.128) is that the right-hand side of each depends only upon the Pohlhausen pressure-gradient parameter $\lambda$.

Given $U_\infty(x)$ and its gradient, a step-by-step numerical procedure generates a value for $\theta$ at each step from which $\lambda$ can be calculated from equations (17.124) and (17.127) (although this involves solving a quintic equation), and all other quantities $(\delta, \delta^*, H, c_f/2,$ etc.) then follow.

The initial conditions for the numerical calculation can be taken either as those for a thin flat plate at zero incidence or, for a shape such as an aerofoil, those for a stagnation point. For a thin flat plate, at $x = 0$, we have $\delta = 0, \theta = 0, \delta^* = 0$, and $\lambda = 0$. In the case of a stagnation point, $U_\infty(0) = 0$, so that according to equation (17.126), for $d\theta/dx$ to remain finite, the right-hand side must also be zero, i.e.

$$2 + \frac{\lambda_0}{6} - (2 + H_0)\,\tilde{\Theta}_0\lambda_0 = 0, \tag{17.129}$$

[164] The curly overbar $\tilde{\ }$ is used here to indicate that $\tilde{\Phi}$ and $\tilde{\Theta}$ are non-dimensional quantities.

the subscript 0 indicating values at the stagnation point, $x = x_0$. After substitution for $\tilde{\Phi}_0$ and $\tilde{\Theta}_0$ from equations (17.123) and (17.124), equation (17.129) can be written in terms of $\lambda_0$ as

$$2 - \frac{116}{315}\lambda_0 + \frac{79}{7560}\lambda_0^2 + \frac{1}{7560}\lambda_0^3 = 0$$

the solution of which is $\lambda_0 = 7.052$. From equations (17.123), (17.124), and (17.125) we then have

$$\tilde{\Theta}_0 = 0.1045, \quad \tilde{\Phi}_0 = 0.2412, \quad H_0 = 2.308$$

and at $x = x_0$, from equation (17.128),

$$\frac{d}{dx}\left(\frac{U_\infty \theta^2}{\nu}\right) = -\tilde{\Theta}_0^2 \lambda_0 = 0.0737 \tag{17.130}$$

which is used to start the numerical integration. In an adverse pressure gradient the calculation cannot proceed beyond the point at which $\lambda = -12$, where from equation (17.122) we see that boundary-layer separation (i.e. $\tau_S = 0$) is predicted to occur. More exact solutions of the laminar boundary-layer equations indicate that separation is found to occur for a $\lambda$-value about 40% lower. We also note that at separation ($\lambda = -12$) equation (17.125) gives $H = 3.5$ compared with $H = 3.95$ from Table 17.2 for the wedge-flow solutions.

A starting condition can also be arrived at through consideration of Falkner and Skan's wedge-flow solutions which showed that, for any value of the exponent $m$

$$\frac{\theta}{x}\sqrt{Re_x} = A(m), \tag{17.131}$$

values of $(\theta/x)\sqrt{Re_x}$ being listed in Table 17.2. From equation (17.131),

$$\frac{U_\infty \theta^2}{\nu} = A^2 x$$

so that

$$\frac{d}{dx}\left(\frac{U_\infty \theta^2}{\nu}\right) = A^2$$

which can be used to initiate the numerical integration if a suitable value can be ascribed to $m$ and hence $A$. For a flat-plate boundary layer, $m = 0$ and, from Table 17.2, $A = 0.6641$. For a stagnation point, $m = 1$, and $A = 0.2923$.

Although numerical integration of equation (17.128) is straightforward, it should not be forgotten that the equation is not exact but an approximation based upon Pohlhausen's quartic velocity profile. We have already seen that for a flat-plate boundary layer there are small but significant differences between the exact results (from Blasius' equation) and those for various assumed profiles, including the quartic. Also, as we have already indicated, Pohlhausen's profile leads to a $-\lambda$ value at separation about 40% too high. It turns out that a simpler approximate analytical procedure, known as **Thwaites' method**, which is based upon the wedge-flow solutions produces results comparable in accuracy to Pohlhausen's method.

### 17.6.3 Thwaites' method of solution

From equation (17.126) we have for Pohlhausen's profile

$$\frac{U_\infty}{\nu} \frac{d\theta^2}{dx} = F_1(\lambda) \tag{17.132}$$

while, from the definitions of $\lambda$,

$$\frac{\delta^2}{\nu} \frac{dU_\infty}{dx} = \lambda \tag{17.98}$$

and $\tilde{\Theta}$,

$$\tilde{\Theta} = \frac{1}{63}\left(\frac{37}{5} - \frac{\lambda}{15} - \frac{\lambda^2}{144}\right), \tag{17.124}$$

we see that

$$\frac{\theta^2}{\nu} \frac{dU_\infty}{dx} = \Theta^2 \lambda = \left[\frac{1}{63}\left(\frac{37}{5} - \frac{\lambda}{15} - \frac{\lambda^2}{144}\right)\right]^2 \lambda = \lambda_\theta(\lambda). \tag{17.133}$$

The pressure-gradient parameter $\lambda_\theta$ was introduced in Subsection 17.4.2 in the comments on the solutions of Falkner and Skan's equation and has the advantage over $\lambda$ that $\theta$, unlike $\delta$, can be calculated exactly. From equations (17.132) and (17.133) we can conclude that, for Pohlhausen's profile,

$$\frac{U_\infty}{\nu} \frac{d\theta^2}{dx} = F_\theta(\lambda_\theta). \tag{17.134}$$

From a range of exact solutions, numerical computations, and the wedge-flow solutions, it is found that equation (17.134) is well represented by the linear relation

$$\frac{U_\infty}{\nu} \frac{d\theta^2}{dx} = F_\theta = 0.45 - 6\lambda_\theta. \tag{17.135}$$

It is important to realise that although equation (17.134) was derived from Pohlhausen's profile (it can be shown to have wider validity) equation (17.135) is an approximate relation which can be used to calculate $\theta(x)$ for any variation of free-stream velocity $U_\infty(x)$.

If we multiply through the momentum-integral equation (17.85) by $U_\infty \theta / \nu$ and rearrange the terms, we find

$$F_\theta = 2T - 2(2 + H)\lambda_\theta \tag{17.136}$$

where $T$ is a skin-friction parameter defined by

$$T = \frac{\theta \tau_S}{\mu U_\infty} = \frac{U_\infty \theta}{\nu} \frac{c_f}{2}. \tag{17.137}$$

Although equation (17.136) bears a superficial resemblance to equation (17.135), neither $T$ nor $H$ is a constant but both are functions of $\lambda_\theta$. Once $\theta(x)$ is known, $\lambda_\theta(x)$ is easily calculated and $\tau_S(x)$ is then found from either a table of values of $T$ vs $\lambda_\theta$ or an empirical correlation such as

$$T = (0.09 + \lambda_\theta)^{0.62}. \tag{17.138}$$

Equation (17.138) shows that the calculation procedure will indicate boundary-layer separation and $\tau_S = 0$ for $\lambda_\theta = -0.09$.

For the wedge-flow solutions, for which $U_\infty = Cx^m$, we found

$$\frac{\theta}{x}\sqrt{Re_x} = \text{constant} = A(m) \tag{17.139}$$

so that

$$F_\theta = \frac{U_\infty}{\nu}\frac{d\theta^2}{dx} = (1-m)A^2 \tag{17.140}$$

and

$$\lambda_\theta = \frac{\theta^2}{\nu}\frac{dU_\infty}{dx} = mA^2. \tag{17.141}$$

Figure 17.9 shows a plot of $F_\theta$ versus $\lambda_\theta$, corresponding to equation (17.135), together with points representing $\lambda_\theta$ and $F_\theta$ calculated from values of $A$ and $m$ taken from Table 17.2, using equation (17.141) for $\lambda_\theta$, and equation (17.140) for $F_\theta$. Equation (17.135) is clearly not the best possible fit to the wedge-flow data but, as already stated, it was based on a range of exact solutions to the boundary-layer equations.

If we rearrange equation (17.135) and multiply through by $\nu U_\infty^5$ we have

$$U_\infty^6 \frac{d\theta^2}{dx} + 6\theta^2 U_\infty^5 \frac{dU_\infty}{dx} = \frac{d}{dx}\left(U_\infty^6 \theta^2\right) = 0.45\nu U_\infty^5$$

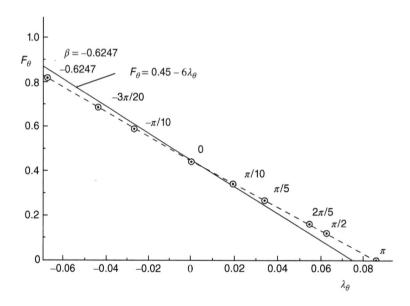

**Figure 17.9** $F_\theta$ versus $\lambda_\theta$ for Thwaites' correlation and for wedge-flow solutions ($\beta$-values shown for each data point)

which can be integrated to give

$$U_\infty^6 \theta^2 - U_{\infty,0}^6 \theta_0^2 = 0.45 \nu \int_0^x U_\infty^5 dx. \qquad (17.142)$$

In equation (17.142), which is referred to as **Thwaites' formula**, $U_{\infty,0}$ is the value of $U_\infty$ at $x = 0$, and $\theta_0$ is the corresponding value of $\theta$.

Numerical evaluation of the integral in Thwaites' formula, known as **quadrature**, is necessary for all but the simplest variations of $U_\infty$. Once $\theta(x)$ has been calculated, $c_f(x)$ can be determined from $c_f/2 = \nu T / (U_\infty \theta)$, with $T$ calculated from equation (17.138).

Equations (17.140) and (17.141) show that, for a wedge flow, $F$ and $\lambda_\theta$ are both functions of $m$. It is easily shown that $T = [(c_f/2) \sqrt{Re_x}] [(\theta/x) \sqrt{Re_x}]$, which must also be a function of $m$, as must $H$. All terms in equation (17.136) are therefore functions of $m$ and so can also be regarded as functions of the pressure-gradient parameter $\lambda_\theta$. A possible interpretation of Thwaites' method is that the velocity profile at any $x$-location within a laminar boundary layer is assumed to have the similarity form, according to Falkner and Skan's equation, corresponding with the $\lambda_\theta$-value at that location.

---

**ILLUSTRATIVE EXAMPLE 17.2**

(a) The free-stream velocity $U_\infty$ for flow around a circular cylinder in crossflow is given by $U_\infty = 2U_0 \sin \phi$, where $U_0$ is the flow velocity far upstream of the cylinder, and the angle $\phi$ is measured from the forward stagnation point. Near the forward stagnation point, the variation of $U_\infty$ can be approximated by $U_\infty = 2U_0\phi$, where $\phi$ is measured in radians. Use Thwaites' formula, equation (17.137), to show that, in the vicinity of the stagnation point,

$$\theta \approx \sqrt{\frac{3\nu R}{80 U_0}}$$

where $R$ is the cylinder radius.

(b) Again using Thwaites' formula, show that, for all values of $\phi$ up to boundary-layer separation,

$$\theta = \frac{1}{\sin^3 \phi} \sqrt{\frac{0.225 \nu R}{U_0} \left( \frac{8}{15} - \frac{5}{8} \cos\phi + \frac{5}{48} \cos3\phi - \frac{1}{80} \cos5\phi \right)}.$$

You will need to make use of the relation

$$\int \sin^5 \phi \, d\phi - \frac{5}{8} \cos\phi + \frac{5}{48} \cos3\phi - \frac{1}{80} \cos5\phi.$$

(c) Use the result of part (b) to find a formula for Thwaites' pressure-gradient parameter $\lambda_\theta$ in terms of $\phi$ and calculate values of $\lambda_\theta$ for $\phi = 30°, 90°, 100°$, and $103.11°$. From these values of $\lambda_\theta$ calculate the shear-stress parameter $T$ at the same values of $\phi$ from the formula

$$T = (0.09 + \lambda_\theta)^{0.62}.$$

Comment on the results.

(d) If $U_0 = 0.5$ m/s, calculate $\theta$, the momentum-thickness Reynolds number, and the wall shear stress at $\phi = 30°$, $90°$, and $100°$ for airflow ($\nu = 1.5 \times 10^{-5}$ m$^2$/s, $\rho = 1.2$ kg/m$^3$) over a cylinder 20 mm in diameter.

## Solution

(a) In the present situation, $U_\infty(0) = 0$ so that Thwaites' formula reduces to

$$U_\infty^6 \theta^2 = 0.45\nu \int_0^x U_\infty^5 dx.$$

The surface distance $x$ from the forward stagnation point is given by $x = \phi R$ and, for small values of $\phi$, the free-stream velocity is given by $U_\infty = 2U_0\phi$. If we substitute for $U_\infty$ and $x$ in Thwaites' formula we have

$$2U_0\phi^6\theta^2 = 0.45\nu R \int_0^\phi \phi^5 d\phi$$

which we can integrate to find

$$2U_0\phi^6\theta^2 = \frac{0.45\nu R\phi^6}{6} \quad or \quad \theta^2 = \frac{3\nu R}{80U_0}.$$

(b) For any angle $\phi$ we have

$$2U_0 \sin^6 \phi \theta^2 = 0.45\nu R \int_0^\phi \sin^5 \phi d\phi$$

$$= 0.45\nu R \left[ \left( -\frac{5}{8}\cos\phi + \frac{5}{48}\cos3\phi - \frac{1}{80}\cos5\phi \right) - \left( -\frac{5}{8} - \frac{5}{48} + \frac{1}{80} \right) \right]$$

so that

$$\theta^2 = \frac{0.225\nu R}{U_0 \sin^6 \phi} \left( \frac{8}{15} - \frac{5}{8}\cos\phi + \frac{5}{48}\cos3\phi - \frac{1}{80}\cos5\phi \right).$$

(c) From the definition of $\lambda_\theta$

$$\lambda_\theta = \frac{\theta^2}{\nu}\frac{dU_\infty}{dx} = \frac{0.225R}{U_0 \sin^6 \phi}\left( \frac{8}{15} - \frac{5}{8}\cos\phi + \frac{5}{48}\cos3\phi - \frac{1}{80}\cos5\phi \right)\frac{dU_\infty}{dx}$$

and, from $U_\infty = 2U_0 \sin \phi$,

$$\frac{dU_\infty}{dx} = \frac{dU_\infty}{d\phi}\frac{d\phi}{dx} = \frac{2U_0 \cos\phi}{R}$$

so that

$$\lambda_\theta = \frac{0.45\cos\phi}{\sin^6 \phi}\left( \frac{8}{15} - \frac{5}{8}\cos\phi + \frac{5}{48}\cos3\phi - \frac{1}{80}\cos5\phi \right).$$

The calculated values of $\lambda_\theta$ and $T$ for different values of $\phi$ are shown in Table E17.2.1.

### Comments:

- At $\phi = 0°$ we have a stagnation-point flow. We showed in part (a) that, for $\phi \to 0$, $\theta^2 \to 3\nu R/(80U_0)$. Also, for $\phi = 0°$, $dU_\infty/dx = 2U_0/R$ so that $\lambda_\theta = 3/40 = 0.075$. For comparison, from Table 17.2 the wedge-flow solutions give $\lambda_\theta = 0.08544$, and $T = 0.3603$.
- At $\phi = 90°$ we have $\lambda_\theta = 0$ so that at this location the boundary layer locally has the form of a flat-plate boundary layer for which the wedge-flow solutions give $T = 0.2205$.

**Table E17.2.1** Calculated    values of $\lambda_\theta$ and $T$ for different values of $\phi$

| $\phi$ | $\lambda_\theta$ | $T$ |
|---|---|---|
| 0° | 0.075 | 0.3272 |
| 30° | 0.07215 | 0.3237 |
| 90° | 0 | 0.2247 |
| 100° | −0.0603 | 0.1131 |
| 103.11° | −0.08999 | 0.0005 |

- At $\phi = 103.11°$ the value of $\lambda_\theta$ is practically equal to −0.09, for which $T = 0$, i.e. the flow has reached the separation point.
- This example suggests that fairly accurate results can be obtained from Thwaites' method for quite a complex flow: one involving a stagnation point, rapid acceleration (in the region $0° < \phi < 90°$), and finally retardation (adverse pressure gradient) beyond $\phi = 90°$ to a separation point.

(d)  $U_0 = 0.5$ m/s, $R = 0.01$ m, $v = 1.5 \times 10^{-5}$ m$^2$/s, and $\rho = 1.2$ kg/m$^3$

By definition $\lambda_\theta = (\theta^2/v)\, dU_\infty/dx$ and we have $U_\infty = 2U_0 \sin\phi$ together with $x = R\phi$ so that $\theta = \sqrt{v\lambda_\theta R/(2U_0 \cos\phi)}$, and $Re_\theta = U_\infty\theta/v = 2U_0\theta \sin\phi/v$.

Note that, when $\phi = 90°$, $\lambda_\theta = 0$ so that $\theta$ must be calculated from the equation in part (b), which leads to $\theta = \sqrt{0.12vR/U_0}$.

By definition $T = \theta\tau_S/(\mu U_\infty)$ so that $\tau_S = 2\mu U_0 \sin\phi\, T/\theta$.

Using the calculated values for $\lambda_\theta$ and $T$ we find the following results (Table E17.2.2)

**Table E17.2.2** Calculated    values for $\theta$,   the momentum-thickness Reynolds number, and the wall shear stress for different values of $\phi$

| $\phi$ | $\theta$ (mm) | $Re_\theta$ | $\tau_S$ (Pa) |
|---|---|---|---|
| 0° | 0.1061 | 0 | 0 |
| 30° | 0.1118 | $3.727 \times 10^3$ | 0.0261 |
| 90° | 0.1897 | $1.265 \times 10^4$ | 0.0213 |
| 100° | 0.2282 | $1.498 \times 10^4$ | 0.0088 |
| 103.11° | 0.2440 | $2.376 \times 10^4$ | 0 |

## 17.7  Aerofoil lift in subsonic flow

The force which arises as a consequence of fluid flow around an object immersed in the fluid may be resolved into two components, one acting in the same direction as the approach flow and the other orthogonal (or perpendicular) to it. The aligned component is referred to as the **drag force** and the orthogonal component as the **lift force**. In Section 12.3 we showed that in supersonic (compressible) gas flow around both an inclined flat plate and around a diamond-shaped aerofoil changes in flow direction brought about by **shockwaves** and **expansion waves** lead to differences in the static pressures acting on the wetted surfaces. The lift and drag forces result from integrating these static pressures over these surfaces.

The lift forces which arise in incompressible fluid flow around an aerofoil are also primarily a consequence of the static-pressure distribution acting on the aerofoil surface but an essential aspect of the flow is the influence of viscosity: in incompressible flow, in the absence of viscosity there is neither lift or drag. The pressure distribution also contributes to the drag force but it is the surface shear stress associated with the viscous boundary layer which normally dominates the drag and also has a direct influence on the lift.

Figure 17.10 shows the cross sections of five profiles which produce lift and drag to varying degrees depending upon the geometrical shape and the angle of attack, $\alpha$. The direction of the approach flow, termed the **relative wind**, is taken to be horizontal so that lift is a vertically upward force[165]. In each case, a major feature is downward deflection of the flow by the aerofoil. Since the approach flow has zero momentum in the vertical direction, and downward momentum immediately downstream of the aerofoil, according to the **linear momentum equation** (Section 9.2) the fluid has been subjected to a downward force. From **Newton's third law**, there must therefore be a force equal in magnitude but opposite in direction acting on the aerofoil: the lift force. A drag force is associated with the reduction in momentum in the approach-flow direction. To this **lift-associated drag force** must be added the drag due to surface friction, termed **profile drag**.

Section (a) is a thin curved plate, much like a guidevane; section (b) is symmetrical about a straight chord line; section (c) is similar to (b) but cambered, i.e. asymmetrical; section (d) has the profile of a compressor blade such as we considered in Chapter 14; and section (e) is typical of a gas-turbine blade. It should be evident that the profiles are not all drawn to the same scale. At zero angle of attack the symmetrical section (b) will generate zero lift whereas section (c) will generate lift for $\alpha = 0$ due to its shape.

In Section 6.3 we pointed out that flow over a stationary aerofoil divides at a forward stagnation point located on its leading edge. On the upper (suction) surface the velocity in the free stream (outside the boundary layer) increases from zero to a maximum, depending upon the aerofoil shape, and then decreases again. From **Bernoulli's equation** we conclude that the static pressure initially decreases then increases. The streamwise **pressure gradient** is thus

---

[165] Lift is defined as the component of aerodynamic force acting on a body which is perpendicular to the relative wind.

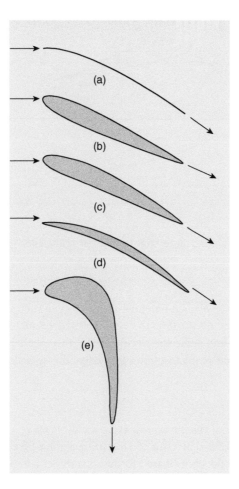

**Figure 17.10** Lifting cross sections: (a) thin curved plate (b) inclined symmetrical aerofoil (c) cambered aerofoil (d) compressor blade (e) gas-turbine blade

initially **favourable** and then **adverse**. As we have seen in this chapter, an adverse pressure gradient can lead to **boundary-layer separation**. Separation on an aerofoil leads to a loss of lift and ultimately **stall**. On the lower surface of the aerofoil there is again a static-pressure decrease beyond the stagnation point but the static pressure remains relatively high compared with the upper (**suction**) surface.

The motion of an inviscid fluid is a well-developed mathematical problem referred to as **potential-flow theory**. Particularly for bluff bodies, potential-flow theory leads to flow patterns which bear little resemblance to real flows due to the neglect of viscosity. Steady potential flow around a body produces no force (i.e. no lift or drag) irrespective of body shape, a result known as **d'Alembert's paradox**. In the case of an aerofoil, potential-flow theory leads to flow patterns similar to that shown in Figure 17.11(a). Particularly in the late 19th and early 20th centuries,

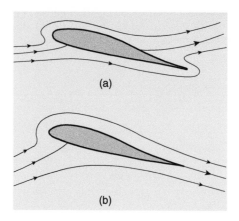

**Figure 17.11** Potential flow around an aerofoil: (a) with zero circulation (b) with circulation imposed to satisfy the Kutta condition

considerable effort was made to adapt potential-flow theory in such a way that it could predict the lift generated by an aerofoil. In 1902 the German mathematician Martin Wilhelm Kutta introduced the idea of adding sufficient **circulation** $\Gamma$ to potential flow around an aerofoil to move the rear stagnation point to the trailing edge, the so-called **Kutta condition**. The result is

$$L = \rho V \Gamma \qquad\qquad (17.143)$$

where $\rho$ is the fluid density, $V$ the flowspeed, and $L$ the lift force.

Aerofoil lift is usually specified in terms of a non-dimensional lift coefficient $C_L$ which varies with angle of attack $\alpha$ and the chord-based Reynolds number $\rho Vc/u$. Typically, $C_L$ increases linearly with $\alpha$ initially but, as the stall condition is approached, $C_L$ reaches a maximum and falls thereafter (see Figure 1.6 in Chapter 1).

We conclude this section by mentioning a fundamentally erroneous, but often stated, explanation for lift. Certain aerofoils have an asymmetric profile, such as sections (c) and (e) in Figure 17.10 and that known as the **Clark Y profile** shown in Figure 17.12, and the distance from the leading edge to the trailing edge is greater over the upper surface than over the lower surface. The erroneous argument is that because the fluid has further travel over the upper surface its velocity must be higher and, according to Bernoulli's equation, the static pressure must be lower. An underlying assumption is that two particles of fluid flowing over the aerofoil, one over the upper surface and the other over the lower surface, must reach the trailing edge at the same time, but there is no reason for this to be so. It is in any case a fact that all particles in contact with the aerofoil's surface are at rest relative to it (the **no-slip condition**), and so never reach the trailing edge. It is also the case that some aerofoils (more generally **lifting surfaces**) have a symmetrical profile but nevertheless generate lift. On geometric considerations alone, the erroneous argument clearly fails for the thin curved plate in Figure 17.10(a) and is barely plausible for the compressor blade in Figure 17.10(d).

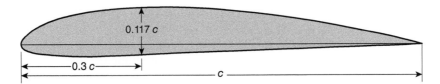

**Figure 17.12** Clark Y aerofoil section, chord length $c$

The brief, essentially two-dimensional, discussion of aerofoil lift given here is greatly simplified and takes no account of **wingspan** or **wing planform**, both of which are of major significance. It should be clear that the principles outlined in this section are equally valid if the aerofoil is 'inverted' to generate **download** (i.e. negative lift).

 17.8 SUMMARY

This chapter started by introducing the concept of a boundary layer and the associated boundary-layer approximations. The laminar boundary-layer equations were then derived from the Navier-Stokes equations. The assumption of velocity-profile similarity was shown to reduce the partial differential boundary-layer equations to ordinary differential equations. Next were discussed the numerical solutions to Blasius' equation, for zero pressure gradient, and the Falkner-Skan equation for wedge flows. von Kármán's momentum integral equation was derived and used to obtain useful results for the zero pressure-gradient boundary layer. Pohlhausen's quartic-profile method was then discussed, followed by the approximate method of Thwaites. The chapter concluded with a qualitative account of the way in which aerodynamic lift is generated.

The student should

- understand the concept of the boundary layer: a thin, near-surface region, in which viscosity plays a dominant role
- be able to state the boundary conditions to which the velocity profile is subject at both the surface over which the boundary layer is developing and in the approach to the free stream
- understand what is meant by entrainment
- understand what is meant by self-similarity of the boundary-layer velocity profiles and the importance of similarity in reducing the boundary-layer equations to an ordinary differential equation
- understand the significance of, and be able to state the definitions of, the displacement $\delta^*$ and momentum-deficit $\theta$ thicknesses, the shape factor $H$, the skin-friction coefficient $c_f/2$, and the Reynolds numbers $Re_x$ and $Re_\theta$
- be able to perform calculations using the tabulated results for $\delta^*, \theta, H$, and $c_f/2$ calculated from the Falkner-Skan equation
- understand the derivation and significance of von Kármán's momentum-integral equation
- be able to solve flat-plate boundary-layer problems using a profile method

- understand how the constants in Pohlhausen's quartic velocity profile are arrived at by satisfying a number of surface and free-stream boundary conditions
- understand how substitution of Pohlhausen's velocity profile into the momentum-integral equation leads to an ordinary differential equation for the calculation of $\theta(x)$ if $U_\infty(x)$ is specified together with $\theta(0)$
- be able to apply Thwaites' approximate formula to the solution of laminar boundary-layer problems
- be able to give a qualitative physical explanation of the manner in which lift is generated by an aerofoil moving steadily through an incompressible viscous fluid

## 17.9  SELF-ASSESSMENT PROBLEMS

**17.1**  Starting with the flat-plate boundary-layer equations

$$\frac{\partial u}{\partial x} + \frac{\partial v}{\partial y} = 0$$

and

$$u\frac{\partial u}{\partial x} + v\frac{\partial u}{\partial y} = v\frac{\partial^2 u}{\partial y^2},$$

show that these two partial differential equations can be reduced to the single ordinary differential equation (Blasius' equation)

$$F''' + FF'' = 0$$

where

$$F = \int_0^\eta f\,d\eta, \quad F'' = \frac{df}{d\eta}, \quad F''' = \frac{d^2 f}{d\eta^2}, \quad \text{and} \quad \eta = y\sqrt{\frac{U_\infty}{vx}}.$$

All symbols have their usual meanings.

**17.2**  (a) The velocity profile within a flat-plate boundary layer may be approximated by $u = A\sin(By)$. Use the boundary conditions $u(\delta) = U_\infty$, and $\partial u/\partial y|_\delta = 0$, to determine the constants $A$ and $B$.

(b) Show that the momentum thickness for the profile of part (a) is given by

$$\frac{\theta}{\delta} = \frac{2}{\pi} - \frac{1}{2}.$$

You will need to make use of the relation

$$\int_0^\phi \sin^2\phi\,d\phi = \frac{1}{2}(\phi - \sin\phi\cos\phi).$$

Show also that

$$\frac{c_f}{2} = \frac{\pi}{2}\frac{v}{U_\infty\delta}.$$

(c) Use the results of part (b) and the momentum-integral equation for a flat-plate boundary layer

$$\frac{d\theta}{dx} = \frac{c_f}{2}$$

to find a differential equation for $\delta$. Solve the equation and then show that

$$\frac{c_f}{2}\sqrt{Re_x} = 0.328.$$

**17.3** The boundary layer which develops over the leading edge of a wing can be assumed to behave like a laminar wedge-flow boundary layer with $m = 1$ for which $F''(0) = 1.233, (\delta^*/x)\sqrt{Re_x} = 0.648$, and $(\theta/x)\sqrt{Re_x} = 0.292$, where $\delta^*$ is the displacement thickness, $\theta$ is the momentum thickness, $x$ is the distance from the stagnation point, $Re_x = U_\infty x/\nu$, $U_\infty$ is the free-stream velocity at distance $x$, $\nu$ is the kinematic velocity of the fluid, and $F''(0)$ is the non-dimensional velocity gradient at the surface, i.e.

$$F''(0) = \sqrt{\frac{\nu x}{U_\infty^3}} \left.\frac{\partial u}{\partial y}\right|_0.$$

For a particular wing, the free-stream velocity gradient $dU_\infty/dx = 750$ m/s. Calculate the shape factor $H = \delta^*/\theta$, the Reynolds number based upon $U_\infty$ and $\theta$, and the wall shear stress at $x = 0.3$ m. Show that the average shear stress over this distance is half the final value. The fluid density can be taken as 1.2 kg/m$^3$, and the kinematic viscosity as $1.5 \times 10^{-5}$ m$^2$/s.
(Answers: 2.22, 619, 35.3 Pa)

**17.4** A laminar boundary layer develops on a flat plate along which the free-stream velocity varies with distance $x$ from the leading edge according to $U_\infty = U_0\sqrt{(1-z)}$, where $z = Cx/L$, $U_0$ is the free-stream velocity at $x = 0$, $L$ is the length of the plate, and $C$ is a positive constant.
(a) Show that the free-stream velocity variation corresponds to a constant adverse pressure gradient.

(b) Assuming that the momentum thickness $\theta = 0$ at $x = 0$, use Thwaites' equation

$$\theta^2 = \frac{0.45\nu}{U_\infty^6} \int_0^x U_\infty^5 dx$$

to show that the variation of $\theta(x)$ is given by

$$\theta = 3\sqrt{\frac{\nu L}{70U_0 C}\left[\left(\frac{U_0}{U_\infty}\right)^6 - \frac{U_\infty}{U_0}\right]}.$$

(c) From the result of part (b), find an equation for $\lambda_\theta$ in terms of $U_0/U_\infty$.

(d) Given that boundary-layer separation occurs for $\lambda_\theta = -0.09$,, find the $x$-location at which separation occurs (in terms of $Cx/L$).

(e) Calculate $\lambda_\theta$ for $Cx/L = 0.1$ and the corresponding value of the shear-stress parameter $T$, using the formula

$$T = (0.09 + \lambda_\theta)^{0.62}.$$

(Answers: (d) 0.2213, (e) −0.0287, 0.1771)

# 18 Turbulent flow

This chapter begins with a brief description of the qualitative character of turbulent flow. This is followed by the decomposition of a turbulent flow into a mean and fluctuating parts, and the derivation of the Reynolds-averaged Navier-Stokes (RANS) equations, the turbulent kinetic-energy equation, and the equation for the transport of the so-called Reynolds stress, which arises from the time averaging. These equations represent the foundation on which turbulence modelling is based. Reduced forms of the RANS equations for two-dimensional boundary-layer and Couette flows are then derived. The Law of the Wall is shown to result from dimensional considerations applied to plane Couette flow. Three separate zones are identified: the viscous sublayer, the fully-turbulent log-law region, and the buffer layer which separates the two. The Law of the Wake for the outer region of a turbulent boundary layer is then presented. A brief discussion follows concerned with the wide spectrum of length, time, and velocity scales, which are an important aspect of any turbulent flow. The log law is then applied to the analysis of turbulent flow through a smooth-walled pipe, followed by consideration of the effect of surface roughness. The calculation of pressure loss in piping systems is presented, largely based on empirical loss coefficients for such components as elbows, tee junctions, and area changes, together with the frictional pressure drop in long sections of pipe. Both the log-law and power-law velocity distributions are used as a basis for the analysis of a flat-plate turbulent boundary layer, and the results compared with empirical skin-friction formulae. The flow over a circular cylinder in crossflow is discussed based upon the experimentally based curve for the drag coefficient $C_D$ versus Reynolds number, $Re$. Values of $C_D$ are given for cylinders of various cross section in crossflow and also for a number of three-dimensional objects.

## 18.1 Transitional and turbulent flow

For any given **shear flow**[166], as the Reynolds number is progressively increased, some regions of the flowfield exhibit an unsteady (i.e. time-dependent) but still relatively orderly, essentially laminar, state, while other zones increasingly show irregular, chaotic, fluctuations in velocity. The latter is called **turbulent flow**, while the intermittent mix of quasi-laminar and turbulent flow is termed **transitional flow**. The **intermittency factor** $\gamma$ at a fixed point in a flow is the fraction of time the flow there is turbulent, so that $\gamma$ ranges from 0 to 1. As a consequence of

---

[166] Any flow in which viscous or turbulent shear stresses play a key role is termed a shear flow. Examples include duct flow, boundary-layer flow, free and wall jets, and wakes.

*Introduction to Engineering Fluid Mechanics.* Marcel Escudier.
© Marcel Escudier 2017. Published 2017 by Oxford University Press.

the velocity differences and overall scales involved, most flows of industrial interest fall into the turbulent-flow category.

It is generally accepted that any flow of a Newtonian fluid, whether laminar, transitional, or turbulent, obeys the **Navier-Stokes equations**. Although turbulent flow is inherently unsteady, a time average can be taken leading to a set of equations for the time-averaged motion. As we show in Section 18.2, these equations closely resemble the laminar-flow equations of Chapter 17 but include additional terms, the so-called **Reynolds stresses**, which arise as a consequence of correlations between the fluctuations in the three velocity components, $u'$, $v'$, and $w'$. Accounting for these correlations is the central problem in the analysis of time-averaged turbulent flows.

In principle the Navier-Stokes equations for the fluctuating motion can be converted to finite-difference, finite-volume, or finite-element form (a process called **discretisation**) and solved numerically, an approach known as **direct numerical simulation (DNS)**. In the case of a turbulent boundary layer, such a simulation has to account for all length scales, ranging from the Kolmogorov length scale[167] $l_\kappa$ to the boundary-layer thickness $\delta$ itself. Since a fluid volume of length $L$, width $W$, and thickness $\delta$ contains $WL\delta/l_\kappa^3$ cells of side length $l_\kappa$, this is the number of cells which would have to be considered in a complete simulation. With typical values $L = W = 1$ m, $\delta = 50$ mm, and $l_\kappa = 50$ $\mu$m, we find that this corresponds to $10^{15}$ (or 1000 trillion) cells, which exceeds the capacity of present-day computers and it is unlikely that DNS will be used for routine engineering calculations in the foreseeable future. Computing power, measured in flop/s, has increased exponentially over the last six decades, roughly according to $flop/s \approx e^{0.5\Delta y}$, where $\Delta y$ is the number of years from 1938[168] at which we have set $flop/s = 1$. The fastest computer as of June 2015 is reported to have achieved close to 34 petaflops, i.e. $10^{15}$ $flop/s$.

## 18.2 Reynolds decomposition, Reynolds averaging, and Reynolds stresses

At any point within a turbulent flow, the three orthogonal components of the velocity, $u$, $v$, and $w$, together with the static pressure $p$, fluctuate apparently randomly with time. In general, it would be necessary to include the density in this list, but we shall limit consideration to constant-density flows. These fluctuations are subject to a number of constraints and so cannot be completely random: the continuity equation 15.7 for the flow of a constant-density fluid shows that $u$, $v$, and $w$ are not independent, while $p$ is related to $u$, $v$, and $w$ through the Navier-Stokes equations 15.29 to 15.31.

Since DNS for practical engineering calculations is at best a prospect for the distant future, it is probable that for the foreseeable future the method of analysing turbulent flow, suggested by Osborne Reynolds in 1895, will be based upon separating all flow quantities into a mean (or time-averaged) part and a fluctuating (time-dependent) part, the latter having an average value

[167] The Kolmogorov length scale, together with other turbulence scales, is discussed in Section 18.4.
[168] The year 1938 is a consequence of extrapolating to zero a linear fit to a graph of computing power (*flop/s*) versus calendar year on log-linear coordinates.

of zero. This is known as **Reynolds decomposition**. If we take the $x$-component of velocity $u$, to illustrate this idea, we have

$$u = \bar{u} + u' \tag{18.1}$$

where $\bar{u}$ is the time average[169] of the fluctuating velocity $u$ and $u'$ is the fluctuating part. The time average is defined by

$$\bar{u} = \frac{1}{T} \int_0^T u \, dt \tag{18.2}$$

where the time interval $T$ over which the average is taken is long compared with the fluctuating time scale. The overbar[170] here and elsewhere signifies a time average. Substitution of $u = \bar{u} + u'$ into equation (18.2) leads immediately to $\overline{u'} = 0$. As we shall see shortly, when we apply this type of averaging process to the Navier-Stokes equations, non-zero terms like $\overline{u'^2}$ and $\overline{u'v'}$ arise.

According to equation (15.7), for an incompressible flow the continuity equation is

$$\frac{\partial u}{\partial x} + \frac{\partial v}{\partial y} + \frac{\partial w}{\partial z} = 0 \tag{18.3}$$

which we can now write as

$$\frac{\partial (\bar{u} + u')}{\partial x} + \frac{\partial (\bar{v} + v')}{\partial y} + \frac{\partial (\bar{w} + w')}{\partial z} = 0. \tag{18.4}$$

If we take the time average of equation (18.4), the three fluctuating terms average to zero and we have

$$\frac{\partial \bar{u}}{\partial x} + \frac{\partial \bar{v}}{\partial y} + \frac{\partial \bar{w}}{\partial z} = 0. \tag{18.5}$$

If we subtract equation (18.5) from equation (18.4) we find

$$\frac{\partial u'}{\partial x} + \frac{\partial v'}{\partial y} + \frac{\partial w'}{\partial z} = 0 \tag{18.6}$$

and we see that the mean and fluctuating parts of the instantaneous velocity components separately satisfy an equation of continuity.

From Subsection 15.1.6 the three components of the Navier-Stokes equations for the flow of a constant- and uniform-property fluid in the absence of body forces are:
$x$-component

$$\rho \frac{Du}{Dt} = \rho \left[ \frac{\partial u}{\partial t} + \frac{\partial u^2}{\partial x} + \frac{\partial (uv)}{\partial y} + \frac{\partial (uw)}{\partial z} \right] = -\frac{\partial p}{\partial x} + \mu \left( \frac{\partial^2 u}{\partial x^2} + \frac{\partial^2 u}{\partial y^2} + \frac{\partial^2 u}{\partial z^2} \right) \tag{18.7}$$

---

[169] To avoid confusion, we shall assume that $\bar{u}$ is independent of time. However, just as we can have an unsteady laminar flow, we can also have a turbulent flow for which $\bar{u}$ varies with time at a frequency much lower than is typical of the turbulent fluctuations.
[170] Angle brackets are sometimes used instead of an overbar to denote a time average, i.e. $u = \bar{u}$.

$y$-component

$$\rho\frac{Dv}{Dt} = \rho\left[\frac{\partial v}{\partial t} + \frac{\partial(vu)}{\partial x} + \frac{\partial v^2}{\partial y} + \frac{\partial(vw)}{\partial z}\right] = -\frac{\partial p}{\partial y} + \mu\left(\frac{\partial^2 v}{\partial x^2} + \frac{\partial^2 v}{\partial y^2} + \frac{\partial^2 v}{\partial z^2}\right) \qquad (18.8)$$

$z$-component

$$\rho\frac{Dw}{Dt} = \rho\left[\frac{\partial w}{\partial t} + \frac{\partial(wu)}{\partial x} + \frac{\partial(wv)}{\partial y} + \frac{\partial w^2}{\partial z}\right] = -\frac{\partial p}{\partial z} + \mu\left(\frac{\partial^2 w}{\partial x^2} + \frac{\partial^2 w}{\partial y^2} + \frac{\partial^2 w}{\partial z^2}\right). \qquad (18.9)$$

In each of these equations we have replaced terms like $v\partial u/\partial y$ by $\partial(uv)/\partial y - u\partial v/\partial y$ and used the continuity equation to eliminate $\partial v/\partial y$[171].

If we now take the time averages of these three equations we have, for a flow which is steady on average

$$\rho\left[\frac{\partial \overline{u^2}}{\partial x} + \frac{\partial \overline{(uv)}}{\partial y} + \frac{\partial \overline{(uw)}}{\partial z}\right] = -\frac{\partial \overline{p}}{\partial x} + \mu\left(\frac{\partial^2 \overline{u}}{\partial x^2} + \frac{\partial^2 \overline{u}}{\partial y^2} + \frac{\partial^2 \overline{u}}{\partial z^2}\right) \qquad (18.10)$$

$$\rho\left[\frac{\partial \overline{(vu)}}{\partial x} + \frac{\partial \overline{v^2}}{\partial y} + \frac{\partial \overline{(vw)}}{\partial z}\right] = -\frac{\partial \overline{p}}{\partial y} + \mu\left(\frac{\partial^2 \overline{v}}{\partial x^2} + \frac{\partial^2 \overline{v}}{\partial y^2} + \frac{\partial^2 \overline{v}}{\partial z^2}\right) \qquad (18.11)$$

$$\rho\left[\frac{\partial \overline{(wu)}}{\partial x} + \frac{\partial \overline{(wv)}}{\partial y} + \frac{\partial \overline{w^2}}{\partial z}\right] = -\frac{\partial \overline{p}}{\partial z} + \mu\left(\frac{\partial^2 \overline{w}}{\partial x^2} + \frac{\partial^2 \overline{w}}{\partial y^2} + \frac{\partial^2 \overline{w}}{\partial z^2}\right). \qquad (18.12)$$

It should now be clear that the difference between the Navier-Stokes equations for a steady laminar flow and for a time-mean-steady turbulent flow arises from the non-linear advective terms on the left-hand side of each of the last three equations.

If we now introduce $u = \overline{u} + u'$, $v = \overline{v} + v'$, and $w = \overline{w} + w'$, we have

$$\overline{u^2} = \overline{u}^2 + \overline{u'u'}, \overline{uv} = \overline{u}\,\overline{v} + \overline{u'v'}, \text{ and } \overline{uw} = \overline{u}\overline{w} + \overline{u'w'}, \text{ etc.} \qquad (18.13)$$

In arriving at these identities we have made use of the fact that $\overline{u'\overline{u}} = \overline{u'}\overline{u} = 0$, $\overline{v'\overline{v}} = \overline{v'}\overline{v} = 0$, etc.

Equation (18.10) may now be written as

$$\rho\left[\frac{\partial}{\partial x}\left(\overline{u}^2 + \overline{u'u'}\right) + \frac{\partial}{\partial y}\left(\overline{u}\,\overline{v} + \overline{u'v'}\right) + \frac{\partial}{\partial z}\left(\overline{u}\,\overline{w} + \overline{u'w'}\right)\right] = -\frac{\partial \overline{p}}{\partial x} + \mu\left(\frac{\partial^2 \overline{u}}{\partial x^2} + \frac{\partial^2 \overline{u}}{\partial y^2} + \frac{\partial^2 \overline{u}}{\partial z^2}\right)$$

which simplifies to

$$\rho\left(\overline{u}\frac{\partial \overline{u}}{\partial x} + \overline{v}\frac{\partial \overline{u}}{\partial y} + \overline{w}\frac{\partial \overline{u}}{\partial z}\right) = -\frac{\partial \overline{p}}{\partial x} + \frac{\partial}{\partial x}\left(\mu\frac{\partial \overline{u}}{\partial x} - \left[\rho\overline{u'u'}\right]\right) + \frac{\partial}{\partial y}\left(\mu\frac{\partial \overline{u}}{\partial y} - \left[\rho\overline{u'v'}\right]\right)$$
$$+ \frac{\partial}{\partial z}\left(\mu\frac{\partial \overline{u}}{\partial z} - \left[\rho\overline{u'w'}\right]\right) \qquad (18.14)$$

equation (18.11) may be written as

$$\rho\left[\frac{\partial}{\partial x}\left(\overline{v}\,\overline{u} + \overline{v'u'}\right) + \frac{\partial}{\partial y}\left(\overline{v}^2 + \overline{v'v'}\right) + \frac{\partial}{\partial z}\left(\overline{v}\overline{w} + \overline{v'w'}\right)\right] = -\frac{\partial \overline{p}}{\partial y} + \mu\left(\frac{\partial^2 \overline{v}}{\partial x^2} + \frac{\partial^2 \overline{v}}{\partial y^2} + \frac{\partial^2 \overline{v}}{\partial z^2}\right)$$

[171] This part of the analysis is the subject of Self-assessment problem 18.1.

which simplifies to

$$\rho\left(\bar{u}\frac{\partial\bar{v}}{\partial x} + \bar{v}\frac{\partial\bar{v}}{\partial y} + \bar{w}\frac{\partial\bar{v}}{\partial z}\right) = -\frac{\partial\bar{p}}{\partial y} + \frac{\partial}{\partial x}\left(\mu\frac{\partial\bar{v}}{\partial x} - \left[\rho\overline{v'u'}\right]\right) + \frac{\partial}{\partial y}\left(\mu\frac{\partial\bar{v}}{\partial y} - \left[\rho\overline{v'v'}\right]\right)$$
$$+ \frac{\partial}{\partial z}\left(\mu\frac{\partial\bar{v}}{\partial z} - \left[\rho\overline{v'w'}\right]\right) \tag{18.15}$$

and equation (18.12) may be written as

$$\rho\left[\frac{\partial}{\partial x}\left(\overline{wu} + \overline{w'u'}\right) + \frac{\partial}{\partial y}\left(\overline{wv} + \overline{w'v'}\right) + \frac{\partial}{\partial z}\left(\overline{w}^2 + \overline{w'w'}\right)\right] = -\frac{\partial\bar{p}}{\partial z} + \mu\left(\frac{\partial^2\overline{w}}{\partial x^2} + \frac{\partial^2\overline{w}}{\partial y^2} + \frac{\partial^2\overline{w}}{\partial z^2}\right)$$

which simplifies to

$$\rho\left(\bar{u}\frac{\partial\overline{w}}{\partial x} + \bar{v}\frac{\partial\overline{w}}{\partial y} + \bar{w}\frac{\partial\overline{w}}{\partial z}\right) = -\frac{\partial\bar{p}}{\partial z} + \frac{\partial}{\partial x}\left(\mu\frac{\partial\overline{w}}{\partial x} - \left[\rho\overline{w'u'}\right]\right) + \frac{\partial}{\partial y}\left(\mu\frac{\partial\overline{w}}{\partial y} - \left[\rho\overline{w'v'}\right]\right)$$
$$+ \frac{\partial}{\partial z}\left(\mu\frac{\partial\overline{w}}{\partial z} - \left[\rho\overline{w'w'}\right]\right). \tag{18.16}$$

Equations (18.14), (18.15), and (18.16) are known as the **Reynolds-averaged Navier-Stokes (or RANS) equations**. It is the terms in square brackets in these equations which distinguish them from their laminar-flow counterparts. In fact, if the velocity fluctuations are zero, the RANS equations reduce to those for steady, constant-property, laminar flow. From now on we shall write $\overline{u'u'}$, $\overline{v'v'}$, and $\overline{w'w'}$ as $\overline{u'^2}$, $\overline{v'^2}$, and $\overline{w'^2}$, respectively. These additional terms, usually shifted to appear on the right-hand sides of the time-averaged Navier-Stokes equations, by comparison with the viscous normal and shear stresses, may be interpreted physically as stresses: $\rho\overline{u'^2}$, $\rho\overline{v'^2}$, and $\rho\overline{w'^2}$ as pressure-like normal stresses, and $\rho\overline{u'v'}$, $\rho\overline{v'w'}$, and $\rho\overline{w'u'}$ as shear stresses. The six stresses are known as the **Reynolds stresses**[172] or **apparent stresses**. Each of the shear stresses arises from the correlation of two orthogonal components of the velocity fluctuation at a given point, a non-zero value of the correlation indicating that the two components are not independent. If the correlation is negative, then the two components are opposite in sign over most of the averaging period.

## 18.3 Turbulent-kinetic-energy equation and Reynolds-stress equation

At any point in a turbulent flow, half the sum of the three normal stresses represents the **specific turbulent kinetic energy**[173] $\bar{k}$, of the fluctuating velocity components

$$\bar{k} = \frac{1}{2}\left(\overline{u'^2} + \overline{v'^2} + \overline{w'^2}\right). \tag{18.17}$$

Since it is often the case that there is a principal flow direction, for example the $x$-direction, it is common to take the value of $\overline{u'^2}$ as a measure of the turbulence intensity. Other measures

---

[172] The quantities $\overline{u'^2}$, $\overline{v'^2}$, $\overline{w'^2}$, $\overline{u'v'}$, $\overline{v'w'}$, and $\overline{w'u'}$ are also often referred to as the Reynolds stresses but, in the absence of the density $\rho$, the term **kinematic Reynolds stresses** is more appropriate.

[173] The specific turbulent kinetic energy is the turbulent kinetic energy per unit mass.

in use include $\sqrt{\overline{u'^2}}/\overline{u}, \sqrt{\overline{u'^2}}/U_\infty$, $\sqrt{k}/\overline{u}$, and $\sqrt{k}/U_\infty$. The normalising velocity, $\overline{u}$ or $U_\infty$, is measured at the same location as $u'$.

In addition to the RANS equations, further exact equations can be derived from the Navier-Stokes equations. For example, for a steady, two-dimensional, constant-property, turbulent boundary layer, neglecting viscous terms other than the dissipation, we have the **turbulent-kinetic-energy equation**

$$\overline{u}\frac{\partial \overline{k}}{\partial x} + \overline{v}\frac{\partial \overline{k}}{\partial y} = -\frac{\partial}{\partial y}\left[\overline{v'\left(\frac{1}{2}k + \frac{p'}{\rho}\right)}\right] + \left(\nu\frac{\partial \overline{u}}{\partial y} - \overline{u'v'}\right)\frac{\partial \overline{u}}{\partial y} - \epsilon \tag{18.18}$$

      I                   II      III       IV

where $\epsilon$ is the average **turbulent-kinetic-energy dissipation rate**[174] given by

$$\epsilon = \nu\left[2\overline{\left(\frac{\partial u'}{\partial x}\right)^2} + 2\overline{\left(\frac{\partial v'}{\partial y}\right)^2} + 2\overline{\left(\frac{\partial w'}{\partial z}\right)^2} + \overline{\left(\frac{\partial u'}{\partial y} + \frac{\partial v'}{\partial x}\right)^2}\right.$$
$$\left. + \overline{\left(\frac{\partial u'}{\partial z} + \frac{\partial w'}{\partial x}\right)^2} + \overline{\left(\frac{\partial v'}{\partial z} + \frac{\partial w'}{\partial y}\right)^2}\right]. \tag{18.19}$$

The combination $\rho\left(\nu\partial\overline{u}/\partial y - \overline{u'v'}\right)$ is the combined viscous-plus-turbulent time-average shear stress $\overline{\tau}$ and, apart from the region close to a solid boundary, it is usually the case that $-\overline{u'v'} \gg \nu\partial\overline{u}/\partial y$. In addition to $\epsilon$, the turbulent kinetic-energy equation has introduced two further turbulence correlations: $\overline{v'k}$ and $\overline{v'p'}$.

The terms in equation (18.18) can be interpreted as follows

  I. Transport of $k$ through advection by the mean flow
  II. Transport of $k$ by velocity fluctuations (turbulent diffusion)
  III. Transport of $k$ by pressure fluctuations
  IV. Rate of production of $k$ by interaction of the shear stress and the mean-velocity gradient

It should be noted that dissipation of kinetic energy also occurs due to the velocity gradients of the time-mean motion (so-called **direct dissipation**).

Another equation commonly considered in two-dimensional turbulent boundary-layer analysis is the **Reynolds-stress equation**

$$\overline{u}\frac{\partial\left(-\overline{u'v'}\right)}{\partial x} + \overline{v}\frac{\partial\left(-\overline{u'v'}\right)}{\partial y} = \overline{v'^2}\frac{\partial\overline{u}}{\partial y} - \frac{\overline{p'}}{\rho}\overline{\left(\frac{\partial u'}{\partial y} + \frac{\partial v'}{\partial x}\right)} + \frac{\partial}{\partial y}\left(\overline{u'v'^2} + \frac{\overline{p'u'}}{\rho}\right)$$
$$+ \nu\frac{\partial^2\left(-\overline{u'v'}\right)}{\partial y^2} + 2\nu\overline{\frac{\partial u'}{\partial y}\frac{\partial v'}{\partial x}} \tag{18.20}$$

which has again introduced further terms.

The turbulent kinetic-energy equation, the Reynolds-stress equation, and other equations derived from or based on the RANS equations are the foundations for the methodology termed turbulence modelling, which we discuss briefly in Section 18.5.

---

[174] The symbol $\epsilon$ represents the lower-case Greek letter epsilon. Epsilon is also represented by $\varepsilon$, which we use for surface-roughness height (see Section 18.9).

## 18.4 Turbulence scales

So far we have made little mention of the structure of a turbulent flow. It is observed experimentally that clumps (or packets) of fluid particles, called **eddies**, form, interact, break up,
and reform throughout a turbulent flow. The length scale (i.e. size) of these eddies varies from
the overall scale of the flow down to a **microscale**, much larger than the **molecular mean free
path**, so that the **continuum hypothesis** still applies (already implied in assuming that the
Navier-Stokes equations apply), where viscosity dominates and turbulent kinetic energy is dissipated into heat. Most of the kinetic energy of a turbulent flow is contained in the **integral
length scales**, which are the largest scales in an energy spectrum, i.e. the distribution of kinetic energy according to length scale or frequency. The largest scales correspond to the lowest
frequency, and vice versa. The kinetic energy in a turbulent flow passes progressively from the
largest energy-bearing eddies to the smallest dissipative eddies in what is termed the **energy
cascade**. It should be evident that, because there are fluctuations in velocity and a wide distribution of length scales, **mixing** within a turbulent flow is much stronger than in a laminar
flow. The practical consequence is enhanced surface shear stress and, where a surface is heated
or cooled, higher rates of heat transfer than in a laminar flow.

The smallest length scale at which turbulence can exist in a flow is the **Kolmogorov length
scale** $l_K$, defined in terms of the kinematic viscosity of the fluid $\nu$ and the rate of dissipation of
turbulent kinetic energy per unit mass $\epsilon$

$$l_K = \left(\frac{\nu^3}{\epsilon}\right)^{1/4}. \tag{18.21}$$

This combination of $\nu$ and $\epsilon$ is arrived at on dimensional grounds since $[\nu] = L^2/T$, and
$[\epsilon] = L^2/T^3$. The **Kolmogorov time and velocity scales**, $\tau_K$ and $v_K$, are similarly defined

$$\tau_K = \left(\frac{\nu}{\epsilon}\right)^{1/2} \tag{18.22}$$

and

$$v_K = (\nu\epsilon)^{1/4}. \tag{18.23}$$

An inevitable consequence of these definitions is that the Reynolds number $v_K l_K/\nu = 1$, indicating that the small-scale motion is quite viscous. In numerical simulations of turbulent flows,
the smallest scale that has to be resolved is usually taken to be of the same order of magnitude
as the Kolmogorov length scale.

At any point in a turbulent flow, fluctuations in velocity and pressure contain energy across
a wide range of frequencies $f$. According to equation (18.17), the turbulent kinetic energy per
unit mass is $k$

$$\bar{k} = \frac{1}{2}\left(\overline{u'^2} + \overline{v'^2} + \overline{w'^2}\right) \tag{18.17}$$

and, if the kinetic energy in the frequency range $f$ to $f + \delta f$ is $E(f)$, then

$$\bar{k} = \int_0^\infty E(f)\, df. \tag{18.24}$$

Rather than frequency, it is usual here to introduce the idea of a **wavenumber** $\kappa$, where

$$\kappa = \frac{2\pi f}{v} = \frac{2\pi}{\lambda} \tag{18.25}$$

and $\lambda = v/f$ is the wavelength, $v$ being the instantaneous velocity at the point of measurement. Equation (18.24) is then written as

$$\bar{k} = \int_0^\infty E(\kappa)\, d\kappa \tag{18.26}$$

and the quantity $E(k)$ is called the **energy spectral density**, or **energy spectrum function**.

The so-called **inertial subrange** corresponds with the non-dissipative **Taylor microscales**, which are intermediate between the largest and smallest scales. Within the inertial subrange Kolmogorov argued that a range of scales exists within which $E(\kappa)$ is dependent upon the dissipation rate $\epsilon$ and wavenumber $\kappa$ but independent of viscosity so that

$$E(\kappa) = F(\epsilon, \kappa). \tag{18.27}$$

It follows from dimensional analysis that

$$E = C_K \epsilon^{2/3} \kappa^{-5/3} \tag{18.28}$$

where $C_K$ is the **Kolmogorov constant**. The inertial subrange covers the wavenumber range $1/l \ll \kappa \ll 1/l_K$, where $l$ is a measure of the largest scale, such as the **integral length scale**.

If it is assumed, as is reasonable, that the turbulence dissipation rate $\epsilon$ is determined by the specific turbulent kinetic energy $k$ and the integral length scale $l$, then dimensional analysis leads to

$$\epsilon \sim \frac{k^{3/2}}{l}. \tag{18.29}$$

For Couette flow, the convective terms on the left-hand side of the **turbulent kinetic-energy equation** (18.18) are zero, and $\bar{u}$ is independent of $x$, so we have

$$-\frac{\partial}{\partial y}\left[ \overline{v'\left(\frac{1}{2}k + \frac{p'}{\rho}\right)} \right] + \left( v\frac{\partial \bar{u}}{\partial y} - \overline{u'v'} \right)\frac{d\bar{u}}{dy} - \epsilon = 0. \tag{18.30}$$

It can be shown that the term involving $\overline{v'k}$ and $\overline{v'p'}$ is negligible compared with the production and dissipation terms, so that equation (18.30) reduces to

$$\epsilon = -\overline{u'v'}\frac{d\bar{u}}{dy} \tag{18.31}$$

where we have neglected the viscous shear stress compared with the Reynolds shear stress.

In Subsection 18.7.2 we show that, for the near-wall region, if $-\rho\overline{u'v'} = \tau_S$, then

$$\frac{y}{u_\tau}\frac{d\bar{u}}{dy} = \frac{1}{k} \tag{18.50}$$

so that from equation (18.31)

$$\epsilon = \frac{u_\tau^3}{\kappa y} \tag{18.32}$$

and we have a good estimate for $\epsilon$ in the log-law region.

From the definitions of the Kolmogorov scales (equations (18.21) to (18.23)) we then have

$$\frac{u_\tau l_k}{\nu} = \left(\kappa y^+\right)^{1/4} \tag{18.33}$$

$$\frac{u_\tau^2 \tau_K}{\nu} = \left(\kappa y^+\right)^{1/2} \tag{18.34}$$

and

$$\frac{v_K}{u_\tau} = \frac{1}{\left(\kappa y^+\right)^{1/4}}. \tag{18.35}$$

In Illustrative Example 18.2 we calculate values for the Kolmogorov scales for a specified pipe flow.

## 18.5  Turbulence modelling

Unfortunately, it is unlikely that physically exact equations will ever be established which take into account the entire range of length and time scales which we have just identified and which link the six Reynolds stresses, and correlations such as $\overline{p'u'}$, $\overline{u'v'^2}$, etc., to $\bar{u}$, $\bar{v}$, $\bar{w}$, $\bar{p}$, and their spatial gradients. Apart from research into DNS and the closely related **large-eddy simulation** (LES), which avoids the need to establish such links, research hitherto has concentrated on a semi-empirical methodology in which the physics of turbulent flow is approximated by partial differential equations for the specific turbulent kinetic energy $k$, the rate of turbulent dissipation $\epsilon$, the Reynolds stresses $\overline{u'v'}$, $\overline{u'^2}$, $\overline{v'^2}$, etc., and various length scales with approximations for the correlations between $u'$, $v'$, $w'$, $p'$, and their gradients. Devising these approximations has become known as **turbulence modelling**.

For the foreseeable future, DNS and LES are likely to remain research topics providing results used to guide the development of, and against which to test the predictions of, turbulence modelling, particularly where experimental data are unavailable. Depending upon the complexity involved and level of accuracy required, engineering applications will rely upon software based upon empirical correlations and turbulence modelling of varying levels of sophistication.

Probably the earliest example of turbulence modelling was Prandtl's **mixing-length hypothesis** based upon the idea that the large-scale random movements of fluid elements in turbulent motion are analogous to the small-scale random motion of molecules in a gas (kinetic theory). From this beginning, with time the following hierarchy of turbulence models has evolved

- zero equation model, such as the mixing-length or eddy-viscosity model
- two-equation model, in which time-averaged equations are solved for the specific turbulent-kinetic energy $k$ and the turbulent-kinetic-energy dissipation rate $\epsilon$
- Reynolds-stress equation model
- algebraic-stress model

A detailed presentation and discussion of turbulence modelling is beyond the scope of this text and from the foregoing the impression could be gained that little progress has been made in devising relatively simple ways to analyse turbulent flows of practical interest. In fact much of what was learned throughout the 20$^\text{th}$ century, based upon simplifications, dimensional analysis, empiricism, and integral methods, forms the basis of current engineering practice and to a large extent is incorporated into turbulence modelling. Some of the ideas involved are discussed in the remainder of this chapter.

## 18.6 Two-dimensional turbulent boundary layers and Couette flow

Although velocity fluctuations in a turbulent flow always occur in the three orthogonal directions, there are many practical situations, just as for laminar flow, where there is no variation of time-averaged quantities in the third direction (usually taken as the $z$-direction). If in addition we introduce the boundary-layer approximations outlined in Chapter 17, the counterpart to equation (17.4) for a turbulent boundary layer is

$$\bar{u}\frac{\partial \bar{u}}{\partial x} + \bar{v}\frac{\partial \bar{u}}{\partial y} = -\frac{1}{\rho}\frac{dp}{dx} + \frac{1}{\rho}\frac{\partial}{\partial y}\left(\mu\frac{\partial \bar{u}}{\partial y} - \rho\overline{u'v'}\right). \tag{18.36}$$

Even though the Reynolds shear stress $\rho\overline{u'v'}$ is the only term remaining from the Reynolds decomposition and time averaging, except in the near-vicinity of a wall, it is usually several orders of magnitude greater than the viscous shear stress, $\mu\partial\bar{u}/\partial y$, and so has to be accounted for. Just as for laminar flow, we can consider fully-developed turbulent flow through a cylindrical channel, for which equation (18.36) reduces to

$$0 = -\frac{1}{\rho}\frac{dp}{dx} + \frac{1}{\rho}\frac{\partial}{\partial y}\left(\mu\frac{d\bar{u}}{dy} - \rho\overline{u'v'}\right). \tag{18.37}$$

A further simplification occurs for **Couette flow** which, as we saw in Section 16.4, is the fully-developed, shear-driven flow between two parallel surfaces where one is moving tangentially with respect to the other. Within such an idealised flow the shear stress is constant (giving rise to the term **constant-stress layer**) so that, if the flow is turbulent, the mean (i.e. time-averaged) shear stress $\bar{\tau}$ is given by

$$\bar{\tau} = \mu\frac{d\bar{u}}{dy} - \rho\overline{u'v'} = \text{constant} = \bar{\tau}_S \tag{18.38}$$

where $\bar{\tau}_S$ is the mean wall shear stress and $y$ is the distance from the stationary surface.

The boundary conditions for equations (18.36), (18.37), and (18.38) are the same as those for laminar flow.

## 18.7 Plane turbulent Couette flow and the Law of the Wall

As we have just seen, for plane turbulent Couette flow the RANS equations reduce to

$$\bar{\tau} = \mu\frac{d\bar{u}}{dy} - \rho\overline{u'v'} = \text{constant} = \bar{\tau}_S. \tag{18.39}$$

Equation (18.39) cannot simply be integrated to give the mean-velocity profile $\bar{u}(y)$ because the dependence of the Reynolds shear stress $-\overline{u'v'}$ on other flow properties is unknown.

We shall return to equation (18.39) shortly but for the time being we introduce the assumption that $\bar{u}$ depends upon $y$, $\bar{\tau}_S$, and the fluid properties $\rho$ and $\mu$, i.e.

$$\bar{u} = f(y, \bar{\tau}_S, \rho, \mu), \tag{18.40}$$

an assumption which can reasonably be applied to the near-wall region of any turbulent shear flow over a smooth surface, i.e. to boundary layers and channel flows. Dimensional analysis leads to

$$\frac{\bar{u}}{u_\tau} = f\left(\frac{u_\tau y}{\nu}\right) \tag{18.41}$$

where $u_\tau = \sqrt{\bar{\tau}_S/\rho}$ has the units of velocity and is termed the **friction velocity** (or **wall-friction velocity**)[175]. Equation (18.41), first postulated by Prandtl, is known as the **Law of the Wall**, or **universal velocity distribution**, and written as

$$u^+ = f(y^+) \tag{18.42}$$

where

$$u^+ \equiv \frac{\bar{u}}{u_\tau} \text{ and } y^+ \equiv \frac{u_\tau y}{\nu} \tag{18.43}$$

are the so-called **wall variables**[176]. It is seen that $y^+$ can be regarded as a turbulence Reynolds number and that $\nu/u_\tau$ is a **viscous length scale**. The Law of the Wall is usually regarded as comprising three parts: a **viscous sublayer**, a **buffer layer**, and a **fully-turbulent region**.

### 18.7.1 Viscous sublayer

From the boundary conditions at a solid surface it must be that $u' = 0$ (the no-slip condition) and $v' = 0$ (impermeable surface) so that in the immediate vicinity of a solid surface both $u'$ and $v'$ decrease and $\mu d\bar{u}/dy \gg \rho\overline{u'v'}$. Equation (18.39) thus reduces to

$$\frac{d\bar{u}}{dy} = \frac{\bar{\tau}_S}{\mu} \tag{18.44}$$

which integrates to give

$$\bar{u} = \frac{\bar{\tau}_S y}{\mu} \tag{18.45}$$

or, in wall variables,

$$u^+ = y^+. \tag{18.46}$$

The region where equation (18.46) is valid is termed the **viscous (or linear) sublayer**[177] and taken to have a thickness $\delta_{SUB}$ given by

---

[175] The symbol $u^*$ is also used to represent the friction velocity and spoken as 'ustar'.
[176] $u^+$ and $y^+$ are spoken as 'uplus' and 'yplus', respectively.
[177] Although the turbulence intensity is small, the flow within the viscous sublayer is not purely laminar, and the term **laminar sublayer** is to be avoided.

$$\frac{u_\tau \delta_{SUB}}{\nu} = \delta_{SUB}^+ = 5. \qquad (18.47)$$

## 18.7.2 Fully-turbulent layer and the log law

Beyond the viscous sublayer, $y^+ > 30$, say, it is argued that direct viscous effects on the turbulent structure and the influence of viscosity on the mean flow is negligible so that the mean-velocity gradient is dependent only upon $y$, $\rho$, and $\overline{\tau}_S$, i.e.

$$\frac{d\overline{u}}{dy} = f(y, \rho, \overline{\tau}_S) \qquad (18.48)$$

or, if we introduce $u_\tau$,

$$\frac{d\overline{u}}{dy} = f(y, u_\tau). \qquad (18.49)$$

The only dimensionally acceptable form of equation (18.49) is

$$\frac{y}{u_\tau} \frac{d\overline{u}}{dy} = \frac{1}{\kappa} \qquad (18.50)$$

or

$$y^+ \frac{du^+}{dy^+} = \frac{1}{\kappa} \qquad (18.51)$$

where $\kappa$ is a constant, known as **von Kármán's constant**.
Equation (18.51) can be integrated to give

$$u^+ = \frac{1}{\kappa} \ln y^+ + B \qquad (18.52)$$

where $B$ is also a universal constant. The velocity distribution represented by equation (18.52) is known as the **log law** and has been confirmed experimentally with the values[178] $\kappa = 0.4$, and $B = 5.5$.

An approximate indication of the sublayer thickness, $\delta_{SUB}$, results from determining the value of $y^+$ at which the sublayer profile, represented by equation (18.46), has the same value of $u^+$ as given by the log law. The result is

$$\frac{u_\tau \delta_{SUB}}{\nu} = \delta_{SUB}^+ = 11. \qquad (18.53)$$

This value is obviously much greater than $\delta_{SUB}^+ = 5$ given by equation (18.47) and which represents the wall distance at which the velocity distribution begins to depart from $u^+ = y^+$.

Although the log law was arrived at primarily using dimensional arguments, it can also be deduced using primitive turbulence modelling. In equation (18.36), the momentum equation for a two-dimensional, turbulent boundary layer, the only additional term, compared with equation (17.4) for a laminar boundary layer, is the Reynolds shear stress $\overline{\tau}_T = -\rho\overline{u'v'}$. A natural first step in attempting to account for $-\rho\overline{u'v'}$, first made by Boussinesq in 1877

---

[178] Slightly different values are sometimes quoted: $\kappa = 0.41$, and $B = 5.0$, and there is evidence that $\kappa$ and $B$ are weakly dependent upon Reynolds number.

(two decades before Reynolds introduced the idea of time averaging), was to assume that this quantity behaved in an analogous way to that for molecular shear, i.e.

$$\overline{\tau}_T = -\rho\overline{u'v'} = \mu_T\frac{\partial\overline{u}}{\partial y},\qquad(18.54)$$

which defines the quantity $\mu_T$ known as the **eddy viscosity**.

As we stated in Section 18.5, the earliest example of turbulence modelling was probably Prandtl's suggestion that, by analogy with the small-scale random motion of molecules in a gas (kinetic theory), the large-scale random movements of fluid elements (i.e. the eddies) in turbulent motion leads to a transverse exchange of momentum. If this exchange occurs over an average distance $l_M$, which has become known as the **mixing length**, then

$$\mu_T = \rho l_M^2\left|\frac{\partial\overline{u}}{\partial y}\right|.\qquad(18.55)$$

One way of quantifying Prandtl's idea is to assume that the **root-mean-square values** of $u'$ and $v'$ are approximated by

$$\sqrt{\overline{u'^2}} \approx \sqrt{\overline{v'^2}} \approx l_M\frac{\partial\overline{u}}{\partial y}\qquad(18.56)$$

so that

$$-\overline{u'v'} \approx \sqrt{\overline{u'^2}}\sqrt{\overline{v'^2}} \approx l_M^2\frac{\partial\overline{u}}{\partial y}\left|\frac{\partial\overline{u}}{\partial y}\right| = \nu_T\left|\frac{\partial\overline{u}}{\partial y}\right|.\qquad(18.57)$$

The modulus sign has been introduced to ensure that the Reynolds shear stress and the velocity gradient have the same sign. In practice, it is found that, for turbulent flows in which the mean velocity is asymmetric about a maximum, there is a small region where $-\overline{u'v'}$ and $\partial\overline{u}/\partial y$ are opposite in sign, but this is of little consequence. By analogy with $\nu$, the quantity $\nu_T$ is termed the **kinematic eddy viscosity**.

In the vicinity of a solid surface over which there is turbulent flow, it is reasonable to assume that $l_M$ is proportional to $y$, i.e.

$$l_M = \kappa y\qquad(18.58)$$

where, at this stage, $\kappa$ is simply a constant although equation (18.60) below shows that it can be identified as **von Kármán's constant**, introduced above.

If we substitute for $l_M$ from equation (18.58) in equation (18.57), and assume that $-\rho\overline{u'v'} = \overline{\tau}_S$, then we have

$$\kappa y\frac{\partial\overline{u}}{\partial y} = \sqrt{\frac{\overline{\tau}_S}{\rho}} = u_\tau\qquad(18.59)$$

which can be rewritten as

$$y^+\frac{du^+}{dy^+} = \frac{1}{\kappa}\qquad(18.60)$$

which is identical to equation (18.51) and so again leads to the **log-law** velocity distribution, equation (18.52).

### 18.7.3 Buffer layer

There is no simple theory covering the near-wall range $5 < y^+ < 30$, which is intermediate between the viscous sublayer and the log-law region, often called the buffer layer. A formula suggested by Spalding (1961), which has equation (18.46) as the asymptote for $y^+ \to 0$, and which asymptotes to equation (18.52) at large $y^+$ is

$$y^+ = u^+ + e^{-\kappa B}\left[e^{\kappa u^+} - 1 - \kappa u^+ - \frac{1}{2}\left(\kappa u^+\right)^2 - \frac{1}{6}\left(\kappa u^+\right)^3\right]. \tag{18.61}$$

The negative terms within square brackets can be regarded as correction terms, which are subtracted from the log-law equation in the form

$$y^+ = e^{-\kappa B}e^{\kappa u^+} = e^{-\kappa B}\left[1 + \kappa u^+ + \frac{1}{2}\left(\kappa u^+\right)^2 + \frac{1}{6}\left(\kappa u^+\right)^3 + \frac{1}{24}\left(\kappa u^+\right)^4 + \dots\right]. \tag{18.62}$$

Equations (18.46), (18.52), and (18.61) are all plotted on semi-logarithmic coordinates in Figure 18.1, together with a power-law equation with $A = 8.75$, and $m = 7$ (see Subsection 18.13.3). Spalding's formula has been shown to be an accurate fit to measured velocity distributions for pipe-flow data. The discrepancy between the power-law equation and the log law for $y^+ < 100$ is exaggerated by the logarithmic scale for the abscissa.

Van Driest (1956) suggested that, to account for the viscous sublayer, equation (18.58) for the mixing length should be modified by a so-called **damping factor**

$$l_M = \kappa y \left(1 - e^{-y^+/C}\right) \tag{18.63}$$

where the empirical constant $C$ is usually given the value 26 if the streamwise pressure gradient is zero. If equation (18.63) is substituted in equation (18.57), numerical integration, for a constant-stress layer, results in a velocity distribution close to that corresponding

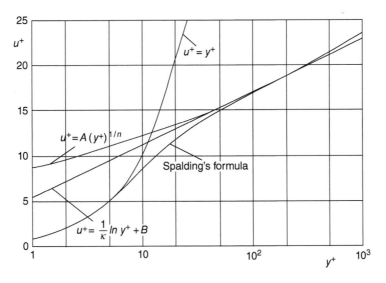

**Figure 18.1** Distributions of mean velocity for near-wall turbulent flow

with Spalding's equation (18.61) but has the advantage that, if $C$ is appropriately modified, pressure-gradient and other effects, such as transpiration, can be incorporated.

### 18.7.4 Outer layer and the Law of the Wake

Coles (1956) observed that, as the free stream is approached, experimentally determined velocity profiles within a turbulent boundary layer increasingly depart from the log law, particularly if there is an adverse pressure gradient, i.e. static pressure increasing with streamwise distance along the surface over which the boundary layer is developing. He showed that a good fit to the data is given by a composite in which a so-called **wake function**[179] $f(\eta)$ is added to the log law

$$u^+ = \frac{1}{\kappa} \ln y^+ + B + \frac{2\Pi}{\kappa} f(\eta) \tag{18.64}$$

where the **wake-strength parameter** $\Pi$ increases in an adverse pressure gradient. The variable $\eta = y/\delta$, $\delta$ being the boundary-layer thickness, and the wake function $f(\eta)$ has a sigmoidal form normalised such that $f(0) = 0$ and $f(1) = 1$. A simple equation which adequately represents the wake function is[180]

$$f(\eta) = 3\eta^2 - 2\eta^3. \tag{18.65}$$

The ratios between $\delta$ and the integral parameters displacement thickness $\delta^*$ (equation ((17.44)) and momentum-deficit thickness $\theta$ (equation (17.61)) were defined in Section 17.3. These ratios, evaluated using equation (18.64) combined with equation (18.65), are given by

$$\frac{\delta^*}{\delta} = \frac{(1+\Pi)}{\kappa} \sqrt{\frac{c_f}{2}} \tag{18.66}$$

and

$$\frac{\theta}{\delta} = \frac{\delta^*}{\delta} - \frac{F(\Pi)}{\kappa^2} \frac{c_f}{2} \tag{18.67}$$

where

$$F(\Pi) = \frac{52}{35}\Pi^2 + \frac{19}{6}\Pi + 2 \tag{18.68}$$

and $c_f$ is the skin-friction coefficient defined by

$$c_f = \frac{\overline{\tau_S}}{\frac{1}{2}\rho U_\infty^2}. \tag{18.69}$$

The shape factor $H$ then follows as

$$\frac{1}{H} = \frac{\theta}{\delta^*} = 1 - \frac{F(\Pi)}{(1+\Pi)\kappa}\sqrt{\frac{c_f}{2}} \tag{18.70}$$

[179] Coles used the term 'wake' because the shape of the function $f(\eta)$ resembles the velocity-defect distribution in a turbulent wake.

[180] Another equation for the wake function which is a good fit to the data is $f(\eta) = sin^2(\pi\eta/2)$ but the cubic form has the advantage that the algebra involved in determining such quantities as $\delta^*$ and $\theta$ is appreciably simpler.

which confirms, as was the case for a laminar boundary layer, that $H > 1$. What is also suggested by equation (18.70) is that, given the dependence on the skin-friction coefficient, $H$ will be Reynolds-number dependent.

Another important result, independent of the form of the wake function, is a consequence of evaluating equation (18.64) at the edge of the boundary layer, $y = \delta$, where $\bar{u} = U_\infty$

$$\sqrt{\frac{2}{c_f}} = \frac{1}{\kappa} \ln \left( \frac{U_\infty \delta}{\nu} \sqrt{\frac{c_f}{2}} \right) + B + \frac{2\Pi}{\kappa} \tag{18.71}$$

wherein we have made use of the identities

$$U_\infty^+ = \sqrt{\frac{2}{c_f}} \quad \text{and} \quad \delta^+ = \frac{u_\tau \delta}{\nu} = \frac{U_\infty \delta}{\nu} \sqrt{\frac{c_f}{2}}. \tag{18.72}$$

For boundary layers subjected to an adverse pressure gradient, experimental measurements increasingly depart from the log law as the pressure-gradient parameter[181]

$$\lambda = \frac{\delta}{\tau_S} \frac{dp}{dx} \tag{18.73}$$

is increased ($\lambda > 10$ corresponds with a strong adverse pressure gradient). For weak favourable pressure gradients ($\lambda < 0$) the wake strength is low so that the log-law equation, equation (18.52), applies throughout the near-wall region, i.e. for a boundary layer this means for $y \le \delta$, while for pipe flow $y \le R$.

---

**ILLUSTRATIVE EXAMPLE 18.1**

Use equation (18.64) with the wake function given by equation (18.65) to show that the ratio of the displacement thickness $\delta^*$ to the boundary-layer thickness $\delta$ for a turbulent boundary layer is given by

$$\frac{\delta^*}{\delta} = \frac{(1 + \Pi)}{\kappa} \sqrt{\frac{c_f}{2}}$$

where $c_f/2$ is the local skin-friction coefficient and $\Pi$ is the wake-strength parameter.

Solution

The definition of the displacement thickness is

$$\delta^* = \int_0^\delta \left( 1 - \frac{\bar{u}}{U_\infty} \right) dy.$$

The mean velocity is approximated by

$$u^+ = \frac{1}{\kappa} \ln y^+ + B + \frac{2\Pi}{\kappa} f(\eta)$$

---

[181] Note that the pressure gradient $\lambda$ defined here is different from those defined for both Poiseuille flow $\lambda_P$, equation (16.52), and a laminar boundary layer, equation (17.97).

where the wake function is given by

$$f(\eta) = 3\eta^2 - 2\eta^3.$$

If we substitute for $\bar{u}$ in the definition of $\delta^*$ we have

$$\delta^* = \delta - \frac{\nu}{u_\tau U_\infty^+} \int_0^{\delta^+} \left( \frac{1}{\kappa} \ln y^+ + B \right) dy^+ - \frac{2\Pi\delta}{\kappa U_\infty^+} \int_0^1 (3\eta^2 - 2\eta^3) \, d\eta$$

$$= \delta - \frac{\nu}{u_\tau U_\infty^+} \left( \frac{\delta^+}{\kappa} \ln \delta^+ - \frac{\delta^+}{\kappa} + B\delta^+ \right) - \frac{\Pi\delta}{\kappa U_\infty^+}$$

$$= \frac{(\Pi + 1)\,\delta}{\kappa U_\infty^+} = \frac{(\Pi + 1)\,\delta}{\kappa} \sqrt{\frac{c_f}{2}}.$$

## 18.8 Fully-developed turbulent flow through a smooth circular pipe

For fully-developed turbulent flow through a smooth circular pipe, velocity-profile measurements show that the wake strength $\Pi$ is small so that equation (18.64) with $\Pi = 0$, i.e. the log-law equation, equation (18.52), is a good approximation to the mean-velocity distribution

$$u^+ = \frac{1}{\kappa} \ln y^+ + B.$$

We can use this equation to calculate a **bulk-average** (or **spatial average**) velocity $\overline{V}$ for fully-developed flow through a pipe of radius $R$ (diameter $D$) through

$$\dot{Q} = \pi R^2 \overline{V} = \int_0^R \bar{u} 2\pi r \, dr \tag{18.74}$$

where $\dot{Q}$ is the volumetric flowrate and $r$ is the radial distance from the pipe centreline. It should be noted that we have neglected not only the wake component of the velocity distribution but also the contribution of the viscous sublayer. The latter approximation is increasingly valid as the pipe Reynolds number increases (see Self-assessment problem 18.2).

Since $r = R - y$, where $y$ is the distance from the pipe wall, equation (18.74) leads to

$$\frac{1}{2}\overline{V}R^2 = R \int_0^R \bar{u} \, dy - \int_0^R \bar{u} \, y \, dy$$

which can be transformed into

$$\frac{u_\tau \overline{V} R^2}{2\nu^2} = R^+ \int_0^{R^+} u^+ dy^+ - \int_0^{R^+} u^+ y^+ dy^+. \tag{18.75}$$

If we substitute for $u^+$ from equation (18.52), we find, after simplification,

$$\overline{V}^+ = \frac{1}{\kappa} \ln R^+ + B - \frac{3}{2\kappa} \tag{18.76}$$

where $\overline{V}^+ = \overline{V}/u_\tau$.

If we define a friction factor $f_F$ as in Section 16.2 (the Fanning friction factor, equation (16.16)), i.e.

$$\frac{f_F}{2} = \frac{\tau_S}{\rho \overline{V}^2} = \left(\frac{u_\tau}{\overline{V}}\right)^2 \tag{18.77}$$

then equation (18.76) leads to

$$\sqrt{\frac{2}{f_F}} = \frac{1}{\kappa} \ln \left( \frac{1}{2} \sqrt{\frac{f_F}{2}} Re_D \right) + B - \frac{3}{2\kappa} \tag{18.78}$$

where $Re_D$ is the Reynolds number based upon $\overline{V}$ and $D$.

This is quite a remarkable result—a friction-factor equation which we have arrived at without direct reference to the equations of motion. With $\kappa = 0.41$, and $B = 5.0$, equation (18.78) gives

$$\sqrt{\frac{2}{f_F}} = 2.439 \ln \left( \sqrt{\frac{f_F}{2}} Re_D \right) - 0.349$$

which is very close to a correlation, valid for $Re_D > 4 \times 10^3$, based upon experimental data[182]

$$\sqrt{\frac{2}{f_F}} = 2.457 \ln \left( \sqrt{\frac{f_F}{2}} Re_D \right) + 0.292. \tag{18.79}$$

---

### ILLUSTRATIVE EXAMPLE 18.2

Air at 25 °C flows through a smooth-walled circular pipe 100 mm in diameter at a bulk velocity of 70 m/s. Calculate the Kolmogorov scales 0.5 mm from the pipe wall, assuming fully-developed turbulent flow. Use the Kármán-Nikuradse formula to calculate the friction factor.

### Solution

$D = 0.1$ m, $\overline{V} = 70$ m/s, $\rho = 1.184$ kg/m$^3$, $\mu = 1.85 \times 10^{-5}$ Pa $\cdot$ s, and $y = 5 \times 10^{-4}$ m.
The Reynolds number $Re_D = 1.184 \times 70 \times 0.1/ \left( 1.85 \times 10^{-5} \right) = 4.48 \times 10^5$, which confirms that the flow is turbulent (i.e. $Re_D > 4 \times 10^3$).
The Kármán-Nikuradse formula is

$$\sqrt{\frac{2}{f_F}} = 2.457 \ln \left( \sqrt{\frac{f_F}{2}} Re_D \right) + 0.292$$

from which the Fanning friction factor $f_F/2 = 1.678 \times 10^{-3}$.
The surface shear stress is then $\tau_S = \rho \overline{V}^2 f_F/2 = 9.734$ Pa so that the friction velocity $u_\tau = \sqrt{\tau_S/\rho} = 2.867$ m/s.
The distance from the surface $y$ in wall units is $y^+ = \rho u_\tau y/\mu = 91.75$.

---

[182] Equation (18.79) is known as the **Kármán-Nikuradse equation**, although according to White (2005) it was originally suggested by Prandtl in 1935. It is sometimes stated in terms of log-base 10 and the Darcy friction factor ($f_D = 8f_F/2$) (see Self-assessment problem 18.9).

The Kolmogorov scales are then calculated as follows
from equation (18.33),

$$\frac{u_\tau l_k}{\nu} = \left(\kappa y^+\right)^{1/4} = 2.477 \text{ so that } l_K = 1.350 \times 10^{-5} \text{ or } 13.5 \ \mu m,$$

from equation (18.34),

$$\frac{u_\tau^2 \tau_K}{\nu} = \left(\kappa y^+\right)^{1/2} = 6.133 \text{ so that } \tau_K = 1.166 \times 10^{-5}s \text{ or } 11.66 \ \mu s$$

and, from equation (18.35),

$$\frac{v_K}{u_\tau} = \frac{1}{\left(\kappa y^+\right)^{1/4}} = 0.404 \text{ so that } v_K = 1.158 \text{ m/s.}$$

Perhaps the most striking thing is how small the length and time scales are at $y^+ = 91.75$, which is just into the log-law region. At the edge of the viscous sublayer (taken as $y^+ = 11$, as we calculated earlier), the values are $l_K = 3.2 \ \mu m$, $\tau_K = 4.04 \ \mu s$, and $v_K = 1.97$ m/s. Self-assessment problem 18.6 concerns the Kolmogorov scales for a boundary layer.

## 18.9 Surface roughness

So far we have dealt with turbulent shear flow over a smooth surface. Even when specially treated, all real surfaces are **hydrodynamically rough** to some degree, i.e. the near-wall flow differs from that for a smooth surface. While small-scale surface roughness has little effect on laminar flow, if the average **height of roughness elements**[183], $\varepsilon$, in turbulent flow is comparable with, or greater than, the thickness of the viscous sublayer $\delta_{SUB}$ then the near-wall (sublayer) velocity distribution and the surface shear stress are affected. If we now include $\varepsilon$ in equation (18.40), we have

$$\bar{u} = f\left(y, \bar{\tau}_S, \rho, \mu, \varepsilon\right), \tag{18.80}$$

so that dimensional analysis leads to

$$\frac{\bar{u}}{u_\tau} = f\left(\frac{u_\tau y}{\nu}, \frac{u_\tau \varepsilon}{\nu}\right) \tag{18.81}$$

or

$$u^+ = f\left(y^+, \varepsilon^+\right) \tag{18.82}$$

where $\varepsilon^+ = u_\tau \varepsilon/\nu$ is the non-dimensional roughness height. Equation (18.47) shows that equation (18.82) can also be written as

$$u^+ = f\left(y^+, \frac{\varepsilon}{\delta_{SUB}}\right) \tag{18.83}$$

which confirms that it is the ratio of the roughness height to the thickness of the viscous sublayer which is important.

---

[183] See footnote 174 regarding the symbol $\varepsilon$.

It should be evident that to represent surface roughness in terms of a simple average rough-ness height is highly simplified and in reality the geometry of the roughness elements plays a role. Roughness may arise from the method of construction or finish of a surface (e.g. riveted, welded, roughly machined, sand blasted, etc.), it may be non-uniform or highly structured (e.g. strips or grooves), it may be a consequence of wear or deposition (e.g. calcium build-up in wa-ter pipes or rust), etc. Roughness is commonly modelled by glueing sand grains of a specified size to a surface.

Experimental studies have shown that the influence of roughness on a near-wall turbulent flow can be categorised as follows

- $\varepsilon^+ < 4$: **hydraulically (or hydrodynamically) smooth**, $\varepsilon < \delta_{SUB}$, and the roughness has no effect on the flow
- $4 < \varepsilon^+ < 60$: transitional-roughness regime
- $\varepsilon^+ > 60$: fully-rough regime where $f_F$ is independent of Reynolds number (see equation (18.90))

Experimental data shows that the effect of roughness on the log law is a downward shift $\Delta B$ dependent upon the magnitude of $\varepsilon^+$, so that

$$u^+ = \frac{1}{\kappa} \ln y^+ + B - \Delta B \left(\varepsilon^+\right) \tag{18.84}$$

where, according to White (2005), based upon **sand-grain roughness** experiments,

$$\Delta B \approx \frac{1}{\kappa} \ln \left(1 + 0.3\varepsilon^+\right). \tag{18.85}$$

If $\Delta B$ from equation (18.85) is substituted into equation (18.84) then, for $0.3\varepsilon^+ \gg 1$, we have

$$u^+ = \frac{1}{\kappa} \ln \left(\frac{y^+}{0.3\varepsilon^+}\right) + B = \frac{1}{\kappa} \ln \left(\frac{y}{\varepsilon}\right) + B - \frac{1}{\kappa} \ln 0.3 = \frac{1}{\kappa} \ln \left(\frac{y}{\varepsilon}\right) + 7.94 \tag{18.86}$$

where we have assumed $\kappa = 0.41$, and $B = 5$.

The principal significance of equation (18.86) is that for large values of the non-dimensional roughness height $\varepsilon^+$ the velocity distribution retains its logarithmic form but loses its dependence on viscosity.

## 18.10 Fully-developed turbulent flow through a rough-surface circular pipe

As for a smooth pipe, the velocity distribution for fully-developed turbulent flow through a pipe with a rough surface can be used to determine an equation for the bulk-average flow velocity, which can be rearranged to give a skin-friction formula. This is left as an exercise for the reader. A useful formula, devised by Colebrook (1939), for pipes with surface roughness representative of commercially available pipes is[184]

---

[184] Equation (18.87) is usually referred to as the **Colebrook-White formula**.

$$\frac{1}{\sqrt{f_D}} = -2log_{10}\left(\frac{2.51}{Re_D\sqrt{f_D}} + \frac{\varepsilon}{3.7D}\right).$$ (18.87)

where $f_D = 8\tau_S/\rho\overline{V}^2$ is the Darcy friction factor. The ratio $\varepsilon/D$ is referred to as **relative roughness** and typically falls within the range $10^{-5} < \varepsilon/D < 0.05$. In terms of the natural logarithm and the Fanning friction factor $f_F = f_D/4$, equation (18.87) transforms to

$$\sqrt{\frac{2}{f_F}} = -2.457\ln\left(\frac{0.887}{Re_D}\sqrt{\frac{2}{f_F}} + \frac{\varepsilon}{3.7D}\right).$$ (18.88)

For a hydraulically smooth pipe, $\varepsilon = 0$, and equation (18.88) reduces to

$$\sqrt{\frac{2}{f_F}} = 2.457\ln\left(Re_D\sqrt{\frac{f_F}{2}}\right) + 0.295$$ (18.89)

which is very close to the empirical Kármán-Nikuradse equation (18.79).

For a fully-rough pipe, where $\varepsilon/D \gg 1/Re_D\sqrt{f_F}$, equation (18.88) leads to

$$\sqrt{\frac{2}{f_F}} = -2.457\ln\left(\frac{\varepsilon}{D}\right) + 3.215$$ (18.90)

and we see that $f_F$ is independent of $Re_D$ and so also of viscosity. The corresponding flow is referred to as **wholly** or **completely turbulent** because the viscous sublayer plays no role.

An equation which is more convenient to use than the Colebrook-White formula is

$$\sqrt{\frac{2}{f_F}} = -2.211\ln\left[\frac{6.9}{Re_D} + \left(\frac{\varepsilon}{3.7D}\right)^{1.11}\right]$$ (18.91)

which is based upon a formula suggested by Haaland (1983).

A diagram in which the Fanning (or Darcy) friction factor is plotted versus pipe Reynolds number $Re_D$ on logarithmic scales for a range of values of the relative roughness height $\varepsilon/D$, calculated from the Colebrook-White equation (18.87), is known as a **Moody chart** (Moody (1944)). Figure 18.2 is a version of the Moody chart for values of $\varepsilon/D$ in the range $10^{-5}$ to 0.05, as well as 0. Equation (16.40) for fully-developed laminar flow through a circular pipe, $f_F Re_D = 16$, is included in Figure 18.2 for reference.

Various sources of roughness were identified in Section 18.9. Typical values for the roughness height $\varepsilon$ are listed in Appendix 5 but should be regarded as no more than guidance to the order of magnitude of the average height of roughness elements likely to be encountered in practice.

The lower limit for the validity of the Colebrook-White formula is usually taken as $Re_D = 4 \times 10^3$, while the upper limit of equation (16.40), for fully-developed laminar pipe flow, is about $Re_D = 2 \times 10^3$. In the **transition region** $2 \times 10^3 < 4 \times 10^3$ (omitted from Figure 18.2) the flow becomes unsteady and there is no simple relationship between $f_F$ and $Re_D$. From a practical point of view, this region is best avoided if a steady flow is required with predictable flow behaviour.

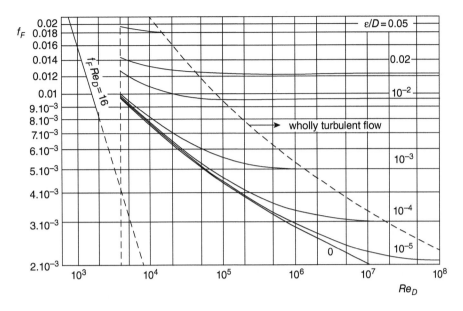

**Figure 18.2** Moody chart: Fanning friction factor $f_F$ versus Reynolds number $Re_D$ for flow through pipes with relative roughness $\varepsilon/D$ (logarithmic scales)

## 18.11 Minor losses in pipe systems

For fluid flow through any practically realistic **pipe system**, in addition to the **stagnation-pressure losses** due to surface friction, termed the **major losses**, account has to be taken of so-called **minor losses**[185], the main causes of which are

- disturbances at the pipe entrance and exit
- sudden increases or decreases in cross section
- gradual increases in cross section (diffusors)
- fully or partially open valves
- bends, elbows, tee junctions, and other pipe fittings

Since pressure can be regarded as a form of energy (see Subsection 7.5.1), the term **pressure loss** is misleading as the mechanical energy concerned is simply converted into heat, and no energy is actually lost. However, the term is well established and so will continue to be used here. From the list above it is apparent that pressure losses are associated primarily with friction, area change, and direction change.

The majority of flows of engineering significance have sufficiently high Reynolds numbers that they are turbulent. Commercial software packages, based upon a wide variety of turbulence models, are now available for the calculation to acceptable levels of accuracy of the flow characteristics of pipe systems, including velocity, pressure, and turbulence-intensity distributions, as well as overall pressure loss. However, if all that is needed is the calculation of overall stagnation-pressure loss from inlet to outlet of a system, it is usually adequate to characterise each component (i.e. **pipe fitting**) in the system using an empirical **loss coefficient** $K$.

---

[185] In spite of the names, it is often the case that in practice the minor losses exceed the minor losses.

For an incompressible fluid of density $\rho$, the definition of $K$ is

$$K = \frac{\Delta p_0}{\frac{1}{2} \rho \overline{V}^2} \tag{18.92}$$

where $\Delta p_0$ is the loss in stagnation pressure across an individual component for a bulk flow velocity $\overline{V}$ at its inlet. If there is no change in cross-sectional area from inlet to outlet of a component, then the changes in static and stagnation pressure are equal.

Although $K$ depends primarily on the basic shape of a component, details of the internal geometry are also important. For example, for a sudden contraction $K$ can be reduced from 0.5 to 0.02 by appropriate rounding of the inlet. In the case of a bend[186], the pressure loss is a consequence of the bend radius (which may not be constant), which gives rise to **secondary flows** (counter-rotating vortices) due to centripetal acceleration. The loss may also be associated with flow separation on the low-radius side and affected by the cross-section shape, internal surface roughness, method of installation (e.g. flanged or threaded), which is often unstated but has a major influence, and the Reynolds number ($K$ typically decreases as the Reynolds number increases). Loss coefficients for all fittings are also affected by the upstream flow conditions, higher losses being associated with a fully-developed upstream flow rather than a uniform flow. For large bends with a rectangular cross section, as is typical for a wind or water tunnel, it is usual to incorporate a **cascade** of **guidevanes** (see Section 10.9) to reduce losses and improve flow quality.

As will be shown in Subsection 18.11.1, for a sudden enlargement a good estimate for the loss coefficient $K_{SE}$ is given by an analysis based upon the linear momentum equation. A similar analysis (Subsection 18.11.2) for a sudden contraction requires a correction factor, however. Subject to the influences mentioned in the previous paragraph, guideline values of $K$ for various elbows and **tee junctions** are listed in Table 18.1.

For more accurate values it is necessary to consult the manufacturer's literature for a given fitting. A regular elbow, a long-radius elbow, a line-flow tee junction, and a branch-flow tee junction are shown schematically in Figure 18.3.

## 18.11.1  Sudden enlargement and Borda-Carnot equation

In Section 10.5 it was shown that the changes in static $p$ and stagnation pressure $p_0$ for flow through a **sudden enlargement** can be calculated by applying the linear momentum equation to the flow, which leads to

$$p_2 - p_1 = \frac{\dot{m}^2}{\rho A_2} \left( \frac{1}{A_1} - \frac{1}{A_2} \right) \tag{18.93}$$

---

[186] The terms **pipe bend** and **pipe elbow** tend to be used interchangeably and inconsistently. Both refer to a component which joins two sections of pipe where there is an angle between the two. It is sometimes said that all elbows are bends but not all bends are elbows. The difference is that the term bend is generic and describes an offset or change in the direction of piping, while an elbow is a component prefabricated to a standard, the bend angle usually being 45°, 90°, or 180°, although any angle is clearly possible. An elbow with an angle of 180° is referred to as a **return bend**. If the nominal (internal) pipe diameter is $D$, the bend radius for a **standard** (or **regular** or **short-radius**) elbow is $1D$, while for a **long-radius elbow** the standard bend radius is $1.5D$. Other common choices for the radius of an elbow are $3D$ and $5D$.

**Table 18.1** Loss coefficients ($K$) for pipe elbows and tee junctions

| Elbows | |
| --- | --- |
| 45° standard radius, flanged | 0.2 |
| 45° standard radius, threaded | 0.4 |
| 45° long radius, flanged | 0.2 |
| 90° standard radius, flanged | 0.3 |
| 90° standard radius, threaded | 1.5 |
| 90° long radius, flanged | 0.2 |
| 90° long radius, threaded | 0.7 |
| 180° standard radius, flanged | 0.2 |
| 180° standard radius, threaded | 1.5 |
| Tee junctions | |
| Line flow, flanged | 0.2 |
| Line flow, threaded | 0.9 |
| Branch flow entering line | 1.3 |
| Line flow entering branch | 1.5 |

and the **Borda-Carnot equation**

$$p_{0,1} - p_{0,2} = \frac{\dot{m}^2}{2\rho} \left( \frac{1}{A_1} - \frac{1}{A_2} \right)^2 \qquad (18.94)$$

where $\rho$ is the (constant) fluid density, $\dot{m}$ is the mass flowrate, $A$ is the cross-sectional area, and the subscripts 1 and 2 refer to the regions upstream and downstream of the enlargement, respectively.

If the upstream diameter is $d$, and the downstream diameter is $D$, it is straightforward to show from equation (18.94) that for a sudden enlargement the loss coefficient $K_{SE}$ is given by

$$K_{SE} = \frac{\Delta p_0}{\frac{1}{2}\rho \overline{V}_1^2} = \left[ 1 - \left( \frac{d}{D} \right)^2 \right]^2. \qquad (18.95)$$

For flow from a duct with area issuing into the surroundings, equation (18.95) with $D \to \infty$ leads to $K_{SE} = 1$, i.e. the stagnation-pressure loss is equal to the dynamic pressure upstream of the contraction.

The variation of $K_{SE}$ with diameter ratio is shown in Figure 18.5.

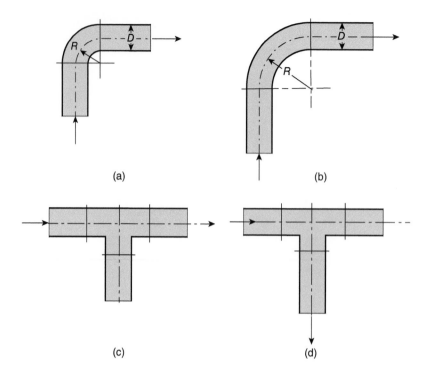

**Figure 18.3** Pipe fittings: (a) regular pipe elbow (b) long-radius pipe elbow (c) line-flow tee junction (d) branch-flow tee junction

### 18.11.2 **Sudden contraction**

As shown in Figure 18.4, flow through a **sudden contraction** is complicated by the fact that the flow separates at the corner of the contraction, in the same way as for an **orifice-plate flowmeter**. Account then has to be taken of the occurrence of a *vena contracta* downstream of the contraction, the cross-sectional area of which ($A_V$) is unknown. If we treat the flow

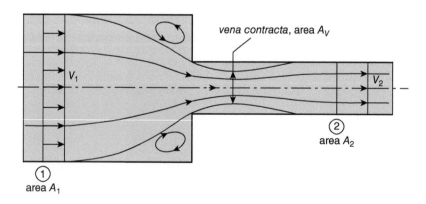

**Figure 18.4** Schematic diagram of flow through a sudden contraction, showing separated flow and the *vena contracta*

between the *vena contracta* and the downstream pipe as though it were a sudden enlargement, and also assume that there is zero loss in stagnation pressure between the upstream region so that $p_{0,V} = p_{0,1}$, then we can use the Borda-Carnot equation, equation (18.94), to calculate the loss in stagnation pressure as

$$p_{0,1} - p_{0,2} = \frac{\dot{m}^2}{2\rho} \left( \frac{1}{A_V} - \frac{1}{A_2} \right)^2.$$
(18.96)

It is usual for a sudden contraction to refer to the bulk-average velocity in the downstream pipe, $\overline{V}_2$, so that

$$\frac{p_{0,1} - p_{0,2}}{\frac{1}{2}\rho\overline{V}_2^2} = \left( \frac{A_2}{A_V} - 1 \right)^2 = K_{SC}$$
(18.97)

where $K_{SC}$ is the loss coefficient for the sudden contraction. Since the ratio $A_V/A_2$ depends upon the overall contraction ratio $A_2/A_1$ so does the loss coefficient. For a contraction with sharp edges, White (2011) recommends the empirical formula

$$K_{SC} = 0.42 \left[ 1 - \left( \frac{d}{D} \right)^2 \right]$$
(18.98)

where $d$ is the diameter of the downstream pipe and $D$ is the diameter of the upstream pipe. Other writers suggest that the coefficient should be 0.5 rather than 0.42.

The variation of $K_{SC}$ with diameter ratio is shown in Figure 18.5. Rounding the contraction edges reduces $K_{SC}$ considerably: by about 50% if the edge radius is 0.06D, and 95% for 0.25D.

### 18.11.3 Total stagnation-pressure loss

For a pipe of diameter $D$ and length $L$ with $N$ fittings, the overall stagnation-pressure loss $\Delta p_{0,OVERALL}$ is given by

$$\Delta p_{0,OVERALL} = \Delta p_L + \sum_{i=1}^{N} \Delta p_{0,i}$$
(18.99)

where $\Delta p_L$ is the pressure loss over the pipe length and $\Delta p_{0,i}$ is the stagnation-pressure loss across the $i^{th}$ fitting.

If $f_F$ is the Fanning friction factor then, assuming fully-developed flow in the pipe with bulk-average velocity $\overline{V}$,

$$\Delta p_L = \frac{4\tau_S L}{D} = 2\rho\overline{V}^2 f_F \frac{L}{D}$$
(18.100)

where $\tau_S$ is the surface shear stress within the pipe. If $\overline{V}$, and hence $Re_D$, are known, the Fanning friction factor can be calculated from the Colebrook-White equation, from Section 18.10, or obtained from the Moody chart.

If the pipe diameter changes between fittings, then the pressure loss in each section has to be calculated separately to account for the changes in $\overline{V}$. Clearly, if a section is too short, the assumption of fully-developed flow (implied by equation (18.100)) becomes invalid.

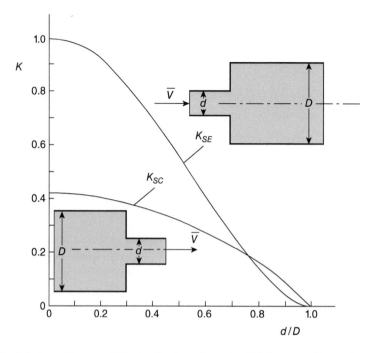

**Figure 18.5** Variation of loss coefficient with diameter ratio $d/D$ for (a) sudden enlargement ($K_{SE}$) and (b) sharp-edge sudden contraction ($K_{SC}$).

For fitting $i$ we have

$$\Delta p_{0,i} = K_i \frac{1}{2} \rho \overline{V}_i^2 \qquad (18.101)$$

so that, finally,

$$\Delta p_{0,OVERALL} = 2\rho \overline{V}^2 f_F \frac{L}{D} + \frac{1}{2}\rho \sum_{i}^{N} K_i \overline{V}_i^2. \qquad (18.102)$$

As we saw in Section 4.3, a static-pressure difference $\Delta p$ can be represented in terms of the height of a column of liquid $h = \Delta p/\rho g$, the height being referred to as the head. This concept can be applied to any fluid but, in the case of gases and vapours, the density of a reference liquid, such as water, an oil, or mercury, has to be introduced. The overall pressure loss expressed in this way is referred to as a **head loss**.

---

**ILLUSTRATIVE EXAMPLE 18.3**

In a chemical plant, paraffin oil (kerosene) with a specific density of 0.804 and dynamic viscosity $1.92 \times 10^{-3}$ Pa·s is pumped between two large containers at a volumetric flowrate of 0.006 m³/s through a pipe with diameter $D = 50$ mm and length $L = 150$ m. The relative roughness is $10^{-4}$. Installed within the pipe are the following: a standard 90° elbow with radius $D$, a long-radius 90° elbow with radius 1.5 $D$, both flanged, and two partially open valves with loss coefficients 2.0 and 5.0, respectively. The flow enters the pipe through a sharp-edged entrance and leaves through a sharp-edged exit. Calculate the overall pressure difference and the pumping power required. Both containers are on the same horizontal level.

## Solution

$D = 0.04$ m, $L = 150$ m, $\dot{Q} = 0.009$ m$^3$/s, $\rho = 804$ kg/m$^3$, $\mu = 1.92 \times 10^{-3}$ Pa·s, $\varepsilon/D = 10^{-4}$, $K_1 = 0.42$ (pipe inlet), $K_2 = 2.0$ (first valve), $K_3 = 0.3$ (flanged 90° standard elbow, $R_3 = D$), $K_4 = 0.2$ (flanged 90° long-radius elbow, $R_4 = 1.5\,D$), and $K_5 = 5.0$ (second valve).

Mean velocity in pipe $\overline{V} = 4\dot{Q}/\pi D^2 = 7.162$ m/s.

Reynolds number $Re_D = \rho \overline{V} D/\mu = 1.2 \times 10^5$.

From the Moody chart a first estimate for the Fanning friction factor is $f_F = 4.5 \times 10^{-3}$.

With $f_F = 4.5 \times 10^{-3}$ as an initial estimate, from the Colebrook-White equation $f_F = 4.475 \times 10^{-3}$.

The sum of the five loss coefficients $\sum K_i = 7.92$.

The overall stagnation-pressure difference $\Delta p_{0,OVERALL}$ is given by

$$\Delta p_{0,TOTAL} = 2\rho \overline{V}^2 f_F \frac{L}{D} + \frac{1}{2}\rho \sum_i^N K_i \overline{V}^2 = 1.547 \text{ MPa}$$

and the required pumping power $P = \dot{Q}\Delta p_{0,OVERALL} = 13.92$ kW.

The electrical power supplied to the pump would need to be about 25% higher, given that pumps are less than 100% efficient.

## 18.12 Momentum-integral equation

In Section 17.5 we showed that, by considering the forces acting on a control volume of infinitesimal width in the streamwise direction, and the flowrates of streamwise momentum into and out of the control volume, we can derive von Kármán's momentum-integral equation for a two-dimensional, constant-property, laminar boundary layer

$$\frac{c_f}{2} = \frac{d\theta}{dx} + (H + 2)\frac{\theta}{U_\infty}\frac{dU_\infty}{dx}. \tag{17.86}$$

This equation can also be derived by formal integration across the boundary layer of the boundary-layer form of the Navier-Stokes equations. The same approach can be applied to the Reynolds-averaged equations for a turbulent boundary layer, with the result

$$\frac{c_f}{2} = \frac{d\theta}{dx} + (H + 2)\frac{\theta}{U_\infty}\frac{dU_\infty}{dx} + \frac{1}{U_\infty^2}\frac{d}{dx}\int_0^\infty \left(\overline{u'^2} - \overline{v'^2}\right) dy \tag{18.103}$$

although it is usual to assume that the integral involving the normal-stress terms is negligible. The momentum-integral equation for a turbulent boundary layer is then identical to that for a laminar boundary layer, although it has to be remembered that the momentum and displacement thicknesses, $\theta$ and $\delta^*$, respectively, have to be calculated from the distribution of the mean velocity $\overline{u}(y)$ for a turbulent boundary layer.

## 18.13  Flat-plate boundary layer

### 18.13.1  Wall-plus-wake velocity profile

From the wall-plus-wake velocity distribution, the skin-friction coefficient $c_f/2$ is given by equation (18.71)

$$\sqrt{\frac{2}{c_f}} = \frac{1}{\kappa} \ln\left(\sqrt{\frac{c_f}{2}} \frac{U_\infty \delta}{\nu}\right) + B + \frac{2\Pi}{\kappa} \qquad (18.71)$$

$\Pi$ being the wake parameter, and $\delta$ the boundary-layer thickness.

In Subsection 18.7.4 we showed that the momentum-deficit thickness $\theta$ corresponding to the combined log-law and law-of-the-wake velocity distribution, equation (18.64), is given by

$$\frac{\theta}{\delta} = \left(\frac{\Pi + 1}{\kappa}\right)\sqrt{\frac{c_f}{2}} - \frac{F(\Pi)}{\kappa^2}\frac{c_f}{2} \qquad (18.104)$$

where

$$F(\Pi) = \frac{52}{35}\Pi^2 + \frac{19}{6}\Pi + 2 \qquad (18.68)$$

so that equations (18.71) and (18.104) can be combined to eliminate $\delta$ and produce the following equation connecting the skin-friction coefficient $c_f$, the momentum-thickness Reynolds number, defined by $Re_\theta = U_\infty \theta / \nu$, and the wake parameter $\Pi$

$$\frac{U_\infty \theta}{\nu} = Re_\theta = \left[\Pi + 1 - F(\Pi)\frac{1}{\kappa}\sqrt{\frac{c_f}{2}}\right]\frac{e^{\kappa\sqrt{2/c_f}}}{\kappa} e^{-(B\kappa + 2\Pi)}. \qquad (18.105)$$

In principle equation (18.105) can be seen as a relationship between the momentum-thickness Reynolds number $Re_\theta$ and $c_f/2$ although the form of the equation is inconvenient, and the value (or $x$-variation) of the wake parameter $\Pi$ is, as yet, unknown.

For a zero-pressure-gradient (flat-plate) boundary layer the momentum-integral equation (18.103), neglecting the $\overline{u'^2} - \overline{v'^2}$ term, reduces to

$$\frac{d\theta}{dx} = \frac{c_f}{2}. \qquad (18.106)$$

Substitution for $\theta$ from equation (18.105) in equation (18.106) leads to an ordinary differential equation which can be solved to give the following equation for the streamwise Reynolds number $Re_x$, if the wake parameter $\Pi$ is assumed to be independent of $x$

$$Re_x = \frac{1}{\kappa^3}\left[(\Pi + 1)\left(\frac{2\kappa^2}{c_f} - 2\kappa\sqrt{\frac{2}{c_f}} + 2\right) - F(\Pi)\left(\kappa\sqrt{\frac{2}{c_f}} - 2\right)\right]e^{\kappa\sqrt{2/c_f}} e^{-(B\kappa + 2\Pi)}. \qquad (18.107)$$

Unfortunately, it is not possible to rearrange the equation such that $c_f$ is an explicit function of $Re_x$ so that, given $Re_x$, an iterative procedure is required to determine $c_f$. Also, although the algebra leading to equation (18.107) is straightforward, it is quite tedious and requires the assumption that $\Pi$ is constant. For a flat-plate boundary layer a value for $\Pi$ of about 0.45 is found to be a good fit to experimental data. With $\Pi = 0$ what remains of the wall-plus-wake equation (18.64) is the log law, but even then the resulting relationships between $c_f$ and both $Re_\theta$ and $Re_x$ are inconvenient for further analysis.

## 18.13.2 Empirical drag laws

A number of purely empirical relations between $c_f$ and $Re_x$, known as **drag laws**, have been proposed, one of the earliest being that suggested by Schultz-Grunow (1940) for the range $10^6 \leq Re_x \leq 10^9$

$$\frac{c_f}{2} = \frac{0.185}{\left(log_{10}Re_x\right)^{2.584}} \qquad (18.108)$$

while, more recently, White (2005) suggested

$$\frac{c_f}{2} = \frac{0.2275}{\ln^2 (0.06 Re_x)}. \qquad (18.109)$$

Both formulae have the merit that $c_f$ is an explicit function of $Re_x$ but, as we shall see in Subsection 18.13.3, it is also useful to express the skin-friction coefficient in terms of the momentum-thickness Reynolds number $Re_\theta = U_\infty\theta/\nu$.

## 18.13.3 Power-law velocity profile

It is straightforward to derive explicit relationships between $c_f$ and both $Re_\theta$ and $Re_x$ if the velocity-profile assumption (for a flat-plate boundary layer) takes the **power-law** form:

$$u^+ = A \left(y^+\right)^m \qquad (18.110)$$

where $A$ and $m$ are constants. Such an assumption is tantamount to assuming $\Pi = 0$: as shown in Figure 18.1, with $A = 8.75$, and $m = 1/7$, equation (18.110) is close to the log law, equation (18.52), for $y^+ < 1500$.

If equation (18.110) is written in the form

$$\frac{\bar{u}}{U_\infty} = \left(\frac{y}{\delta}\right)^m = \xi^m, \qquad (18.111)$$

where $\xi = y/\delta$, from the definition of $\theta$, equation (17.46), we have

$$\frac{\theta}{\delta} = \frac{1}{\delta} \int_0^\delta \frac{\bar{u}}{U_\infty} \left(1 - \frac{\bar{u}}{U_\infty}\right) dy = \int_0^1 \xi^m \left(1 - \xi^m\right) d\xi = \frac{m}{(m+1)(2m+1)}. \qquad (18.112)$$

With $m = 1/7$ equation (18.112) gives $\theta/\delta = 7/72 = 0.0972$, i.e. $\theta \ll \delta$.

According to the momentum-integral equation for a flat-plate boundary layer,

$$\frac{d\theta}{dx} = \frac{c_f}{2} \qquad (18.106)$$

so that

$$\frac{c_f}{2} = \frac{m}{(m+1)(2m+1)} \frac{d\delta}{dx} = \frac{m}{(m+1)(2m+1)} \frac{dRe_\delta}{dRe_x} \qquad (18.113)$$

where $Re_x = U_\infty x/\nu$, and $Re_\delta = U_\infty\delta/\nu$.

From equation (18.110) at $y = \delta$

$$U_\infty^+ = A\left(\delta^+\right)^m$$

which can be transformed into a skin-friction equation in terms of $Re_\delta$

$$\frac{c_f}{2} = A^{-2/(m+1)}Re_\delta^{-2m/(m+1)}.$$ (18.114)

With $A = 8.75$, and $m = 1/7$, this becomes

$$\frac{c_f}{2} = 0.0225\,Re_\delta^{-0.25}.$$ (18.115)

Substitution for $\delta$ from equation (18.112) allows equation (18.114) to be transformed into the required relationship between $c_f/2$ and $Re_\theta$

$$\frac{c_f}{2} = A^{-2/(m+1)}\left[\frac{m}{(m+1)(2m+1)}\right]^{2m/(m+1)}Re_\theta^{-2m/(m+1)}.$$ (18.116)

With the values for $A$ and $m$ used above we have

$$\frac{c_f}{2} = 0.0125\,Re_\theta^{-0.25}.$$ (18.117)

If equation (18.114) is used to eliminate $Re_\delta$ from equation (18.113) to give a differential equation for $c_f/2$, we find, after integration

$$\frac{c_f}{2} = \left[A^{-2/(3m+1)}\frac{m}{(2m+1)(3m+1)}\right]^{2m/(3m+1)}Re_x^{-2m/(3m+1)}$$ (18.118)

and, with $A = 8.75$, and $m = 1/7$, we have

$$\frac{c_f}{2} = 0.0288\,Re_x^{-0.2}.$$ (18.119)

The equations for $c_f/2$ derived here from the power-law equation are less accurate than the empirical equations based directly on experimental data. They are, however, very convenient for analytical studies.

It may be remarked that this analysis of a flat-plate turbulent boundary layer is similar in a number of ways to the profile-method of analysis for a laminar boundary layer presented in Section 17.6.

---

### ILLUSTRATIVE EXAMPLE 18.4

(a) Calculate the value of $c_f/2$ given by White's equation, equation (18.109), for a flat-plate turbulent boundary layer with $Re_x = 10^9$.
(b) Solve equation (18.107) with $\Pi = 0.45$, $\kappa = 0.41$, and $B = 5.0$ to find the value of $Re_x$ corresponding to the value of $c_f/2$ found in part (a).
(c) Calculate the value of $c_f/2$ given by White's equation with the value of $Re_x$ obtained in part (b). Comment on the results.

## Solution

(a) White's equation (equation (18.109)) is

$$\frac{c_f}{2} = \frac{0.2275}{\ln^2(0.06Re_x)}.$$

With $Re_x = 10^9$ this equation gives $c_f/2 = 0.2275/\ln^2(0.06 \times 10^9) = 7.092 \times 10^{-4}$.

(b) Equation (18.107) is

$$Re_x = \frac{1}{\kappa^3}\left[(\varPi+1)\left(\frac{2\kappa^2}{c_f} - 2\kappa\sqrt{\frac{2}{c_f}} + 2\right) - F(\varPi)\left(\kappa\sqrt{\frac{2}{c_f}} - 2\right)\right]e^{\kappa\sqrt{2/c_f}}e^{-(B\kappa+2\varPi)}$$

with

$$F(\varPi) = \frac{52}{35}\varPi^2 + \frac{19}{6}\varPi + 2.$$

With $\varPi = 0.45$ we find $F(\varPi) = 3.726$. It is convenient to substitute $\beta = \kappa\sqrt{2/c_f}$ so that

$$Re_x = \frac{1}{0.41^3}\left[1.45(\beta^2 - 2\beta + 2) - 3.726(\beta - 2)\right]e^{\beta}e^{-(5\times0.41+0.9)}.$$

With $c_f/2 = 7.092 \times 10^{-4}$ we have $\beta = 15.40$ and, finally, $Re_x = 9.289 \times 10^8$.

(c) If we substitute the value of $Re_x$ from part (b) into White's equation we find $c_f/2 = 7.151 \times 10^{-4}$.

## Comments:

The value for $Re_x$ found in part (b) is within 7% of $10^9$, the value used to determine $c_f/2$ from White's equation, which initially suggests a significant discrepancy. However, this value of $Re_x$ leads to a value of $c_f/2$ within 1% of that given by White's equation, showing that the dependence of $c_f/2$ on $Re_x$ is very weak and also that the log law plus wake function leads to a skin-friction equation which is of comparable accuracy with experimental data as represented by White's empirical equation.

---

### 18.13.4 Flat-plate boundary-layer transition

It is usually the case, unless special measures are used to 'trip' the laminar boundary layer which develops from the leading edge of a flat plate, that transition to turbulent flow occurs following the growth of instabilities within the boundary layer. The streamwise distance over which transition occurs is usually relatively short and it is sufficient to assume that transition occurs instantaneously once a critical Reynolds number is reached.

For a zero pressure-gradient boundary layer on a smooth surface, an appropriate value for the critical streamwise Reynolds number $Re_{x,C} = U_\infty x/\nu$ is $3\times10^6$ if the free-stream turbulence level $\bar{k}_\infty < 1.5 \times 10^{-6}U_\infty^2$, where $\bar{k}_\infty$ is the turbulent kinetic-energy per unit mass in the free stream (i.e. for $y > \delta$). Experiments show that the value of $Re_{x,C}$ decreases monotonically for higher values of $\bar{k}_\infty$, becoming negligible for $\bar{k}_\infty > 10^{-3}U_\infty^2$. Surface roughness also leads to a decrease in $Re_{x,C}$, a value of $5 \times 10^5$ being appropriate for a typical industrial material where low friction is important.

If we assume that the boundary layer remains laminar up to the point of transition, $x_C$, then, based upon the result for zero-pressure gradient tabulated in Table 17.2 in Chapter 17, the momentum thickness $\theta_C$ corresponding with $x_C$ is given by

$$\frac{\theta_C}{x_C}\sqrt{Re_{x,C}} = 0.6641$$

from which we conclude, if $Re_{x,C} = 3 \times 10^6$, then

$$Re_{\theta,C} \approx 1150. \tag{18.120}$$

It is sometimes more convenient to present results as functions of the **momentum-thickness Reynolds number** $Re_\theta$ rather than the streamwise Reynolds number $Re_x$.

Once the transition-onset location $x_C$ has been determined, the continuing development of the (now turbulent) boundary layer can be calculated as follows. If it is assumed that the transition region is short compared with the downstream stretch of turbulent boundary layer, then it is reasonable to assume that both the mass flowrate per unit width within the boundary layer $\dot{m}'$ given by

$$\dot{m}' = \rho \int_0^\delta u \, dy$$

and the corresponding momentum flowrate per unit width $\dot{M}'$ given by

$$\dot{M}' = \rho \int_0^\delta u^2 \, dy$$

remain unchanged across the transition region. From this it can be concluded that the momentum thickness also remains unchanged, since

$$\theta = \int_0^\delta \frac{u}{U_\infty}\left(1 - \frac{u}{U_\infty}\right) dy = \frac{\dot{m}'}{\rho U_\infty} - \frac{\dot{M}'}{\rho U_\infty^2}. \tag{18.121}$$

Assuming that the momentum thickness remains unchanged leads to the interesting conclusion that the boundary-layer thickness also changes only slightly. We found in Subsection 17.3.2 that, for a flat-plate laminar boundary layer,

$$\frac{\theta_L}{\delta} = 0.09393 \tag{17.47}$$

and, assuming the $1/7^{\text{th}}$ power-law velocity profile for a turbulent boundary layer,

$$\frac{\theta_T}{\delta} = \frac{7}{72} = 0.0972 \tag{18.122}$$

so that, at the location of transition,

$$\frac{\theta_T}{\theta_L} = 1.035.$$

This value of $\theta_T/\theta_L$ is subject to many uncertainties: the value of $\delta_L$ depends upon how close to unity is the value of $u/U_\infty$ at which it is decided that the edge of the boundary layer has

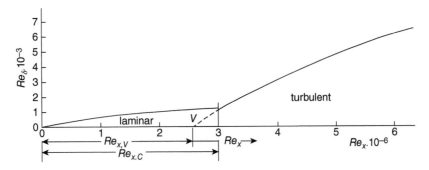

**Figure 18.6** Schematic diagram of transition of a flat-plate boundary layer (note that there is a factor of 250 between the $Re_x$- and $Re_\delta$-scales)

been reached (according to Table 17.1 in Chapter 17, the value of $y/\Delta$ which leads to equation (17.47) corresponds to $u/U_\infty = 0.99994$), and the value of $\delta_T$ is a consequence of the choice of the simple power-law representation of the velocity profile. Nevertheless, it is clear that the boundary-layer thickness at has to adjust to accomodate the redistribution of momentum which occurs during transition.

The continuing development of the (now turbulent) boundary layer can be calculated assuming that it originates from a **virtual origin** such that the momentum thickness of the turbulent boundary layer at $x_C$ is equal to the momentum thickness of the laminar boundary layer at that location. The flow configuration is illustrated in Figure 18.6, with the virtual origin marked by $V$. The more rapid growth rate of the turbulent boundary layer is clearly seen. The consideration here is a good example of the merit in specifying a transition criterion in terms of the momentum-thickness Reynolds number.

---

**ILLUSTRATIVE EXAMPLE 18.5**

A flat plate 20 m long and 5 m wide is placed in an airflow at 20 °C having a velocity of 40 m/s. The streamwise pressure gradient is zero. Assume that the boundary layer on the plate transitions from laminar to turbulent instantaneously at a location where the streamwise Reynolds number $Re_x = 3 \times 10^6$.

   (i) Calculate the streamwise location $x_C$ at which transition occurs and the corresponding values of the momentum-thickness Reynolds number $Re_{\theta,C}$, the momentum thickness $\theta_C$, the surface shear stress $\tau_{S,C}$, and the drag force $D_L$ for the laminar-flow section.

   (ii) Calculate the location $x_V$ of the virtual origin of the turbulent boundary layer, the surface shear stress at the location $x_C$ for the turbulent boundary layer, the surface shear stress at the end of the plate $\tau_{S,E}$, and the total drag force $D_T$ over the turbulent-flow section. Hence calculate the drag force $D_E$ exerted by the flow on the entire plate.

  (iii) What would be the drag force if the flow were entirely laminar from the leading edge to the trailing edge of the plate?

  (iv) What would be the drag force if the flow were entirely turbulent from the leading edge of the plate?

## Solution

$L = 20$ m, $W = 5$ m, $\mu = 1.82 \times 10^{-5}$ Pa·s, $\rho = 1.204$ kg/m³, $U_\infty = 40$m/s, and $Re_C = 3 \times 10^6$. To distinguish, where appropriate, the laminar section from the turbulent section of the boundary layer we shall use the subscripts $L$ and $T$, respectively, while subscript $E$ will denote the end of the plate, and $C$ the location of transition.

(i) From $Re_C = 3 \times 10^6$, we have

$$x_C = \mu \, Re_C/(\rho U_\infty) = 1.82 \times 10^{-5} \times 3 \times 10^6/(1.204 \times 40) = 1.134 \; m.$$

For a flat-plate laminar boundary layer, from Table 17.2, $Re_\theta = 0.6641\sqrt{Re_x}$, so that $Re_{\theta,C} = 0.6641\sqrt{Re_{x,C}}$. We thus find $Re_{\theta,C} = 0.6641 \times \sqrt{3 \times 10^6} = 1150.3$, so that $\theta_C = \mu Re_{\theta,C}/(\rho U_\infty) = 4.347 \times 10^{-4}$ m, or 0.435 mm.
From Table 17.2, for a zero-pressure-gradient laminar boundary layer $(c_{f,L}/2)Re_\theta = 0.2205$ so that at $x = x_C$ we have $c_{f,L}/2 = 0.2205/1150.3 = 1.917 \times 10^{-4}$. Since $c_f/2 = \tau_S/\rho U_\infty^2$, we find $\tau_{S,C} = 1.204 \times 40^2 \times 1.917 \times 10^{-4} = 0.369$ Pa.
From equation (17.89), the drag force exerted over the length $x_C$ is given by $D_C = \rho U_\infty^2 \theta_C W = 4.19$ N. The same result is arrived at by noting that, since $\tau_S \propto x^{-1/2}$ for a laminar, flat-plate boundary layer, the average wall shear stress over any length $x = 2\tau_{S,x}$, and so $D_C = 2\tau_{S,C} W x_C$.

(ii) It is assumed that the momentum thickness for the turbulent boundary layer at the location of transition is unchanged from that for preceding the laminar boundary layer, i.e. $\theta_C = 0.435$ mm, and $Re_{\theta,C} = 1150.3$.
We shall use the symbol $\chi$ to represent streamwise distance from the virtual origin, as shown in Figure 18.6, i.e. $\chi = x - x_V$. The location of the virtual origin $x_V$ is then found as follows. According to equations (18.117) and (18.119), for the turbulent boundary layer at the transition location,

$$\frac{c_{f,T}}{2} = 0.0125 \, Re_{\theta,C}^{-0.25} = 0.0288 \, Re_{\chi,C}^{-0.2}$$

from which we find

$$\frac{c_{f,T}}{2} = 2.146 \times 10^{-3} \text{ and } Re_{\chi,C} = 4.349 \times 10^5$$

so that

$$\chi_C = \frac{\mu}{\rho U_\infty} Re_{\chi,C} = 0.164 \text{ m and } \tau_{S,C} = \rho U_\infty^2 \frac{c_{f,T}}{2} = 4.135 \text{ Pa.}$$

Given that $\chi_C = x_C - x_V$ (see Figure 18.6), we have $x_V = x_C - \chi_C = 1.134 - 0.164 = 0.970$ m. The distance from the virtual origin to the end of the plate $\chi_E = L - x_V = 19.03$ m. The corresponding Reynolds number $Re_{\chi,E} = \rho U_\infty \chi_E/\mu = 5.036 \times 10^7$, and the skin-friction coefficient, from equation (18.119), $c_{f,T}/2 = 0.0288 \, Re_{\chi,E}^{-0.2}$, is $c_{f,T}/2 = 8.298 \times 10^{-4}$. The shear stress at the end of the plate is then $\tau_{S,E} = \rho U_\infty^2 c_{f,T}/2 = 1.599$ Pa.
The drag force over the turbulent section $D_T$ is obtained from

$$D_T = W \int_{\chi_C}^{\chi_E} \tau_{S,T} dx = \mu U_\infty W \int_{Re_{\chi,C}}^{Re_{\chi,E}} \frac{c_{f,T}}{2} dRe_\chi = 0.0288 \, \mu U_\infty W \int_{Re_{\chi,C}}^{Re_{\chi,E}} Re_\chi^{-0.2} dRe_\chi$$

$$= 0.0360 \, \mu U_\infty W \left( Re_{\chi,E}^{0.8} - Re_{\chi,C}^{0.8} \right) = 185.9 \text{ N.}$$

The overall drag force on the plate is then $D_E = D_C + D_T = 190.1$ N.

(iii) If the boundary layer were laminar over the entire length of the plate, the Reynolds number $Re_E$ would have the value $5.292 \times 10^7$, and the drag force would be given by

$$D = \mu U_\infty W \int_0^{Re_L} \frac{c_{f,L}}{2} dRe_x = 0.3321 \, \mu U_\infty W \int_0^{Re_L} Re_x^{-0.5} dRe_x$$

$$= 0.6642 \, \mu U_\infty W Re_E^{0.5} = 17.59 \text{ N}.$$

(iv) If the boundary layer were turbulent over the entire length of the plate, the drag force would be given by

$$D = \mu U_\infty W \int_0^{Re_L} \frac{c_{f,T}}{2} dRe_x = 0.0288 \, \mu U_\infty W \int_0^{Re_L} Re_x^{-0.2} dRe_x$$

$$= 0.036 \, \mu U_\infty W Re_E^{0.8} = 197.8 \text{ N}.$$

**Comments:**

(a) At the transition location $x_C$ we see that the shear stress for the laminar boundary layer is 0.369 Pa, whereas for the turbulent boundary layer the value is 4.14 Pa, i.e. an increase by an order of magnitude. These values should be regarded as indicative rather than 100% accurate, but it is clearly the case that the shear stress in a turbulent boundary layer is far in excess of that for a laminar boundary layer.

(b) Assuming the boundary layer to be turbulent over the entire plate would lead to an error in the overall drag of only +4%. In general, if the drag calculated assuming the flow is entirely turbulent is an order of magnitude (or more) greater than the drag due to laminar flow up to the transition location, then neglect of the laminar-flow contribution is justified.

## 18.14 Boundary layers with streamwise pressure gradient

In Section 18.13, concerned with the flat-plate boundary layer, it was found that an explicit relationship between $c_f$ and $Re_x$ resulted from the assumption of a power-law form for the velocity profile. The approach effectively neglects the outer-region wake contribution to the velocity profile, i.e. the wake parameter $\Pi = 0$, a simplification which cannot be justified where there is a streamwise pressure gradient, particularly an adverse gradient. Furthermore, in the latter case experiments show that $\Pi$ can reach values as high as 100. Under such circumstances it is clear that the variation of $\Pi(x)$ has to be accounted for.

It is reasonable to assume that the momentum-integral equation (18.103) is still valid

$$\frac{c_f}{2} = \frac{d\theta}{dx} + (H + 2) \frac{\theta}{U_\infty} \frac{dU_\infty}{dx} \qquad (18.103)$$

where the normal-Reynolds-stress term has been neglected as before.

Equation (18.105) is also still valid

$$\frac{U_\infty \theta}{\nu} = Re_\theta = \left[ \Pi + 1 - F(\Pi) \frac{1}{\kappa} \sqrt{\frac{c_f}{2}} \right] \frac{e^\kappa \sqrt{2/c_f}}{\kappa} e^{-(B\kappa + 2\Pi)}. \qquad (18.105)$$

We also found earlier that the shape factor $H$ corresponding to the combined log law plus law-of-the wake velocity distribution is given by

$$\frac{1}{H} = \frac{\theta}{\delta^*} = 1 - \frac{F(\Pi)}{(1 + \Pi)\kappa}\sqrt{\frac{c_f}{2}}. \tag{18.70}$$

Equations (18.105) and (18.70) can, in principle, be substituted into equation (18.103), leading to an equation involving $dc_f/dx$, $d\Pi/dx$, $c_f$, and $\Pi$, as well as the known (specified) quantities $\nu$, $\kappa$, $U_\infty$, and $dU_\infty/dx$. However, to proceed further requires additional information which cannot be obtained by manipulating any of the existing equations or derived from the Navier-Stokes equations. Various empirical equations have been proposed to provide this information, including an empirical **entrainment function**, in which the rate of entrainment into the boundary layer is related to the wake strength. Another approach is based upon an integral equation for kinetic-energy dissipation. With these empirical equations, solutions can be obtained using numerical integration. Such empirical approaches have now been superseded by much more sophisticated, and general, methods based upon the partial differential equations for the transport of turbulent kinetic energy, the rate of dissipation of turbulent kinetic energy, etc.

Qualitatively the influence of an adverse pressure gradient is similar to that for a laminar boundary layer, as discussed in Chapter 17, i.e. decreasing surface shear stress and, if the pressure-gradient parameter is sufficiently strong, boundary-layer separation. If the pressure gradient is favourable, there is a tendency for the turbulence intensity to reduce and ultimately to approach a laminar-like state, a process termed **laminarisation**, or **relaminarisation**.

## 18.15 **Bluff-body drag**

The force exerted by a flowing fluid on an immersed object is known as the **drag force**, or just **drag**. That part of the drag force due entirely to the surface shear stress acting on the object is called the **skin-friction drag**, or just **friction drag**. For a thin flat plate aligned with the flow, the drag is due entirely to friction drag. **Pressure drag**, or **form drag**, is the net force arising from the static pressure acting on an object's surface. For a thin flat plate normal to the flow, the drag is due entirely to pressure drag. For any object the sum of the form drag and the friction drag is called the **profile drag**. Additional contributions to profile drag come from **wave drag**, which in the case of marine craft arises from surface waves and in compressible flow from shockwaves, and the drag associated with lift known as **induced drag**.

Solid objects can be categorised as **streamlined** or **bluff** depending upon whether their shape is such that the flow over them remains attached, with accompanying low drag, or separates, with associated high drag. The flow over **low-drag aerofoils**, for example, may remain laminar but for most bodies of engineering interest the boundary layers will be turbulent over much of the surface as will be the region of flow downstream known as the **wake**. Although computer software has been developed which allows full details of the flow over complex shapes to be calculated quite accurately, for many engineering purposes it is sufficient to characterise the overall drag force $D$ exerted on a body through a drag coefficient $C_D$ defined by

$$C_D = \frac{D}{\frac{1}{2}\rho V^2 A} \tag{18.123}$$

where $\rho$ is the fluid density, $V$ is the flow velocity upstream of the body, and $A$ is an appropriate area, usually the frontal projected (or silhouette) area of the body. It is to be expected that for any given shape, $C_D$ will depend on the Reynolds number $Re$ defined by

$$Re = \frac{\rho V L}{\mu}, \tag{18.124}$$

the Mach number $M$,

$$M = \frac{V}{c}, \tag{18.125}$$

surface roughness, etc. In these equations $\mu$ is the dynamic viscosity of the fluid, $c$ is the soundspeed, and $L$ is a characteristic length of the body. For the most part we shall restrict attention to incompressible flow, which corresponds with $M$ less than about 0.5.

The variation of $C_D$ with $Re$ for a long, smooth-surface, circular cylinder in crossflow is shown in Figure 18.7. The curve shown is based upon a number of experimental investigations carried out in the early-to-mid $20^{th}$ century. The early data are included in the paper by Roshko (1961), who extended the range of conditions covered to $Re \approx 10^7$. The original experimental data exhibits considerable scatter, particularly for $Re > 4 \times 10^5$. Several flow regimes have been identified, some of which are evident from the $C_D$ versus $Re$ curve.

The flow at very low Reynolds numbers is initially steady, symmetric, and laminar without separation but as $Re \to 4$ a closed separation bubble appears attached to the downstream face of the cylinder. The vortices within the bubble grow, become unstable, are eventually ($Re > 50$) shed from alternate sides of the cylinder, and are advected downstream. The pattern of vortices

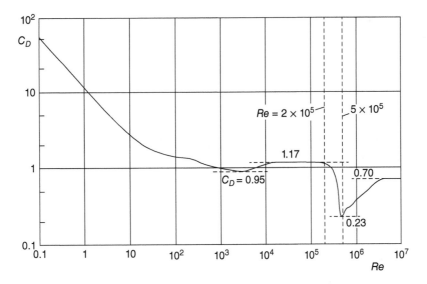

**Figure 18.7** Drag coefficient $C_D$ versus Reynolds number $Re$ for a smooth circular cylinder in crossflow (logarithmic scales)

of alternating rotation is known as a **Kármán vortex street**. The frequency $f$ corresponding to successive vortices passing a fixed point satisfies the equation

$$St = \frac{fD}{V} \approx 0.2 \tag{18.126}$$

for $40 < Re < 60 - 100$. The vortices are stable until $200 < Re < 400$, when they become unstable. The drag coefficient decreases progressively with increasing $Re$, reaching a first minimum of $C_D \approx 0.95$ at $Re \approx 2000$.

Once $Re > 400$ the vortices are already turbulent as they detach from the cylinder although the boundary layer on the cylinder surface remains laminar until $Re \approx 2 \times 10^5$ with separation occurring slightly before $90°$ measured from the forward stagnation line. A wide turbulent wake is created downstream of the cylinder, within which the static pressure is much lower than the free-stream pressure. From the minimum at $Re \approx 2000$ the drag coefficient increases slightly until $Re \approx 10^4$ and then remains constant at about 1.2 until $Re \approx 2 \times 10^5$. At this Reynolds number the boundary layer on the cylinder transitions from laminar to turbulent and remains attached until about $120°$ from the stagnation line. The turbulent wake is now narrower than before, the static pressure is close to its free-stream value, and the drag coefficient much lower with a minimum value $C_D \approx 0.23$ at $Re \approx 5 \times 10^5$. The sudden drop in $C_D$ at $Re \approx 2 \times 10^5$ is referred to as the **drag crisis**, and the corresponding value of $Re$ as the **critical Reynolds number**, $Re_C$. If $Re < Re_C$ the flow is said to be **subcritical**, and **supercritical**[187] if $Re > Re_C$. Beyond this point the drag coefficient increases progressively until it appears to plateau at a value of 0.7 at $Re \approx 4 \times 10^6$. Although there are no data for $Re > 10^7$, from a practical point of view this is unlikely to cause problems: for a 50 mm-diameter cylinder with flow of air at 50 °C the velocity corresponding with $Re = 10^7$ would be 3000 m/s (i.e. for air at STP a Mach number close to 9, and for a 1 m diameter cylinder it would be 150 m/s ($M = 0.44$).

The variation of $C_D$ with $Re$ for a smooth sphere is qualitatively similar to that for a circular cylinder although the plateau value in the range $10^3 < Re < 2 \times 10^5$ is much lower at about 0.4. The critical Reynolds number for a sphere or cylinder is reduced significantly by surface roughness, trip wires, and free-stream turbulence. For example, at $Re \approx 10^5$ the drag coefficient for a dimpled golf ball is less than 50% of the value for a smooth sphere.

Once the Mach number exceeds about 0.4 compressibility effects become important and $C_D$ begins to increase, initially gradually but dramatically so in the vicinity of $M = 1$ as shockwaves arise.

Drag coefficients reported in the literature for long cylinders of various cross section in crossflow are given in Figure 18.8, and for various three-dimensional objects in Figure 18.9. In all cases the approach flow is from left to right. It is a great simplification to present $C_D$ values as a single number but, as we have seen for a smooth circular cylinder, $C_D$ is approximately constant in the range $10^4 < Re < 2 \times 10^5$, and the same is true for a similar $Re$ range for most of the shapes in these two figures. The values listed are an average from several sources, some of which for a given object differ by as much as 20%. In these two figures, each object is considered in isolation, whereas in practice it is often the case that there are two or more

---

[187] The terms 'subcritical' and 'supercritical' as used here should not be confused with the same terms used for open-channel flow.

|  | |  | | | | | | | | |
|---|---|---|---|---|---|---|---|---|---|---|
| square | 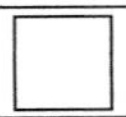 R/H | 0 | 0.02 | 0.17 | 0.33 | | | | | |
| | $C_D$ | 2.15 | 2.0 | 1.2 | 1.0 | | | | | |
| square |  $C_D$ | 1.6 | | | | | | | | |
| rectangle | 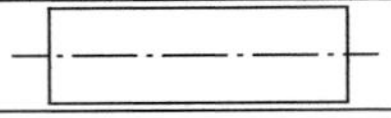 L/H | ≤ 0.4 | 0.4 | 0.5 | 0.65 | 1.0 | 1.2 | 2.0 | 3.0 | 6.0 |
| | $C_D$ | 1.9 | 2.4 | 2.4 | 2.9 | 2.2 | 2.1 | 1.7 | 1.3 | 0.9 |
| rectangle with rounded nose R/H = 1 | 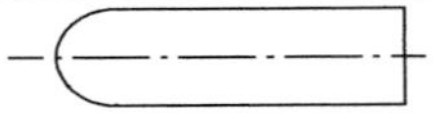 L/H | 0.5 | 1.0 | 2.0 | 4.0 | 6.0 | | | | |
| | $C_D$ | 1.15 | 0.90 | 0.70 | 0.68 | 0.64 | | | | |
| rectangle with rounded nose and taper |  $C_D$ | 0.15 | | | | | | | | |
| semi circle | 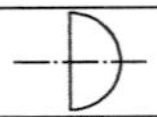 $C_D$ | 1.9 | | | | | | | | |
| semi-circular shell | 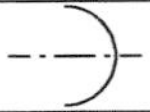 $C_D$ | 2.3 | | | | | | | | |
| semi-circular shell | 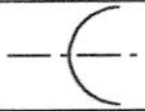 $C_D$ | 1.1 | | | | | | | | |
| ellipse | 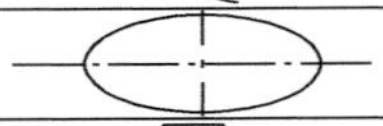 L/H | 1.0 | 2.0 | 4.0 | 8.0 | | | | | |
| | $C_D$ | 0.3 | 0.2 | 0.15 | 0.1 | | | | | |
| hexagon | 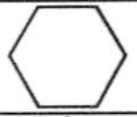 $C_D$ | 1.0 | | | | | | | | |
| hexagon | 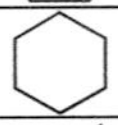 $C_D$ | 0.7 | | | | | | | | |
| equilateral triangle |  R/H | 0 | 0.02 | 0.08 | 0.25 | | | | | |
| | $C_D$ | 1.5 | 1.2 | 1.3 | 1.1 | | | | | |
| equilateral triangle | 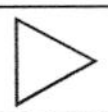 R/H | 0 | 0.02 | 0.08 | 0.25 | | | | | |
| | $C_D$ | 2.15 | 2.0 | 1.9 | 1.3 | | | | | |

**Figure 18.8** Typical drag coefficients for long cylinders in crossflow

|  | | | | | | | | | |
|---|---|---|---|---|---|---|---|---|---|
| cube | $C_D$ | 1.06 | | | | | | | |
| cube | 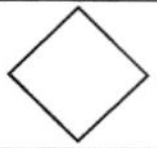 $C_D$ | 0.81 | | | | | | | |
| rectangular plate normal to flow |  $B/H$ | 1 | 5 | 10 | 20 | ∞ | | | |
| | $C_D$ | 1.18 | 1.2 | 1.3 | 1.5 | 2.0 | | | |
| circular cylinder normal to flow | 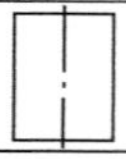 $H/D$ | 1 | 2 | 3 | 5 | 10 | 20 | 40 | ∞ |
| | $C_D$ | 0.64 | 0.68 | 0.74 | 0.74 | 0.82 | 0.91 | 0.98 | 1.2 |
| circular cylinder aligned with flow | 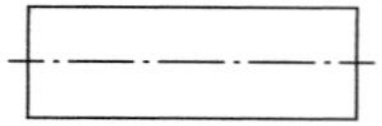 $L/D$ | ≪1 | 0.5 | 1 | 2 | 4 | 8 | | |
| | $C_D$ | 1.16 | 1.13 | 0.92 | 0.84 | 0.86 | 0.99 | | |
| solid hemisphere |  $C_D$ | 1.17 | | | | | | | |
| solid hemisphere |  $C_D$ | 0.42 | | | | | | | |
| hollow hemisphere |  $C_D$ | 1.41 | | | | | | | |
| hollow hemisphere |  $C_D$ | 0.39 | | | | | | | |
| ellipsoid | 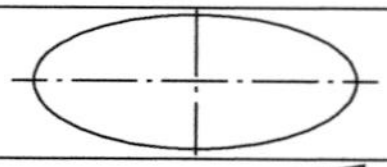 $L/D$ | 0.75 | 1 | 2 | 4 | 8 | | | |
| | $C_D$ | 0.2 | 0.2 | 0.13 | 0.1 | 0.08 | | | |
| cone | 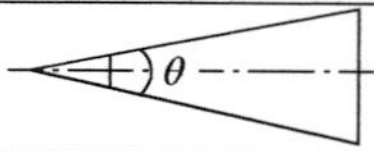 $\theta$ | 10° | 20° | 30° | 40° | 60° | 75° | | |
| | $C_D$ | 0.3 | 0.4 | 0.55 | 0.65 | 0.8 | 1.05 | | |
| streamlined body | 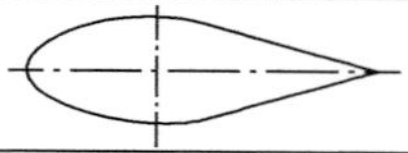 $C_D$ | 0.04 | | | | | | | |

**Figure 18.9** Typical drag coefficients for various three-dimensional objects

objects in close proximity. The situation is then more complex and it is not realistic to list all the possible shape combinations.

In Figure 18.8 the streamwise length of each cylinder is $L$, its height is $H$, and, where the corners of an object have been rounded, $R$ denotes the corner radius. The projected area $A$ is given by $BH$, where $B (\gg H)$ is the cylinder span.

The benefit of rounding corners is clearly evident from the entries for square and triangular cross sections. Even more striking is the drag reduction associated with rounding to $R = 0.5\ H$ the nose of a cylinder of rectangular cross section. Tapering the base of a rectangular cross section with such a rounded nose reduces the drag coefficient to 0.15, the level for an elliptical cross section with aspect ratio 4:1.

In Figure 18.9, for the cube, $H$ is the side length; for the rectangular plate, $H$ is the height, and $B$ is the span; for the axisymmetric three-dimensional objects, $D$ is the maximum diameter; for the cylindrical cylinder aligned with the flow, $L$ is the length; for the cylindrical cylinder normal to the flow, $L$ is the height; and for the cone, $\theta$ is the total included angle. The corresponding projected areas $A$ are $H^2, \sqrt{2}H^2, BH, HD$, and $\pi D^2/4$.

 18.16 SUMMARY

In this chapter we have outlined the principal characteristics of a turbulent flow and shown how Reynolds time-averaging procedure, applied to the Navier-Stokes equations, leads to a set of equations similar to those governing laminar flow but including additional terms which arise from correlations between fluctuating velocity components and velocity-pressure correlations. We showed that dimensional considerations applied to the kinematic viscosity, and the rate of dissipation $\epsilon$ of specific turbulent kinetic energy $k$ leads to the Kolmogorov time, length, and verlocity scales which characterise the smallest eddies of turbulent motion. The complex nature of turbulent motion has led to an empirical methodology called turbulence modelling in which the correlation terms in the Reynolds-averaged Navier-Stokes equations and in the transport equations for $k, \epsilon$, and other turbulence quantities are modelled. We show that limited, but useful, results for fully-developed turbulent channel flow and zero-pressure-gradient boundary layers can be deduced by treating the turbulent shear flow in the immediate vicinity of a solid surface as a Couette flow which leads to the Law of the Wall and the logarithmic velocity variation termed the log law. We discuss the characterisation of surface roughness, and its effect on both the velocity distribution and surface shear stress. It is shown that the distribution of mean velocity within a turbulent boundary layer can be represented by a linear combination of the near-wall log law and an outer-layer Law of the Wake.

The student should be able to

- give a qualitative description of turbulent flow
- understand Reynolds' decomposition and the derivation of the Reynolds-averaged Navier-Stokes equations
- use dimensional analysis to derive the Kolmogorov length, time, and velocity scales
- explain why turbulence modelling is necessary
- use dimensional analysis to show that the velocity distribution in the near-wall region of a turbulent shear flow has the universal form $u^+ = f(y^+)$
- show that within the viscous sublayer $u^+ = y^+$
- show that if viscosity has no direct influence on the velocity distribution then $u^+ = \ln y^+/\kappa + B$

- understand the qualitative behaviour of a boundary-layer velocity profile described by the combination of the log law and the law of the wake and the role of the wake parameter $\Pi$
- derive an equation for the dependence of the Fanning friction factor on the Reynolds number for fully-developed turbulent flow through a pipe assuming that the velocity distribution follows the log law
- apply the individual loss coefficients for various pipe fittings to calculate the pressure loss through a piping system
- understand the influence of surface roughness on the near-wall velocity distribution in a turbulent flow
- derive an equation for the dependence of the skin-friction coefficient on the momentum-thickness Reynolds number $Re_\theta$ for a flat-plate turbulent boundary layer, assuming that the velocity distribution follows the log law plus Law of the Wake combination
- derive an equation for the dependence of the skin-friction coefficient on the streamwise Reynolds number $Re_x$ for a flat-plate turbulent boundary layer, assuming that the velocity distribution follows the log law plus Law of the Wake combination
- derive an equation for the dependence of the skin-friction coefficient on the momentum-thickness Reynolds number $Re_\theta$ for a flat-plate turbulent boundary layer, assuming that the velocity distribution has a power-law form
- derive an equation for the dependence of the skin-friction coefficient on the streamwise Reynolds number $Re_x$ for a flat-plate turbulent boundary layer, assuming that the velocity distribution has a power-law form
- make use of drag coefficients to calculate the drag force acting on an object of given shape

## 18.17  SELF-ASSESSMENT PROBLEMS

**18.1**  Show that, for an incompressible flow,

$$u\frac{\partial u}{\partial x} + v\frac{\partial u}{\partial y} + w\frac{\partial u}{\partial z} = \frac{\partial u^2}{\partial x} + \frac{\partial(uv)}{\partial y} + \frac{\partial(uw)}{\partial z}$$

where the symbols have their usual meaning.

**18.2**  Show that, for fully-developed turbulent flow through a smooth-wall circular pipe, neglect of the contribution of the viscous sublayer to the mean-velocity distribution is valid if

$$\sqrt{\frac{2}{f_F}}\frac{1}{Re_D} \ll 1$$

where $f_F = 2\tau_S/\rho\overline{V}^2$ is the Fanning friction factor, $Re_D = \rho\overline{V}D/\mu$ is the pipe Reynolds number, $\tau_S$ is the surface shear stress, $\overline{V}$ is the bulk-mean velocity, $D$ is the pipe diameter, $\rho$ is the fluid density, and $\mu$ is the dynamic viscosity of the fluid.

**18.3**  Water at 10 °C is pumped through a pipeline which consists of a 50 mm diameter pipe 25 m long followed by an 80 mm diameter pipe also 25 m long. The relative roughness for both pipes is 0.001. The two sections of pipe are connected by a sharp-edged sudden expansion. Entry to the 50 mm pipe and exit from the 80 mm pipe are both sudden with sharp edges. A threaded 90° standard elbow is installed in the 50 mm pipe, and a threaded 90° long-radius elbow in the 80 mm pipe. The flow is controlled by a valve in the 80 mm pipe, for which the loss coefficient is 6.0 when the valve is fully open. If the flowrate is 0.011 $m^3/s$ calculate the overall loss in stagnation pressure and the pumping power required to maintain the flow. (Answers: 0.235 MPa, 2.583 kW)

**18.4** Assume that for fully-developed turbulent flow through a rough-wall circular pipe the velocity distribution is given by the equation

$$u^+ = \frac{1}{\kappa} \ln y^+ + B - \frac{1}{\kappa} \ln \left(1 + 0.3\varepsilon^+\right).$$

Show that the bulk-mean velocity $\overline{V}$ for a pipe of diameter $D$ is given by

$$\overline{V}^+ = \frac{1}{\kappa} \ln \left( \frac{R^+}{1 + 0.3\varepsilon^+} \right) + B - \frac{3}{2\kappa}$$

and hence that the Fanning friction factor $f_F$ is related to the Reynolds number $Re_D$ and the relative roughness $\varepsilon/D$ through the formula

$$\sqrt{\frac{2}{f_F}} = \frac{1}{\kappa} \ln \left[ \frac{\sqrt{\frac{f_F}{2}} Re_D}{2 \left( 1 + 0.3 \frac{\varepsilon}{D} \sqrt{\frac{f_F}{2}} \right)} \right] + B - \frac{3}{2\kappa}.$$

**18.5** Show that, for a boundary layer with $\Pi = 0$, the streamwise variation of the skin-friction coefficient is given by

$$Re_x = \frac{1}{\kappa^3} \left( \beta^2 - 4\beta + 6 \right) e^{(\beta - B\kappa)}$$

where $\beta = \kappa \sqrt{2/f_F}$.
If $\kappa = 0.4$, $B = 5.5$, and $c_f/2 = 7.15 \times 10^{-4}$, find the value of $Re_x$ using this formula. Use the calculated value of $Re_x$ to calculate $c_f/2$ from White's formula:

$$\frac{c_f}{2} = \frac{0.2275}{\ln^2 (0.06 Re_x)}.$$

(Answer: $1.531 \times 10^9, 6.767 \times 10^{-4}$)

**Comments:**

The two values of $c_f/2$ are within about 5.5% of each other.
Note that with $\Pi = 0$ it has been assumed that the log law applies throughout the boundary layer. This is exactly the same assumption made in the analysis of fully-developed turbulent pipe flow. The key difference here is that the boundary-layer thickness increases with streamwise distance whereas the pipe diameter is fixed.

**18.6** Assume that for fully-developed turbulent flow through a parallel-wall channel of height $2H$ the velocity distribution is given by the log law

$$u^+ = \frac{1}{\kappa} \ln y^+ + B.$$

Show that the bulk-mean velocity $\overline{V}$ is given by

$$V^+ = \frac{1}{\kappa} \ln H^+ + B - \frac{1}{\kappa}$$

and hence that the Fanning friction factor $f_F$ is related to the Reynolds number $Re_H = 2\overline{V}H/\nu$ through the formula

$$\sqrt{\frac{2}{f_F}} = \frac{1}{\kappa} \ln \left( \frac{1}{2} \sqrt{\frac{f_F}{2}} Re_H \right) + B - \frac{1}{\kappa}.$$

**18.7**  Use the power-law equations

$$\frac{c_f}{2} = 0.0288 \, Re_x^{-0.2}$$

and

$$\frac{c_f}{2} = 0.0225 \, Re_\delta^{-0.25}$$

to calculate the skin-friction coefficient $c_F/2$, the wall shear stress $\tau_S$, the friction velocity $u_\tau$, and the boundary-layer thickness $\delta$ at the end of a flat plate 20 m long if the free-stream velocity is 70 m/s and the flowing fluid is air at 25 °C. Assume the boundary layer is turbulent starting at the leading edge of the plate.
Calculate the Kolmogorov scales given by

$$l_K = \left( \frac{\nu^3}{\epsilon} \right)^{1/4}, \quad v_K = (\nu\epsilon)^{1/4}, \quad \text{and} \quad \tau_K = \left( \frac{\nu}{\epsilon} \right)^{1/2}$$

and the turbulent-energy dissipation rate given by

$$\epsilon = \frac{u_\tau^3}{\kappa y},$$

where the von Kármán constant $\kappa = 0.41$, at a distance 1 mm from the plate surface.
Calculate the ratios $l_K/\delta$, $v_K/U_\infty$, and $\tau_K U_\infty/\delta$.
(Answers: $7.395 \times 10^{-4}$; 4.290 Pa; 1.904 m/s; 191.3 mm; 21.82 $\mu$m; 0.725 m/s; 30.1 $\mu$s; $1.682 \times 10^4$ m$^3$/s$^2$; $1.184 \times 10^{-4}$; 0.014; 0.0110)

**18.8**  Write the log law for a boundary layer in terms of $u/U_\infty$, $U_\infty y/\nu$, $c_f/2$, and the log-law constants $B$ and $\kappa$. Suggest how a value for the friction factor could be obtained from an experimentally determined velocity distribution, $u(y)$.

**18.9**  Show that the following form of the Kármán-Nikuradse formula for turbulent flow through a smooth pipe,

$$\frac{1}{\sqrt{f_D}} = -2 \log_{10} \left( \frac{2.51}{Re_D \sqrt{f_D}} \right)$$

can be transformed into the form

$$\sqrt{\frac{2}{f_F}} = 2.457 \ln \left( \sqrt{\frac{f_F}{2}} Re_D \right) + 0.292$$

where $f_F = \tau_S/\rho\overline{V}^2$ is the Fanning friction factor, $f_D = 4f_F$ is the Darcy friction factor, and $Re_D = \rho\overline{V}D/\mu$ is the Reynolds number.

# Appendix 1
# Principal contributors to fluid mechanics

The entries here are for engineers, mathematicians, and other scientists whose names appear in the main text. They are listed in chronological order according to year of birth. Many made major contributions to other areas of science and mathematics but, for the most part, it is their work in fluid mechanics which is outlined here.

**Archimedes** (ca 287–ca 212 BC) A Greek mathematical scientist remembered for discovering the principle concerning buoyancy, which bears his name. According to legend, as he was entering a public bath, the concept came to him as a way of determining the amount of gold used in the fabrication of the crown of the Hieronn II, the king of Syracuse. Archimedes is also credited with a number of inventions, including the water screw, and for recognising that a fluid is a continuous substance.

Sextus Julius **Frontinus** (40 AD–103 AD) A Roman military engineer responsible for the control and maintenance of the aqueducts of Ancient Rome. He wrote technical treatises, including *Deaqaducte*. It is thought he was aware of the continuity equation.

Leonardo **da Vinci** (1452–1519) An Italian polymath who was fascinated by the phenomenon of flight. Among his many inventions, he designed flying machines (ornithopters), parachutes, and elementary hydraulic machines. He is credited with formulating a one-dimensional form of the continuity equation, and made observations of various fluid flows, including open-channel flow, eddies downstream of obstacles in a water flow, and flow through contracted weirs.

Evangelista **Torricelli** (1608–1647) An Italian mathematician and philosopher who established the principle of a barometer and was the first to create a sustained vacuum. He showed that the flowrate of a liquid through an opening in the wall of a container is proportional to the square root of the height of the liquid above the opening, a result formulated as Torricelli's theorem. The obsolete pressure unit torr is named after him.

Blaise **Pascal** (1623–1662) A French mathematician and physicist who developed the theory of hydrostatics. A key contribution was the recognition that the pressure at any point within a fluid is the same in all directions, now termed Pascal's law. He also recognised the connection between atmospheric pressure and the height of a barometric column. The pascal (Pa) is the SI unit of pressure.

Sir Isaac **Newton** (1642–1727) It is impossible to do justice in just a few lines to the English scientist who discovered the three laws of motion which form the basis of all classical mechanics, including fluid mechanics, and are contained in his *Philosophiae naturalis principia mathematica*, first published in 1687. His specific contributions to fluid mechanics, to be found in Book II of the *Principia*, include the first derivation of the velocity-squared drag law, the sine-squared law, which accounts for the influence on drag of a plate in a fluid stream of the angle of attack, the Newtonian shear-stress law, and the concept of a fluid property now called viscosity. Fluids for which the viscosity is independent of the shear stress are called Newtonian, while those for which the viscosity is affected by shearing and other aspects of the motion are non-Newtonian. The newton (N) is the SI unit of force.

Henri de **Pitot** (1695–1771) A French civil engineer who devised an instrument, the forerunner of the Pitot-static tube, to measure the velocity distribution in a water channel and the speed of a boat.

Daniel **Bernoulli** (1700–1782) A Dutch-born Swiss mathematician who laid the foundations of hydrodynamics in a treatise published in 1738 entitled *Hydrodynamica, sive de viribus et motibus fluidorum commentarii.* He recognised that there is a relationship between pressure and velocity, now embodied in the equation known as Bernoulli's equation, and also discovered some of the basic energy relationships that apply to a liquid.

Leonhard **Euler** (1707–1783) A Swiss mathematician, one of the most prolific of all time, who made major contributions to mechanics, dynamics, and hydrodynamics. He was the first to recognise that pressure in a moving fluid is strictly a point property. He formulated equations of fluid motion and introduced the concept of centrifugal machinery. The analysis of most problems in fluid mechanics is based upon the Eulerian method, which is concerned with the entire flowfield at any position and time. The Euler equations are a set of partial differential equations for the flow of an inviscid fluid. The Euler number is the ratio of a pressure difference to dynamic pressure. Euler's turbomachine equation expresses the relationship between torque and the change in the flowrate of moment of momentum for a turbomachine.

Jean le Rond **d'Alembert** (1717–1783) A French mathematician and physicist who was the first to derive an expression for mass conservation in fluid flow. He introduced the concept of a moving fluid element, the volume and velocity of which varied from point to point in a flow, aspects of fluid flow encompassed in what is now called the continuity equation. d'Alembert's paradox states that a body moving through a perfect fluid experiences zero drag.

Jean-Charles de **Borda** (1733–1799) A French naval engineer who analysed ballistics problems and performed experiments using a whirling-arm apparatus. Borda observed that the combined drag of two bodies in close proximity is different from the sum of the drag of each body taken individually, a phenomenon termed the Borda effect. He introduced the concept of a stream tube. The Borda-Carnot equation connects the reduction in stagnation pressure across a sudden expansion to the mass flowrate, fluid density, and the cross-sectional areas.

Joseph-Louis **Lagrange** (1736–1813) An Italian-born, French mathematician who made contributions to number theory and applied mechanics. He was the first to integrate Euler's equation for the motion of an irrotational, compressible fluid, obtaining a result which reduces to Bernoulli's equation for an incompressible fluid. The Lagrangian method of analysis of fluid motion follows individual fluid particles.

Giovanni Battista **Venturi** (1746–1822) An Italian physicist who carried out experiments on liquid flow through convergent-divergent tubes, including sudden enlargements. He also investigated open-channel flows and the hydraulic jump. The Venturi-tube flowmeter, the open-channel form of which is called a Venturi flume, was developed by Clemens Herschel.

Lazare Nicolas Marguerite **Carnot** (1753–1823) A French military engineer, mathematician, physicist, and politician. The Borda-Carnot equation is named after him and Jean-Charles de Borda. The Carnot cycle is named after Carnot's son, Nicolas Léonard Sadi **Carnot** (1796–1832), also a military engineer, who was a pioneer in the development of thermodynamics.

Amedeo Carlo **Avogadro** (1776–1856) An Italian scientist best known for his contributions to molecular theory. The number of molecules in 1 kmol of any substance is a fundamental physical constant known as the Avogadro number.

(Claude-) Louis (-Marie-Henri) **Navier** (1785–1836) A French physicist and civil engineer who was the first to formulate the Navier-Stokes equations, which govern viscous flow of a Newtonian fluid. Navier's formulation was flawed in accounting for shear stress without reference to viscosity. Navier was the first to precisely define the concept of mechanical work and recognised for practical contributions to the design of bridges.

Augustin Louis de **Cauchy** (1789–1857) A French engineer and mathematician who contributed to the analysis of wave motion, an extension of Navier's equations, and conformal transformation for the analysis of irrotational flow. The general partial differential equations for the conservation of linear momentum applied to fluid flow are known as Cauchy's equations. The Cauchy-Riemann equations relate the partial spatial derivatives of the stream function and velocity potential (or the corresponding velocity components) in incompressible, two-dimensional irrotational flow. The Cauchy number preceded the Mach number, of which it is the square, as the basic parameter of compressible flow.

Jean Léonard Marie **Poiseuille** (1797–1869) A French physician and physiologist trained in physics and mathematics, who performed experiments on liquid flow through capillary tubes. Poiseuille flow is the name now given to fully-developed laminar flow of a constant-property fluid through a cylindrical channel. Such flows are characterised by the non-dimensional Poiseuille number.

Gotthilf Heinrich Ludwig **Hagen** (1797–1884) A German physicist and hydraulic engineer who carried out pipe-flow experiments from which he concluded that some kind of transition occurs when an unstable laminar flow is disturbed by heating. The Hagen-Poiseuille equation governs fully-developed laminar flow of an incompressible Newtonian fluid through a circular pipe.

Henry Philibert Gaspard **Darcy** (1803–1858) A French engineer who made important contributions to hydraulics and flow through porous media. Darcy improved the design of the Pitot tube by adding static-pressure tappings some distance from the tip of the tube. He performed and correlated the results of experiments on the pressure drop of water flowing through smooth- and rough-wall pipes. The non-dimensional pressure drop in such pipes, commonly used in hydraulics, is called the Darcy (or Darcy-Weisbach) friction factor. The equation governing low Reynolds-number flow through a porous medium is called Darcy's law.

Julius Ludwig **Weisbach** (1806–1871) A German mathematician and engineer whose contributions to hydraulics include an equation for flow over a weir, and the Darcy-Weisbach friction factor.

William **Froude** (1810–1879) An English engineer, hydrodynamicist, and naval architect who pioneered the use of towing tanks with small-scale models in the design of ships and the prediction of their wave and boundary resistance. The Froude number is the non-dimensional group used to characterise free-surface flows.

James Prescott **Joule** (1818–1889) An English brewer and physicist who established the relationship between thermal energy and work, which led to the first law of thermodynamics. The SI-derived unit of energy is the joule (J). Joule's law relates the heat dissipated by an electrical resistance to the current flowing through it.

George Gabriel **Stokes** (1819–1903) An Irish physicist and mathematician who independently of, and more rigorously than, Navier formulated the Navier-Stokes equations for the flow of an incompressible viscous fluid. Critically, Stokes formally incorporated the dynamic viscosity in a set of constitutive equations for the normal and shear stresses in Cauchy's equations of motion. Stokes' name is associated with many fluid-flow phenomena, including low-Reynolds-number viscous flow, which is referred to as Stokes flow, and the governing equation as Stokes equation; Stokes number is the non-dimensional group which characterises the behaviour of a small particle immersed in a fluid. The term also refers to the non-dimensional group which characterises oscillatory viscous flow past a solid object; the square root of this Stokes number is the Wommersley number. Stokes' assumption is that the thermodynamic and mechanical pressure at any point in a fluid are equal; Stokes' drift is the term given to the average translation velocity in the direction of wave propagation of a fluid or floating particle due to pure wave motion of a liquid; Stokes stream function describes the streamlines in axisymmetric three-dimensional flow; and the stokes is the obsolete cgs unit of kinematic viscosity. Stokes also contributed to the theories of sound and light waves.

William John Macquorn **Rankine** (1820–1872) A Scottish engineer who established an absolute scale for temperature, in which one degree Rankine is equal to one degree Fahrenheit, and a thermodynamic cycle used as a standard of efficiency for steam power. Rankine also established the equations of continuity, momentum, and energy, which govern changes across a shockwave, but failed to recognise the impossibility of a rarefaction shock. These equations, later published independently by Pierre Henri Hugoniot, are known as the Rankine-Hugoniot relations.

Hermann Ludwig Ferdinand **von Helmholtz** (1821–1894) A German surgeon and physicist who studied wave motion, free-surface stratified flow, viscous flow, irrotational flow, and cavitation. A Helmholtz resonator is a rigid cavity with a short, narrow neck which resonates when excitation is applied at the opening. Helmholtz distinguished between rotational and irrotational flows, and is credited with introducing the concepts of vorticity, a term he coined, vortex filaments, and vortex sheets. He also introduced the concept of a surface of discontinuity in an inviscid fluid flow. The Kelvin-Helmholtz instability, sometimes referred to as Helmholtz instability, is the instability of the interface between immiscible, inviscid fluids due to shearing.

William Thomson, Lord **Kelvin** (1824–1907) A Northern-Ireland-born British scientist who made significant contributions to numerous topics, including theories of sound, vortex motion, capillary waves, and flow instability. The analogy between electrical and acoustic quantities is called the Kelvin system. Kelvin was the first to derive an equation for the wavespeed of shallow-water waves which includes the effects of surface tension and finite depth. He was also the first to derive the exact kinematic condition for a free surface. The Kelvin-Helmholtz instability is the instability of the interface between immiscible, inviscid fluids due to shearing. In recognition of devising the absolute temperature scale, the SI unit of absolute temperature is the kelvin (K). He also formulated in detail the energy relationships of the second law of thermodynamics. That the limiting thermal efficiency of a heat engine is 100% is embodied in the Kelvin-Planck statement of the second law. Kelvin is also credited with introducing the word turbulence to describe the unsteady flow which occurs once the critical Reynolds number is exceeded. A model for the stress-strain relationship of a viscoelastic fluid is known as the Kelvin-Voigt model, sometimes referred to as either the Kelvin model or the Voigt model.

Lester Allen **Pelton** (1829–1908) An American mining engineer at the time of the California gold rush (ca 1850) who refined the design of the cup-shaped buckets then in use for impulse turbines, now called Pelton turbines.

Peter Guthrie **Tait** (1831–1901) A Scottish mathematical physicist who collaborated with Lord Kelvin. Tait contributed to many areas of physics, including the early development of the kinetic theory of gases and the density of ozone. His experimental research on the compressibility of water, glass, and mercury led to the formulation of the Tait equation of state, which correlates liquid density and pressure.

John Thomas **Fanning** (1837–1911) An American architect and hydraulic engineer who designed the municipal water system for the city of Manchester, New Hampshire. The non-dimensional Fanning friction factor is widely used to characterise frictional pressure drop in pipes and other duct forms.

Ernst **Mach** (1838–1916) An Austrian experimental physicist who was a pioneer in the field of supersonic aerodynamics and demonstrated the existence of supersonic flow and shock-waves, which he photographed using shadowgraphy. The Mach-Zehnder interferometer is a more sensitive device used to visualise density variations in compressible flow. The Mach number is the non-dimensional group used to characterise compressible flow. A Mach wave is a weak wave due to an infinitesimal disturbance in steady supersonic flow. The angle between the wavefront and the incident-flow direction is the Mach angle. A point disturbance in a supersonic flow produces a Mach cone. The diagram which illustrates the propagation of sound waves originating from a point source in a flow is known as Mach's construction.

Josiah Willard **Gibbs** (1839–1903) An American mechanical engineer (the first American to be awarded a PhD in engineering) who became a theoretical physicist credited as a co-founder of statistical mechanics, together with James Clerk Maxwell and Ludwig Boltzmann. He also devised vector calculus and made contributions to physical optics, including birefringence. The set of fundamental equations relating specific internal energy, specific enthalpy, Helmholtz function, and Gibbs function to temperature, pressure, specific volume, and specific entropy are known as the Gibbs relations.

Osborne **Reynolds** (1842–1912) A Northern-Ireland-born British mathematician, the first professor of engineering at what is now the University of Manchester. Reynolds was the first to suggest the decomposition of a turbulent flow into a mean and a fluctuating part and made major contributions to the theory of turbulent flow. The apparent shear stress that arises due to the statistical correlation between orthogonal fluctuations in flow velocity is termed the Reynolds stress. Reynolds also deduced that the transition from laminar to turbulent flow in pipe flow depended upon the non-dimensional combination of variables now called the Reynolds number. He was also the first to demonstrate the phenomenon of cavitation and suggested that the eddy diffusivities for momentum and heat transfer are equal, a conjecture known as Reynolds analogy.

John William Strut, Lord **Rayleigh** (1842–1919) An English scientist who developed the theory of sound, discovered argon, investigated the density of gases, analysed bubble collapse (the basis for studies of cavitation damage), investigated wave phenomena, proposed the technique of dimensional analysis, and contributed the exponent method now known as Rayleigh's method. He appears to have been the first to recognise aerodynamic heating at hypersonic speeds. Rayleigh showed that both viscosity and heat conduction were essential in determining the structure of a shockwave and that a rarefaction shock violated the second law of thermodynamics. Rayleigh's supersonic Pitot formula relates the ratio of static to stagnation pressure to the Mach number. Rayleigh flow is the term given to compressible pipe flow with surface heat transfer, and Rayleigh line to the curve of specific enthalpy versus specific entropy for such a flow as the Rayleigh line. Plateau-Rayleigh instability, or just Rayleigh instability, is the name given to the instability of a falling stream of liquid due to surface tension, which leads to breakup of the stream and the formation of droplets. Rayleigh-Taylor instability is the instability of the interface between two immiscible liquid layers, the liquid of higher density being above that of lower density. The Rayleigh number is the non-dimensional parameter which arises in viscous fluid flows where buoyancy is important.

Joseph Valentin **Boussinesq** (1842–1929) A French mathematician and physicist who contributed to the study of turbulence and hydrodynamics, in addition to vibration, light, and heat. He is credited with introducing the concept of an eddy viscosity to model the

Reynolds shear stress. The Boussinesq approximation refers to the assumption that fluid properties are independent of temperature, with the exception of the density in the gravitational body-force term in the momentum equations. Another Boussinesq approximation concerns the propagation of water waves, including solitary waves.

Clemens **Herschel** (1842–1930) An American hydraulic engineer best known for developing the Venturi-tube flowmeter. He also designed a flume for testing water wheels and translated a manuscript by Sextus Julius Frontinus, which he titled 'Frontinus and the Water Supply of the City of Rome'. The Herschel-Bulkley model for a yield-stress fluid is named after Herschel's son, Winslow H. Herschel.

Ludwig Eduard **Boltzmann** (1844–1906) An Austrian theoretical physicist, mathematician, and philosopher who developed statistical mechanics. The ratio of the universal gas constant to the Avogadro number defines the Boltzmann constant.

Carl Gustav Patrik de **Laval** (1845–1913) A Swedish engineer who designed a convergent-divergent nozzle, now called a Laval nozzle, to drive a steam turbine. He is better known for the centrifugal separators he designed to separate out the cream in milk.

Nicolai Egorovich **Zhukovsky** (1847–1921) A Russian mathematician, regarded as the founder of hydrodynamics and aerodynamics in Russia, who made contributions to the understanding of lift, hydraulic shock, and vortex theory. Conformal mapping of the complex plane is known as the Zhukovsky transformation, and its application leads to the design of Zhukovsky aerofoils. An equation for the lift on an aerofoil, established independently by Zhukovsky and Martin Kutta, is known as the Kutta-Zhukovsky equation.

Čeněk (or Vincenc) **Strouhal** (1850–1922) A Czech physicist who showed that the frequency of oscillations (so-called æolian tones) in flow over a taught wire is related to the flow velocity. The frequency is characterised by the non-dimensional Strouhal number.

Pierre Henri **Hugoniot** (1851–1887) A French mechanical engineer and ballistician who contributed to the theory of shockwaves. The Rankine-Hugoniot equation relates changes in thermodynamic properties across a shockwave. An equation relating changes in cross-sectional area and fluid velocity to the Mach number for flow through a convergent-divergent nozzle is termed the Hugoniot equation.

Maurice Marie Alfred **Couette** (1858–1943) A French physicist who studied the frictional effects of fluids and devised a concentric-cylinder apparatus for the measurement of viscosity. Viscous flow brought about by the relative tangential movement of two surfaces is termed Couette flow. Taylor-Couette flow is the name given to flow of a viscous fluid in the annular gap between differentially rotating cylinders.

William **Sutherland** (1859–1911) A Scottish-born, Australian theoretical physicist who contributed to the understanding of gas kinetics, the surface tension of aqueous solutions, the behaviour of gases at low temperatures, Brownian motion, and the viscosity of gases. Sutherland's formula is a widely used equation for the temperature dependence of the viscosity of a gas.

Edgar **Buckingham** (1867–1940) An American physicist who made fundamental contributions to dimensional analysis, including the equation defining the number of independent non-dimensional groups, now called Buckingham's Π theorem. His research also included the flow of non-Newtonian fluids: the Buckingham-Reiner equation governs fullydeveloped laminar pipe flow of a Bingham plastic.

Martin Wilhelm **Kutta** (1867–1944) A German mathematician who established the Kutta condition, which requires that in potential-flow theory the circulation for flow over a body with a sharp trailing edge is such as to fix the rear stagnation point at the trailing edge. He is also credited with deriving the Zhukovsky-Kutta equation for aerofoil lift. Kutta also devised the Runge-Kutta method for the numerical solution of ordinary differential equations.

Martin **Knudsen** (1871–1949) A Danish physicist who helped develop the kinetic theory of gases. The Knudsen number is a non-dimensional group used to characterise rarefied gas flows.

Moritz **Weber** (1871–1951) A German professor of naval mechanics who is credited with formalising the use of non-dimensional groups as the basis for similarity studies. The Weber number is the non-dimensional capillarity parameter involving surface tension.

Ludwig **Prandtl** (1875–1953, German) As with Newton, it is impossible in a few lines to do justice to Prandtl, regarded as the father of modern fluid mechanics. He was a mechanical engineer who introduced probably the most significant concept in all fluid mechanics: that of the boundary layer. Together with his more mathematically inclined student, Heinrich Blasius, Prandtl developed boundary-layer theory, which underlies much of fluid dynamics. Prandtl's mixing-length hypothesis is often considered to be the first example of turbulence modelling. He made many other contributions to the understanding of turbulent flow, including the first formulation of the universal law of the wall, and he also contributed to the understanding of low-speed and finite-wing lifting-line theory as well as compressible-flow theory. Prandtl's name is also given to the two kinds of turbulent secondary flow. In the first decade of the 20$^{th}$ century Prandtl designed, and carried out experiments on, a small-scale supersonic (Mach number ca 1.5) nozzle and published photographs of the shock and expansion-wave patterns he observed. Together with his PhD student Theodor Meyer, Prandtl developed the theory of oblique shocks and expansion waves. Prandtl's relation for a normal shock is a remarkably simple equation between the velocities on either side of a shock and the critical soundspeed. He established, but never published, a rule for correcting low-speed aerofoil lift coefficients to account for compressibility effects at high subsonic speeds. The correction was published independently by the British aerodynamicist Hermann Glauert and is now known as the Prandtl-Glauert rule. Together with Adolf Busemann, another student, Prandtl employed the method of characteristics to establish the correct shape of a supersonic nozzle, the method still used to design such nozzles and rocket engines. The Prandtl number is the ratio of momentum diffusivity to thermal diffusivity, a key non-dimensional parameter in convective heat transfer.

Eugene Cook **Bingham** (1878–1945) An American chemist who made significant contributions to rheology, a term he co-coined with Markus Reiner. The Bingham number is a non-dimensional group used to characterise the flow of a yield-stress liquid.

Carl Wilhelm **Oseen** (1879–1944) A Swedish theoretical physicist who attempted to extend to higher Reynolds numbers Stokes' analysis of flow at very low Reynolds number.

Lewis Ferry **Moody** (1880–1953) An American, the first professor of hydraulics at Princeton, who developed the so-called Moody chart, a graphical representation of the Colebook-White equation, which accounts for the effect of surface roughness on the frictional resistance of turbulent pipe flow. He is also known for work on draft-tube design.

Karl **Hiemenz** (born ca 1880) A German PhD student of Prandtl who carried out experiments involving water flow over a circular cylinder in crossflow. Hiemenz observed that the flow was unsteady and is famous for his daily comment to von Kármán, 'It always oscillates.' This led von Kármán to explain theoretically the phenomenon of vortex shedding. The viscous flow in the vicinity of a stagnation point which obeys the Falkner-Skan equation is known as Hiemenz flow.

Theodor von **Kármán** (1881–1963, Hungarian-born American) He was a major contributor to many aspects of applied mechanics, including elasticity and above all fluid mechanics.

He explained theoretically the tendency for periodic (i.e. fixed-frequency) disturbances to arise in the wake of an object such as a circular cylinder immersed in a steady flow. The alternating pattern of contra-rotating vortices in the wake is known as a Kármán vortex street. The constant $\kappa$ in the log law which describes the velocity distribution in near-wall turbulent flow is known as von Kármán's constant and originates from a mixing-length model of turbulence. Von Kármán's momentum-integral equation has been the basis for many attempts to calculate boundary-layer development, including Pohlhausen's method. The Kármán-Schoenherr formula for the total frictional drag on a surface is employed by naval architects. Von Kármán also made numerous contributions to compressible-flow theory and was the first to realise that at an altitude of 100 km the density of the atmosphere is too low to support aeronautical flight. The so-called Kármán line delineates the boundary between the earth's atmosphere and outer space.

Percy Williams **Bridgman** (1882–1961) An American who was awarded the Nobel prize for physics in 1946. He was a major contributor to the development of dimensional analysis and in 1922 published a book concerned with the logic of dimensional reasoning.

Gino Girolamo **Fanno** (1882–1962) An Italian mechanical engineer who studied compressible duct flow with wall friction, now known as Fanno flow.

Theodor **Meyer** (1882–1972) A German mathematician who was a PhD student and major collaborator of Prandtl. In his PhD thesis in 1908 Meyer developed the first theory for calculating the properties of oblique shock and expansion waves in supersonic flow. A Prandtl-Meyer expansion is the centred expansion wave which arises in supersonic flow around a sharp corner, and the key parameter in the theoretical analysis is the Prandtl-Meyer function. Meyer and Prandtl are also credited with developing oblique-shock theory. In his thesis he also published a Schlieren photograph of the Mach waves generated within a supersonic nozzle with roughened internal surfaces.

(Paul Richard) Heinrich **Blasius** (1883–1970) A German fluid dynamicist, one of the first PhD students of Prandtl, who developed laminar boundary-layer theory. Blasius' equation is the ordinary differential equation which governs the velocity distribution within a zero-pressure-gradient laminar boundary layer. Blasius' formula is an empirical friction factor-Reynolds number correlation for turbulent flow through a smooth-wall pipe.

Henri Marie **Coandă** (1886–1972) A Romanian aeronautical engineer who designed and constructed aircraft in the early 20$^{th}$ century. He discovered the influence on a free jet of a nearby surface, now called the Coandă effect.

Sir Geoffrey Ingram **Taylor** (1886–1975, English) A physicist and mathematician, frequently referred to as G. I. Taylor or simply G. I., who made numerous contributions to fluid mechanics and other areas of applied mechanics. He is credited with making the first measurements of the pressure distribution over a wing in steady flight. He showed, independently and almost simultaneously with Rayleigh, that only compression shockwaves are physically possible. In his second published paper, he presented an analysis for the internal structure of a shockwave and its thickness. The differential equations governing supersonic flow over a cone were integrated numerically by Taylor and J. W. Maccoll. Taylor is probably best known for his work on the stability of axisymmetric Couette flow in the annular gap between differentially rotating cylinders, also known as Taylor-Couette flow, on shear-augmented dispersion (Taylor dispersion), and on the statistical analysis of turbulent flow. The Taylor number is the non-dimensional parameter used to characterise the stability of Taylor-Couette flow. The circumferential vortices which occur once a critical value of the Taylor number is exceeded are called Taylor vortices. The Taylor-Proudman theorem and the Taylor column refer to phenomena occurring in rotating flow. Rayleigh-Taylor instability is the instability of the interface between two immiscible liquid layers, the liquid of higher density being above that of lower density. Taylor made numerous significant contributions to the understanding of turbulent flow, including the idea of a mixing length, a decade before Prandtl. He also proposed the idea of approximating turbulent transfer rates of momentum, heat, and water vapour in the atmosphere by eddy diffusivities.

Markus **Reiner** (1886–1976) An Austro-Hungarian-born Israeli who was a major contributor to the field of rheology: the Buckingham-Reiner equation is named after him and Edgar Buckingham. Together with Eugene Bingham he is credited with coining the term 'rheology'.

Karl **Pohlhausen** (1892–1980) A German PhD student of Prandtl and collaborator of von Kármán who made significant contributions to the development of boundary-layer theory. He is primarily remembered for the Kármán-Pohlhausen calculation procedure for laminar boundary-layer development, and the Pohlhausen pressure-gradient parameter which arises. He was the first to demonstrate that von Kármán's momentum-integral equation can be derived directly by integration from the boundary-layer equations.

Johann **Nikuradse** (1896–1979) A Russian-born, German PhD student of Prandtl best known for his pioneering experimental research into the influence of surface roughness on turbulent pipe flow. The Kármán-Nikuradse formula is a friction factor-Reynolds number correlation for turbulent flow through smooth pipes.

Jakob **Ackeret** (1898–1981) A Swiss mechanical engineer who contributed to supersonic fluid dynamics, including the design in 1935 of a supersonic wind tunnel, the development of linearised supersonic-flow theory (postdoctoral work carried out in Prandtl's laboratory in Göttingen), and in 1929 proposed the name Mach number for the non-dimensional group which is used to characterise compressible flow. He also carried out research on cavitation and on variable-pitch propellers for ships and aircraft, and designed a closed-circuit gas turbine.

Cedric Masey **White** (1898–1993) A British civil engineer who was the PhD supervisor of the British engineer Cyril Frank **Colebrook** (1910–1997) and with whom he developed the empirical Colebook-White equation for the calculation of frictional pressure drop for transitional and turbulent flow through smooth and rough pipes. The equation is the basis for the Moody chart.

Andrei Nikolaevich **Kolmogorov** (1903–1987) A Russian mathematician who made major contributions to several areas of mathematics and physics, including probability theory, stochastic processes, topology, and turbulence theory. The microscales of turbulence are called the Kolmogorov scales, and the $E \propto k^{-5/3}$ form of the turbulence energy spectrum is also named after him.

Antonio **Ferri** (1912–1975) An Italian aeronautical engineer and former Chief of the Gas Dynamics Section at NACA Langley Field who made significant contributions to the theory of supersonic and hypersonic flow, particularly supersonic and hypersonic jet engines, supersonic combustion, and aerodynamic heating. His book *Elements of Aerodynamics of Supersonic Flows* includes original contributions to the theory of oblique shocks, and a general discussion of conical flow.

Edward Reginald **van Driest** (1913–2005) An American engineer who carried out research on boundary-layer transition under von Kármán. He suggested a widely used formula for the variation of the mixing length in the buffer region of a turbulent boundary layer. He also studied under Ackeret and made contributions to the understanding of high-speed aerodynamics, particularly the effect of roughness on transition under supersonic and hypersonic conditions. Other work concerned isotropic turbulence and the design of rockets and missiles.

David Carl **Ipsen** (1921–2015) An American engineer who made a major contribution to dimensional analysis, particularly the method of sequential elimination of dimensions, now called Ipsen's method.

Sir Bryan **Thwaites** (1923–) An English applied mathematician who invented the Thwaites flap and devised an approximate approach to the calculation of laminar boundary-layer development, known as Thwaites' method.

Dudley Brian **Spalding** (1923–2016) An English mechanical engineer whose early research was concerned primarily with combustion, heat, and mass transfer. Spalding is one of the pioneers and key developers of computational fluid mechanics and turbulence modelling. The CFD code PHOENICS was developed under his direction. Spalding's law-of-the-wall formula covers the viscous sublayer, the buffer region, and the logarithmic layer. The Spalding transfer number is a non-dimensional thermodynamic parameter which characterises the rate of diffusion of fuel from a burning fuel droplet. A non-dimensional group which arises in convective heat transfer for pipe flow is also called the Spalding number.

Pierre Jean **Carreau** (1939–) A Canadian chemical engineer who has made numerous contributions to the understanding of polymer rheology and the design of mixing systems. The five-parameter Carreau-Yasuda equation accurately describes the flow curve for many polymer solutions, polymer melts, and other non-Newtonian liquids.

# Appendix 2
# Physical properties of selected gases and liquids, and other data

Gaps have been left in a number of tables where information could not be found for some properties of a particular fluid. In a few instances, linear interpolation was used to find the value of a property at a particular temperature.

The data in **Tables A.1** to **A.5** have been compiled primarily from Kaye and Laby (1973) and White (2011).

The data for gases in **Table A.6** have been compiled primarily from the online Air Liquide Gas Encyclopedia (https://encyclopedia.airliquide.com).

**Table A.1** The atomic weights and molecular weights of some common elements and molecules, and their symbols and formulae

| Atom | | Atomic weight | Molecule | | Molecular weight, $\mathcal{M}$ |
|------|------|------|------|------|------|
| Carbon | C | 12.011 | Air[184] | | 28.96 |
| Chlorine | Cl | 38.453 | Butane | $C_4H_{10}$ | 58.12 |
| Fluorine | F | 18.998 | Chlorine | $Cl_2$ | 76.91 |
| Helium | He | 4.003 | Ethane | $C_2H_6$ | 30.07 |
| Hydrogen | H | 1.008 | Hydrogen | $H_2$ | 2.02 |
| Nitrogen | N | 14.007 | Hydrogen sulphide | $H_2S$ | 34.08 |
| Oxygen | O | 15.999 | Methane | $CH_4$ | 16.04 |
| Sulphur | S | 32.066 | Nitrogen | $N_2$ | 28.01 |
| | | | Nitric oxide | NO | 30.01 |
| | | | Nitrogen dioxide | $NO_2$ | 46.01 |
| | | | Oxygen | $O_2$ | 32.00 |
| | | | Pentane | $C_5H_{12}$ | 72.15 |
| | | | Propane | $C_3H_8$ | 44.10 |
| | | | Sulphur hexafluoride | $SF_6$ | 146.0 |
| | | | Water | $H_2O$ | 18.02 |

[184] Although air is a mixture of gases (major constituents 78.08% nitrogen, 20.95% oxygen, and 0.93% argon, and 0.04% carbon dioxide by volume) and not a compound, it is normally treated as a gas with a molecular weight of 28.96.

**Table A.2** Some universal constants and their symbols and values

| | | |
|---|---|---|
| Avogadro number (or constant) | $N_A$ | $6.02214 \times 10^{26}$ molecules/kmol |
| Boltzmann constant | $k_B$ | $1.3807 \times 10^{-23}$ J/K ($= \mathcal{R}/N_A$) |
| Speed of light in vacuum | $c_0$ | $2.9979 \times 10^8$ m/s |
| Standard acceleration due to gravity | $g$ | $9.80665$ m/s$^2$ |
| Universal (or molar) gas constant | $\mathcal{R}$ | $8.3145$ kJ/kmol.K ($= \mathcal{M}R$) |

**Table A.3** Physical properties of pure water at 1 atm

| $t$ (°C) | $\rho$ (kg/m$^3$) | $\mu$ (Pa · s) | $\nu^{185}$ (m$^2$/s) | $\sigma^{186}$ (N/m) | $p_V$ (kPa) | $c$ (m/s) |
|---|---|---|---|---|---|---|
| 0 | 999.8 | 1.787 E–3 | 1.787 E–6 | 0.0757 | 0.6107 | 1402 |
| 4 | 1000.0 | 1.573 E–3 | 1.573 E–6 | 0.0749 | 0.8130 | 1422 |
| 10 | 999.7 | 1.304 E–3 | 1.304 E–6 | 0.0742 | 1.2276 | 1447 |
| 20 | 998.2 | 1.002 E–3 | 1.004 E–6 | 0.07275 | 2.3384 | 1482 |
| 30 | 995.7 | 7.982 E–4 | 8.016 E–7 | 0.0712 | 4.2451 | 1509 |
| 40 | 992.2 | 6.540 E–4 | 6.591 E–7 | 0.0696 | 7.3812 | 1529 |
| 50 | 988.0 | 5.477 E–4 | 5.544 E–7 | 0.0679 | 12.345 | 1543 |
| 60 | 983.2 | 4.674 E–4 | 4.754 E–7 | 0.0662 | 19.933 | 1551 |
| 70 | 977.8 | 4.048 E–4 | 4.140 E–7 | 0.0644 | 31.177 | 1555 |
| 80 | 971.8 | 3.554 E–4 | 3.657 E–7 | 0.0626 | 47.375 | 1555 |
| 90 | 965.3 | 3.155 E–4 | 3.268 E–7 | 0.0608 | 71.120 | 1550 |
| 100 | 958.4 | 2.829 E–4 | 2.952 E–7 | 0.0588 | 1.0133 E+2 | 1542 |

[185] Calculated from the definition $\nu = \mu/\rho$.
[186] In contact with air.

In **Tables A.2** to **A.5**, the symbols have the following meanings and units

$c$   speed of sound, for perfect gas $c = \sqrt{\gamma R T}$ (m/s)
$K$   bulk modulus (Pa)
$\mathcal{M}$   molecular weight (kg/kmol)
$p$   pressure (Pa)
$p_V$   vapour pressure (Pa)
$R$   specific gas constant, $R = \mathcal{R}/M$ (m$^2$/s$^2$ · K or J/kg · K)
$t$   temperature (°C)
$T$   absolute temperature (K)

$t_B$  boiling point (°C)
$\gamma$  ratio of specific heats for perfect gas, $\gamma = C_P/C_V$
$\mu$  dynamic viscosity (Pa · s)
$\nu$  kinematic viscosity, $\nu = \mu/\rho$ (m$^2$/s)
$\rho$  density (kg/m$^3$)
$\sigma$  surface tension (N/m)

**Table A.4** Physical properties of air at 1 atm

| $t$ (°C) | $\rho$ (kg/m$^3$) | $\mu$ ($\mu$Pa · s) | $\nu$ (m$^2$/s) | $c$ (m/s) |
|---|---|---|---|---|
| −40 | 1.514 | 0.157 | 1.04 E−5 | 306.2 |
| −20 | 1.395 | 0.163 | 1.17 E−5 | 319.1 |
| 0 | 1.292 | 0.171 | 1.32 E−5 | 331.4 |
| 5 | 1.269 | 0.173 | 1.36 E−5 | 334.4 |
| 10 | 1.247 | 0.176 | 1.41 E−5 | 337.4 |
| 15 | 1.225 | 0.180 | 1.47 E−5 | 340.4 |
| 20 | 1.204 | 0.182 | 1.51 E−5 | 343.4 |
| 25 | 1.184 | 0.185 | 1.56 E−5 | 346.3 |
| 30 | 1.165 | 0.186 | 1.60 E−5 | 349.1 |
| 40 | 1.127 | 0.187 | 1.66 E−5 | 354.7 |
| 50 | 1.109 | 0.195 | 1.76 E−5 | 360.3 |
| 60 | 1.060 | 0.197 | 1.86 E−5 | 365.7 |
| 70 | 1.029 | 0.203 | 1.97 E−5 | 371.2 |
| 80 | 0.9996 | 0.207 | 2.07 E−5 | 376.6 |
| 90 | 0.9721 | 0.214 | 2.20 E−5 | 381.7 |
| 100 | 0.9461 | 0.217 | 2.29 E−5 | 386.0 |
| 150 | 0.8343 | 0.238 | 2.85 E−5 | 412.3 |
| 200 | 0.7461 | 0.253 | 3.39 E−5 | 434.5 |
| 250 | 0.6748 | 0.275 | 4.08 E−5 | 458.4 |
| 300 | 0.6159 | 0.298 | 4.84 E−5 | 476.3 |
| 400 | 0.5243 | 0.332 | 6.34 E−5 | 514.1 |
| 500 | 0.8343 | 0.364 | 7.97 E−5 | 548.8 |
| 1000 | 0.2772 | 0.504 | 1.82E−4 | 694.8 |

**Table A.5** Physical properties of some common liquids at 20 °C and 1 atm

| Liquid[187] | $\mathcal{M}$ | $\rho$ (kg/m³) | $\mu^{188}$ (Pa·s) | $\sigma^{189}$ (N/m) | $t_B$ (°C) | $p_V$ (Pa) | $K$ (Pa) | $c$ (m/s) |
|---|---|---|---|---|---|---|---|---|
| Benzene $C_6H_6$ | 78.1 | 879 | 6.03 E−4 | 0.0289(A) | 80.1 | 1.01 E+4 | 1.4 E+9 | 1320 |
| Carbon tetrachloride $CCl_4$ | 153.8 | 1632 | 9.12 E−4 | 0.0270(V) | 76.7 | 1.20 E+4 | 9.65 E+8 | 940 |
| Castor oil $C_{57}H_{110}O_9$ | 938 | 960 | 0.7 | | 313 | | | 1500 |
| Ethanol $C_2H_5OH$ | 46.1 | 789 | 1.08 E−3 | 0.0224(V) | 78.3 | 5.7 E+3 | 9.0 E+8 | 1162 |
| Ethylene glycol $C_2H_6O_2$ | 62.1 | 1110 | 0.0199 | | 197.3 | | | 1339 |
| Freon 12 $CCl_2F_2$ | 126.9 | 1330 | 2.63 E−4 | | −29.8 | | | |
| Glycerol $C_3H_8O_3$ | 92.1 | 1261 | 1.41 | 0.0634(A) | 182 | 1.4 E−2 | 4.39 E+8 | 1860 |
| Mercury Hg | 200.6 | 13546 | 1.53 E−3 | 0.489(V) | 357 | 1.1 E−3 | 2.55 E+10 | 1454 |
| Methanol $CH_3OH$ | 32.0 | 791 | 5.43 E−4 | 0.0225(A) | 64.7 | 1.34 E+4 | 8.3 E+8 | 1121 |
| Olive oil | | 900 | 0.0607 | | | | 1.6 E+9 | 1440 |
| Paraffin oil (kerosene) | | 804 | 1.92 E−3 | 0.028 | | 3.11 E+3 | 1.6 E+9 | 1315 |
| Petrol | 100 | 680 | 2.92 E−4 | 0.0216 | | 5.51 E+4 | 9.58 E+8 | |
| Pure water $H_2O$ | 18 | 998.2 | 1.00 E−3 | 0.0728 | | 2.34 E+3 | 2.19 E+9 | 1482 |
| SAE 10 W oil | | 870 | 0.104 | 0.036 | | | 1.31 E+9 | |
| SAE 30 W oil | | 891 | 0.29 | 0.035 | | | 1.38 E+9 | |
| SAE 150 W oil | | 902 | 0.86 | | | | | |
| Sea water (3.5% salinity) | | 1025 | 1.08 E−3 | 0.0728 | 100.6 | 2.34 E+3 | 2.33 E+9 | 1522 |

[187] The principal constituent of castor oil is ricinoleic acid. Castor oil, petrol, paraffin oil (kerosene), and SAE oil are all generic terms for blends of miscible oils, either natural hydrocarbons or synthetic. Since there are wide variations in the composition of these blends, the values of properties listed are representative.

[188] For a liquid, there is a significant decrease in dynamic viscosity with increase in temperature (see Section 2.8 and Figure 2.5). Below 100 bar, viscosity is practically independent of pressure.

[189] $A$ = against air; $V$ = against own vapour.

**Table A.6** Physical properties of some common gases at 1 atm[190]

| Gas | $\mathcal{M}$ | $\rho$ | $\mu$ | $R$[191] | $\gamma$ | $c$[192] |
|---|---|---|---|---|---|---|
| Dry air[193] | 28.96 | 1.225 | 1.721 E–5 | 287.0 | 1.402 | 340.4 |
| Argon (Ar) | 39.95 | 1.690 | 2.102 E–5 | 208.1 | 1.670 | 316.4 |
| $n$-Butane ($C_4H_{10}$) | 58.12 | 2.544 | 6.769 E–6 | 143.1 | 1.105 | 213.4 |
| Carbon dioxide ($CO_2$) | 44.01 | 1.871 | 1.371 E–5 | 188.9 | 1.294 | 265.3 |
| Carbon monoxide (CO) | 28.01 | 1.185 | 1.652 E–5 | 296.8 | 1.401 | 346.1 |
| Chlorine ($Cl_2$) | 70.91 | 3.04 | 1.245 E–5 | 117.3 | 1.355 | 214.0 |
| Ethane ($C_2H_6$) | 30.07 | 1.282 | 8.613 E–6 | 276.5 | 1.194 | 308.4 |
| Ethylene ($C_2H_4$) | 28.05 | 1.194 | 9.47 E–6 | 296.4 | 1.246 | 326.1 |
| Helium (He) | 4.003 | 0.169 | 1.870 E–5 | 2077 | 1.667 | 998.6 |
| Hydrogen ($H_2$) | 2.016 | 0.0852 | 8.397 E–6 | 4124 | 1.405 | 1292 |
| Methane ($CH_4$) | 16.04 | 0.680 | 1.025 E–5 | 518.4 | 1.306 | 441.6 |
| Neon (Ne) | 20.18 | 0.853 | 2.938 E–5 | 412.0 | 1.667 | 444.7 |
| Nitric oxide (NO) | 30.01 | 1.340 | 1.780 E–5 | 277.1 | 1.394 | 333.5 |
| Nitrogen ($N_2$) | 28.01 | 1.185 | 1.663 E–5 | 296.8 | 1.401 | 346.1 |
| Nitrogen dioxide ($NO_2$) | 46.01 | 1.947 | 1.331 E–10 | 180.7 | | |
| Nitrous oxide ($N_2O$) | 44.01 | 1.872 | 1.363 E–5 | 188.9 | 1.280 | 263.9 |
| Oxygen ($O_2$) | 32.00 | 1.354 | 1.914 E–5 | 259.8 | 1.397 | 323.3 |
| Sulphur dioxide ($SO_2$) | 64.06 | 2.763 | 1.180 E–5 | 129.8 | 1.281 | 218.8 |
| Sulphur hexafluoride ($SF_6$) | 146.06 | 6.256 | 1.377 E–5 | 56.93 | 1.098 | 134.2 |
| Water vapour ($H_2O$) | 18.00 | 461.4 | 9.7 E–6 | | | |

[190] The temperatures are 15 °C for $\rho$, $R$, and $c$, 0 °C for $\mu$, and 25 °C for $\gamma$.
[191] Calculated from $R = \mathcal{R}/M$.
[192] Calculated from $c = \sqrt{\gamma R T}$.
[193] Major constituents of dry air by volume: 78.08% $N_2$, 20.95% $O_2$, 0.93% Ar, 0.04% $CO_2$.

**Table A.7** Physical properties of the 1976 Standard Atmosphere; values determined using the Digital Dutch 1976 Standard Atmosphere Calculator (http://www.digitaldutch.com/atmoscalc/)

**(0) Troposphere:** 0 to 11 km, 15 °C to –56.5 °C, lapse rate $\Gamma$ = 6.5 °C/km

| $z'_G$ (m) | $T$ (K) | $p$ (Pa) | $\rho$ (kg/m$^3$) | $z'_G$ (m) | $T$ (K) | $p$ (Pa) | $\rho$ (kg/m$^3$) |
|---|---|---|---|---|---|---|---|
| 0 | 288.15 | 1.01325 E+5 | 1.2250 | 6000 | 249.15 | 4.7181 E+4 | 0.6597 |
| 500 | 284.90 | 9.5461 E+4 | 1.1673 | 6 500 | 245.90 | 4.4035 E+4 | 0.6238 |
| 1 000 | 281.65 | 8.9875 E+4 | 1.1116 | 7 000 | 242.65 | 4.1061 E+4 | 0.5895 |
| 1 500 | 278.40 | 8.4556 E+4 | 1.0581 | 7 500 | 239.40 | 3.8251 E+4 | 0.5562 |
| 2 000 | 275.15 | 7.9495 E+4 | 1.0065 | 8 000 | 236.15 | 3.5600 E+4 | 0.5252 |
| 2 500 | 271.90 | 7.4683 E+4 | 0.9569 | 8 500 | 232.90 | 3.3099 E+4 | 0.4951 |
| 3 000 | 268.65 | 7.0109 E+4 | 0.9091 | 9 000 | 229.65 | 3.0743 E+4 | 0.4663 |
| 3 500 | 265.40 | 6.5764 E+4 | 0.8632 | 9 500 | 226.40 | 2.8524 E+4 | 0.4389 |
| 4 000 | 262.15 | 6.1640 E+4 | 0.8191 | 10 000 | 223.15 | 2.6436 E+4 | 0.4127 |
| 4 500 | 258.90 | 5.7728 E+4 | 0.7768 | 10 500 | 219.90 | 2.4474 E+4 | 0.3877 |
| 5 000 | 255.65 | 5.4020 E+4 | 0.7361 | 11 000 | 216.65 | 2.2632 E+4 | 0.3639 |
| 5 500 | 252.40 | 5.0507 E+4 | 0.6971 | | | | |

**(1) Tropopause:** 11 km to 20 km, isothermal at –56.5 °C (216.65 K)

| $z'_G$ (m) | $T$ (K) | $p$ (Pa) | $\rho$ (kg/m$^3$) | $z'_G$ (m) | $T$ (K) | $p$ (Pa) | $\rho$ (kg/m$^3$) |
|---|---|---|---|---|---|---|---|
| 11 000 | 216.65 | 2.2632 E+4 | 0.3639 | 16 000 | 216.65 | 1.0288 E+4 | 0.1654 |
| 11 500 | 216.65 | 2.0916 E+4 | 0.3363 | 16 500 | 216.65 | 9.508 E+3 | 0.1529 |
| 12 000 | 216.65 | 1.9330 E+4 | 0.3108 | 17 000 | 216.65 | 8.787 E+3 | 0.1413 |
| 12 500 | 216.65 | 1.7865 E+4 | 0.2873 | 17 500 | 216.65 | 8.121 E+3 | 0.1306 |
| 13 000 | 216.65 | 1.6510 E+4 | 0.2655 | 18 000 | 216.65 | 7.505 E+3 | 0.1207 |
| 13 500 | 216.65 | 1.5259 E+4 | 0.2454 | 18 500 | 216.65 | 6.936 E+3 | 0.1115 |
| 14 000 | 216.65 | 1.4102 E+4 | 0.2268 | 19 000 | 216.65 | 6.410 E+3 | 0.1031 |
| 14 500 | 216.65 | 1.3033 E+4 | 0.2096 | 19 500 | 216.65 | 5.924 E+3 | 0.0953 |
| 15 000 | 216.65 | 1.2045 E+4 | 0.1937 | 20 000 | 216.65 | 5.475 E+3 | 0.0880 |
| 15 500 | 216.65 | 1.1113 E+4 | 0.1790 | | | | |

**(2) Lower stratosphere:** 20 to 32 km, −56.5 °C to −44.5 °C, lapse rate $\Gamma = -1$ °C/km

| | | | | | | | |
|---|---|---|---|---|---|---|---|
| 20 000 | 216.65 | 5.475 E+3 | 0.0880 | 28 000 | 224.65 | 1.586 E+3 | 0.0246 |
| 22 000 | 218.65 | 4.000 E+3 | 0.0637 | 30 000 | 226.65 | 1.172 E+3 | 0.0180 |
| 24 000 | 220.65 | 2.930 E+3 | 0.0463 | 32 000 | 228.65 | 8.680 E+2 | 0.0132 |
| 26 000 | 222.65 | 2.153 E+3 | 0.0337 | | | | |

**(3) Upper stratosphere:** 32 to 47 km, −44.5 °C to −2.5 °C, lapse rate $\Gamma = -2.8$ °C/km

| | | | | | | | |
|---|---|---|---|---|---|---|---|
| 32 000 | 228.65 | 8.680 E+2 | 0.0132 | 47 000 | 270.65 | 1.109 E+2 | 0.0014 |
| 40 000 | 251.05 | 2.775 E+2 | 0.0039 | | | | |

**(4) Stratopause:** 47 km to 51 km, isothermal at −2.5 °C

| | | | | | | | |
|---|---|---|---|---|---|---|---|
| 47 000 | 270.65 | 1.109 E+2 | 0.0014 | 51 000 | 270.65 | 66.9 | 0.0009 |
| 50 000 | 270.65 | 75.9 | 0.00098 | | | | |

**(5) Lower mesosphere:** 51 to 71 km, −2.5 °C to −58.5 °C, lapse rate $\Gamma = 2.8$ °C/km

| | | | | | | | |
|---|---|---|---|---|---|---|---|
| 51 000 | 270.65 | 66.9 | 0.0009 | 70 000 | 217.45 | 4.63 | 0.0001 |
| 60 000 | 245.45 | 20.3 | 0.0003 | 71 000 | 214.65 | 3.96 | 0.0001 |

**(6) Upper mesosphere:** 71 km to 84.852 km, −58.5 °C to −86.2 °C, lapse rate $\Gamma = 2.0$ °C/km

| | | | | | | | |
|---|---|---|---|---|---|---|---|
| 71 000 | 214.65 | 3.96 | 0.0001 | 84 852 | 186.95 | 0.373 | 0.0000 |
| 80 000 | 196.65 | 0.886 | 0.0000 | | | | |

**(7) Mesopause:** 84.852 km, −86.2 °C

In **Section 4.13** the relationship between the geopotential altitude $z'_G$ and the geometric altitude was shown to be

$$z'_G = \frac{R_E z'}{R_E + z'},$$

$R_E = 6356$ being the radius of the earth. The calculated values of $z'$ corresponding to $z'_G$ for each layer are as follows

    **(0) Troposphere:** 0 to 11.019 km
    **(1) Tropopause:** 11.019 to 20.063 km

(2) **Lower stratosphere:** 20.063 to 32.162 km
(3) **Upper stratosphere:** 32.162 to 47.350 km
(4) **Stratopause:** 47.350 to 51.413 km
(5) **Lower mesosphere:** 51.413 to 71.802 km
(6) **Upper mesosphere:** 71.802 to 86.000 km
(7) **Mesopause:** 86.000 km

Not included in the 1976 Standard Atmosphere are the following regions of the upper atmosphere

**Thermosphere:** 85 km to 600 km, –120 °C to 2000 °C
**Thermopause:** 600 km, 2000 °C
**Exosphere:** 600 km to 10,000 km

# Appendix 3
## Areas, centroid locations, and second moments of area for some common shapes

$C$ ———⬤——— $C$   denotes centroid and axis of rotation

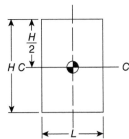

(a) Rectangle

$$A = L\,H$$

$$I_C = \frac{A\,H^2}{12} = \frac{L\,H^3}{12}$$

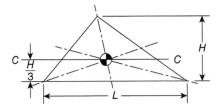
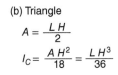

(b) Triangle

$$A = \frac{L\,H}{2}$$

$$I_C = \frac{A\,H^2}{18} = \frac{L\,H^3}{36}$$

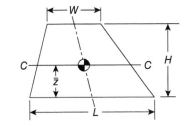

(c) Trapezium

$$A = \frac{(W + L)\,H}{2}$$

$$\bar{z} = \frac{(2W + L)\,H}{3(W + L)}$$

$$I_C = \frac{(W^2 + 4W\,L + L^2)\,H^3}{36(W + L)}$$

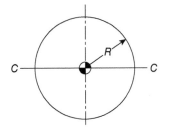

(d) Circle

$$A = \pi\,R^2$$

$$I_C = \frac{A\,R^2}{4} = \frac{\pi\,R^4}{4}$$

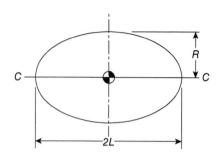

(e) Ellipse

$$A = \pi R L$$

$$I_C = \frac{AR^2}{4} = \frac{\pi R^3 L}{4}$$

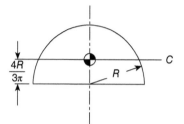

(f) Semicircle

$$A = \frac{\pi R^2}{2}$$

$$I_C = \left(\frac{1}{4} - \frac{16}{9\pi^2}\right) A R^2$$

$$= \left(\frac{\pi}{4} - \frac{16}{9\pi}\right) \frac{R^4}{2}$$

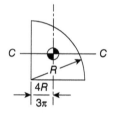

(g) Quadrant of circle

$$A = \frac{\pi R^2}{4}$$

$$I_C = \left(\frac{1}{4} - \frac{16}{9\pi^2}\right) A R^2$$

$$= \left(\frac{\pi}{4} - \frac{16}{9\pi}\right) \frac{R^4}{2}$$

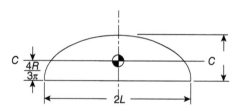

(h) Semiellipse

$$A = \frac{\pi R L}{2}$$

$$I_C = \left(\frac{1}{4} - \frac{16}{9\pi^2}\right) A R^2$$

$$= \left(\frac{\pi}{4} - \frac{16}{9\pi}\right) \frac{R^3 L}{2}$$

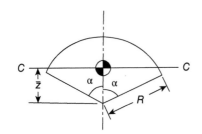

(i) Circular sector

$$A = \alpha R^2$$

$$\bar{z} = \frac{2\,R\,\sin\alpha}{3\,\alpha}$$

$$I_C = \left(\frac{\alpha + \sin\alpha\,\cos\alpha}{4} - \frac{4\sin^2\alpha}{9\,\alpha}\right) R^4$$

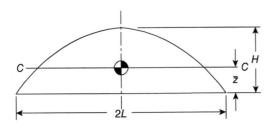

(j) Sine wave

$$A = \frac{4\,L\,H}{\pi}$$

$$\bar{z} = \frac{\pi\,H}{8}$$

$$I_C = \left(\frac{2}{9} - \frac{\pi^2}{64}\right) A^2$$

$$= \left(\frac{8}{9\pi} - \frac{\pi}{16}\right) L H^3$$

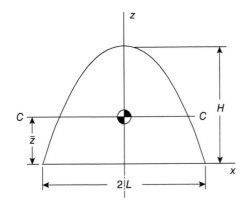

(k) Segment of $n^{\text{th}}$ degree

$$z = H\left[1 - \left(\frac{x}{L}\right)^n\right]$$

$$A = \frac{n\,L\,H}{n+1}$$

$$\bar{z} = \frac{n\,H}{2n+1}$$

$$I_C = \frac{n^2(n+1)\,H^2 A}{(2n+1)^2\,(3n+1)}$$

$$= \frac{n^3 L\,H^3}{(2n+1)^2\,(3n+1)}$$

# Appendix 4
# Differential equations for compressible pipe flow

In **Chapter 13** we discussed the trends with increasing or decreasing Mach number for Fanno flow, for isothermal pipe flow with friction, and for Rayleigh flow based upon the differential equations for these three pipe flows. In this appendix we show how the differential equations can be derived from the basic equations presented in **Chapter 13**. The key results are highlighted in colour. Where necessary we shall make use of the relations $C_P = \gamma R/(\gamma - 1)$, $c^2 = \gamma RT$, $M = V/c$, $\rho V = G =$ constant, and $p = \rho RT$.

## (a) Fanno flow: Adiabatic pipe flow with friction

For adiabatic flow, $\dot{q}' = 0$ so that $T_0 =$ constant, and the energy equation derived in **Chapter 13**, equation (13.3), simplifies to

$$C_P \frac{dT}{dx} + V \frac{dV}{dx} = 0. \tag{A1}$$

If we introduce the relations $C_P = \gamma R/(\gamma - 1)$, $c^2 = \gamma RT$, and $M = V/c$, after some algebraic manipulation, we find

$$\frac{1}{T} \frac{dT}{dx} = \frac{1}{c^2} \frac{dc^2}{dx} = - \left[ \frac{\left(\frac{\gamma - 1}{2}\right) M^2}{1 + \left(\frac{\gamma - 1}{2}\right) M^2} \right] \frac{1}{M^2} \frac{dM^2}{dx}. \tag{A2}$$

From equation (A1) we have

$$\frac{1}{V} \frac{dV}{dx} = - \frac{1}{(\gamma - 1)M^2} \frac{1}{T} \frac{dT}{dx} \tag{A3}$$

so that, after substitution for $(1/T)\, dT/dx$ from equation (A2),

$$\frac{1}{V} \frac{dV}{dx} = \left\{ \frac{1}{2\left[1 + \left(\frac{\gamma - 1}{2}\right) M^2\right]} \right\} \frac{1}{M^2} \frac{dM^2}{dx}. \tag{A4}$$

From $\rho V = G =$ constant, we have

$$\frac{1}{\rho} \frac{d\rho}{dx} = - \frac{1}{V} \frac{dV}{dx} \tag{A5}$$

so that

$$\frac{1}{\rho} \frac{d\rho}{dx} = - \left\{ \frac{1}{2\left[1 + \left(\frac{\gamma - 1}{2}\right) M^2\right]} \right\} \frac{1}{M^2} \frac{dM^2}{dx}. \tag{A6}$$

From $p = \rho RT$, we have

$$\frac{1}{p}\frac{dp}{dx} = \frac{1}{\rho}\frac{d\rho}{dx} + \frac{1}{T}\frac{dT}{dx} \tag{A7}$$

and, after substitution for $(1/\rho)\, d\rho/dx$ from equation (A6) and for $(1/T)\, dT/dx$ from equation (A2), we find

$$\frac{1}{p}\frac{dp}{dx} = -\frac{1}{2}\left[\frac{1 + (\gamma - 1)M^2}{1 + \left(\frac{\gamma - 1}{2}\right)M^2}\right]\frac{1}{M^2}\frac{dM^2}{dx}. \tag{A8}$$

The stagnation pressure $p_0$ is given by an equation derived in **Chapter 11**,

$$p_0 = p\left[1 + \left(\frac{\gamma - 1}{2}\right)M^2\right]^{\gamma/(\gamma - 1)} \tag{11.22}$$

from which

$$\frac{1}{p_0}\frac{dp_0}{dx} = \frac{1}{p}\frac{dp}{dx} + \frac{\gamma M^2}{2\left[1 + \left(\frac{\gamma - 1}{2}\right)M^2\right]}\frac{1}{M^2}\frac{dM^2}{dx}. \tag{A9}$$

After substitution for $(1/p)\, dp/dx$ from equation (A8) in equation (A9), and some algebraic manipulation, we find

$$\frac{1}{p_0}\frac{dp_0}{dx} = -\frac{1 - M^2}{2\left[1 + \left(\frac{\gamma - 1}{2}\right)M^2\right]}\frac{1}{M^2}\frac{dM^2}{dx}. \tag{A10}$$

The entropy variation, from equation (11.1) in **Chapter 11**, is

$$T\frac{ds}{dx} = C_P\frac{dT}{dx} - \frac{1}{\rho}\frac{dp}{dx} \tag{A11}$$

which can be rewritten as

$$\frac{1}{R}\frac{ds}{dx} = \frac{\gamma}{(\gamma - 1)}\frac{1}{T}\frac{dT}{dx} - \frac{1}{p}\frac{dp}{dx}. \tag{A12}$$

After substitution in equation (A12) for $(1/T)\, dT/dx$ from equation (A2) and for $(1/p)\, dp/dx$ from equation (A8), we have

$$\frac{1}{R}\frac{ds}{dx} = \frac{1 - M^2}{2\left[1 + \left(\frac{\gamma - 1}{2}\right)M^2\right]}\frac{1}{M^2}\frac{dM^2}{dx}. \tag{A13}$$

We can see from equations (A10) and (A13) that

$$\frac{1}{R}\frac{ds}{dx} = -\frac{1}{p_0}\frac{dp_0}{dx}, \tag{A14}$$

a result which could have been derived directly from equation (A11) since the flow is adiabatic (i.e. $T_0 = $ constant).

The final form for each of the differential equations is arrived at by substituting for $\left(1/M^2\right)/dM^2/dx$ from equation (13.14), with $f_F$ replaced by $\overline{f_F}$, in equation (A2) for $T$, (A4) for $V$, in equation (A6) for $\rho$, in equation (A8) for $p$, in equation (A10) for $p_0$, and in equation (A13) for $s$

$$\frac{1}{M^2}\frac{dM^2}{dx} = \frac{\gamma M^2 \left[1 + \left(\frac{\gamma-1}{2}\right) M^2\right]}{1 - M^2} \frac{4\overline{f_F}}{D}. \tag{A15}$$

The resulting equations are given in Subsection 13.2.2, as equations (13.18) to (13.22).

## (b) Isothermal pipe flow with friction

Because the static temperature $T$ is now constant, so is the soundspeed $c$, and a convenient starting point is the definition of the Mach number $M$

$$M = \frac{V}{c} \tag{A16}$$

from which

$$\frac{1}{V}\frac{dV}{dx} = \frac{1}{2M^2}\frac{dM^2}{dx}. \tag{A17}$$

Equation (A5) is still valid, so that

$$\frac{1}{\rho}\frac{d\rho}{dx} = -\frac{1}{V}\frac{dV}{dx} = -\frac{1}{2M^2}\frac{dM^2}{dx}. \tag{A18}$$

From $p = \rho RT$, we have

$$\frac{1}{p}\frac{dp}{dx} = \frac{1}{\rho}\frac{d\rho}{dx} = -\frac{1}{2M^2}\frac{dM^2}{dx}. \tag{A19}$$

The stagnation temperature $T_0$, as defined in Chapter 11, is defined by equation (11.13)

$$T_0 = T + \frac{V^2}{2C_P} \tag{11.13}$$

so that

$$\frac{dT_0}{dx} = \frac{V^2}{C_P}\frac{1}{V}\frac{dV}{dx}. \tag{A20}$$

After division by $T_0$ this gives

$$\frac{1}{T_0}\frac{dT_0}{dx} = \left(\frac{\frac{V^2}{C_P}}{T + \frac{V^2}{2C_P}}\right)\frac{1}{V}\frac{dV}{dx} = \frac{(\gamma-1)M^2}{\left[1 + \left(\frac{\gamma-1}{2}\right)M^2\right]}\frac{1}{V}\frac{dV}{dx}$$

and, after substitution for $(1/V)\,dV/dx$ from equation (A17),

$$\frac{1}{T_0}\frac{dT_0}{dx} = \frac{(\gamma-1)M^2}{2\left[1 + \left(\frac{\gamma-1}{2}\right)M^2\right]}\frac{1}{M^2}\frac{dM^2}{dx}. \tag{A21}$$

Equation (A9), derived from the definition of the stagnation pressure $p_0$, is still valid

$$\frac{1}{p_0}\frac{dp_0}{dx} = \frac{1}{p}\frac{dp}{dx} + \frac{\gamma M^2}{2\left[1 + \left(\frac{\gamma-1}{2}\right)M^2\right]}\frac{1}{M^2}\frac{dM^2}{dx} \tag{A9}$$

and after substitution for $(1/p)\,dp/dx$ from equation (A19) in equation (A9) we find

$$\frac{1}{p_0}\frac{dp_0}{dx} = -\frac{\left[1 - \left(\frac{\gamma+1}{2}\right)M^2\right]}{2\left[1 + \left(\frac{\gamma-1}{2}\right)M^2\right]}\frac{1}{M^2}\frac{dM^2}{dx}. \tag{A22}$$

From equation (A11), with $T = $ constant,

$$\frac{ds}{dx} = -\frac{1}{\rho T}\frac{dp}{dx} \tag{A23}$$

which leads to

$$\frac{1}{R}\frac{ds}{dx} = -\frac{1}{\rho RT}\frac{dp}{dx} = -\frac{1}{p}\frac{dp}{dx} = \frac{1}{2M^2}\frac{dM^2}{dx}. \tag{A24}$$

The final form for each of the differential equations is arrived at by substituting for $(1/M^2)\,dM^2/dx$ from equation (13.39), again replacing $f_F$ with $\overline{f_F}$, in equation (A17) for $V$, in equation (A18) for $\rho$, in equation (A19) for $p$, in equation (A21) for $T_0$, in equation (A22) for $p_0$, and in equation (A24) for $s$

$$\frac{1}{M^2}\frac{dM^2}{dx} = -\frac{\gamma M^2}{\left(\gamma M^2 - 1\right)}\frac{4\overline{f_F}}{D}. \tag{13.39}$$

The resulting equations are given in **Subsection 13.3.1** as equations (13.44) to (13.47).

## (c) Rayleigh flow: Frictionless pipe flow with surface heat transfer

If we differentiate the perfect-gas equation

$$p = \rho R T \tag{2.9}$$

we have

$$\frac{1}{T}\frac{dT}{dx} = \frac{1}{p}\frac{dp}{dx} - \frac{1}{\rho}\frac{d\rho}{dx}. \tag{A25}$$

From the mass-conservation equation derived in **Chapter 6**, equation (6.1), with $A = $ constant,

$$\rho V = \text{constant}$$

from which

$$\frac{1}{\rho}\frac{d\rho}{dx} = -\frac{1}{V}\frac{dV}{dx}. \tag{A5}$$

Equations (A25) and (A5) combine to give

$$\frac{1}{T}\frac{dT}{dx} = \frac{1}{p}\frac{dp}{dx} + \frac{1}{V}\frac{dV}{dx}. \tag{A26}$$

Since the flow is frictionless, the momentum equation derived in **Chapter 13**, equation (13.1), reduces to

$$\frac{dp}{dx} + \rho V \frac{dV}{dx} = 0 \tag{A27}$$

from which

$$\frac{1}{p}\frac{dp}{dx} = -\frac{\rho V^2}{p}\frac{1}{V}\frac{dV}{dx} = -\gamma M^2 \frac{1}{V}\frac{dV}{dx}. \tag{A28}$$

Substitution from equation (A28) in equation (A26) then gives

$$\frac{1}{T}\frac{dT}{dx} = \left(1 - \gamma M^2\right)\frac{1}{V}\frac{dV}{dx}. \tag{A29}$$

From $M = V/c$ we have

$$\frac{1}{M}\frac{dM}{dx} = \frac{1}{2M^2}\frac{dM^2}{dx} = \frac{1}{V}\frac{dV}{dx} - \frac{1}{c}\frac{dc}{dx} = \frac{1}{V}\frac{dV}{dx} - \frac{1}{2T}\frac{dT}{dx}. \tag{A30}$$

Equations (A29) and (A30) can be combined to eliminate $dV/dx$, with the result

$$\frac{1}{T}\frac{dT}{dx} = -\left(\frac{\gamma M^2 - 1}{\gamma M^2 + 1}\right)\frac{1}{M^2}\frac{dM^2}{dx}. \tag{A31}$$

From equations (A25), (A30), and (A31), we thus have

$$\frac{1}{V}\frac{dV}{dx} = -\frac{1}{\rho}\frac{d\rho}{dx} = \left(\frac{1}{\gamma M^2 + 1}\right)\frac{1}{M^2}\frac{dM^2}{dx} \tag{A32}$$

and, from equation (A28),

$$\frac{1}{p}\frac{dp}{dx} = -\left(\frac{\gamma M^2}{\gamma M^2 + 1}\right)\frac{1}{M^2}\frac{dM^2}{dx}. \tag{A33}$$

Equation (A9), derived from the definition of the stagnation pressure $p_0$, is still valid

$$\frac{1}{p_0}\frac{dp_0}{dx} = \frac{1}{p}\frac{dp}{dx} + \frac{\gamma M^2}{2\left[1 + \left(\frac{\gamma-1}{2}\right)M^2\right]}\frac{1}{M^2}\frac{dM^2}{dx}. \tag{A9}$$

Combining equations (A9) and (A33) then gives

$$\frac{1}{p_0}\frac{dp_0}{dx} = -\frac{\gamma M^2\left(1 - M^2\right)}{2\left(\gamma M^2 + 1\right)\left[1 + \left(\frac{\gamma-1}{2}\right)M^2\right]}\frac{1}{M^2}\frac{dM^2}{dx}. \tag{A34}$$

From equation (11.14) for the stagnation temperature $T_0$,

$$T_0 = T\left[1 + \left(\frac{\gamma-1}{2}\right)M^2\right] \tag{11.14}$$

we have

$$\frac{1}{T_0}\frac{dT_0}{dx} = \frac{1}{T}\frac{dT}{dx} + \frac{\left(\frac{\gamma-1}{2}\right)M^2}{\left[1 + \left(\frac{\gamma-1}{2}\right)M^2\right]}\frac{1}{M^2}\frac{dM^2}{dx}. \tag{A35}$$

After substitution for $(1/T)\,dT/dx$ from equation (A31) in equation (A35) we find

$$\frac{1}{T_0}\frac{dT_0}{dx} = \frac{1-M^2}{\left(\gamma M^2 + 1\right)\left[1 + \left(\frac{\gamma-1}{2}\right)M^2\right]}\frac{1}{M^2}\frac{dM^2}{dx}. \tag{A36}$$

The entropy variation from equation (A11) is

$$T\frac{ds}{dx} = C_P\frac{dT}{dx} - \frac{1}{\rho}\frac{dp}{dx} \tag{A11}$$

which leads to

$$\frac{1}{R}\frac{ds}{dx} = \frac{\gamma}{(\gamma-1)}\frac{1}{T}\frac{dT}{dx} - \frac{1}{p}\frac{dp}{dx}. \tag{A37}$$

After substitution for $(1/T)\,dT/dx$ from equation (A31) and for $(1/p)\,dp/dx$ from equation (A33) we find

$$\frac{1}{R}\frac{ds}{dx} = \frac{\gamma\left(1-M^2\right)}{(\gamma-1)\left(\gamma M^2+1\right)}\frac{1}{M^2}\frac{dM^2}{dx}. \tag{A38}$$

The final step is to introduce the energy-conservation equation derived in **Chapter 13**, equation (13.3)

$$\dot{q}' = \dot{m}C_P\frac{dT_0}{dx} \tag{13.3}$$

from which we have

$$\frac{1}{T_0}\frac{dT_0}{dx} = \frac{\dot{q}'}{\dot{m}C_P T_0} \tag{A25}$$

so that, from equation (A36),

$$\frac{1}{M^2}\frac{dM^2}{dx} = \frac{\left(\gamma M^2+1\right)\left[1+\left(\frac{\gamma-1}{2}\right)M^2\right]}{\left(1-M^2\right)}\frac{\dot{q}'}{\dot{m}C_P T_0}. \tag{A39}$$

The final form for each of the differential equations is arrived at by substituting for $(1/M^2)\,dM^2/dx$ from equation (A39) in equation (A32) for $V$ and $\rho$, in equation (A33) for $p$, in equation (A36) for $T_0$, in equation (A34) for $p_0$, and in equation (A38) for $s$. The resulting equations are given in **Subsection 13.4.1** as equations (13.68) to (13.74).

# Appendix 5
# Roughness heights

The values here have been collected from a wide range of sources and should be regarded, at best, as a rough (!) guide as there are very large variations (100%+ in some instances) in quoted values. Also, in several instances the description of a surface is too vague and subjective to be of real value, e.g. worn cast iron or moderately corroded carbon steel.

| | |
|---|---|
| glass | 0.3 $\mu$m |
| drawn aluminium, brass, copper, or lead | 1–2 $\mu$m |
| PVC | 1.5–7 $\mu$m |
| fibreglass | 5 $\mu$m |
| flexible rubber tubing | 6–7 $\mu$m |
| stretched steel | 15 $\mu$m |
| stainless steel | 15–30 $\mu$m |
| carbon steel | 20–50 $\mu$m |
| sheet metal | 20–100 $\mu$m |
| wrought iron | 45 $\mu$m |
| commercial or welded steel pipe | 45–90 $\mu$m |
| slightly corroded carbon steel | 50–150 $\mu$m |
| galvanised iron or steel | 25–150 $\mu$m |
| sheet or asphalted cast iron | 0.10–0.15 mm |
| moderately corroded carbon steel | 0.15–1 mm |
| rusted steel | 0.15–4 mm |
| cast iron | 0.25–1 mm |
| smoothed cement | 0.3–0.5 mm |
| concrete | 0.3–3 mm |
| reinforced rubber tubing | 0.3–4 mm |
| coarse concrete | 0.3–5 mm |
| worn cast iron | 0.8–1.5 mm |
| riveted steel | 0.9–9 mm |
| badly corroded carbon steel | 1–3 mm |
| rusted cast iron | 1.5–2.5 mm |

# Bibliography

There is a very wide choice of text and reference books concerned with fluid mechanics. The majority are by American authors and tend to be more mathematical than those by British writers. The small selection below includes those which the author of this book has found to be the most useful over many years. Those by Bird et al., Chapman and Walker, John and Keith, Shapiro, Thompson, and White (2005) deal with more advanced topics than the others. Also included are the books by Anderson and by Rouse and Ince concerning the history of fluid mechanics, and a few books and research papers referred to in the main text.

## History

John D. Anderson 'A History of Aerodynamics and Its Impact on Flying Machines' Cambridge University Press, 1998 ISBN 0-521-66955-3

Hunter Rouse and Simon Ince 'History of Hydraulics' Dover Publications, Inc., 1963

## General textbooks

Theodore Allen, Jr, and Richard L. Ditsworth 'Fluid Mechanics' McGraw-Hill, Inc, 1972

G. K. Batchelor 'An Introduction to Fluid Dynamics' Cambridge University Press, 2000 ISBN 10 0521663962

R. Byron Bird, Warren E. Stewart, and Edwin N. Lightfoot 'Transport Phenomena' John Wiley $2^{nd}$ Edition, 2001 ISBN-10:9752843670

George Emanuel 'Analytical Fluid Dynamics' CRC Press $2^{nd}$ Edition, 2000 ISBN 10-0849391148

T. E. Faber 'Fluid Dynamics for Physicists' Cambridge University Press, 1995 ISBN 0-521-42969-2

James A. Fay 'Introduction to Fluid Mechanics' MIT Press, 1994 ISBN 0-262-06165-1

Rolf H. Sabersky, Allan J. Acosta, Edward G. Hauptmann, and E. M. Gates 'Fluid Flow A First Course in Fluid Mechanics' Prentice-Hall, Inc $4^{th}$ Edition, 1999 ISBN 0-13-576372-X

Frank Mangrem White 'Fluid Mechanics' McGraw-Hill $7^{th}$ Edition, 2011 ISBN 978-007-131121-1

## Compressible fluid flow

Alan J. Chapman and William F. Walker 'Introductory Gas Dynamics' Holt, Reinhart and Winston, Inc., 1971 ISBN 10-003-077035-1

Antonio Ferri 'Elements of Aerodynamics of Supersonic Flows' Macmillan, 1949

James E. A. John and Theo Keith 'Gas Dynamics' Prentice Hall 3$^{rd}$ Edition, 2006 ISBN 13:9780131206687

H. W. Liepmann and A. Roshko 'Elements of Gas Dynamics' John Wiley & Sons, Inc, 1957

Ascher H. Shapiro 'The Dynamics and Thermodynamics of Compressible Fluid Flow' (two volumes) The Ronald Press Company, 1953

Philip A. Thompson 'Compressible-Fluid Dynamics' McGraw-Hill Inc., 1972 ISBN 07-064405-5

## Turbomachinery

N. A. Cumpsty 'Compressor Aerodynamics' Longman Scientific & Technical, 1989 ISBN 0-582-01364-X

Nicholas Cumpsty 'Jet Propulsion A Simple Guide to the Aerodynamic and Thermodynamic Design and Performance of Jet Engines' Cambridge University Press, 1997 ISBN 0-521-59674-2

S. L. Dixon 'Fluid Mechanics and Thermodynamics of Turbomachinery' Elsevier-Butterworth-Heinemann 5$^{th}$ Edition, 1998 ISBN 0-7506-7870-4

David Japikse and Nicholas C. Baines 'Introduction to Turbomachinery' Concepts ETI, Inc and Oxford University Press, 1994 ISBN 0-933283-06-7

## Viscous fluid flow

L. Rosenhead (Ed) 'Laminar Boundary Layers' Oxford University Press, 1963

Hermann Schlichting and Klaus Gersten 'Boundary Layer Theory' Springer 8$^{th}$ Edition, 2000 ISBN 978-3540662709

R. K. Shah and A. L. London 'Laminar Flow Forced Convection in Ducts' Academic Press, 1978

Frank Mangrem White 'Viscous Fluid Flow' McGraw-Hill 3$^{rd}$ Edition, 2005 ISBN 13-978-0071244930

## Non-Newtonian fluid flow

R. Byron Bird, Robert C. Armstrong, and Ole Hassager 'Dynamics of Polymeric Fluids Volume 1 Fluid Mechanics' John Wiley & Sons 2$^{nd}$ Edition, 1987 ISBN 0-471-80245-X

## Research-paper references

C. F. Colebrook 'Turbulent flow in pipes, with particular reference to the transition region between smooth and rough pipe laws' Trans. Inst. Civ. Eng, $\underline{11}$, pp. 133–156 (1939)

D. E. Coles 'The law of the wake in the turbulent boundary layer' J. Fluid Mech, $\underline{1}$, pp. 191–226 (1956)

E. R. van Driest 'On turbulent flow near a wall' J. Aero. Sci., <u>23</u>, pp. 1007–1011 (1956)

S. E. Haaland 'Simple and explicit formulas for the friction factor in turbulent pipe flow' J. Fluids Eng, <u>105,</u> pp. 89–90 (1983)

A. N. Kolmogorov 'The local structure of turbulence in incompressible viscous fluid for very large Reynolds number' Proc. Roy. Soc. Lond, <u>A434</u>, pp. 9–13 (1941)

L. F. Moody 'Friction factors for pipe flow' ASME Trans. <u>66</u>, pp. 671–684 (1944)

L. Prandtl 'Über Flüssigkeitsbewegung bei sehr kleiner Reibung' Proc. III Int Math Congress, Heidelberg (1904)

O. Reynolds 'On the dynamical theory of incompressible viscous fluids and the determination of the criterion' Phil Trans Roy Soc Lond <u>56</u>, pp. 40–45 (1895)

A. Roshko 'Experiments on the flow past a circular cylinder at very high Reynolds number' J. Fluid Mech, <u>10</u>, pp. 345–356 (1961)

F. Schultz-Grunow 'Neues Reibungswiderstandsgesetz für glatte Platten' Luftfahrtforschung <u>17</u>, pp. 239–246 (1940); translated as 'New frictional law for smooth plates', NACA TM 986 (1941)

D. B. Spalding 'A single formula for the "law of the wall"' J. Appl. Mech. <u>28</u>, pp. 455–458 (1961)

G. I. Taylor 'The conditions necessary for discontinuous motion in gases' Proc Roy Soc Lond A84, pp 371–377 (1910)

G. I. Taylor 'Stability of a viscous liquid contained between two rotating cylinders' Phil Trans Roy Soc Lond <u>A215</u>, pp. 1–26 (1915)

## Reference books and websites

Ira H. Abbott and Albert E. von Doenhoff **'Theory of Wing Sections'** Dover Publications, 1959 ISBN -13:978-0-486-60586-9

AMES Research Staff 'Equations, tables, and charts for compressible flow' NACA Report 1135, 1953

Robert D. Blevins **'Applied Fluid Dynamics Handbook'** Van Nostrand Reinhold Co. Inc., 1984 ISBN 0-442-21296-8

G. W. C. Kaye and T. H. Laby **'Tables of Physical and Chemical Constants** And Some Mathematical Functions' Longman Group Ltd 14<sup>th</sup> Edition, 1973, reprinted with corrections 1978, ISBN 0-582-46326-2; the 16<sup>th</sup> edition (published 1995) is now available free at Kaye and Laby online (http://www.kayelaby.npl.co.uk/)

G. F. C. Rogers and Y. R Mayhew **'Thermodynamic and Transport Properties of Fluids'** Wiley-Blackwell 5<sup>th</sup> Edition, 1994 ISBN-13:978-0631197034

B. N. Taylor and A. Thompson (Eds) 'The International System of Units (SI)' NIST Special Publication 330, National Institute of Standards and Technology, 2008 (http://physics.nist.gov/Pubs/SP330/sp330.pdf)

# Index